Helical Piles

A Practical Guide to Design and Installation

Helical Piles

A Practical Guide to Design and Installation

Howard A. Perko, Ph.D., P.E.

WILEY
John Wiley & Sons, Inc.

This book is printed on acid-free paper. ♾

Published by John Wiley & Sons, Inc., Hoboken, New Jersey
Published simultaneously in Canada

Library of Congress Cataloging-in-Publication Data:

Perko, Howard A.
Helical piles : a practical guide to design and installation / Howard A. Perko.
p. cm.
Includes bibliographical references and index.
ISBN 978-0-470-40479-9 (cloth)
1. Steel piling. I. Title.
TA786.P47 2009
624.1'54—dc22 2009019343

Printed in the United States of America

SKY10161554_070726

Contents

Foreword

Helical piles offer a versatile and efficient alternative to conventional deep foundations or anchors in a wide variety of applications. This technology has enjoyed an increased awareness and use by engineers in recent years, a trend which is due at least in part to the efforts of Howard Perko and the members of the Deep Foundation Institute's Helical Foundations and Tie-Backs Committee.

With this greater implementation of helical piles comes an increased need for a comprehensive guide to the current state of knowledge regarding the appropriate methods of design and installation. Howard's book is a much needed resource to meet that need and will serve as the authoritative and comprehensive reference on helical piles.

The fundamental mechanisms by which helical piles develop resistance to load are described in a manner consistent with basic principles of soil mechanics. Along with the thorough description of installation methods and equipment that is provided, the concepts used for design and quality control/quality assurance follow logically. The section on corrosion and life expectancy is particularly important now as applications of helical piles expand into greater use with permanent structures with longer intended service periods. Applications for helical piles are described which may prove novel to many engineers and open opportunities for innovation and development of more cost-effective solutions.

In summary, this text provides a valuable reference on an emerging technology that should serve as an important resource for any practicing engineer or constructor involved in the design or construction of foundation or earth support systems.

Dan Brown, Ph.D., P.E.
Dept. of Civil Engineering, Auburn University
Dan Brown and Associates, PLLC

Foreword

Preface

Helical piles have been used in construction for over 200 years. Today, there are over 50 helical pile manufacturing companies in at least twelve countries on four continents. There may be more than 2,000 helical pile installation contractors in the United States alone.

In the past, helical piles were an interesting alternative that some geotechnical engineers would take into consideration in special cases. Fifteen years ago, helical piles were barely mentioned in undergraduate and graduate civil engineering studies. Now helical piles are well known by most practicing engineers and should be considered an essential part of any graduate course in foundation engineering. Helical piles have gained in popularity to the extent that they are used more frequently than other deep foundations in some geographic locations. Even owners and developers are beginning to request helical piles.

At the time of this writing, an average of 1,500 people per week visit the trade Web site www.helicalpierworld.com. Over 100 technical papers and numerous articles have been written about helical piles. There are 163 U.S. patents pertaining to helical piles. The Helical Foundations and Tie-Backs committee of the Deep Foundation Institute (DFI), a professional trade organization, formed in 2001 and has been one of the largest DFI committees.

Helical piles were adopted into the International Building Code in 2009. Helical piles most certainly have a bright future in geotechnical engineering and foundations construction. Yet most of the information about these systems is contained in proprietary manuals published by helical pile manufacturing companies. An unbiased and universally applicable text dedicated to the design and installation of helical piles is needed to compile the current state of knowledge and practice in the industry. The goal of this book is to satisfy that need.

Acknowledgements

Several professionals in industry helped by proofreading and editing various chapters of this book, contributing images, and providing general feedback. Their contributions are gratefully acknowledged.

Most notably, my good friends and current employers, Bill Bonekemper and Brian Dwyer of Magnum Piering, Inc., contributed in a number of ways, from photographs and load test results, to unwavering moral support. Also, I recognize my former supervisors, Ron McOmber, chairman of the board of CTL|Thompson, and Robert Thompson, founder of CTL|Thompson, who supported the book by providing load test data, laboratory swell test data, and company resources for printing and computing as well as for their review of Chapter 9 and all the guidance through the years.

Support for this book also was provided by helical pile manufacturing representatives, researchers, professionals, and faculty members. Gary Seider and Don Deardorff of Chance Civil Construction/Hubbell, Inc. reviewed Chapters 1 and 2, provided a copy of HeliCap software, and contributed several images. Tony Jacobsen and Justin Porter of Grip-Tite Manufacturing Co., LLC reviewed Chapters 1 and 2 and provided helpful discussions. Darin Willis of Ram Jack Systems Distribution, LLC reviewed Chapters 4,6, 8, and 10 and provided load test data, several images, and a copy of RamJack Foundation Solutions software. Jeff Tully of Earth Contact Products, Inc. provided project photographs and reviewed Chapters 1 and 2. Steve Petres and Wei-Chung Lin of Dixie/MacLean Power provided load test data, a description of their inventions, and an image for Chapter 16. John Pack, senior engineer with IMR in Wheat Ridge, Colorado, reviewed Chapter 9. Mamdouh Nasr of Shaw Construction in Dubai provided load test results and images from finite element analysis of helical piles. Dr. Amy Cerato, assistant professor at Oklahoma University, provided copies of her presentations and papers for reference and incorporation into Chapter 5. Rich Davis of www.helicalpierworld.com provided industry statistics.Gary Bowen, an independent consultant from Mill City, Oregon, is appreciated for providing his unpublished paper on capacity-to-torque ratios and also his assistance with review Chapter 6. Eileen Dornfest, senior geologist and project manager, with Tetratech in Fort Collins, CO, is appreciated for reviewing Chapter 9. I would also like to thank my friends at Mueser Rutledge Consulting Engineers, especially Peter Deming, Sitotaw Fantaye, and Kathleen Schulze, for selecting me to work with on drafting the New York City helical pile building code and for Sitotaw Fantaye's precursory review of Chapters 4, 5, 6, 10, and 11. As a special note, I thank Dr. John Nelson, retired professor from Colorado State University, author of one of the most well regarded books on expansive soils, and my Ph.D. advisor for helping to develop my technical writing skills, teaching me most of what I know about expansive soils, and for his review of Chapter 9.

A number of colleagues from CTL|Thompson, Inc. helped with the book, often in their spare time. Robin Dornfest, Chip Leadbetter, and one of my early mentors, Frank J. Holiday, reviewed and edited Chapter 3. Chief structural engineer James Cherry helped indirectly through the years by teaching a soil engineer about foundation design and structural analysis as well as directly contributing by reading and editing Chapters 7 and 12. Staff members Becky Young and Antoinette Roberts prepared the database of swell tests contained in Chapter 9. Chief environmental engineer Tom Norman and business development manager Timiry Kreiger assisted with the environmental benefits of helical piles contained in Chapter 1.

I also recognize three well-known experts in the helical pile industry: Bob Hoyt, independent engineering consultant, Sam Clemence, distinguished professor from Syracuse University, and Al Lutenegger, distinguished professor from the University of Massachusetts Amherst. I thank them for their many contributions to the industry. Their work provided a foundation for this book. I owe much of my professional growth to listening to their lectures and reading their papers. I am honored to have had many personal conversations with them through the years. I am also thankful for Sam Clemence's review of Chapters 4, 5, and 6.

Chapter 1

Introduction

Helical piles are a valuable component in the geotechnical tool belt. From an engineering/architecture standpoint, they can be adapted to support many different types of structures with a number of problematic subsurface conditions. From an owner/developer standpoint, their rapid installation often can result in overall cost savings. From a contractor perspective, they are easy to install and capacity can be verified to a high degree of certainty. From the public perspective, they are perhaps one of the most interesting, innovative, and environmentally friendly deep foundation solutions available today.

This book contains an introduction, a primer on installation and basic geotechnics, advanced topics in helical pile engineering, practical design applications, and other topics. The introduction starts with basic features and components of helical piles. The reason for all the different terms, such as "helical pier," "helix pier," "screw pile," "torque anchor," and others, is explained through a discussion of terminology. This introductory chapter contains the story of Alexander Mitchell and the invention of the helical pile. Next a brief history of helical pile use is told through an analysis of U.S. patents. Then many modern applications are discussed with the goal of introducing how the helical pile might be applied to everyday projects.

The installation of helical piles is fairly straightforward; however, as with any process, there are a number of tricks of the trade based on years of experience in the installation of helical piles. Many of these tricks are revealed in Chapter 2 along with guidelines for proper installation procedures and equipment. The installation chapter is generally organized as a standard prescription specification with some basic how-to information. Chapter 3 is on basic geotechnics. It contains an overview of some of the basic concepts in soil and rock mechanics that are important for designers and installers of helical piles. These topics include interpretation of exploratory boring logs, soil and rock classification, and shear strength. The soil and rock conditions that are particularly conducive to helical pile use and those conditions that prohibit helical pile use are discussed.

The engineering of helical piles is broken into seven concepts, which comprise the main technical chapters of this book: Chapter 4 on bearing capacity, Chapter 5 on pullout capacity, Chapter 6 on capacity to torque ratio, Chapter 7 on axial load testing, Chapter 8 on reliability and sizing, Chapter 9 on expansive soil resistance, Chapter 10 on lateral load resistance, and Chapter 11 on corrosion and design life expectancy. These engineering concepts are applied to the practical design of foundations in Chapter 12, earth retention systems in Chapter 13, and underpinning systems in Chapter 14. These technical and design chapters are organized as a handy reference with guide capacity charts, design examples, sample calculations, many references, and real test data.

The book concludes with chapters on nontechnical topics: Chapter 15 on foundation economics, Chapter 16 on proprietary systems, and Chapter 17 on current building codes regarding helical piles. Contained in the appendices are a list of common symbols and abbreviations used in design and construction, a fairly complete list of all U.S. helical pile patents, data from over 275 load tests, a list of the nomenclature used throughout the book, and a glossary of terms pertaining to helical piles. It is intended that this book will appeal primarily to foundation contractors, foundation inspectors, practicing engineers, and architects. It may also serve as a useful supplementary reference to graduate students and university professors in the academic departments of engineering, architecture, and construction.

1.1 BASIC FEATURES

Helical piles are manufactured steel foundations that are rotated into the ground to support structures. The basic components of a helical pile include the lead, extensions, helical bearing plates, and pile cap as detailed in Figure 1.1. The lead section is the first section to enter the ground. It has a tapered pilot point and typically one or multiple helical bearing plates. Extension sections are used to advance the lead section deeper into the ground until the desired bearing stratum is reached. Extension sections can have additional helical bearing plates but often are comprised of a central shaft and couplings only. The couplings generally consist of bolted male and female sleeves. The central shaft is commonly a solid square bar or a hollow tubular round section.

Helical piles have been used in projects throughout the world. Uses for helical piles include foundations for houses, commercial buildings, light poles, pedestrian bridges, and sound walls to name a few. Helical piles also are used as underpinning elements for repair of failed foundations or to augment existing foundations for support of new loads. Helical piles can be installed horizontally or at any angle and can support tensile in addition to compressive loads. As a tensile member, they are used for retaining wall systems, utility guy anchors, membrane roof systems, pipeline buoyancy control, transmission towers, and many other structures.

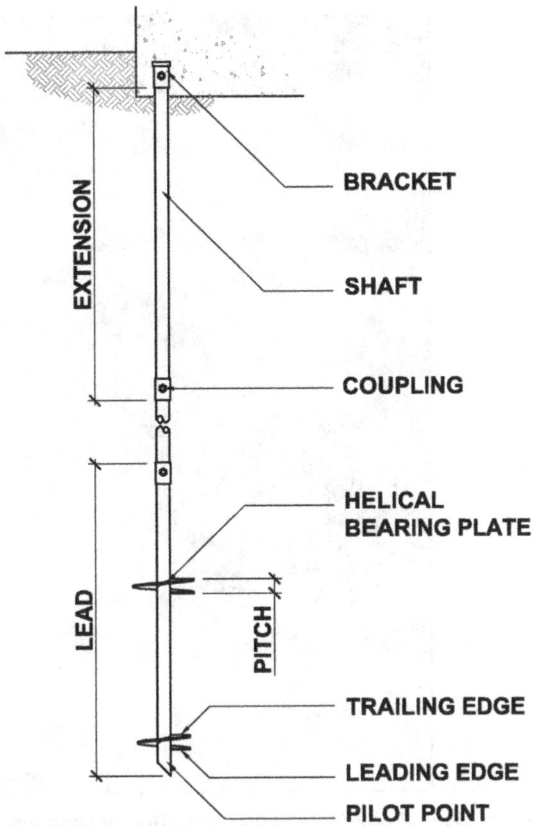

Figure 1.1 Basic helical pile

Helical piles offer unique advantages over other foundation types. Helical pile installation is unaffected by caving soils and groundwater. Installation machinery has more maneuverability than pile-driving and pier-drilling rigs. Installation can even be done with portable, hand-operated equipment in limited access areas such as inside crawl spaces of existing buildings. A photograph of a limited access rig working inside the basement of a commercial building is shown in Figure 1.2. Helical pile installation does not produce drill spoil, excessive vibrations, or disruptive noise. Installation of a new foundation system consisting of 20 helical piles is conducted in typically less than a few hours. Loading can be immediately performed without waiting for concrete to set. Helical piles can be removed and reinstalled for temporary applications, if a pile is installed in an incorrect location or if plans change. A summary of these and other advantages of helical piles is given in Table 1.1. Helical piles are practical, versatile, innovative, and economical deep foundations. Helical piles are an excellent addition to the variety of deep foundation alternatives available to the practitioner.

Figure 1.2 Helical pile installation in limited access area (Courtesy of Earth Contact Products, Inc.)

Table 1.1 Benefits of Helical Piles

Benefits of Helical Piles
Resist scour and undermining for bridge applications
Can be removed for temporary applications
Are easily transported to remote sites
Torque is a strong verification of capacity
Can be installed through groundwater without casing
Typically require less time to install
Can be installed at a batter angle for added lateral resistance
Can be installed with smaller more accessible equipment
Are installed with low noise and minimal vibrations
Can be grouted in place after installation
Can be galvanized for corrosion resistance
Eliminate concrete curing and formwork
Do not produce drill spoil
Minimize disturbance to environmentally sensitive sites
Reduce the number of truck trips to a site
Are cost effective

1.2 TERMINOLOGY

There is often some question as to whether a helical foundation should be considered a pile or a pier. In some parts of the United States, especially the coastal areas, the terms "pile" and "pier" are used with reference to different foundations based on their length. As defined in the International Building Code (2006), a "pile" has a length equal to or greater than 12 diameters. A "pier" has a length shorter than 12 diameters. In other parts of the United States, specifically Rocky Mountain regions, the terms "pile" and "pier" are defined by the installation process. A pier is drilled into the ground, whereas a pile is driven into the ground. Some European foundation engineering textbooks explain that a pier is a type of pile with a portion that extends aboveground, as in the case of marina piers. Geographic differences in definitions of the same terms often create considerable confusion at national and international meetings and conferences. Before attempting a technical discussion, definitions should be clearly stated and agreed on.

The original device that is the precursor to the modern-day helical pile was termed the "screw pile." Sometime later, the phrase "helical anchor" became more common, probably because the major application from 1920 through 1980 was for tension. In about 1985, one of the largest manufacturer's of helical anchors, the AB Chance Company, trademarked the name "helical pier" in order to promote bearing or compression applications. In the last 20 years, other manufacturers attempting to avoid the trade name have promoted terms such as "helix pier," "screw pier," "helical foundation," "torque anchor," and others. The Canadian building code uses the phrase "augered steel pile." The terms "heli-coils" and even "he-lickers" are heard in isolated regions.

Given that most helical piles are typically installed to depths greater than 12 diameters and the trade name issues, the Helical Foundations and Tie-Backs committee of the Deep Foundation Institute decided in 2005 to henceforth use the phrase "helical pile." This is the name that will be used throughout this text. "Helical pile" is defined below. Other terms related to helical piles and foundations in general are defined in the Glossary of Terms.

> *Helical Pile* (noun) "A manufactured steel foundation consisting of one or more helix-shaped bearing plates affixed to a central shaft that is rotated into the ground to support structures."

Since they can resist both compression and tension, helical piles can be used as a foundation or as an anchor. The phrase "helical pile" is generally used for compression applications, whereas the phrase "helical anchor" is reserved for tension applications. The devices themselves are the same. The phrase "helical pile" is used herein for the general case unless the distinction between applications is a necessary clarification.

1.3 INVENTION

The first recorded use of a helical pile was in 1836 by a blind brickmaker and civil engineer named Alexander Mitchell. Mitchell was born in Ireland on April 13, 1780, and attended Belfast Academy. He lost his sight gradually from age 6 to age 21. Being blind limited Mitchell's career options, so he took up brick making during the day and studied mechanics, mathematics, science, and building construction in his leisure. One of the problems that puzzled Mitchell was how to better found marine structures on weak soils, such as sand reefs, mudflats, and river estuary banks. At the age of 52, Mitchell devised a solution to this problem, the helical pile. The author Irwin Ross (Hendrickson, 1984 pp. 332–333) describes Mitchell's moment of invention in this way:

> Necessity is often cited as the mother of invention, but in the case of Mitchell's invention it may be said that it was incubated by his love for mankind and actually discovered by accident.
>
> In the early 1830s, there were many storms. During the long October and November nights, at the beginning of this period, Mitchell lay in bed listening to the raging storms outside, which violently shook the window sashes, made the slates drum, howled in the chimney, and seemed at the retreat of every gust a requiem for those poor mariners whose dead bodies he pictured being swept on the crest of an angry sea.
>
> Mitchell lay thinking. He could only sleep in brief snatches. Something had to be done, and he resolved to do it. Many original ideas occurred to him regarding lighthouse foundations on sandy beds, but in practice they proved to be unsuccessful.
>
> One day in 1832, when experimenting with a sail which he had made to enable a boat to sail in the teeth of the wind by means of a broad-flanged screw in the water and a canvas-covered screw in the air, he happened to place the water screw on the ground, and a great gust of wind, violently propelling the aerial canvas screw, embedded that water screw firmly in the ground.
>
> Mitchell tugged at the connecting spindle, and then his nimble fingers traveled toward the earth, his sense of touch disclosing what had taken place. He sprang upright and danced around his discovery with delight. He had discovered the principle of the screw pile.
>
> One evening he hired a boat, and with his son John as boatman, he steered his course to a sandy bank in Belfast Lough, where he planted a miniature screw pile. He then returned home, no one being any wiser about his experiment. Very early the next morning, before the working world was astir, they rowed out again, examined the pile, and found it firmly fixed where they had placed it, although the sea that night had been a bit rough. This was a moment of great satisfaction to both father and son.

In 1833, Mitchell patented his invention in London. Mitchell called the device a "screw pile" and its first uses were for ship moorings. A diagram of Mitchell's screw pile is shown in Figure 1.3. The pile was turned into the ground by human and animal power using a large wood handle wheel called a capstan. Screw piles on the order

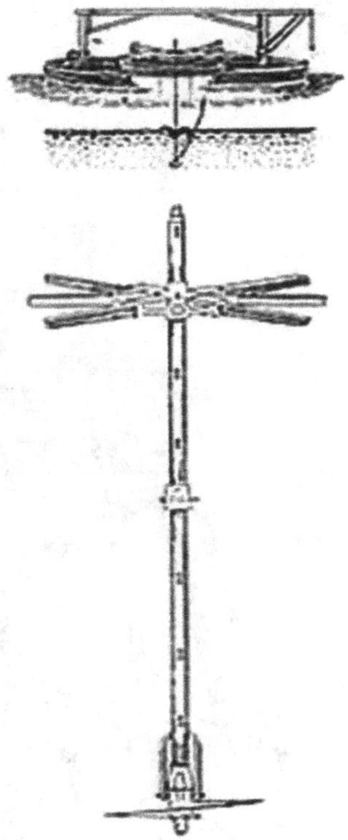

Figure 1.3 Mitchell screw pile

of 20 feet [6 m] long with 5-inch- [127-mm-] diameter shafts required as many as 30 men to work the capstan. Horses and donkeys were sometimes employed as well as water jets.

In 1838, Mitchell used screw piles for the foundation of the Maplin Sands Lighthouse on a very unstable bank near the entrance of the river Thames in England. A profile view of the Maplin Sands Lighthouse is shown in Figure 1.4. The foundation consisted of nine wrought-iron screw piles arranged in the form of an octagon with one screw pile in the center. Each pile had a 4-foot [1.2 m] diameter helix at the base of a 5-inch [127 mm] diameter shaft. All nine piles were installed to a depth of 22 feet [6.7 m], or 12 feet [3.7 m] below the mud line, by human power in nine consecutive days. The tops of the piles were interconnected to provide lateral bracing (Lutenegger, 2003).

Author Irwin Ross (Hendrickson, 1984, pp. 332–333) explained how valuable the invention of the helical pile was to lighthouse construction.

> The erection of lighthouses on this principle caused the technical world to wonder. This invention, which has been the means of saving thousands of lives and preventing the loss of millions of dollars worth of shipping, has enabled lighthouses and beacons to be built

Figure 1.4 Maplin Sands lighthouse

on coasts where the nature of the foreshore and land formations forbade the erection of conventional structures. The screw pile has been used in the construction of lighthouses and beacons all over the world, and it earned for Mitchell and his family a large sum.

. . .Although Mitchell was blind, he never failed to visit his jobs, even in the most exposed positions, during rough weather. In examining the work, he always crawled on his hands and knees over the entire surface, testing the workmanship by his sense of touch. . . .On many occasions he stayed out the whole day, with a few sandwiches and a flask, cheering his men at their work and leading them in sea songs as they marched around on the raft driving the screws.

In 1853, Eugenius Birch started using Mitchell's screw pile technology to support seaside piers throughout England. The first of these was the Margate Pier. From 1862 to 1872, 18 seaside piers were constructed on screw piles. Photographs of three of these piers, the Eastbourne Pier, Bournemouth Pier, and the Palace Pier are shown in Figure 1.5. As can be seen in the figure, each bridge pier consisted of a series of interconnected columns. Each of these columns was supported on a screw pile. The piers themselves supported the weight of pedestrians, carts, buildings, and ancillary structures. The foundations had to support tidal forces, wind loads, and occasional ice flows. Screw piles also were used to support Blankenberg Pier in Belgium in 1895 (Lutenegger, 2003).

During the expansion of the British Empire, screw piles were used to support new bridges in many countries on many continents. Technical articles were published in *The Engineering and Building Record* in 1890 and in *Engineering News* in 1892 regarding bridges supported on screw piling. Excerpts from these journal articles are shown in Figure 1.6. The foundations for the bridges shown look very similar to those used to support seaside piers. Screw piles were installed in groups and occasionally at a batter angle. Pier shafts were braced with horizontal and diagonal members above the mud line. Notably, concrete is absent from the construction of these foundations. As a result of British expansion, screw piles were soon being applied around the world (Lutenegger, 2003).

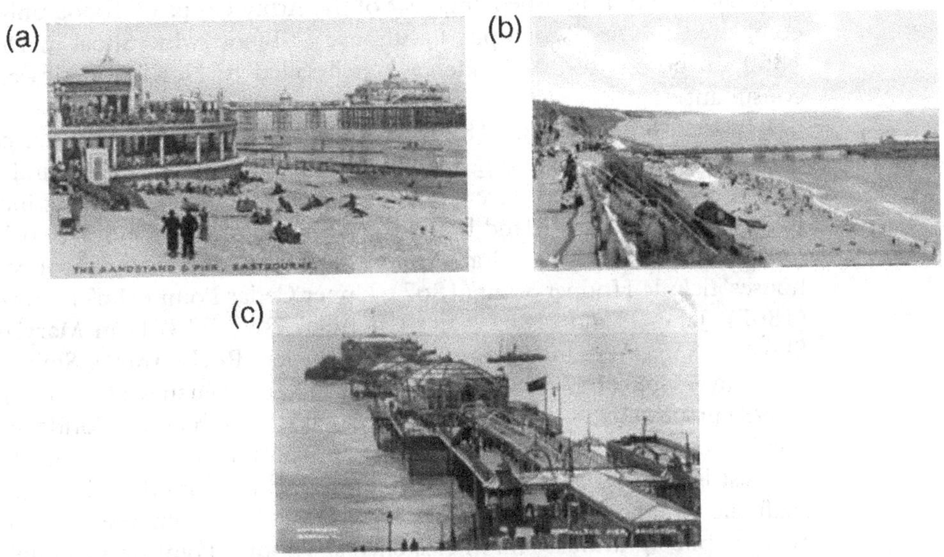

Figure 1.5 Oceanside piers supported by helical piles: (a) Eastbourne Pier; (b) Bournemouth Pier; (c) Palace Pier

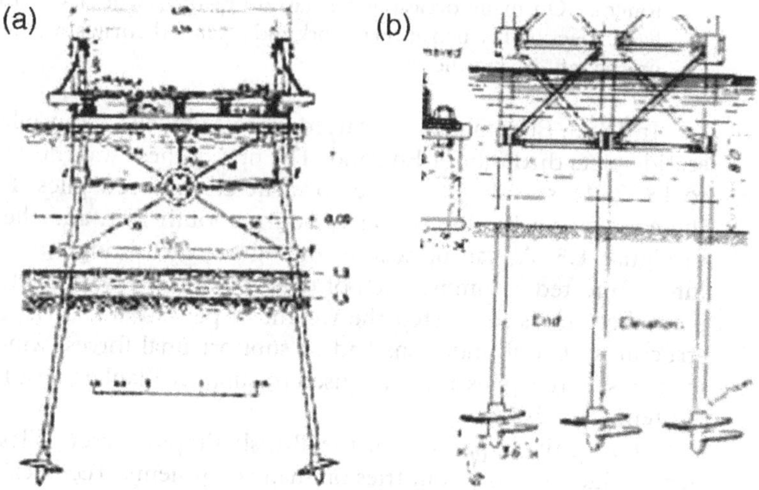

Figure 1.6 Early helical pile supported bridges.
"Screw Pile Bridge over the Wumme River," Engineering and Building Record, April 5, 1890; "Screw Piles for Bridge Piers," Engineering News, August 4, 1892.

There is some controversy as to the first known use of a helical pile in the United States. According to Lutenegger (2003), Captain William H. Swift constructed the first U.S. lighthouse on screw piles in 1843 at Black Rock Harbor in Connecticut. According to the National Historic Landmark Registry (NPS, 2007), Major Hartman Bache, a distinguished engineer of the Army Corps of Topographical Engineers, completed the first screw pile lighthouse at Brandywine Shoal in Delaware Bay in 1850. In both cases, Alexander Mitchell sailed to North America and served as a consultant.

In the 1850s through 1890s, more than 100 lighthouses were constructed on helical pile foundations along the East Coast of the United States and along the Gulf of Mexico. Examples of screw pile lighthouses in North Carolina include Roanoke River (1867), Harbor Island Bar (1867), Southwest Point Royal Shoal (1867), Long Point Shoal (1867), and Brant Island (1867). Other examples of screw pile lighthouses include Hooper Strait (1867), Upper Cedar Point (1867), Lower Cedar Point (1867), Janes Island (1867), and Choptank River (1871) in Maryland and White Shoals (1855), Windmill Point (1869), Bowlers Rock (1869), Smith Point (1868), York River Spit (1870), Wolf Trap (1870), Tue Marshes (1875), and Pages Rock (1893) in Virginia. Screw pile lighthouses also were built in Florida at Sand Key and Sombrero Key. Many of the lighthouse foundations in the Northeast were required to resist lateral loads from ice flows and performed considerably better than straight shaft pile foundations. Most historic lighthouses have been destroyed or disassembled. A screw pile lighthouse still in existence is Thomas Point Shoal Light Station (NPS, 2007).

"I'm glad we installed that helical pile foundation before the glacier hit."

The first technical paper written on helical piles was "On Submarine Foundations; particularly Screw-Pile and Moorings," by Alexander Mitchell, which was published in the *Civil Engineer and Architects Journal* in 1848. In this paper, Mitchell stated that helical piles could be employed to support an imposed weight or resist an upward strain. He further stated that a helical pile's holding power depends on the area of the helical bearing plate, the nature of the ground into which it is inserted, and the depth to which it is forced beneath the surface.

From about 1900 to 1950, the use of helical piles declined. During this time, there were major developments in mechanical pile-driving and drilling equipment. Deep foundations, such as Raymond drilled foundations, belled piers, and Franki piles, were developed. With the development of modern hydraulic torque motors, advances in manufacturing, and new galvanizing techniques, the modern helical pile evolved primarily for anchor applications until around 1980 when engineer Stan Rupiper designed the first compression application in the U.S. using modern helical piles (Rupiper, 2000).

Radio conversation of a U.S. naval ship with Canadian authorities off the coast of Newfoundland in October 1995.

CANADIANS: “Please divert your course 15 degrees to the north to avoid a collision.”

AMERICANS: “Recommend YOU divert your course 15 degrees to the south to avoid a collision.”

CANADIANS: “Negative. You will have to divert your course 15 degrees to the north to avoid a collision.”

AMERICANS: “This is the captain of a US Navy ship. I say again, divert YOUR course”

CANADIANS: “No, I say again, you divert your course”

AMERICANS: “This is the Aircraft Carrier USS LINCOLN, the second largest ship in the United States Atlantic Fleet. We are accompanied with three Destroyers, three Cruisers and numerous support vessels. I DEMAND that you change your course 15 degrees south, or counter-measures will be undertaken to ensure the safety of this ship”

CANADIANS: “This is a LIGHTHOUSE on a helical foundation. Your call.”

1.4 EARLY U.S. PATENTS

There are more than 160 U.S. patents for different devices and methods related to helical piles (see Chapter 16 and Appendix B). One of the earliest patents filed shortly after the first lighthouse was constructed in the U.S. on helical piles was by T.W.H. Moseley. Moseley's patent described pipe sections coupled together with flanges. The lead pipe section was tapered with a spiral section of screw threads and an optional spade point as shown in Figure 1.7. Another aspect of the invention, shown in Figure 1.8, consisted of a wooden pile driven through the center of the screw pile and concrete filling the annular space. The screw portion of the pile is shown installed below the mud line. The bottom most flange rests at the mud line. Historic documents indicate that

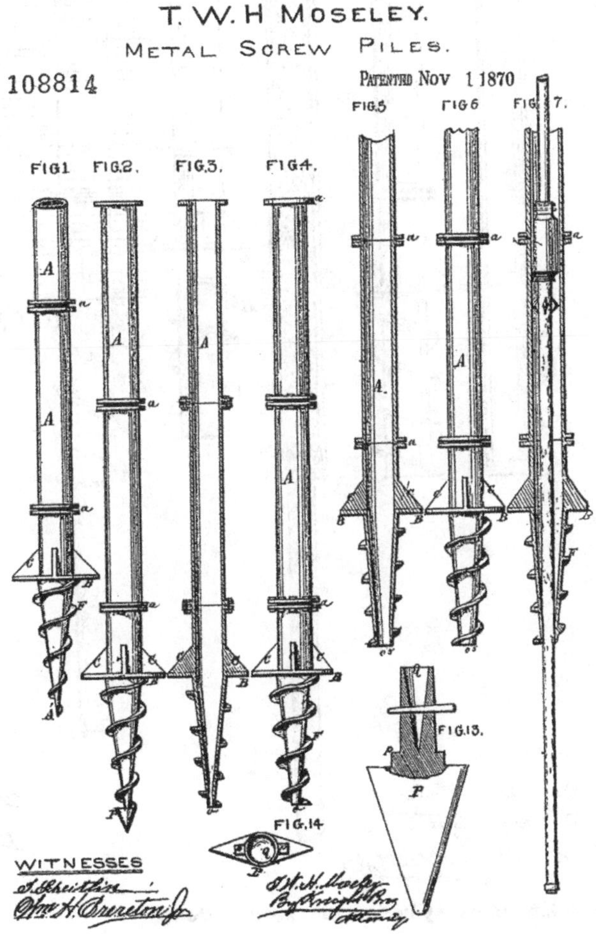

Figure 1.7 Moseley helical pile patent

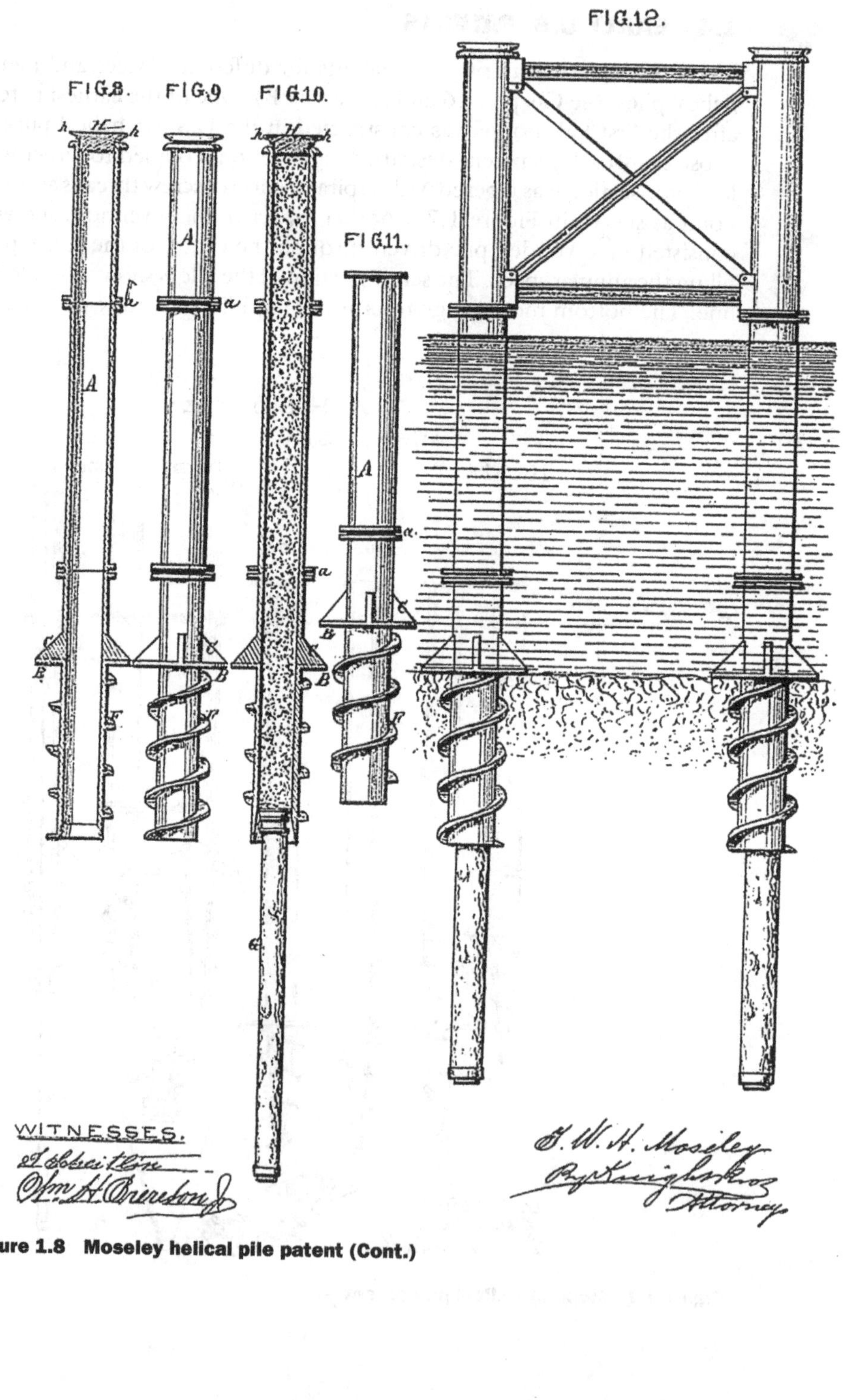

Figure 1.8 Moseley helical pile patent (Cont.)

this method of combining driven piles with screw piles was used to construct a number of marine structures in the 1800s (NPS, 2007).

Although Moseley described using concrete to fill the inside of a helical pile, the first use of pressurized grouting on the exterior of a helical pile was by Franz Dyche in 1952. As shown in Figure 1.9, Dyche explained that a lubricating fluid or grout could be pumped through openings at each screw flight spaced along the helical pile lead section. Dyche's helical pile consisted of a lead section with bearing plates in a continual spiral over the length of the lead. The lead section could be extended in depth by one or more tubular extensions. A guy wire or other anchor cable could be attached to a flange at the top of the lead section. The installation tooling could be removed after the appropriate depth is obtained. It was determined later by others that group effects within soil make the continuous spiral unnecessary and that single helical bearing plates spaced along the length of a lead can match the capacity of a continuous spiral in soil.

One of the first U.S. patents on helical ground anchors can be credited to A.S. Ballard of Iowa, who in 1860 patented what he called an earth borer. In later patents, Ballard's device is referred to as an earth anchor. The device, shown in Figure 1.10, had two helix-shaped plates with a solid steel shaft and conical pilot point. The helical plates are riveted to a cross bar attached to the shaft. Ballard's patent was followed by forty variations in helical anchors over the next one hundred years. One variation, which occurred 15 years after Ballard's patent issue date, was a similar anchoring device by Clarke. Clarke's device, shown in Figure 1.11, differed from that of Ballard in that the pitch of the helical plates was increased and the installation tool was made detachable so that a section of pipe with guy wire eyelet could be inserted after anchor installation.

Patents have been filed for helical anchors with different shaped installation tools including L-shape, S-shape, square, round, and cruciform shaft sockets. Many patents for helical anchors regard special spade-shaped and corkscrew pilot points for penetrating difficult soils. There also are many patents regarding the shape of the helix and its cutting edge. Most of these early patents for helical anchors are more than 25 years old and are now public domain.

Many of the U.S. patents for helical piles involve different methods for supporting structures. An example, depicted in Figure 1.12, involves the hold down of pipelines for buoyancy control. When a partially full pipeline is submersed below open water or in groundwater, it is subject to a significant upward force due to buoyancy. In the example, Hollander describes a method of simultaneously installing two helical anchors rotating in opposite directions using a crane mounted drilling apparatus. The opposite direction of rotation of the anchors during installation eliminated any net rotation force on the suspended drills. This method of anchor installation for buoyancy control patented in 1969 is still used today.

Another notable application of helical piles is for underpinning existing structures. Underpinning is used to repair failed foundations or to support new loads. In 1991, Hamilton and others from the A.B. Chance Company patented a method of installing a steel underpinning bracket under an existing foundation and screwing a helical pile

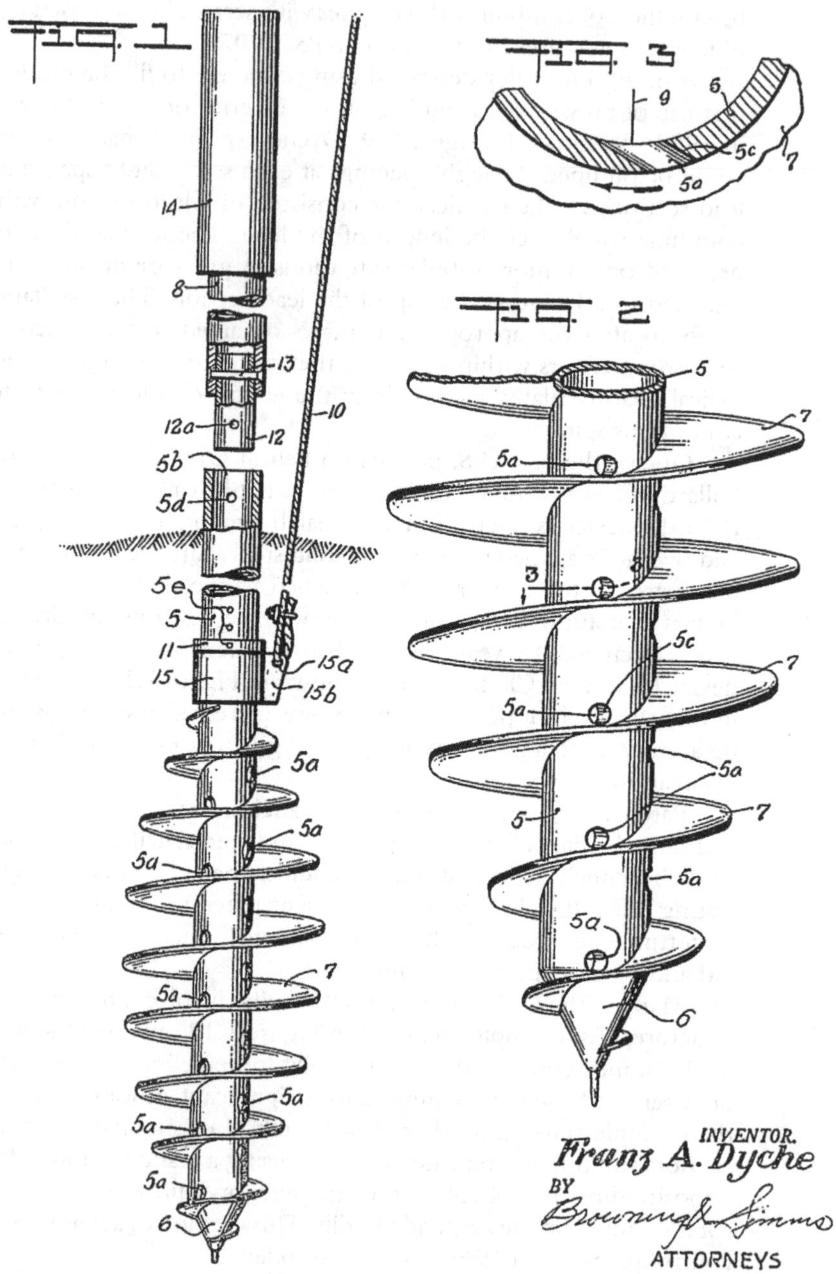

Figure 1.9 First helical pile grouting method

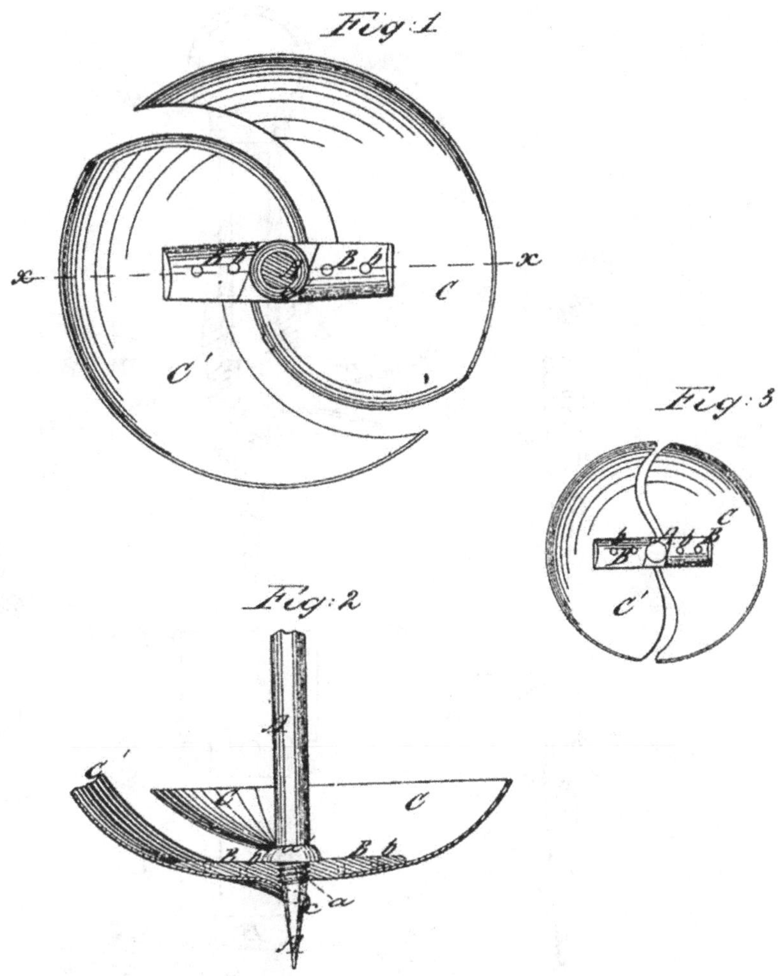

Figure 1.10 Ballard earth-borer device

at a slight angle directly adjacent to the bracket as pictured in Figure 1.13. The helical pile and bracket are used to lift and permanently support the foundation. A legal battle ensued between the patent holders and helical pile installers led by Richard Ruiz of Fast Steel, a competing helical pile manufacturer. Ruiz challenged the originality and novelty of the patent claims. After many appeals, the claims of Hamilton's patent were overturned. It is no longer proprietary to underpin existing foundations using helical piles. A flurry of patents regarding different underpinning brackets followed in the last decade. Despite the loss of their patent rights, much credit is owed to Hamilton and the A.B. Chance Company for advancing the state of the art with respect to helical piling for underpinning.

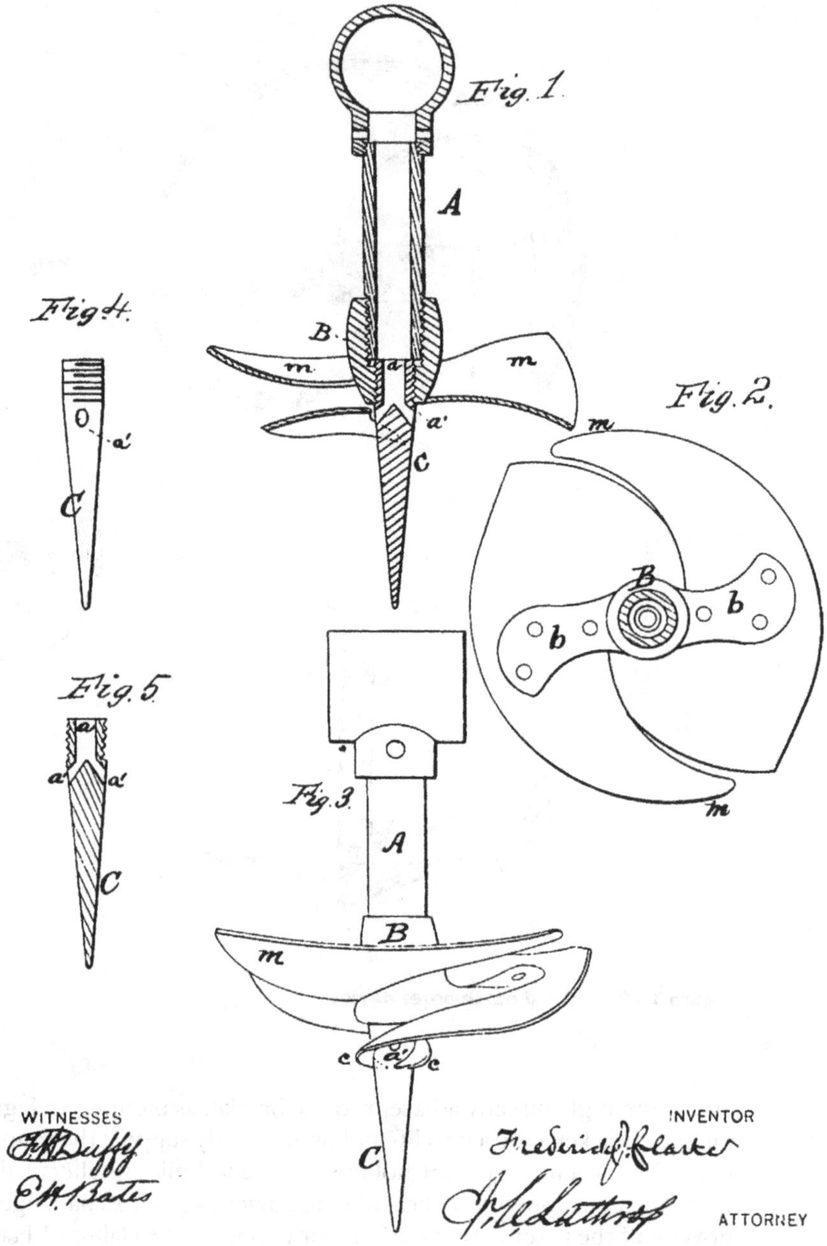

Figure 1.11 Clarke anchor device

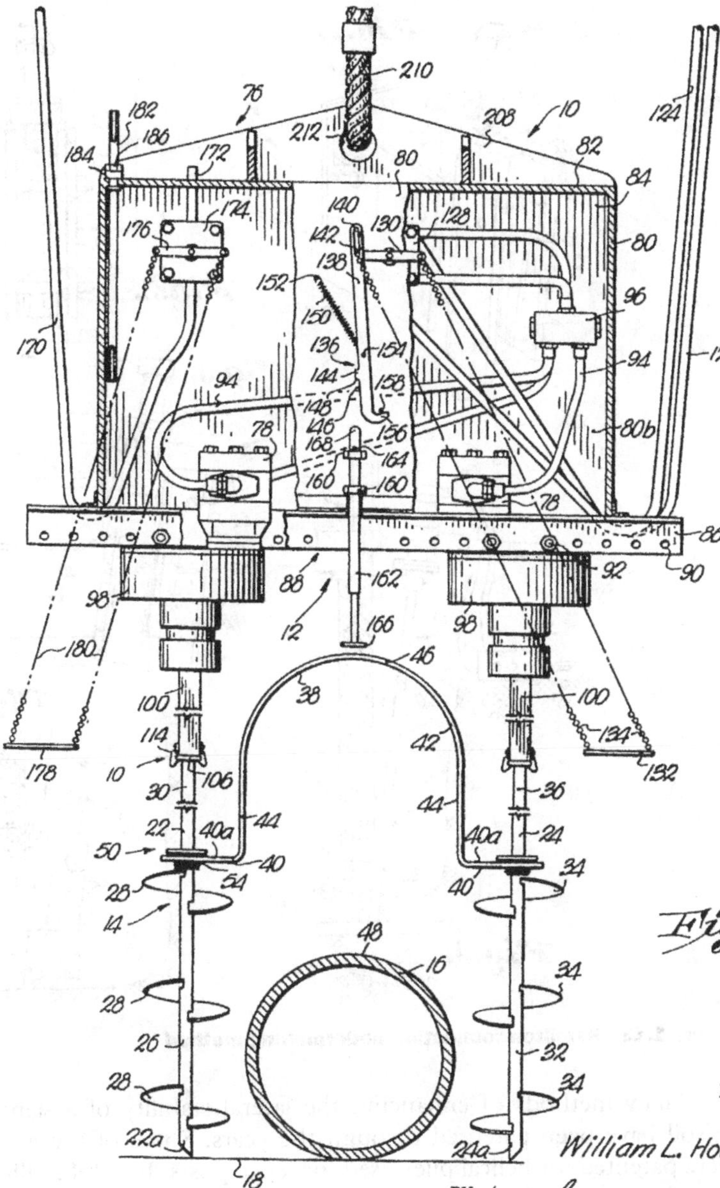

Fig. 1.

INVENTOR

William L. Hollander

BY *Hovey, Schmidt, Johnson & Hovey*

ATTORNEYS

Figure 1.12 Hollander pipeline anchor installation method

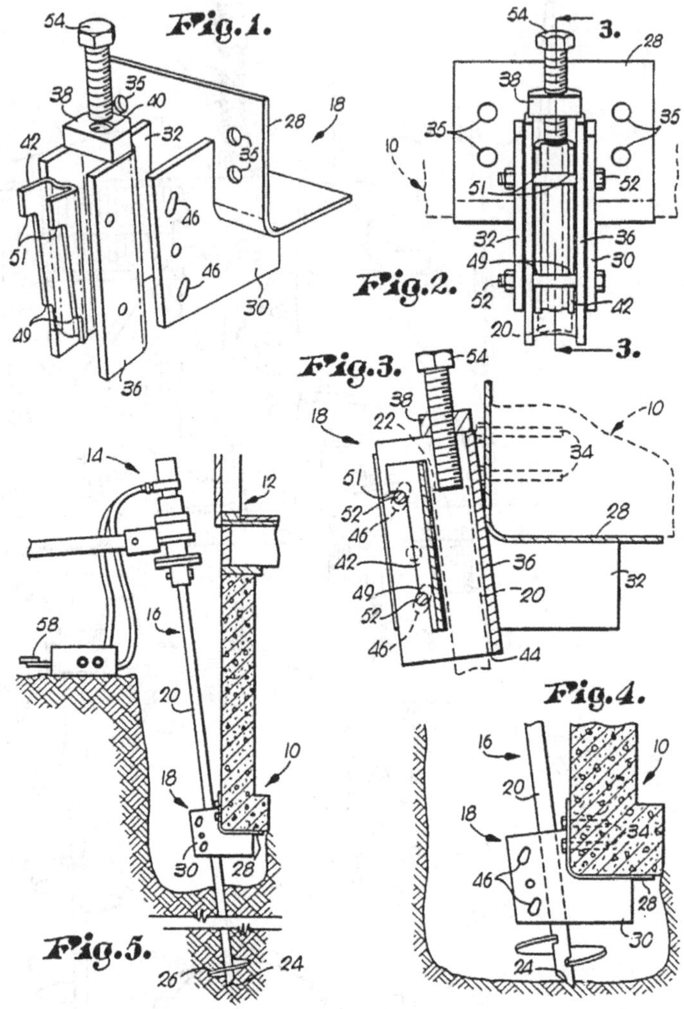

Figure 1.13 Hamilton foundation underpinning method

Many methods of enhancing the lateral stability of a slender helical pile shaft in soil have been patented through the years. Some of the earlier known methods were patented for helical piles used for fence posts. In 1898, Oliver patented a screw-type fence post with a shallow X-shaped lateral stabilizer where the pile meets the ground surface. A year later, Alter patented a screw-type fence post with large-diameter, shallow, cylindrical, lateral stabilizer also near the ground surface. Another example of a lateral stabilizer used with piles similar in appearance to the modern helical pile is shown in Figure 1.14. In 1961, Galloway and Galloway patented this method of placing three triangular plates on a swivel located on the trailing end of a helical pile. The plates or fins are drawn into the ground by the bracket on the end of the helical

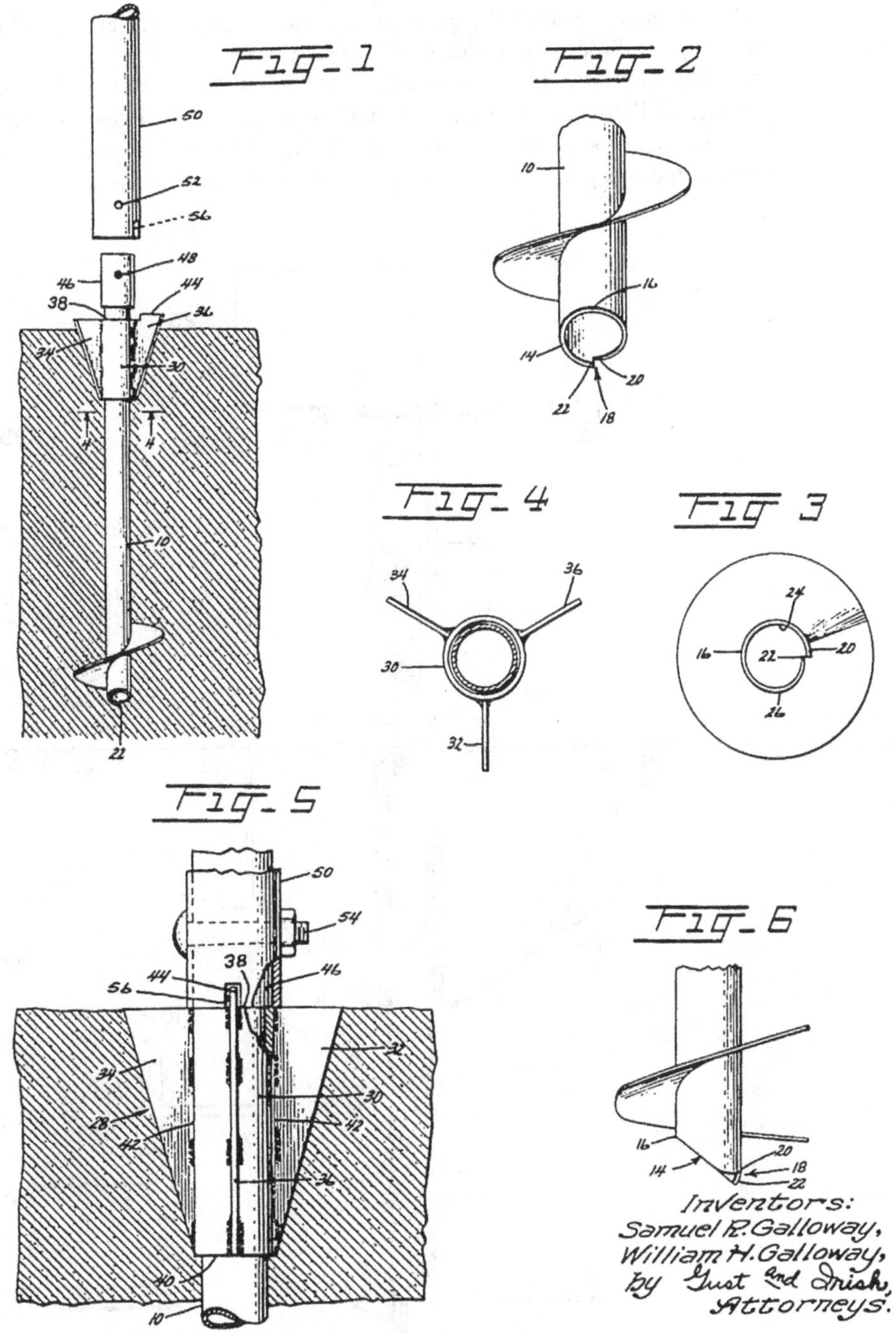

Figure 1.14 Galloway lateral stability device

pile as it advances into the ground. The helical pile with lateral stability enhancer can be coupled directly to a post or other structure.

The Galloway patent was followed in 1989 with the slightly different variation shown in Figure 1.15. In this variation, trapezoidal plates are attached to a square tubular sleeve slipped over the central shaft of a helical pile. The stabilizer sleeve is connected to a pile bracket using an adjustable threaded bar. Any number of structures

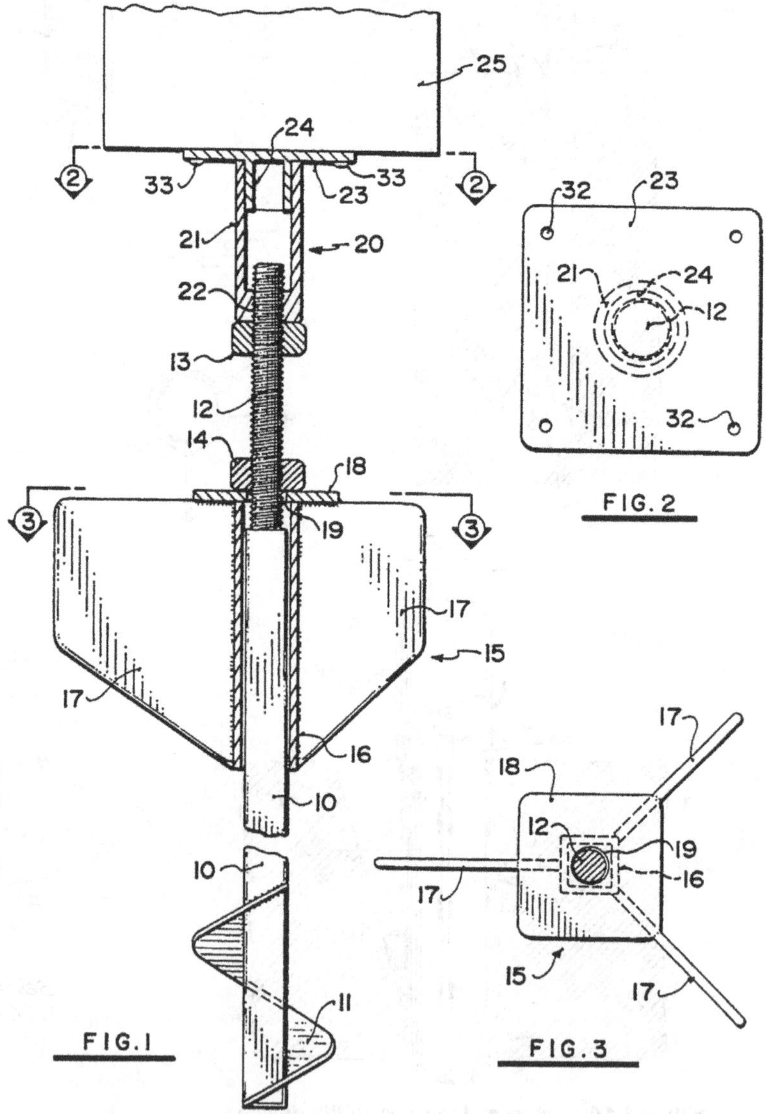

Figure 1.15 McFeetors lateral stability device

could be supported on the thread bar connection. Those familiar with the practice can see that there are many other approaches that could be taken to enhance lateral stability of slender shaft helical piles as are discussed later in this text.

Of course, another way to enhance the lateral resistance of a helical pile is to make the shaft larger. Many helical pile manufacturers currently produce relatively short, large diameter, helical lightpole bases. These products generally consist of a cylindrical or tapered polygonal shaft with helical bearing plate located at the bottom. The helical bearing plate is affixed to a short pilot point for centralizing the base. The top of the pile is fixed to a base plate with bolt hole pattern. Soil is forced aside most lightpole bases so that the central shaft remains empty during installation. In this way, an electrical conduit can be fed through the hollow center of the pile.

1.5 PERIODS OF USE

Much can be gleaned about the history of helical piles from studying the many patents filed through time. A plot of the number of U.S. patents filed regarding helical piles is shown in Figure 1.16. These patents can be grouped generally into four categories, or historical eras. As discussed in Section 1.2, the first uses of helical piles were for ship moorings, lighthouses, and other marine structures. The period from the invention of the screw pile to 1875, when these uses were most common, can generally be termed the "Marine Era." Very few of the earliest patents from this era could be found. Patent 30,175 from 1860 and patents 101,379 and 108,814 from 1870 refer to improvements in prior art, which indicates earlier patents could exist.

A majority of the early patents in Appendix B, beginning with Mudgett in 1878 and ending with Mullet in 1931, involve fence post applications. Known developments in irrigation and plant/soil science during the same general time frame combined with the series of fencing related helical pile patents are reasons for naming this period the "Agricultural Era."

The next group of patents, beginning in about 1920 and spanning into the 1980s, primarily regard guy anchors, tower legs, utility enclosures, and pipelines. This period can be termed the "Utility Era." Historically, this period of time also corresponds to a number of significant infrastructure projects in the United States including many large dams, the interstate highway system, power plants, aqueducts, and great cross-continental electrical transmission projects.

The last group of patents, issued from roughly 1985 until the present, generally concern many types of buildings and other construction applications. Several patents relate to mobile homes, retaining walls, underpinning, sound walls, and special types of helical piles for foundations. This era can be termed the "Construction Era." The Construction Era spans the residential housing boom in the United States.

The number of patents and proprietary systems on the market today should not dissuade the engineer, architect, and contractor from using helical piles. Rather, one should conclude from the vast history of U.S. patents that the helical pile and its many variations and applications, with a few exceptions, are public domain. The helical

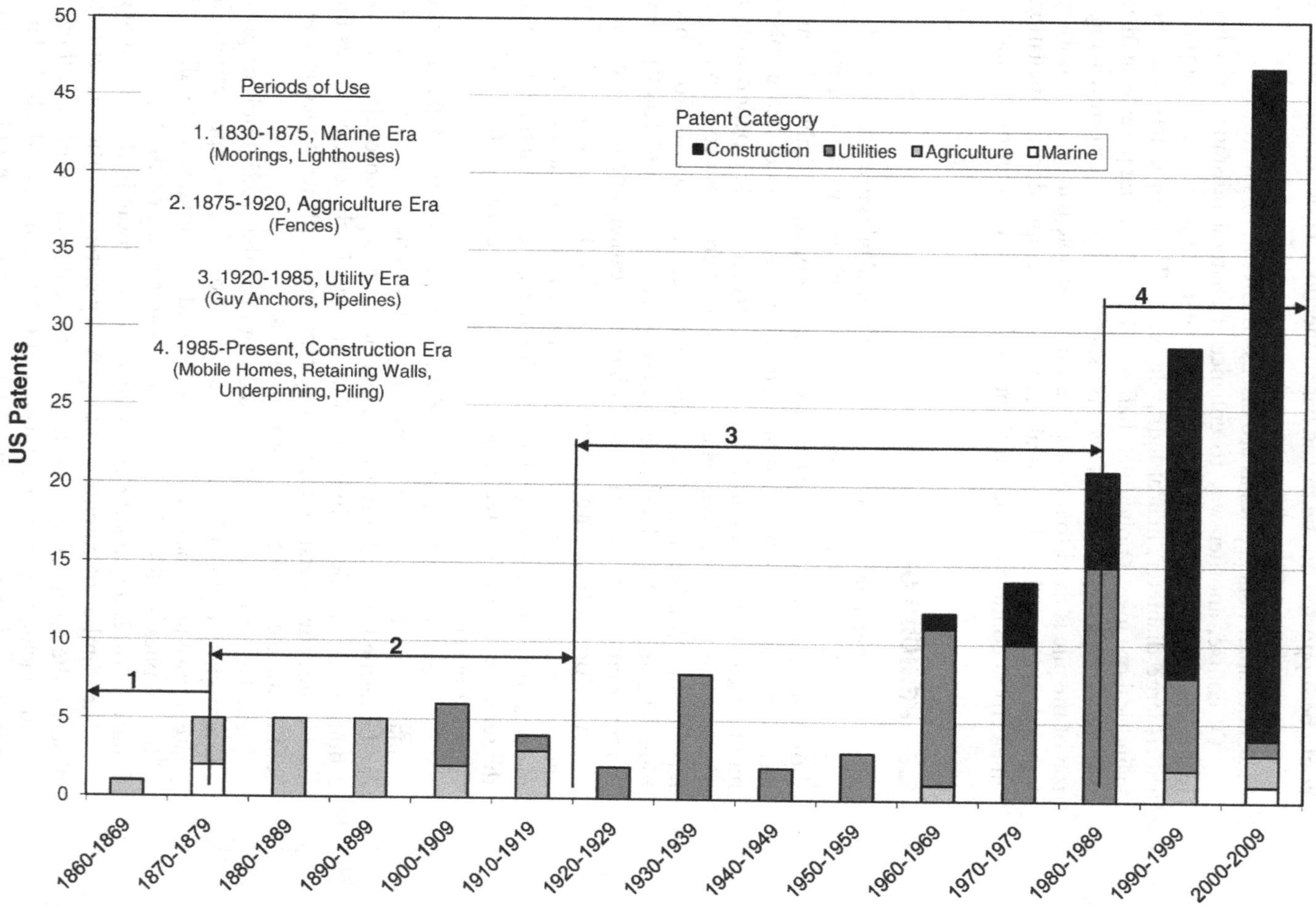

Figure 1.16 U.S. helical pile patents

pile can be used widely and diversely without fear of infringement on older patents. Newer innovations in helical piling also should be looked at with enthusiasm as these technological advancements can be drawn on when specific situations merit.

1.6 MODERN APPLICATIONS

Helical piles have many modern applications. In the electrical utility market, helical piles are used as guy wire anchors and foundations for transmission towers. For example, Figure 1.17 shows three square-shaft helical anchors embedded into the ground at a batter angle and attached to five high-tension guy wires. An example transmission tower foundation is shown in Figure 1.18. The tower in this image is founded on a cast-in-place concrete pile cap over several helical piles. A single helical pile can support design tensile loads typically on the order of 25 tons [222 kN]. An equivalent mass of concrete used in ballast for a transmission tower or guy wire would measure 8 feet [5.5 m] square ×5 feet [1.5 m] thick. Using helical piles and anchors can reduce the amount of concrete required and result in cost savings especially in remote sites.

In residential construction, helical piles are used for new foundations, additions, decks, and gazebos in addition to repair of existing foundations. Helical piles are being installed for an addition to single-story mountain home in Figure 1.19. Helical piles were selected as the foundation for the addition in this image due to the remoteness of the site, uncontrolled fill on the slope, difficult access, and economics. Small and maneuverable installation equipment and low mobilization cost make helical piles ideal for sites with limited access, such as narrow lots and backyards. An article in the *Journal of Light Construction* claimed that these factors combined with the speed of installation make helical piles more economical than footing-type foundations for residential additions (Soth and Sailer, 2004).

A residential deck supported on helical piles is shown in Figure 1.20. The tops of the helical piles can be seen extending from the ground surface. The helical piles

Figure 1.17 Utility guy wire anchors (Courtesy of Hubbell, Inc.)

Figure 1.18 Utility tower foundations (Courtesy of Hubbell, Inc.)

are attached to wooden posts supporting the deck by a simple U-shaped bracket. The helical piles under the deck extend only 6 feet into the ground. Helical piles were used for this deck in northern Minnesota due to the depth of frost and pervasiveness of frost heave in this area. Uniform-diameter concrete piers that bottom below frost are often heaved out of the ground by successive freeze-thaw cycling. One of the unique features of the helical pile is its resistance to frost heave and expansive soils. The slender central shaft limits the upward stresses due to soil heave, while the helical bearing plates resist uplift. Entire subdivisions with hundreds of homes and decks have been founded on helical piles in areas of frost-susceptible or expansive soils.

There are almost unlimited possibilities with helical piles in commercial construction. The lightweight and low impact of installation equipment has made helical piles an attractive alternative in environmentally sensitive wetland areas. Many miles of nature walks have been supported on helical piles. An example nature walk is shown in Figure 1.21. Every 8 feet of this nature walk is supported on a cross member spanning between two helical piles embedded deep in the soft wetland soils. Nature walks can be constructed using helical pile installation equipment supported on completed portions of the walkway so that the equipment does not disturb sensitive natural areas.

Figure 1.19 Residential addition

Some nature walks are installed when the ground is frozen during the winter season to minimize the impact of installation equipment.

The ability to install helical piles without vibration in low-headroom areas within existing buildings has resulted in their use inside many commercial buildings where new loads are planned. The photograph in Figure 1.22 shows a stadium in South Carolina where helical piles were used to support the loads of a new weightlifting and locker room addition. Helical piles were installed to refusal on bedrock at depths of 30 to 40 feet [9 to 12 m]. Project specifications called for vertical design loads of 25 tons [222 kN] at each pile location and a maximum deflection of 1/2 inch [13 mm]. Two load tests were performed on the helical piles used for the stadium project, and the measured loads and deflections met project specifications.

Another example of how helical piles have been used inside existing buildings is to support mezzanines or additional floors. Given the many advantages of helical piles including speed and ease of installation, construction of foundations inside retail or warehouse buildings can be done during off hours without disruption for the proprietor. A photograph of a new mezzanine foundation under construction is shown in

Figure 1.20 Deck and gazebo foundations (Courtesy of Magnum Piering, Inc.)

Figure 1.23. As can be seen in the figure, the work area can be sectioned off from the other areas of the building with a dust and visual barrier. Compact, low-noise equipment can be used to conduct the work.

Helical piles have been used to support staircase and elevator additions for satisfying new commercial building egress requirements for a change of use. Helical piles also have been used to support heavy manufacturing equipment within commercial buildings. The slender helical pile shaft has a high dampening ratio for resisting machine vibrations.

Helical piles can be combined in a group to carry larger loads of commercial construction. The International Building Code, Chapter 18, states that the tops of all types of piles need to be laterally braced. A common way to accomplish this is to use a minimum of three piles in a group to support column loads. Three helical piles can support design loads on the order of 75 to 600 tons [670 to 2,5,340 kN]. In this way, helical piles have been used in a variety of low- to high-rise commercial construction projects.

Another feature that makes helical piles attractive is the ability to install in almost any weather condition. Figure 1.24 shows installation of helical piles being conducted in the rain for a Skyline Chili restaurant. This project was originally designed for driven wood piling with a design capacity of 25 tons [222 kN] per pile. A contractor bid the project using helical piles as an alternative and was found to be more economical. One 25-ton [222 kN] helical pile was substituted at each driven pile location. It turned

Figure 1.21 Nature walk construction (Courtesy of Magnum Piering, Inc.)

out that adverse weather conditions would have delayed pile driving by several weeks. Thirty-five helical piles were installed in two days during the adverse weather and the project continued on schedule.

Another application of helical piles is for underground structures and excavation shoring. The project depicted in Figure 1.25 shows an excavation and pile foundation for an underground MRI research facility at Ohio State University. Helical anchors and shotcrete were used to support the staged excavation on this project, while closely spaced helical piles were used to construct the foundation. The MRI facility has 5-foot-[1.5-m-] thick reinforced concrete walls and ceiling for radiation shielding. Each of the piles is required to support a design load of 25 tons [222 kN]. The excavation was made inside an existing university building. Groundwater and soft soils were encountered at the base of the excavation. Lightweight, tracked machinery was used to install the helical piles. (Perko, 2005)

Figure 1.22 Stadium locker room addition

Another application of helical anchors and shotcrete for excavation shoring is shown in Figure 1.26. This shoring system was constructed for the basement of a retail building. The excavation was made in several stages. Reinforcing steel and manufactured drain boards were placed over the excavated soil after helical anchor installation. Several layers of shotcrete were applied over the reinforcing steel until a smooth uniform broom finish was achieved.

Helical anchors are often used as tie-backs in a variety of other shoring systems, including sheetpiling and soldier piling. They also can be used as soil nails by spacing helical bearing plates along the entire length of the shaft. Helical anchors were used to support a majority of the earth-retaining walls in Ford Field in Detroit, Michigan. Helical soil nails have been used coast to coast in the United States. With small, lightweight equipment and the short bond length of helical soil nails and helical tie-backs, many have been able to deal with site access restrictions and limitations of rights-of-way or property boundaries. A photograph of helical anchors being installed to tie back a soldier beam and lagging system for a medical building is shown in Figure 1.27. The final excavation was approximately 18 feet [5.5 m] deep. Two horizontal walers were held in place by virtue of helical anchors spaced at roughly 5 to 6 feet [1.5 to 2 m] on-center.

Figure 1.23 New mezzanine foundation (Courtesy of Earth Contact Products, Inc.)

1.7 ENVIRONMENTAL SUSTAINABILITY

Helical pile foundations are an environmentally conscientious and sustainable construction practice. The construction of a helical pile foundation consumes less raw material and requires fewer truck trips compared to other types of deep foundations. Substitution of helical piles for other deep foundations almost always reduces the carbon footprint of a foundation. Helical piles also can reduce disturbance in sensitive natural areas.

The unique configuration of helical piles consisting of large bearing surfaces and slender shafts is an efficient use of raw materials. The construction of helical piles requires on the order of 65 percent less raw materials by weight to construct compared to driven steel piles and 95 percent less raw material by weight compared to drilled shafts or augercast piles.

Helical pile foundations require fewer truck trips to and from a construction site. Installation of a helical foundation system requires the piles be shipped from the supplier to the site and mobilization of the installation machine. Construction of a drilled shaft foundation requires shipments of reinforcing steel and concrete as well as mobilization of a drill rig and often a concrete pump truck. As can be seen in Table 1.2, it takes fewer truck trips, to and from a construction site, to install a helical pile

Figure 1.24 Adverse weather installation (Courtesy of Magnum Piering, Inc.)

foundation system compared to other deep foundation systems. Fewer truck trips mean less traffic, less pollution, and less wear-and-tear on roads, streets, and highways.

Helical piles reduce the overall carbon footprint of a project in many ways. Even though helical piles are typically shipped long distances (e.g., from national supplier to construction site), the fact that helical pile foundations require less raw material by weight and fewer truck trips means that overall energy consumption for material transportation often can be much less. For a recent project, it was determined that shipping approximately 350 helical piles from Cincinnati, Ohio, to Denver, Colorado, consumed on the order of 40 percent less fuel than would be required to transport concrete and reinforcing steel from local suppliers to the site for the construction of a drilled shaft foundation with equivalent capacity and performance (Perko, 2008a). The omission of concrete for the foundation piles also reduced pollution because the production of cement is one of the leading producers of carbon emissions. On many occasions, helical piles can be installed with smaller equipment with better fuel

Figure 1.25 Helical pile foundation for underground MRI facility (Courtesy of Magnum Piering, Inc.)

Figure 1.26 Shotcrete and helical anchor shoring system

Figure 1.27 Soldier beam and lagging system with helical tie-backs (Courtesy of Earth Contact Products, Inc.)

Table 1.2 Required Truck Trips

Foundation Option	Number of Trips to/from Site	Trip Description
50 helical piles	1	truck & trailer (installation machine)
	2	flatbed tractor-trailers (helical piles)
	3	
50 drilled shafts	14	concrete trucks
	1	pump truck
	1	flatbed tractor trailer (reinforcing steel)
	1	drill rig
	17	
50 driven H-piles	2	crane delivery & pickup
	4	flatbed tractor trailers (H-Piles)
	1	pile-driving rig
	7	

economy in a shorter time period than other deep foundations. Fuel savings and less air pollution during installation of helical piles reduce the carbon footprint still further.

Helical piles make excellent low-impact foundations for projects that are located in environmentally sensitive areas, such as wetlands, prairies, or historical sites. Lightweight installation equipment minimizes disturbance, making less impact on fragile ecosystems. Structures can be constructed over marshland by keeping the machine on the constructed sections and reaching out to install the helical piles. Alternately, construction can be done during the winter season by installing helical piles from frozen ground. Overall, helical piles may be one of the most environmentally friendly deep foundation systems.

Chapter 2

Installation

Installation of helical piles is fairly simple provided proper equipment and procedures are maintained to produce consistent results. The simplicity of installation has been a catalyst to the growing popularity of helical piles. Nonetheless, installation of helical piles, like any deep foundation, can have its share of challenges due in large part to constantly changing and variable subsurface conditions. This chapter contains some of the basic equipment and procedures for helical pile installation. Injected within the basic information are some tips and suggestions to improve the installation process.

2.1 EQUIPMENT

The helical pile shaft is turned into the ground by application of torsion using a truck-mounted auger or hydraulic torque motor attached to a backhoe, fork lift, front-end loader, skid-steer loader, derrick truck, or other hydraulic machine. A photograph showing example installation equipment is shown in Figure 2.1. The principal component of the equipment is the hydraulic torque motor, which is used to apply torsion (or rotational force) to the top of the helical pile. Helical piles should be installed with high-torque, low-speed torque motors, which allow the helical bearing plates to advance with minimal soil disturbance. Torque motors commonly used for helical pile installation produce a torque of 4,500 to 80,000 foot-pounds (ft-lbs) [6,000 to 100,000 N-m], or higher. The torque motor should have clockwise and counterclockwise rotation capability and should be adjustable with respect to revolutions per minute during installation. Percussion drilling equipment is not appropriate. The torque motor should have a torque capacity equal to or greater than the minimum installation torque required for a project. It is also beneficial if the maximum torque delivered by a hydraulic torque motor is equal the maximum installation torque of the helical pile to prevent overstressing.

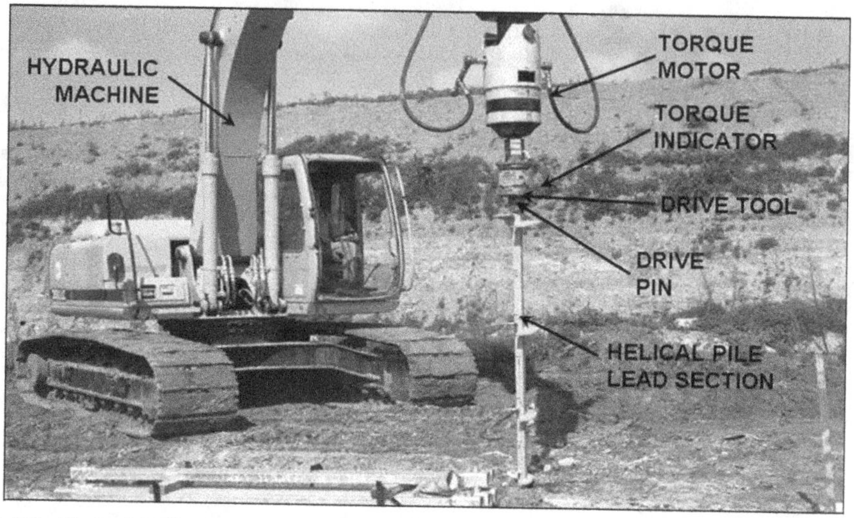

Figure 2.1 Typical helical pile installation equipment (Courtesy of Hubbell, Inc.)

Almost any hydraulic machine can be used to drive a torque motor. The hydraulic machine needs to be matched with the size of the torque motor. Torque motors generally have a minimum hydraulic flow rate and cannot be run from a small electric hydraulic power pack. In general, the higher the hydraulic flow rate, the faster the motor rotates. Typical rotation rates are between 10 and 30 revolutions per minute (rpm). Torque motors also have a recommended operating pressure range. Refer to torque motor manufacturers' technical literature for minimum hydraulic flow and operating pressure information in order to size the hydraulic machinery. Hydraulic machinery should be capable of applying crowd and torque simultaneously to ensure normal advancement of helical piles. The equipment also should be capable of maintaining proper pile alignment and position. The connection between the torque motor and the hydraulic machine should have a maximum of two pivots oriented 90 degrees from each other. Additional pivot points promote wobbling.

The connection between the torque motor and helical pile should be in-line, straight, and rigid, and should consist of a hexagonal, square, or round adapter and helical shaft socket. This is typically accomplished using a manufactured drive tool. The central shaft of the helical pile is best attached to the drive tool by a high-strength, smooth tapered pin with average diameter equal to the helical pile bolt hole size. High-strength hitch pins, the unthreaded portion of bolts, or nondeformed reinforcing steel bars are sometimes used for the drive pin. The drive pin should be as approved by the helical pile manufacturer, maintained in good condition, and safe to operate at all times. The drive pin should be regularly inspected for wear and deformation. The pin should be replaced with an identical pin when worn or damaged. Some helical piles require more than one drive pin.

A very basic, convenient, and useful aspect of most helical piles is that the capacity can be verified from the installation torque. The relationship between capacity and torque is experimentally and theoretically well established, as discussed in subsequent chapters of this book. A torque indicator should be used to measure torque during installation. The torque indicator can be an integral part of the installation equipment or externally mounted device placed in-line with the installation tooling. Most torque indicators are capable of measurements in increments with a precision of 250 to 500 ft-lbs [500 to 1,000 N-m]. Torque indicators should be calibrated prior to start of installation work. Most torque indicators have to be calibrated at an appropriately equipped test facility. Indicators that measure torque as a function of hydraulic pressure should be calibrated with the designated hydraulic machine and torque motor at normal operating temperatures. Torque indicators may be recalibrated if, in the opinion of the engineer, reasonable doubt exists as to the accuracy of the torque measurements. More information on torque measurement is contained in Section 2.5.

Helical piles also can be installed using torque motors operated by hand as shown in Figure 2.2. Hand installation equipment requires a long reaction bar. The reaction bar prevents rotation of the torque motor. The high torque produced by torque motors cannot be resisted by hand. A 6-foot [1.8 m] reaction bar will produce a 1,000 lb [4.45 kN] force at the end to prevent rotation at 6,000 ft-lbs [8,140 N-m]. Hence, care should be taken with the bracing of the reaction arm. Do not brace against an object that can become dislodged or damaged under high loads. It is important to note that the men pictured in Figure 2.2 are shown testing the stability of the reaction bar. It is inadvisable to stand on or hold the reaction bar during application of torque.

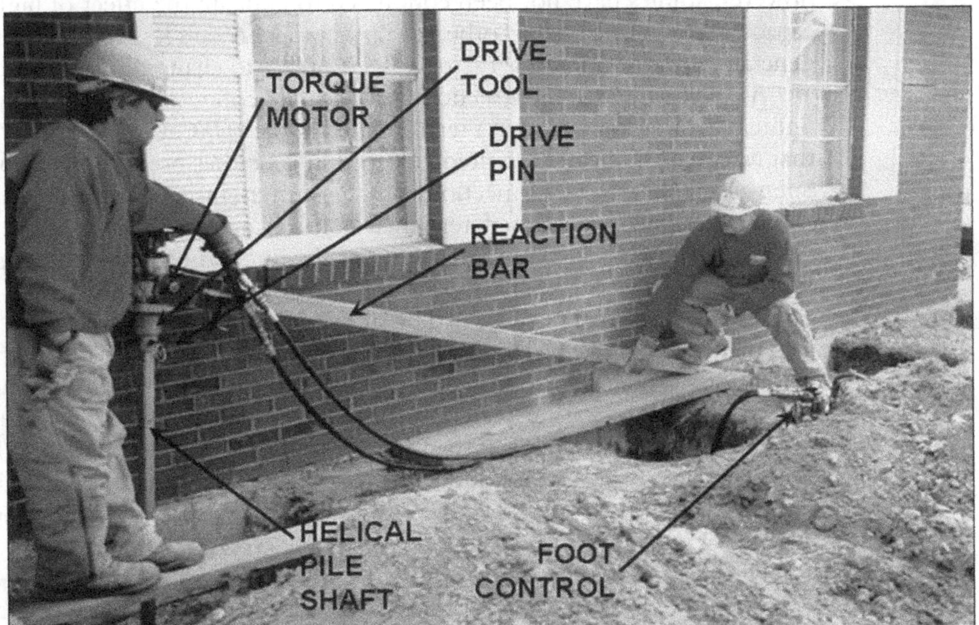

Figure 2.2 Typical hand installation equipment (Courtesy of Hubbell, Inc.)

One of the drawbacks of hand installation is that crowd cannot be applied very effectively to the top of the helical pile. Multiple helical bearing plates can help improve pile advancement when crowd is insufficient. However, multiple helical bearing plates require additional excavation when underpinning structures. Some proprietary systems can be employed to improve hand installation efficiency. Most manufacturers supply short lead and extension sections so that hand installation can be done even in crawl spaces with 2 to 4 feet [0.6 to 1.2 m] of headroom.

2.2 GENERAL PROCEDURES

General procedures for helical pile installation are shown in Figure 2.3. Installation begins by attaching the helical pile lead section to the torque motor using a drive tool and drive pin. The lead section should be positioned and aligned at the desired location and inclination. Next, crowd should be applied to force the pilot point into the ground, then plumbness and alignment of the torque motor should be checked before rotation begins. Advancement continues by adding extension sections as necessary. Plumbness should be checked periodically during installation. Installation torque and depth should be recorded at select intervals. As each extension is stopped just above the ground surface and a new extension is added, it is advantageous to stop so that the operator can directly observe the bolted connection in order to adjust alignment.

All sections should be advanced into the soil in a smooth, continuous manner at a rate of rotation typically less than 30 rpm. Installation can be done at a faster rate; however, studies have not been conducted to evaluate the effect of higher speeds on capacity to torque ratios or other pile properties. A rate less than 30 rpm also allows generally sufficient time for the operator to react to changing ground conditions.

As each new extension section is added, the connection bolts are typically snug-tightened. Snug-tightened is a term defined in the *AISC Manual of Steel Construction* that essentially indicates a bolted connection is neither pretensioned nor slip-critical. Snug-tightened bolted connections simplify design, installation, and inspection. They eliminate the need for a predetermined bolt torque or specific tension. Essentially, a snug-tightened specification indicates that the nut and bolt are well seated.

Constant axial force (crowd) should be applied while rotating helical piles into the ground. The crowd applied should be sufficient to ensure that the helical pile advances into the ground a distance equal to at least 80 percent of the blade pitch during each revolution. The amount of force required varies with soil conditions and the configuration of helical bearing plates. Insufficient crowd can result in augering wherein the pile advances at much less than the pitch during each revolution. When augering occurs, torque drops significantly and correlations between torque and capacity are no longer valid. Augering can adversely affect tensile capacity but does not necessarily indicate reduced bearing capacity.

Helical piles are generally advanced until the termination criteria are satisfied. Termination criteria for helical piles involve achieving the required final installation torque and obtaining the minimum depth, if any . . . Minimum depth generally corresponds to the planned bearing stratum. Minimum depth can be influenced by frost

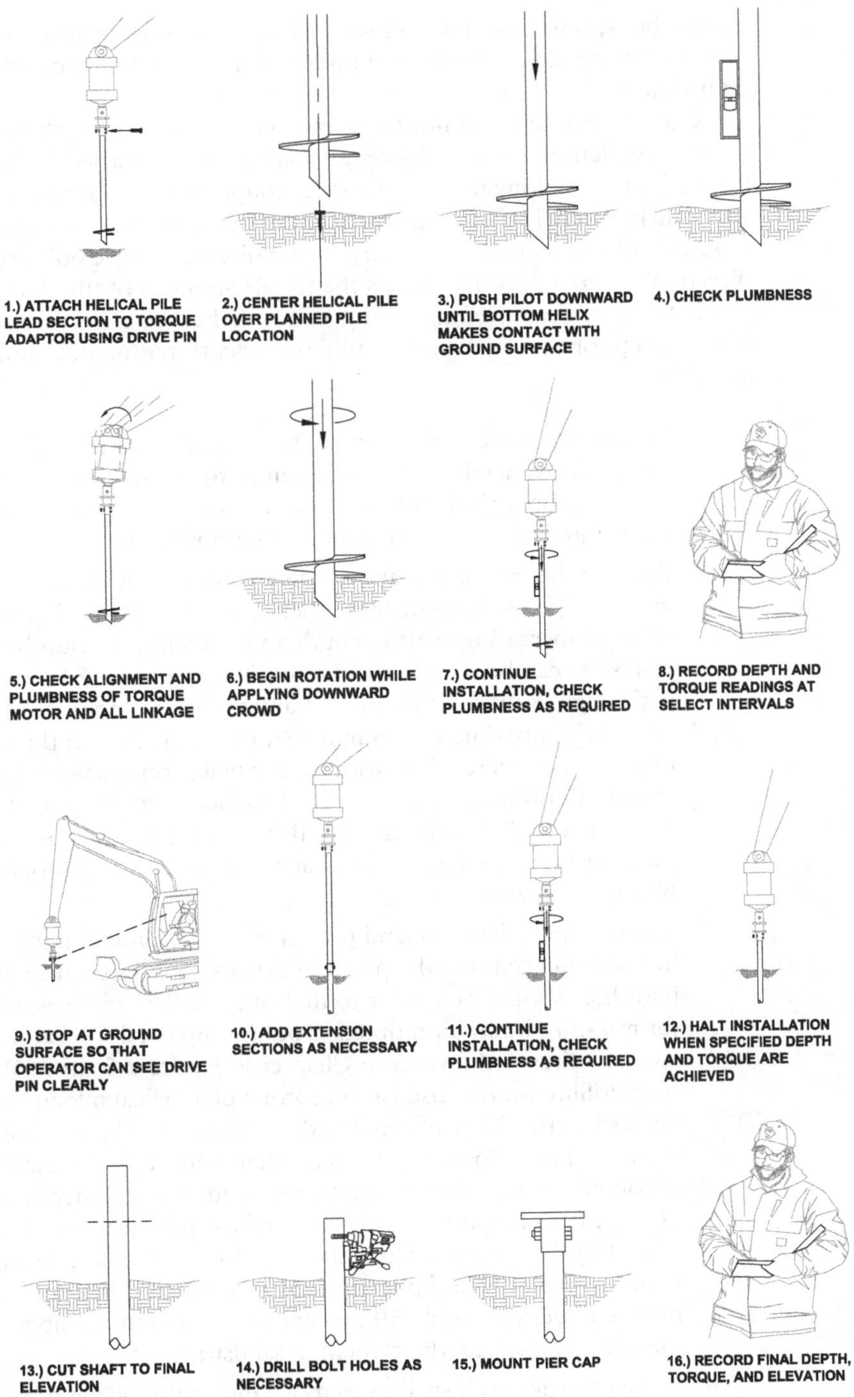

Figure 2.3 General installation procedures

susceptibility, unknown fill, soft soils, collapsible soils, expansive soils, or liquefiable soils. In tension applications, a minimum depth may be specified to ensure a certain embedment.

Care should be taken not to exceed the torsional strength rating of a helical pile during installation. Torsional strength ratings are published by helical pile manufacturers. Bolt hole elongation of the shaft coupling at the drive tool should be limited to 1/4 inch [6 mm] or less. Helical anchors with bolt hole damage exceeding this criterion should be uninstalled, removed, and discarded since bolt hole elongation in the direction of applied torque affects the tensile strength of the shaft.

If the torsional strength rating of the helical pile has been reached or augering occurs prior to achieving the minimum depth required, a number of options are available.

1. Reverse the direction of torque, back out the helical pile a distance of 1 to 2 feet [approximately 1 m], and attempt to reinstall by decreasing crowd and augering through the obstruction. For difficult ground conditions, this procedure may have to be repeated several times.
2. Remove the helical pile and install a new one with higher strength shaft and fewer and/or smaller-diameter helical bearing plates. Figure 2.4 shows the effect of increasing shaft strength and reducing the number of helical bearing plates. As can be seen in the figure, the 2 7/8-inch (73 mm-) diameter shaft with single 8-inch- (203 mm-) diameter helix encountered refusal at a blow count of approximately 86 blows/ft (50/7 inches) in this test. The 3-inch- (76-mm-) diameter shaft with similar helix configuration encountered refusal at a blow count of approximately 120 blows/ft (50/5 inches), whereas that same shaft with 8-inch- and 10-inch- (203- and 254-mm-) diameter dual-cutting-edge helix refused at a blow count of approximately 150 blows/ft (50/4 inches).
3. Remove the helical pile and predrill a small-diameter pilot hole in the same location and reinstall the pile. On expansive soil sites, the diameter of the pilot hole should be similar to that of the pile shaft so as not to create a path for moisture infiltration unless grout is employed. Precautions for expansive soils are discussed further in Chapter 9. Figure 2.5 shows the effect of pilot hole drilling on the installation torque of a helical pile installed in claystone bedrock with a layer of cemented sandstone. In this example, a helical pile with a 3-inch- (76-mm-) diameter shaft and single 12-inch (305-mm) helix encountered refusal in the cemented sandstone at a depth of about 36 feet (11 m). Project specifications required the pile penetrate to a minimum depth of 45 feet due to anticipated effects of expansive soils. As can be seen in the figure, a similar helical pile installed in a 4-inch- (102-mm-) diameter pilot hole resulted in roughly 50 percent less torque in the upper soils and allowed the pile to penetrate the cemented sandstone.
4. If the obstruction is shallow, remove the helical pile and dislodge the obstruction by surface excavation. Backfill and compact the resulting

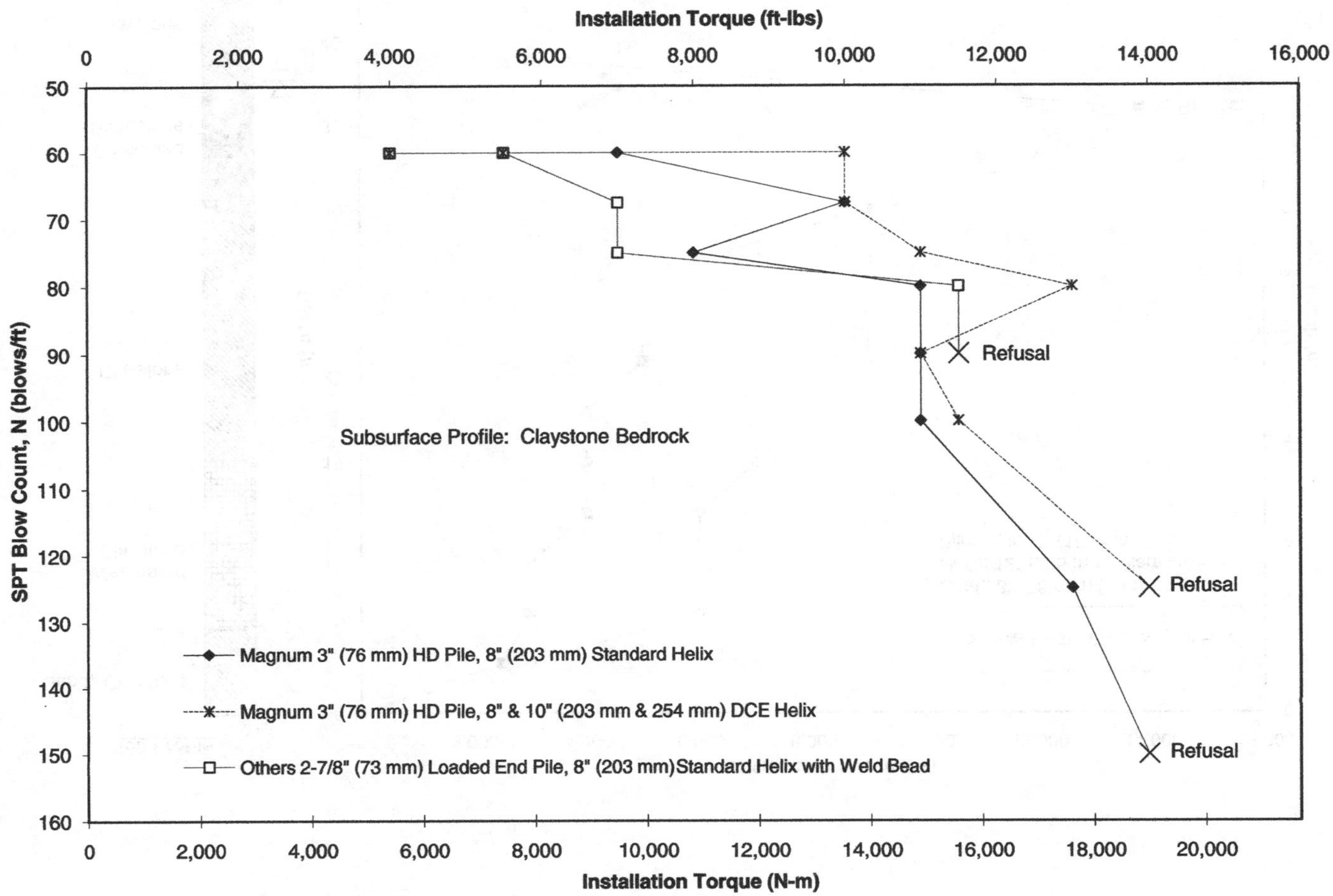

Figure 2.4 Refusal of different helical pile configurations

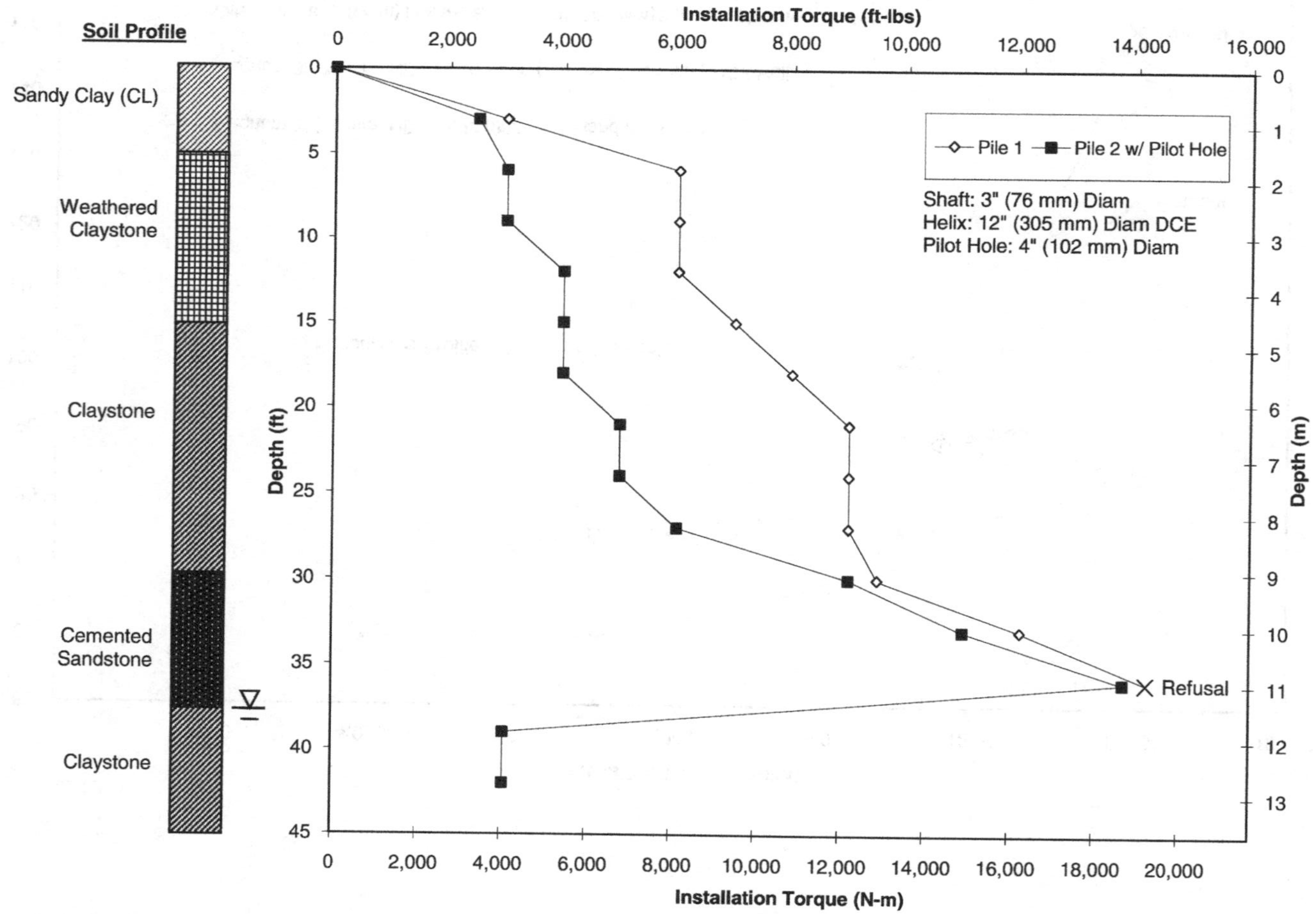

Figure 2.5 Effect of pilot hole on installation torque

Figure 2.6 Removal of shallow obstruction (Courtesy of Magnum Piering, Inc.)

excavation and reinstall the anchor/pier. An extreme example of a near-surface obstruction after removal is shown in Figure 2.6.

5. Remove the helical pile and relocate it a short distance to either side of the installation location.
6. Terminate the installation at the depth obtained and reevaluate the capacity and functionality of the pile. Installation of additional helical piles may be required.
7. Remove the helical pile and sever the uppermost helical bearing plate from the lead section if more than one helical bearing plate is in use, or decrease the diameter of the helical bearing plates by cutting with a band saw. Reinstall the pile with revised helical bearing plate configuration.
8. Remove the helical pile and use a tapered helix such as the seashell cuts shown in Figure 2.7. According to Atlas Systems, Inc., these helix shapes help to improve penetration in difficult soils. The installer is cautioned that certain cuts may be proprietary. Other effective types of helix shapes include the dual-cutting-edge helix by Magnum Piering, Inc. and the cam action helix by Dixie Electrical Manufacturing, Inc.

Modification of helical bearing plates, using a different pile or helix configuration, moving a pile, or reevaluation of capacity should be subject to review and acceptance of the engineer and owner. If the final installation torque is not achieved within a reasonable depth, a number of options are available.

1. Until the maximum depth is achieved (if any), install the helical pile deeper using additional extension sections.
2. Add an extension section with helical bearing plates in order to increase torque and bearing.
3. Remove the helical pile and install a new one with additional and/or larger-diameter helical bearing plates.
4. Decrease the rated load capacity of the helical pile and install additional helical piles at locations specified by the engineer.

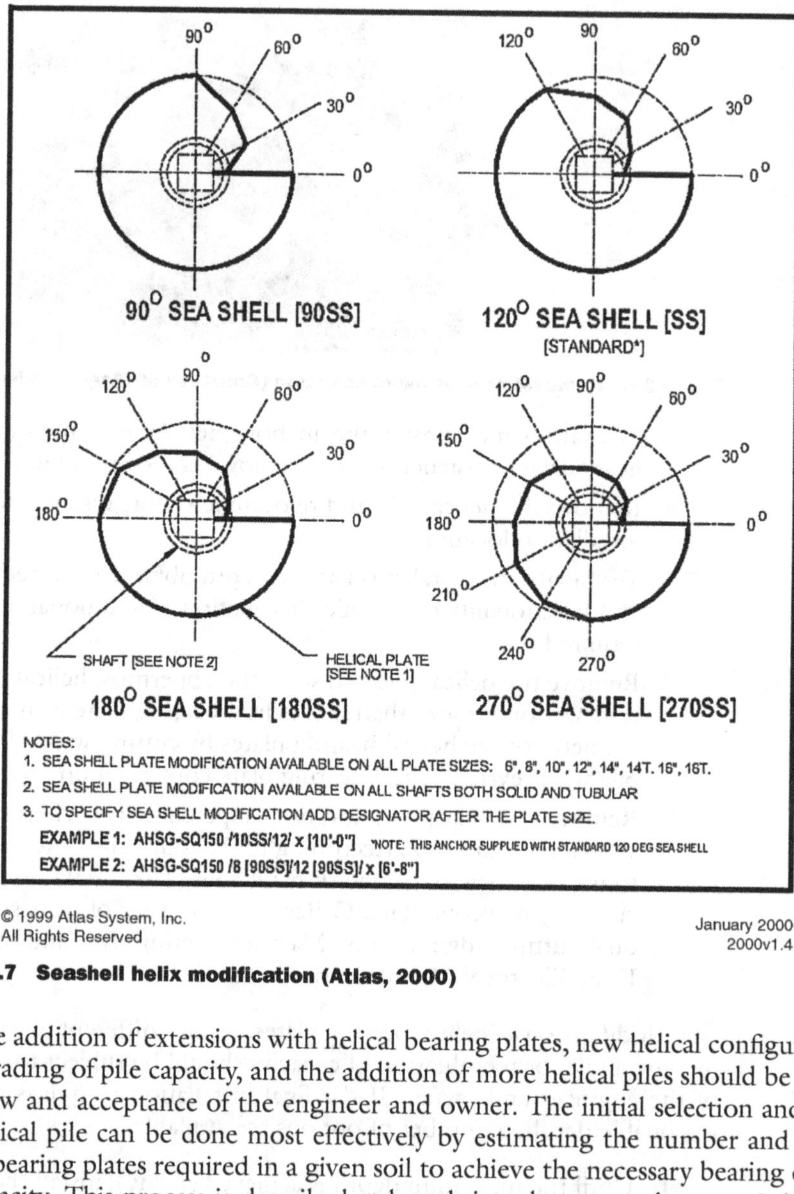

Figure 2.7 Seashell helix modification (Atlas, 2000)

The addition of extensions with helical bearing plates, new helical configurations, downgrading of pile capacity, and the addition of more helical piles should be subject to review and acceptance of the engineer and owner. The initial selection and sizing of a helical pile can be done most effectively by estimating the number and area of helical bearing plates required in a given soil to achieve the necessary bearing or pull-out capacity. This process is described at length in subsequent chapters. Subsurface exploration is required for sizing in this manner.

Where subsurface information is unavailable, a test pile program similar to that shown in Figure 2.8 can be done to evaluate the installation torque achieved with different helix configurations and thereby assess the likely holding capacity. In the study shown in the figure, helical piles were installed with two, three, and four 12-inch- (305-mm-) diameter helical bearing plates. In general, installation torque

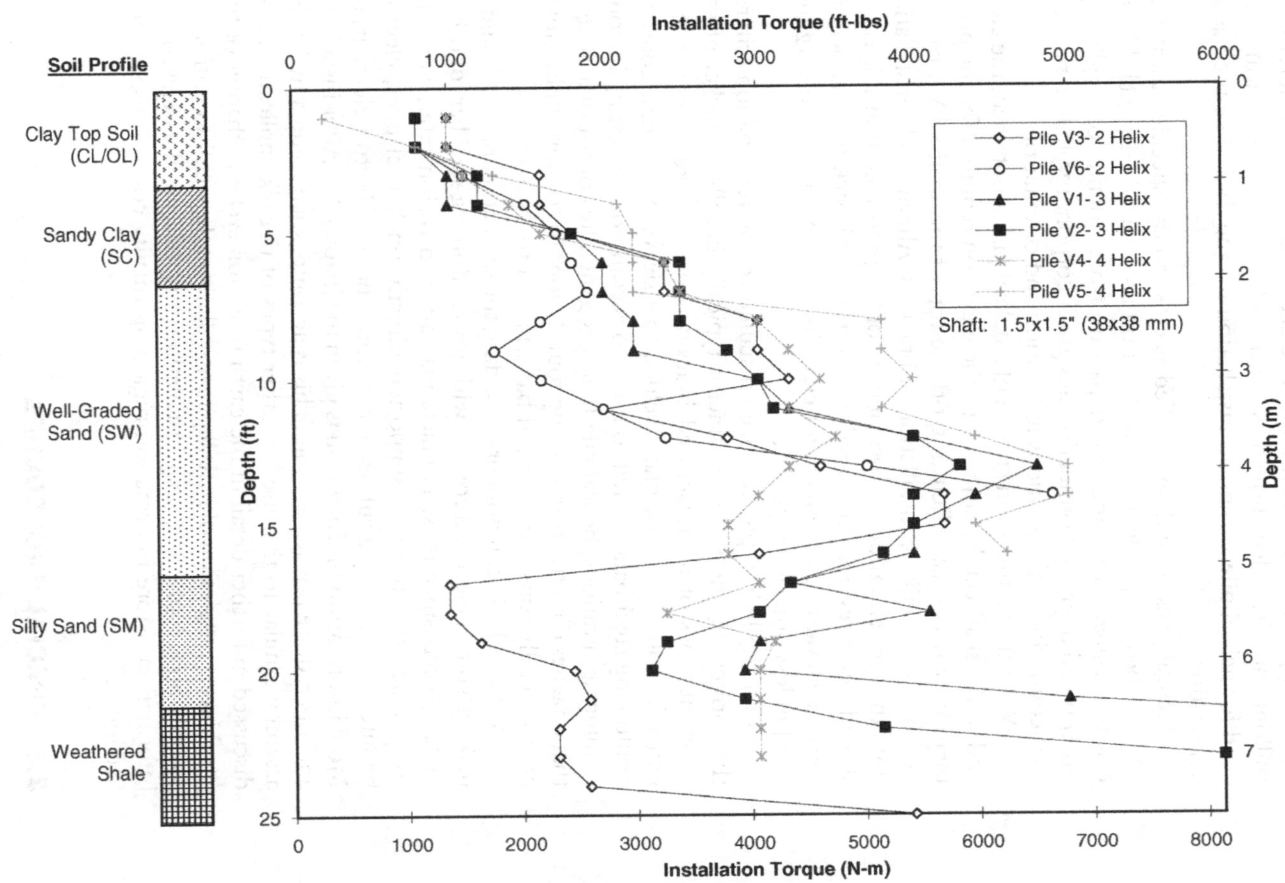

Figure 2.8 Installation torque comparison for various helical piles (Cerato, 2007)

increased with greater number of helical bearing plates. However, as can be seen, installation torque can vary significantly and is not absolute. Whenever possible, bearing or pullout capacity calculations based on traditional soil boring information should be used for helical pile sizing and installation torque should be used as a method of field verification.

Helical piles should be installed as close to the specified location and orientation angle as possible. Tolerances for location are typically ±1 inch [25 mm] unless otherwise specified. Tolerance for departure from orientation angle is generally on the order of ±5 degrees. Foundation designs incorporating helical piles should be devised to account for small variations in location and orientation.

When the termination criteria of a helical pile is obtained, the elevation of the top end of the shaft can be adjusted to the elevation required for the project. This adjustment usually consists of cutting off the top of the shaft with a band saw and drilling new holes to facilitate installation of brackets. Alternatively, installation may continue until the final elevation and orientation of the predrilled bolthole is in alignment. One should never reverse the direction of torque and back out the helical pile to achieve the final elevation. Tolerances for elevation are typically +1 to −1/2 inch [+25 to −13 mm] unless otherwise specified.

After the helical pile has been installed to the termination criteria and cut off to the correct elevation, the contractor typically installs a bracket of some type. Many different types of manufactured brackets exist. Brackets for new construction piles generally consist of a flat plate welded to a sleeve. Some new construction caps have reinforcing steel bars welded to them to accommodate casting in concrete. In underpinning applications, helical pile brackets generally consist of a plate or angle bracket that attaches directly to an existing foundation. In tension applications, helical anchor caps generally consist of a thread bar attached to a sleeve.

All helical pile components including the shaft and bracket should be isolated from making a direct electrical contact with any reinforcing steel bars or other nongalvanized metal objects since these contacts may alter corrosion rates. For applications where steel structures are to be supported directly on helical piles, electrical contact can be interrupted using a rubberized coating, nonconductive element, or Teflon buffer. More information on corrosion is given in Chapter 11. After installation, some helical anchors require posttensioning. This is accomplished using a hydraulic jack and chair assembly similar to that used in other types of tieback applications. Posttensioning is discussed in further detail in the earth retention systems chapter of this book.

Upon completion of helical pile installation, the quality control inspector typically records final depth and final installation torque as well as elevation and plumbness information. More information regarding quality control inspection is contained in Section 2.7.

2.3 SPECIAL PROCEDURES

A number of installation tips are listed in the excerpt. Most are self-explanatory. This section is dedicated to special procedures for particularly challenging site and ground conditions. The special procedures are told through a number of real examples.

Installation Tips

Tip: Clear Utilities

Always locate and clear underground utilities and structures prior to helical pile installation. Pot-hole to accurately locate utilities close to planned piles.

Tip: Reduce Pile Deviations

Use a drive tool that is closely matched to the helical pile shaft size and long enough to properly engage the pile. Short or loose-fitting drive tools can cause excessive wobble during rotation.

Tip: Repeat Lifting of Portable Torque Motors

Use a block-and-tackle system for repeat lifting of portable torque motors, especially when working with short sections in confined areas.

Tip: Controlling Plumbness

For better control of plumbness, an installation assistant should observe the helical pile at a vantage point that is perpendicular to that of the operator.

Tip: Increase Crowd

If needed, more crowd sometimes can be achieved by securing the installation machine to a helical anchor installed as a temporary reaction.

Tip: Difficult Couplings

Correct dimpled round shaft boltholes by removing the coupling bolts and rotating the drive tool several revolutions prior to decoupling. When necessary, use the machine to force the next extension.

Tip: Stubborn Bolt Holes

If necessary, align stubborn boltholes using a drift pin.

Tip: Toxic Fumes

Cutting, welding, or grinding galvanized coatings can result in the production of toxic gases. Perform these activities only with proper ventilation and safety equipment.

(Continued)

Tip: Improve Placement Accuracy

Position lead section over marker and, without rotation, use downward pressure to push pilot point into the ground until first helix rests at ground surface. Check and adjust plumbness, and then begin rotation while maintaining constant downward pressure.

Tip: Removing Helical Piles

When removing helical piles, it is important to limit the tensile force applied to the pile. Some hydraulic machines can apply excessive tensile loads to helical piles. Since they are pulling against the earth, the force is not limited by the weight of the machine.

Tip: Drilling Bolt Holes

The effectiveness of most drill bits is enhanced with low speed and high pressure. Drill pressure can be increased using a pry bar. If available, an electromagnetic drill press (Figure 2.9) is very effective.

Tip: Thru Bolts

Each side of a tubular pile shaft can be drilled independently with perfect alignment for a thru bolt by using a tubular jig with prealigned guide holes. The jig is slid over and securely clamped to the top of a pile shaft.

Tip: Limit Torque

To avoid overstressing helical piles during installation, it is sometimes feasible to install an adjustable pressure-limiting device along the inflow hydraulic line.

Several projects have been completed where rig access was impossible. One such site was a four-story, narrow apartment building. Three stories of the building were cut into a hillside. The back wall of the building was constructed with insufficient lateral restraint, and it began moving. A repair plan was devised to enhance lateral stability by installing a number of helical anchors through the back wall and deep into the hillside. Rig access to the second and third floors was impossible. In order to save time, improve crowd, and simplify installation, the contractor used a large excavator with torque motor. Working from the outside of the front of the building, the contractor incorporated extra-long extensions to reach across the building through

Figure 2.9 Electromagnetic drill press

open windows. A worker inside the building guided installation using handheld radios to communicate with the operator.

Another impossible site was a 40-year-old bridge abutment that was originally designed as a floating vault in very soft clay soils. The vault failed, filled with water, and started to settle. The toe of the vault could not be accessed easily due to the soft soils, shallow water, and environmentally sensitive coastal wetlands. The contractor worked overhead from the bridge deck and installed helical piles with long extensions to augment the existing abutment. The helical piles were bracketed to the outside of the abutment vault. The vault was successfully lifted to near its former elevation. The helical piles continue to provide support to this day.

The ability of helical piles to be installed in low-headroom conditions has been mentioned previously. An example of extreme headroom conditions was a duplex residential structure with 3-foot [0.9-m] crawl space and 12-inch [305-mm] drop beams. The building was constructed on soft soils and started to settle over time. The outside of the structure could be underpinned using conventional equipment. The repair plan also called for several helical piles installed under drop beams within the crawl space. This provided only 2 feet [0.6 m] of clearance. The contractor used a portable torque motor that required only 1 foot of clearance. The contractor also hand-dug a 1-foot- [0.3-m-] deep hole under the grade beam at each pile location. Then using special 2-foot [610-mm] leads and extensions provided by the manufacturer, the contractor was able to tediously install 30-foot [9-m] deep piles that would support design loads of 16 kips [71 kN]. A photograph of the installation is shown in Figure 2.10. A come-along supported by the floor joists was used to lift and lower the torque motor.

Figure 2.10 Low-headroom installation (Courtesy of Magnum Piering, Inc.)

Sometimes a helical pile contractor encounters projects with extremely challenging ground conditions. One example was a university building constructed inadvertently over a sinkhole. Several contractors attempted in vain to perform compaction grouting. After pumping hundreds of yards of grout under the structure, movements continued year after year. It was decided to attempt conventional underpinning using helical piles. Soil borings revealed several layers of 2,000 pounds per square inch (psi) [14 MPa] grout up to 12 inches [305 mm] thick at various depths beneath the structure. Special helical piles were manufactured to penetrate the grout. The manufacturer reinforced the helical bearing plates with tungsten carbide weld bead. Installation took extra time, but eventually the contractor was able to grind and auger through all grout lenses with the special helical bearing plates. All helical piles were advanced to a stable bearing stratum at a depth of approximately 40 to 50 feet [12 to 15 m]. The project was successful.

Another example of extreme ground conditions was a subdivision of homes built on an old ravine that had been filled with material. The upper material appeared to be suitable engineered fill. It was determined later, after much settlement and damage to the homes, that the fill was up to 60 feet [18 m] deep in the center of the subdivision. The deeper material contained construction debris, logs, branches, entire trees, and even a buried car or two. The contractor used a drill rig to penetrate the debris-laden fill and provide a pilot hole equal to the diameter of the helical pile shaft. Helical piles with special dual-cutting-edge helices were then inserted in each pilot hole until the depth of fill was penetrated and appropriate torque and bearing was achieved below the fill. With the assistance of the pilot holes, the special dual-cutting-edge helical bearing plates were able to penetrate the massive obstructions and eventually achieve bearing.

2.4 INSTALLATION SAFETY

Proper helical pile installation requires at least a two-person crew: an operator who runs the hydraulic machine and a spotter who typically handles the pile sections and performs the coupling and decoupling activities. Perhaps more important, the spotter also takes plumbness measurements, corrects positioning, checks alignment of the torque motor, and observes and helps direct the installation. One of the most important factors for safe installation of helical piles is clear communication between the operator and the spotter. Often verbal commands are difficult to hear over the noise of the hydraulic machine. It is best if the spotter combines hand signals with verbal cues.

Several different standards exist for hand signals, including those for excavation, trucking, and crane operation. There are many similarities between these standards. However, all of them have shortfalls with respect to the unique aspects of helical pile installation. Figure 2.11 shows some standard hand signals based on years of experience that could be incorporated for helical pile installation.

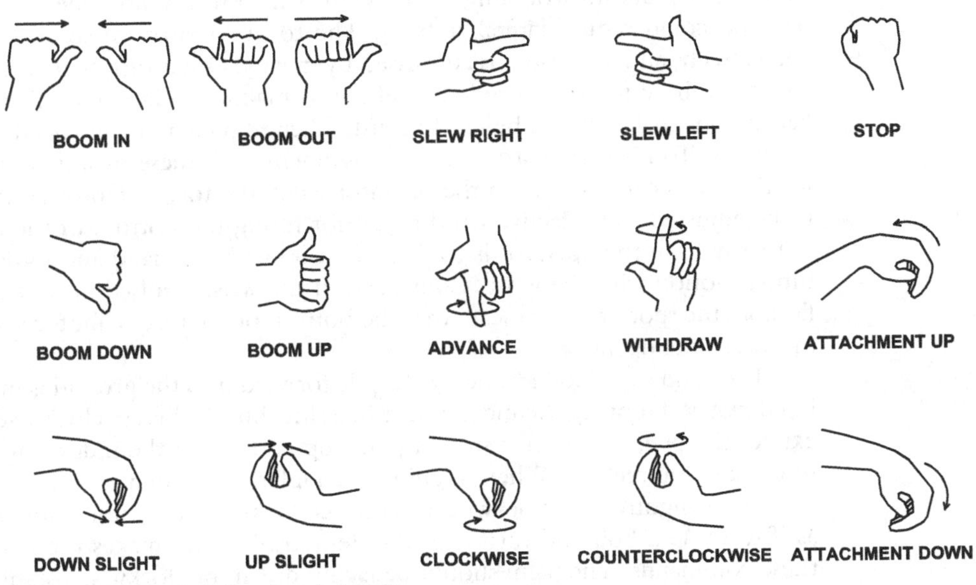

Figure 2.11 Installation hand signals

Often the operator is unable to determine position accurately in a direction perpendicular to his or her vantage point. The spotter is most effective when he or she stands stationary at a vantage point that is perpendicular to that of the operator. The spotter must be able to see the operator at all times. If the spotter cannot see the operator, then the operator cannot see him or her either. This creates an unsafe situation and interrupts proper communication.

The hand signal to "boom in," or move the torque motor toward the operator, is displayed by making a fist with each hand, palms forward, with thumbs extended toward each other. Likewise, to make the signal to "boom out," or move the torque motor away from the operator, fists are made with palms in and thumbs turned away from each other.

To cause the boom to move or slew left or right with respect to the operator's viewpoint, the spotter needs only point in that direction. To move right or left slightly, the spotter should point in that direction and make a quick but abrupt motion, as if to tap the air in that direction. The hand signal to boom down or up is communicated by making a fist and pointing the thumb upward or downward, respectively. To move the torque motor up slightly, the spotter should hold his or her hand with the palm upward and move the thumb and fingers in a pinching manner. The signal to move the torque motor down slightly is similar except with the palm facing downward.

Prior to starting and periodically during advancement, it is appropriate for the spotter to check alignment of the torque motor. The torque motor should be positioned in a straight line with the axis of the helical pile shaft. Otherwise, it is possible to induce significant wobbling of the shaft, which will wallow the hole made by the pile and could reduce lateral stability. The torque motor on some installation rigs hangs freely, and alignment is addressed by gravity. Other rigs, such as excavators and backhoes, have to be controlled carefully to maintain alignment. The advantage of backhoes is that they can be used to install helical piles at a controlled batter angle if necessary. To maintain torque motor alignment with these rigs, it is often necessary for the spotter to indicate to the operator when the torque motor attachment needs to be adjusted. An effective hand signal for tilting the bottom of the torque motor out away from the operator is to form a scoop with the hand and signal an outward tilting motion with the wrist. Conversely, if the wrist is jolted inward in a shoveling fashion, the spotter can indicate that the bottom of the torque motor should be tilted in toward the operator.

The hand signal for advancing the pile forward into the ground is made by pointing down and moving the index finger in a slow but deliberate clockwise rotation. To extract the pile, the spotter should point up and rotate the index finger in a counterclockwise direction. When aligning a coupler, it is often necessary to rotate the drive tool slightly in a clockwise or counterclockwise direction. Moving the fingers as if to grasp a bolt and turn it in the desired direction makes the hand signals for these commands. The palm should be facing down for clockwise rotation and facing up for counterclockwise rotation. Finally, the hand signal for stopping all rotation and movement is a fist with thumb collapsed tightly over the index finger. If the spotter really likes the helical pile installation, he or she can bang the closed fist against the heart. If the spotter does not like the installation, there is one more universal hand signal that is left to the reader's imagination.

As with any heavy machinery, accidents can happen when an operator is unaware of the presence of a person or object. When an inspector, engineer, or other person enters the job site, he or she should always make eye contact with the operator prior to approaching. Likewise, eye contact is important when parking a vehicle or placing an object anywhere near heavy equipment.

The Bureau of Labor Statistics identified 253 heavy equipment–related deaths on U.S. construction sites in the excavation work industry for the years 1992 through 2002. Heavy equipment operators and construction laborers made up 63 percent of the heavy equipment deaths. Backhoes and trucks were involved in half the deaths. Rollovers were the main cause of operator death. For workers on foot and maintenance workers, being struck by heavy equipment, especially while backing up, and being struck by loads were the major causes of death. Ensuring adequate rollover protection for heavy equipment, requiring seat belts, adoption of a lock-out/tag-out standard, establishing restricted access zones around heavy equipment, and requiring spotters for workers who must be near heavy equipment would reduce the risk of heavy equipment deaths in construction (McCann, 2006).

Serious injury or death can occur if a helical pile encounters an underground electric, high-pressure water, or gas line. Severing a buried fiber optic cable or other communications lines can be an extremely expensive proposition. It is imperative that all underground utilities and structures be located and cleared prior to pile installation. Most areas have free locate services for finding and marking public underground utilities. Service companies are available for locating private utilities. Site plans often show utilities on and around a construction site, but these should not be relied on for final utility locations. Helical piles should not be placed within three helix diameters of a buried utility or the uncertainty in the locate plus 1 foot (0.3 m), whichever is greater. If a helical pile must be installed closer to a utility, potholing should be done with water, air, or vacuum in order to find the line and allow the helical pile to be placed reliably alongside or below the utility.

Be conscious of overhead power lines. Serious injury or death can result if a helical pile rig is boomed up into or near a power line. There are cases of electrocution from arcing of electricity from high-voltage lines to drill rigs and other high-mast equipment that get too close to a line.

One of the more dangerous pieces of equipment used for helical pile installation is the reaction bar that resists rotation of hand-operated torque motors. Prior to starting rotation, the reaction bar should be securely braced against an immovable structure. Double-check to be sure it is braced in the correct direction. Do not stand on or otherwise hold the reaction bar during installation. Do not stand on the straining side of the reaction bar. The best policy is to stay clear of the reaction bar in case the operator switches direction inadvertently.

There is inherent danger with all hydraulic equipment. Hydraulic hoses, lines, and hose couplings can break. Damaged and worn hydraulic components should be replaced immediately when detected. All original safety equipment on heavy machinery should be kept in good working condition. Never exceed the pressure ratings of any of the hydraulic equipment. It is best to stay clear of hydraulic machines, lines, and equipment to the extent possible.

Do not exceed the maximum torque rating of a helical pile. Replace bent, worn, or otherwise damaged drive pins. Exceeding the torsional capacity of a helical pile or breaking a drive pin sometimes can result in debris being thrown into the air. Eye and ear protection should be worn at all times. Hard hats, steel-toe shoes, and gloves can help reduce injuries. Many helical pile sections weigh in excess of 75 pounds [34 kg] or more. The Occupational Safety and Health Administration (OSHA) does not provide standard values for the maximum weight that a worker can lift. However, insurance policies and job site safety programs may specify maximum weight. Some common policies allow lifting up to 50 pounds keeping the weight in close and using the legs, team lifting up to 50 pounds per person, and using hauling/lifting equipment including carts, and requiring use of dollies, cranes, and winches for heavier loads whenever practicable.

Stubborn boltholes in couplings can be aligned using a drift pin and hammer. The amount of toque used for helical pile installation is sufficient to sever an appendage. Never place a finger in a bolthole. Keep loose clothing and hair away from rotating components. Coupling sleeves can be another source of danger. Keep fingers away from the coupling when it is being made. Guide the coupling by holding the sections well above or below the coupling opening.

Many helical pile projects require excavation. Contractors should ensure safe excavations at all times. More than 30 construction workers are killed each year in the United States in trenching and excavation-related incidents. Many more suffer injuries and near misses. Placing a spoil pile near the edge of a slope and standing water contribute to roughly 35 percent of excavation cave-ins. Although cave-ins are the major event leading to injury, being struck by a falling object caused 5 percent of injuries in recent years. Workers falling into an excavation result in 8 percent of all accidents. A cubic yard of soil can weigh between 3,000 and 4,000 pounds [1,400 and 1,800 kg] depending on soil type and moisture content. Excavation safety should be taken very seriously at all times (Plog, et al., 2006).

OSHA establishes standard practices for trench and excavation safety (OSHA, 2008). According to this standard, it is permissible to make short-term excavations with a maximum 3.5-foot (1.1-m) vertical cut provided the soil above that is sloped at 3/4:1 if the total depth of excavation is less than 8 feet (2.4 m) or 1:1 if the total depth of excavation is less than 12 feet (3.7 m). For further clarification, the two basic geometries for short-term excavations are shown in Figure 2.12. OSHA defines "short term" as less than 24 hours.

For longer-term excavations, the contractor should employ a competent person who can identify soil and rock types and assist in planning and monitoring excavations. OSHA defines stable rock and three major soil categories as shown in the gray shaded box. These definitions are adapted from OSHA (2008), NCDOL (2008), and general rules of thumb based on geotechnical relationships given in Chapter 3.

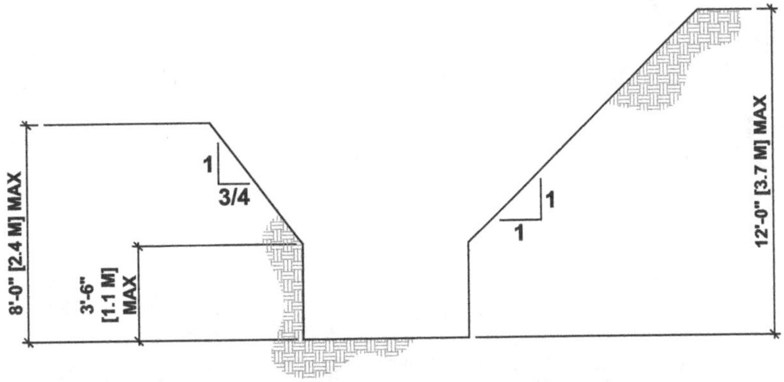

Figure 2.12 Short-term excavations (Modified from NCDOL, 2008)

OSHA Soil and Rock Types

Stable Rock. Massive, competent bedrock that will remain intact while exposed with vertical sides; strike and dip of fracture patterns are not parallel to and do not slope toward the face of the excavation, respectively.

Type A Soils. Stiff to very stiff, fine-grain soils (ML, MH, CL, CH) with an unconfined compressive strength of at least 1.5 tsf (144 kPa), stable weathered bedrock, and medium dense to very dense clayey sands (SC); soils of this category typically have an SPT blow count of at least 12 blows/ft and a moisture content at or below the plastic limit; no fissures, seepage, or previous disturbance; fractures or layers dip into the excavation at less than 4:1 (horizontal:vertical) slope.

Type B Soils. Dense, angular, undisturbed coarse-grain soil with a high degree of particle interlocking and a steep angle of repose (GP, GW, SP, SW, SM), less stable weathered bedrock, and soft to medium stiff or disturbed fine-grained soils (ML, MH, CL, CH, SC) with an unconfined compressive strength between 0.5 tsf (48 kPa) and 1.5 tsf (144 kPa); fine-grain soils of this category typically have a SPT blow count between 3 and 12 blows/ft and a moisture content between the liquid limit and the plastic limit; no seepage; fractures or layers dip into the excavation at less than 4:1 (horizontal:vertical) slope.

Type C Soils. Loose, rounded, or flowing coarse-grain soils (GP, GW, SP, SW, SM) and very soft or submerged fine-grained soils (ML, MH, CL, CH, SC) with an unconfined compressive strength less than 0.5 tsf (48 kPa); soils of this category either have a blow count less than 3 and a moisture content exceeding the liquid limit, have water actively seeping from the face, are practically cohesionless, or have fractures or layers dipping into the excavation at greater than 4:1 (horizontal:vertical) slope.

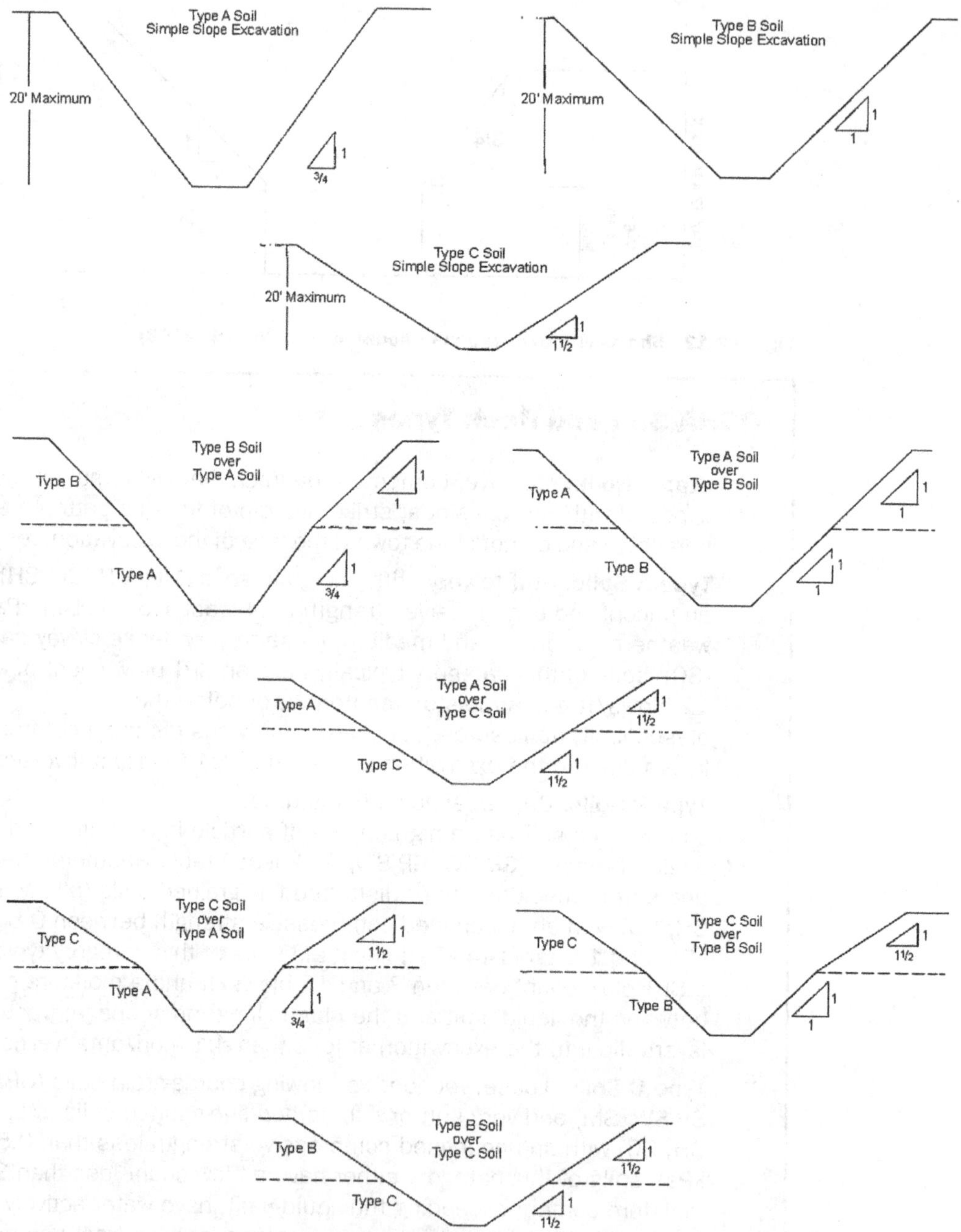

Figure 2.13 Slope configurations for excavations in various soil types

Vertical excavations in stable rock are allowed by OSHA. The slope configurations allowed in uniform and layered soil systems are shown in Figure 2.13. In cohesive soils, benching also can be used to form these slope configurations. As a general rule, benches can have a maximum height of 4 feet (1.2 m) and must be set back or located below the projected maximum slope for a given soil type. Sloping, shoring, or benching that is not in strict accordance with OSHA standards must be designed by a licensed professional engineer. OSHA excavation guidelines are limited to excavations with a maximum depth of 20 feet (6.1 m). Deeper excavations or nonconforming excavations must be designed by a licensed professional engineer. Where soil is interlayered, the classification must be based on the weakest soil layer. In two-layer systems with a more stable layer below a less stable layer, each layer may be classified individually.

It has been shown that proper planning, supervision, compliance with regulations, daily inspections by a competent person, review of accident case histories with workers, and worker training are critical factors for improving excavation and job site safety. All OSHA standards and safety precautions should be followed. This section provides only a brief overview of OSHA criteria and basic safety precautions for helical pile installation. It is neither universally applicable nor comprehensive, and it is not a substitute for a well-conceived safety plan created by qualified safety personnel and construction safety experts. If there is a contradiction between this text and any local safety codes, OSHA standards, job site safety plans, case law, or other regulations, the latter should supersede information contained in this book.

2.5 TORQUE MEASUREMENT

The torque required to advance a helical pile can be correlated to soil shear strength and capacity. These relationships are discussed at length in Chapter 6. Torque should be measured on all projects since it provides such an important verification of capacity.

There are several methods available to measure torque during helical pile installation. One of the methods is a shear pin indicator, which consists of two circular plates attached together by a central hub. One of the plates is affixed to a coupling for attachment to a helical pile. The other plate has a socket for receiving the round or hexagonal kelly bar from a torque motor. The plates have a series of 10 or 20 holes arranged along the perimeter. A number of calibrated shear pins can be placed in the holes. Without the pins, the two plates are free to spin with respect to each other. The pins shear at a certain known torque, providing a one-time measurement after which the device has to be reloaded with new shear pins. A photograph of a shear pin indicator is shown in Figure 2.14. The advantages of this device are that it is less expensive than other indicators and provides a fairly accurate direct measurement of torque. Another advantage is that it can be used to limit the maximum installation torque in order to protect a helical pile from being overstressed and damaged during installation when using a torque motor that exceeds the torsional capacity of the pile. The obvious disadvantage of the device is that it does not provide a continuous log

Figure 2.14 Shear pin indicator (Courtesy of Magnum Piering, Inc.)

of torque during installation. Shear pins can get stuck in the device and be difficult to remove, and it is somewhat cumbersome to keep reloading with shear pins.

Another method of torque measurement is the mechanical dial gauge shown in Figure 2.15. This device has an internal spring-actuated strain transducer. Torque is read directly from a calibrated dial gauge attached to the outer housing of the unit. The main advantage of this unit is that it provides a fairly accurate, continuous, direct torque reading. It also does not require electric power. The main disadvantage of the unit is that the dial gauge rotates with the pier and can be difficult to read. It cannot be read easily from within the operator station on a hydraulic machine, and a peak reading can be missed if it occurs while the unit is facing away from the observer. According to Downey (2008), the device can be damaged under excessive tension or bending, so some care must be exercised during use.

There are several different types of electronic load cells (transducers) for torque measurement. Load cells contain strain gauges that detect elastic deformation of the steel housing under torsion. The strain gauges are essentially electrical resistors that are fed current from a power supply. Strain is measured based on the principle that a change in length translates to a change in wire diameter and resistance. A data acquisition device detects the change in resistance and displays torque digitally. Electronic load cells can be very accurate and provide a direct measurement of torque. The digital display can be placed in the operator station of a hydraulic machine. On more sophisticated devices, torque can be recorded on a laptop or other electronic

Figure 2.15 Mechanical dial indicator (Courtesy of Hubbell, Inc.)

media. Disadvantages of electronic load cells are that they can be expensive, there are few devices that have been perfected, they are not readily available to the installation contractor, and some are very fragile.

In an early adaptation, the electronic load cell was placed in-line between the torque motor and the helical pile. Power supply and data acquisition cables were on a swivel that had to be constantly monitored to prevent the cables from wrapping around the pile during rotation. This early version was cumbersome to use and easily damaged. Modern technology has led to advances in load cells. A schematic diagram of a more modern electronic load cell is pictured in Figure 2.16. The device is placed between the hydraulic machine and the torque motor. The advantage of this device is that it is not rotating during installation and therefore can be hooked directly to the machine's power supply, and cables can be run into the operator station for display. Two disadvantages of this device are that it may increase wobble of the torque motor and the cables can be damaged easily. Another more modern version is the

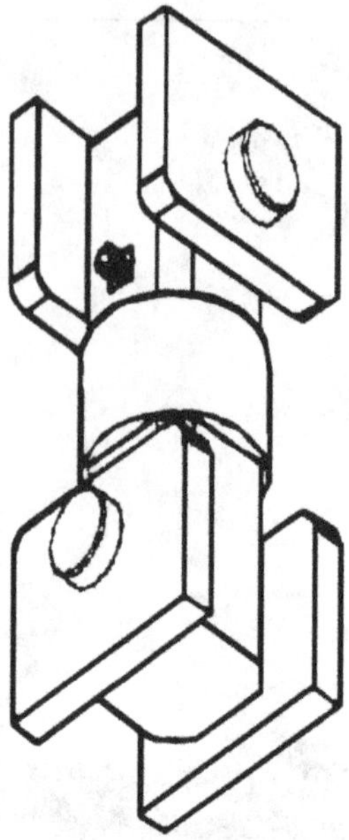

Figure 2.16 Electronic load cell (Courtesy of Pier-Tech, Inc.)

battery-powered, wireless device pictured in Figure 2.17. This device, called Torque-Master, is placed in-line between the torque motor and the pile shaft and is available through Magnum Piering, Inc. The rechargeable battery lasts several days. The wireless transmitter has a range of several hundred feet [30 to 60 m] and comes with one or more receivers for displaying and recording torque.

Torque can be measured indirectly by monitoring the pressure drop across a hydraulic torque motor. The advantage of this method is that torque can be assessed using a pair of very inexpensive pressure dial gauges. The pressure drop is simply determined by reading the input pressure and subtracting the back pressure. The latter remains fairly constant throughout most of the installation so that monitoring both gauges is practicable. It is possible to miss a spike in readings when trying to monitor two gauges. An example of the correlation between pressure drop across a motor and the output torque is shown in Figure 2.18. Each motor has a unique correlation between torque and capacity. Manufacturers often provide theoretical relationships. These should be verified at least annually. Torque motors can lose efficiency with wear, so torque correlation curves can change over time. The location of gauges, length of hydraulic hose, and fittings located between the gauges and torque motor also can affect the accuracy of readings. Frequent (e.g., annual) calibration is required to obtain accurate torque readings due to equipment wear. Torque correlation curves also may change with different hydraulic fluid viscosities, flow velocities, and operating temperatures.

Figure 2.17 TorqueMaster device (Courtesy of Magnum Piering, Inc.)

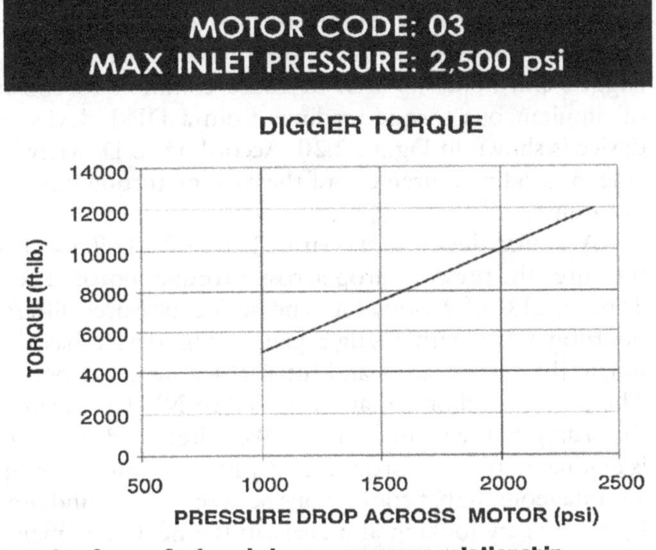

Figure 2.18 Example of manufacturer's torque-pressure relationship

Figure 2.19 DP-1 Torque indicator

Another somewhat ingenious torque measuring device is the DP-1 gauge shown in Figure 2.19. This device consists of a long slender housing with hose fittings, bolt plates, and a pressure gauge. The input and output hoses from the torque motor are spliced to opposite sides of the device. The dial gauge provides a direct reading of the differential hydraulic pressure. The device works by mechanically subtracting the back pressure from the input pressure through the use of opposing hydraulic cylinders. The disadvantages of this device are that it is difficult to read from the operator station of a hydraulic machine and the torque reading is subject to the inaccuracies described earlier for pressure to torque correlations. The advantages of the device are that it is fairly simple to use and reasonably affordable. It is also fairly rugged and holds up well to abuse common to construction sites. A comparison of simultaneous torque readings from a DP-1 device and a mechanical dial gauge device is shown in Figure 2.20. According to Downey (2008), the DP-1 device provides a good measurement of the average torque but may miss sharp drops or spikes in torque.

A similar device is shown in Figure 2.21. This device, named TruTorque, also measures the pressure drop across a torque motor. The device is available from TruTorque, LLC of Austin, TX. The device measures differential pressure using multiple bourdon tubes with a single gauge. The device has the advantage in that it is less fragile than transducers and internal leakage, if it occurs, is immediately detectable. The device is calibrated and certified to NIST standards for a specific torque motor. According to the manufacturer's Web site, the dial gauge on the TruTorque indicator is graduated to read torque directly in foot-pounds or equivalent metric units. This is advantageous in that conversions between torque and pressure do not need to be made by hand or by looking at a chart in the field. The main disadvantage is the possible inaccuracy inherent in indirect measurement of torque using differential pressure

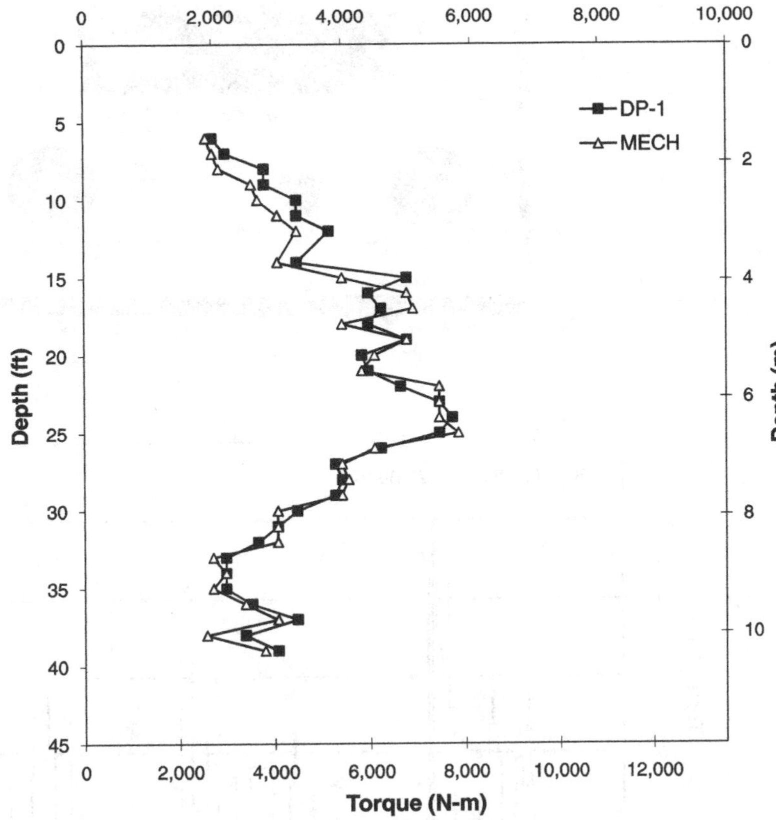

Figure 2.20 Comparison between DP-1 and mechanical torque measurements (Downey, 2008)

Figure 2.21 TruTorque indicator (Courtesy of TruTorque, LLC)

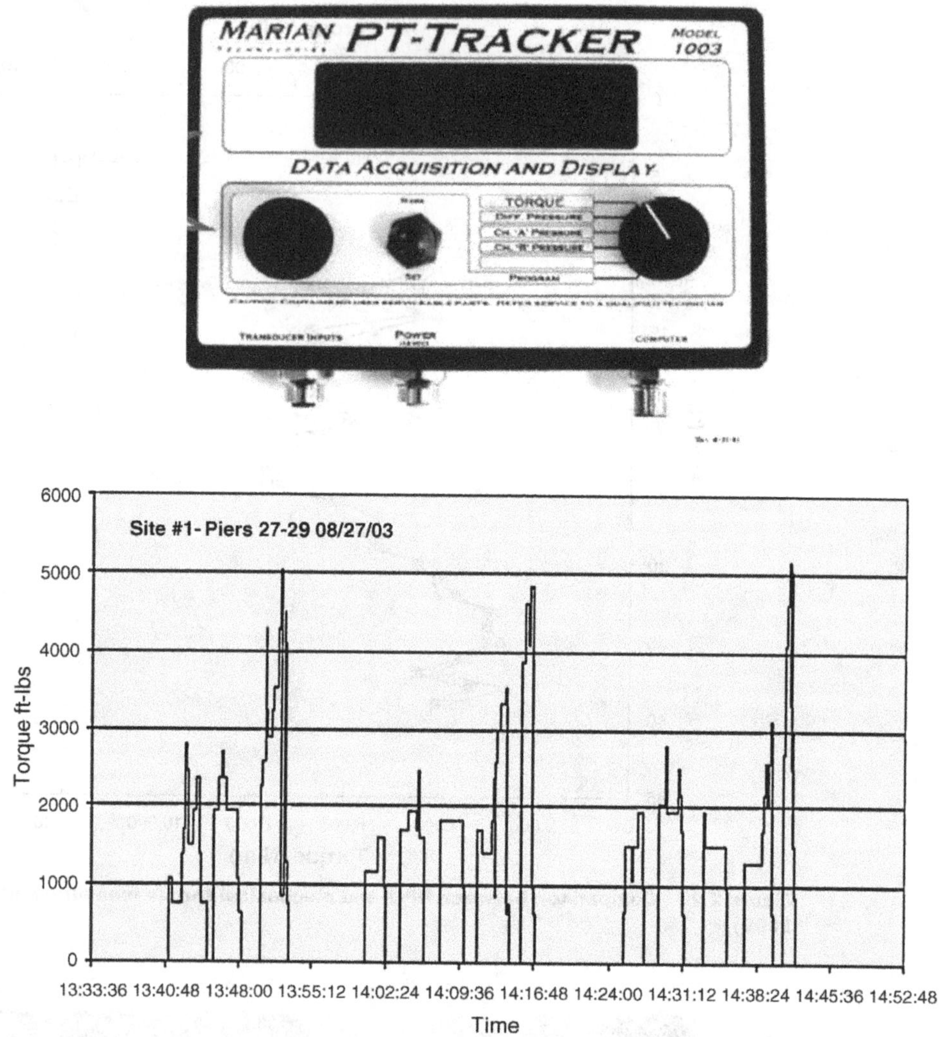

Figure 2.22 Electronic pressure indicator (Courtesy of Marian Technologies, Inc.)

measuring devices. The author has no experience with this device, its dependability, and how it is calibrated for different machines and torque motors, but it appears promising.

A sophisticated way to measure the pressure differential across a torque motor is the use of pressure transducers as in the PT-Tracker by Marian Technologies, Inc. shown in Figure 2.22. This device uses a pair of electronic pressure transducers to measure the input and back pressure. The inventors of this device provide software for a laptop that converts the pressure differential to a torque reading using torque correlation curves supplied by torque motor manufacturers. The theoretical calibration curves from almost all readily available torque motors are preloaded into the software. An installer also can purchase a PT-Tracker with digital readout wherein the theoretical torque correlations are hard-coded onto firmware within the device itself. In all cases, the operator needs only to select the torque motor to obtain torque readings. If a computer is used, the torque readings can be displayed over time as shown in Figure 2.22, providing a useful continuous log of installation activity. Still, this device is subject to the limitations and inaccuracies of correlating pressure with torque as discussed previously.

Perhaps the simplest way to measure torque is to install a helical pile to the maximum or stall torque provided by a particular torque motor. For example, the torque is generally known when a certain rated torque motor stalls during installation of a helical pile. As described, torque motors can experience decreased efficiency with time and wear, so this method should be approached with some caution. Torque motors can be as much as 30 to 40 percent or more inefficient. Stall torque should be measured occasionally using another calibrated device to check efficiency. Stall torque, although simple, is used fairly frequently on small projects with success.

One thread of advice that can be offered through years of installation and inspection of various helical pile systems, different ground conditions, and a multitude of installation equipment is that the most reliable torque measurements are obtained from redundant sources. The conscientious installation contractor and inspector should rely on multiple observations. Helical piles themselves have very consistent responses to torque such as twisting of the shaft or slight elongation of a bolthole at the maximum installation torque. These observations, when coupled with stall torque, the pressure drop across a torque motor, and if possible readings from a mechanical or electric torque indicator, provide indisputable evidence that the correct torque was achieved.

Several examples of twisted square-shaft helical anchor shafts are shown in Figure 2.23. The number of twists per unit length of shaft can be estimated for each anchor using a tape measure. Torque can be estimated from a relationship between shaft twist and torque similar to that shown in Figure 2.24. This relationship is for SS5 anchors manufactured by the A.B. Chance Company. Other manufacturers should be able to provide a relationship between torque and shaft twist for their products upon request.

2.6 TORQUE CALIBRATION

Torque measurement devices should be calibrated at least annually and whenever torque readings are questionable. An example calibration for an electronic load cell torque measuring device is shown in Figure 2.25. These data were obtained from running three load tests on the TorqueMaster device in a calibrated Tinius-Olson electromechanical torsion load bench. The coefficient of determination (i.e., R-squared value) is 0.9998. Standard uncertainty may be taken as the statistically estimated standard deviation for a normal distribution. Standard deviation can be determined using a simple spreadsheet and calculating correlations between measured torque and actual

Figure 2.23 Twisted helical anchor shafts (Downey, 2008)

torque. The standard deviation of the correlation is 0.024 and therefore the uncertainty of measurement is ±2.4 percent.

An example calibration for a set of pressure gauges used to measure the pressure drop across a torque motor is shown in Figure 2.26. These data were obtained by installing three helical piles and simultaneously measuring torque with a calibrated electronic load cell and a set of pressure gauges. The torque motor used in this example had two speeds: low torque and high torque. The coefficients of determination and standard deviations in low and high speeds are 0.9587 and 0.9855 and 0.040 and 0.175, respectively. The use of dual pressure gauges to measure torque in this example has an uncertainty of measurement of ±4.0 percent in low torque and ±17.5 percent in high torque. Percent uncertainty appears to be greatest at low torque readings for this motor and hydraulic machine.

An example calibration for a differential pressure gauge used to measure the pressure drop across a torque motor is shown in Figure 2.27. These data were obtained using similar methods as the last calibration except that pressure differential was measured directly using a DP-1 device. The torque motor in this example had a single speed. The coefficient of determination and standard deviation are 0.9906 and 0.140, respectively. The DP-1 in this example has an uncertainty of measurement of ±14.0 percent. One would expect that most of the uncertainty of measurement associated with using hydraulic pressure—whether with pairs of pressure gauges or with a differential pressure device—is a function of the torque motor type, age, and wear as well as the speed of installation, hydraulic machine type, hoses and fittings, temperature, hydraulic fluid, and other factors.

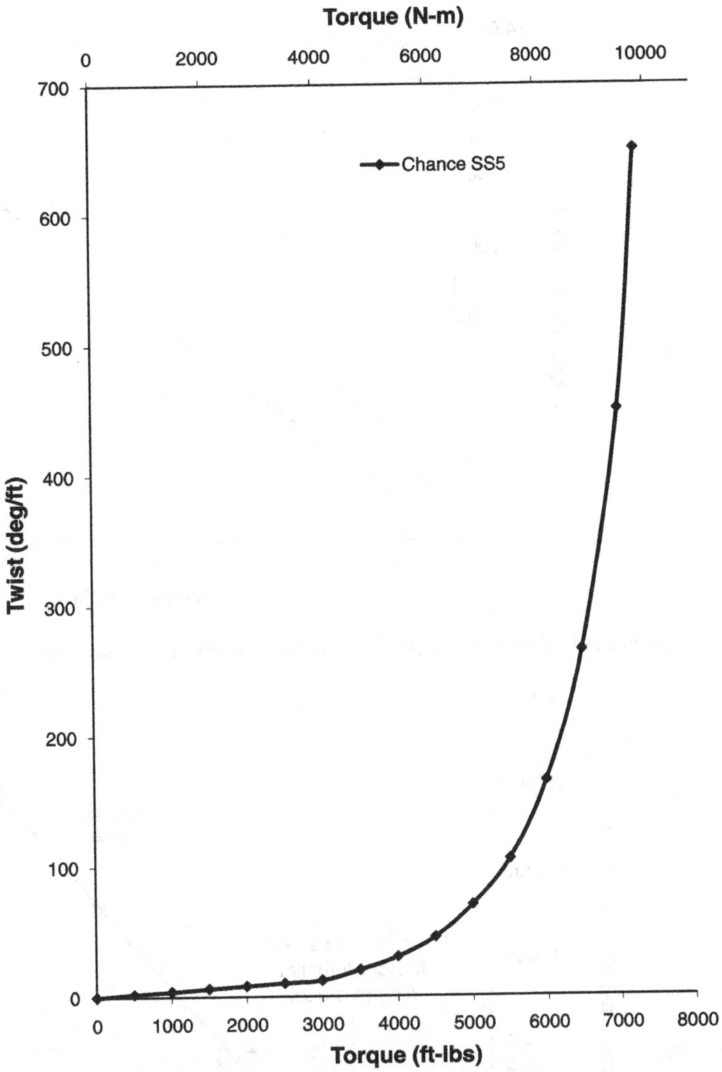

Figure 2.24 Example shaft twist to torque relationship (Downey, 2008)

When the uncertainties of the example calibrations are compared, one can conclude that load cells are more accurate than differential pressure methods. This is to be expected based on the discussions in Section 2.5. For the preceding examples, the load cell had an uncertainty of ±2.4 percent. The differential pressure gauges had uncertainties ranging from ±4.0 percent to ±17.5 percent with an average of ±11.8 percent. This amount of uncertainty still may be within reason for helical pile torque measurement, depending on the criticality of the project and especially considering the variability of underground conditions.

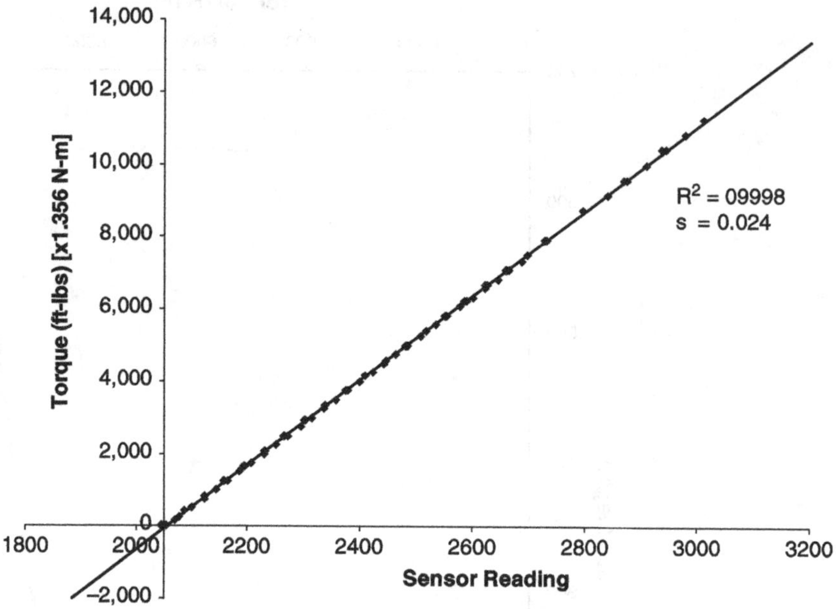

Figure 2.25 Example torque calibration for electronic load cell

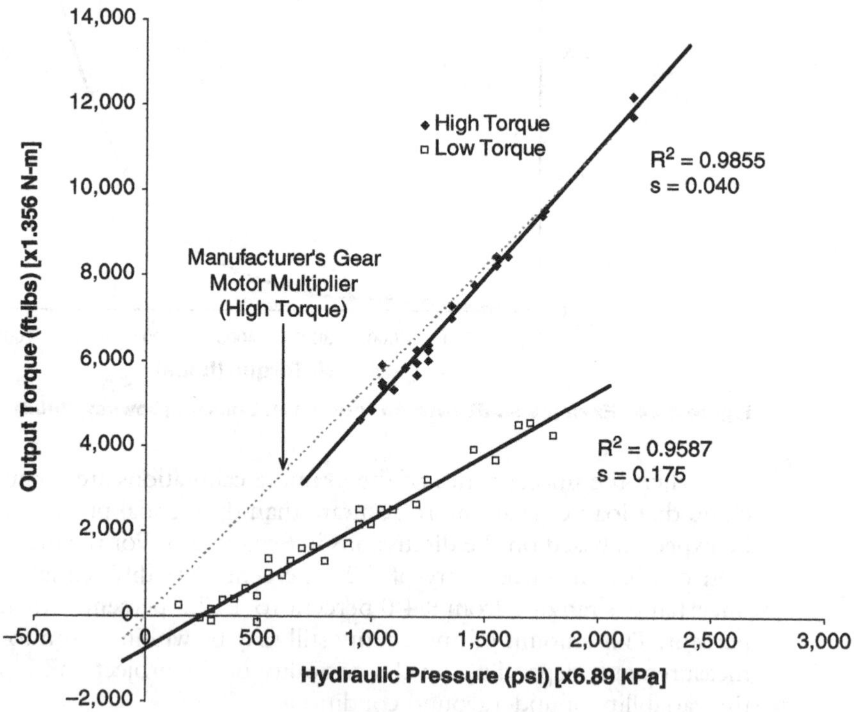

Figure 2.26 Example torque calibration for pressure gauges

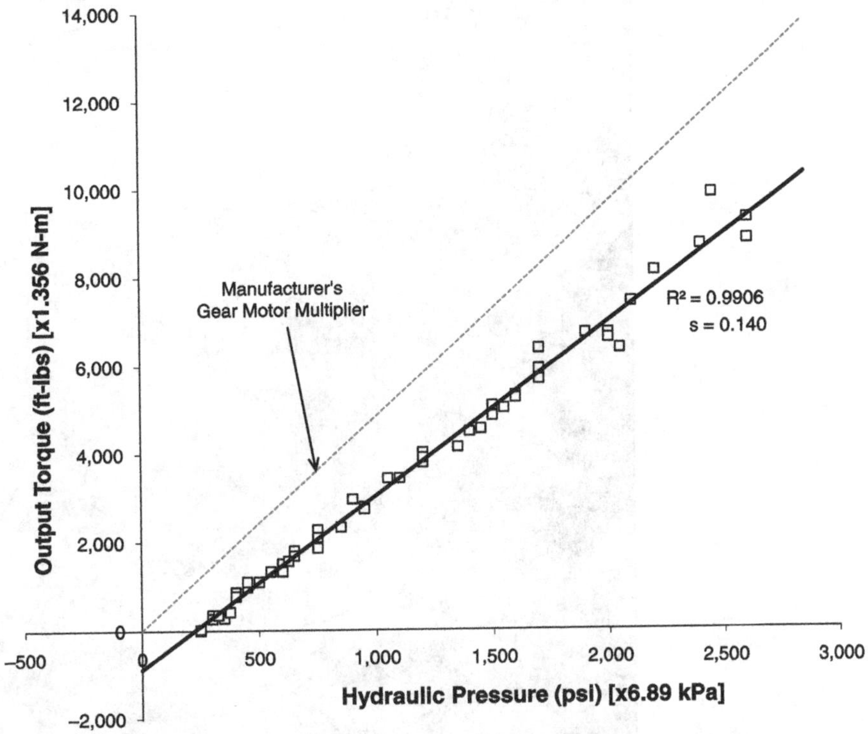

Figure 2.27 Example calibration for differential pressure device

Many torque measurement methods rely on torque motor manufacturer's gear motor multiplier (GMM). Manufacturers use semi-empirical equations to determine this multiplier. Motor GMM varies with planetary gear ratio, motor displacement, and motor efficiency. The correlation between torque and capacity for the two different motors tested in the example calibrations is shown by the dashed gray line in Figures 2.26 and 2.27. As can be seen, the correlation based on GMM is fairly close for one of the motors but is more than 40 percent off for the other motor. Deardorff (2007) found that GMM changes significantly with gear motor age (hours used), replacement of seals, rebuilding of the planetary gears, or replacement of the hydraulic motor. Therefore, it is recommended that gear motors be certified on an annual basis or whenever changes occur to alter their performance.

2.7 FIELD INSPECTION

Owners and engineers often require inspection of helical pile installations (Figure 2.28). The inspector's job is to verify that helical piles are being installed in general accord with project plans and specifications. In some locales, this inspection is full time, wherein every helical pile installation is observed. In other locales, it is common

Figure 2.28 Helical pile inspection

practice to observe only a sample (10 to 20 percent) of helical pile installations. In still other locales, the installation contractor provides oversight and self-inspection. Upon project completion, the contractor provides a record of final installation torque and depth to the owner and engineer. The frequency of inspection is the choice of the design team, local practice, and applicable building codes. Requirements for inspection depend on the difficulty of installation, the irregularity of subsurface conditions, the importance of the structure, and project budgets. Inspection requirements typically are specified in project documents.

In any case, it is important to keep a record of depth and torque during all helical pile installations. At a minimum, the final installation torque and final depth should always be recorded. Torque is an important field verification of capacity that is discussed at length in subsequent chapters. Torque readings with depth can provide a useful log of subsurface conditions and soil strength. Although useful, logs of torque and depth are usually not intended to replace a soil boring done by a competent geotechnical

Company Logo

Helical Pile Installation Observation Record

PROJECT NAME ______________________ PROJECT NUMBER ______________________

PROJECT LOCATION ______________________ DATE OF INSTALLATION ______________________

PILE DESIGNATION								
GRID LOCATION								
ELEVATION	GROUND SURFACE							
	PLANNED TOP OF PILE							
	DEVIATION							
POSITION	HORIZONTAL DEVIATION							
	INCLINATION ANGLE							
	INCLINATION DEVIATION							
LENGTH	REQUIRED MINIMUM							
	TOTAL INSTALLED							
	ABOVE GROUND							
LEAD GEOMETRY	LENGTH							
	NUMBER OF HELICES							
	HELIX DIAMETER							
EXTENSIONS	LENGTHS							
	HELICES							
	CUT-OFF LENGTH							
TORQUE MEASURMENTS	DEPTH (FT)							
	5							
	10							
	15							
	20							
	25							
	30							
	35							
	40							
	45							
	50							
TERMINATION CRITERIA	REQUIRED TORQUE							
	FINAL TORQUE							
	OTHER							
SUPPORT STRATUM	PLANNED TYPE							
	REQUIRED PENETRATION							
	MEASURED PENETRATION							

PILE MANUFACTURER ______________________ INSTALLATION CONTRACTOR ______________________

INSTALLATION EQUIPMENT ______________________

METHOD OF TORQUE MEASUREMENT ______________________

FIELD REPRESENTATIVE ______________________ REVIEWER ______________________

Figure 2.29 Example helical pile inspection form

firm; rather they are used to complement such an investigation. Logs of torque and depth can be compared with soil boring information to further enhance project quality assurance. If the general trend of torque and depth does not compare well with the trend of soil consistency found in soil borings, the inspector and installation contractor may recommend additional subsurface exploration with conventional soil borings to better evaluate the discrepancy. In some cases, changed conditions may be revealed that affect the overall design of the project.

A typical helical pile inspection record is shown in Figure 2.29. Installation records generally contain the pile designation, a log of depth and torque, termination criteria, and final pile length. Piles are typically numbered on as-built plans. Proper installation records should contain the hydraulic equipment, torque motor, and torque indicator make and model along with the installation contractors name, crew chief, inspection company, and inspector's name. Recent calibration certificates should be obtained for the torque indicator. In most cases, the brand and type of helical pile is recorded along with the configuration of the lead section and number and length of extension sections used. Installation records also may contain pile deviations with respect to horizontal location, elevation, and inclination (plumbness). Planned length, required capacity, and planned bearing stratum might be documented.

Another aspect of helical pile inspection that is up-and-coming is the verification of product identification numbers for comparison with International Code Council Evaluation Services, Inc. (ICC-ES) evaluation reports (ER). The popularity of helical piles has resulted in the emergence of numerous helical pile manufacturing companies. Unfortunately, not all companies have the same quality or reputation. The verification of product designation and quality at construction sites is becoming an important aspect of proper inspection. ICC-ES Acceptance Criteria No. 10 (AC10) requires that products used in construction under the jurisdiction of the *International Building Code* (IBC) or the *Uniform Building Code* (UBC) shall be clearly marked with a product identification number that is traceable back to the manufacturer's quality control program. This tracking ensures the product was manufactured to strict specifications and was checked for quality of construction, quality of materials, form, fit, and function prior to shipment to the construction site. More information about the ICC-ES and acceptance criteria for helical pile foundations is contained in Chapter 17.

The helical pile inspector is an important member of the construction team. He or she should be a representative of the foundation engineering company. If a problem occurs during installation, it is beneficial and often well worth the cost of full-time inspection to have such a representative present so that an immediate solution can be determined. Independent third-party verification of helical pile installations is also very important to the owner for quality assurance and can be beneficial to all parties involved in case a dispute arises.

Chapter 3

Basic Geotechnics

A soil report with borings and penetration testing is almost always necessary for proper design and application of helical piles. Soil reports are much easier to understand for those with a background in basic geotechnical engineering, including soil properties and exploration techniques. In this section, geotechnical topics particularly relevant to helical pile foundations are discussed. The experienced geotechnical engineer may elect to skip to the end of the chapter, where some useful index properties can be found. Others may find the simple definitions and explanations within the entire chapter valuable.

This chapter contains an overview of exploratory drilling, penetration testing, and soil sampling procedures. Soil classification, gradation, and plasticity are explained. Some guidelines are presented for selecting sites and subsurface conditions appropriate for helical pile use. The chapter concludes with a description of soil shear strength parameters, methods of determination, and some reference values. Shear strength will be used in subsequent chapters to calculate helical pile bearing, pullout, and lateral capacity.

3.1 SUBSURFACE EXPLORATION

The degree of subsurface exploration required for a project depends on the size and importance of the structure, sensitivity of the structure to foundation movement, local building codes, and local construction practices. In areas where stable soils predominate, subsurface exploration sometimes is omitted and foundations are based on local building codes. In areas where potentially unstable soils predominate, subsurface exploration is performed on building sites, even for small residential additions. Subsurface exploration should always be performed in areas where unknown fill, expansive soils, collapsible soils, soft clays, liquefiable soils, landslides, or other geologic hazards

"The soil report said to use helical piles, but I've used footings on at least 30 other projects like this."

are suspected. Table 3.1 provides guidelines for subsurface exploration. In relatively hazard-free areas, a minimum of one boring is suggested for a single-family residence, a set of five contiguous track home lots, each communication tower, every 15,000 sf [1,400 m^2] of warehouse, and every 400 lf [120 m] of retaining wall. The suggested minimum depth of borings is between 15 and 20 feet [5 and 6 m]. Where geologic hazards are suspected, the suggested minimum number and depth of borings is about two times the aforementioned numbers. Local codes and practices may require more or less exploration. In most cases, the number and depth of borings are determined by a local engineer based on experience and the standard of practice in the area at that time.

Depending on subsurface conditions, exploration can be conducted by drilling with solid-stem auger, hollow-stem auger, with and without drilling fluids as well as various dry and wet coring methods. Subsurface samples can be taken by driving split barrels, pushing thin-wall tubes, collecting auger cuttings, or coring. Subsurface exploration also can be done using exploratory test pits or trenches excavated with a backhoe. A resourceful geotechnical engineer or geologist can use just about any interaction with the soil to gain some information about the subsurface (Perko, 2006a). For example, measuring the torque and observing the advancement of a helical pile can provide information about soil shear strength and consistency.

This section focuses on one of the most common methods of subsurface investigation, exploratory drilling and drive sampling. A photograph of a typical exploratory

Table 3.1 Subsurface Exploration Guide (Modified from Atlas Systems, Inc. (2000))

	Suspected Geologic Hazard		*Relatively Hazard-free Area*	
	Number of Borings	Depth of Borings	Number of Borings	Depth of Borings
Single-family residence	2	25 ft [8 m] or 10 ft [3 m] into bearing stratum	1	15 ft [5 m]
Tract homes	1/Lot	*(as above)*	1/5 Lots	*(as above)*
Warehouse/ manufacturing building	2/20,000 sf [2/2,000 m^2]	*(as above)*	1/15,000 sf [1/1,400 m^2]	*(as above)*
Multistory commercial building	2/5,000 sf [2/500 m^2]	Height of building < 100 ft [30 m]	1/10,000 sf [1/1,000 m^2]	20 ft [6 m]
Communication tower	1/Anchor Cap & 1/Tower Base	35 ft [11 m] or 20 ft [6 m] into bearing stratum	1/Tower Base	20 ft [6 m] into Bearing Stratum
Earth retention project	1/200 lf [1/60 m]	20 ft [6 m] below toe of slope or BOE	1/400 lf [1/120 m]	10 ft [3 m] below toe of slope or BOE

BOE = bottom of excavation

drill rig is shown in Figure 3.1. In the image, a 6-inch-[152 mm] diameter, hollow-stem, continuous-flight auger is being used with a mobile drill. A typical drill crew consists of an operator and a laborer. A geotechnical engineer, geologist, or technician will watch the drilling and classify the soil and bedrock types exposed in the borings by observing the auger cuttings and drive samples taken at various depths. The subsurface profile and sample classifications are described and graphically represented on a "boring log."

An example boring log is shown in Figure 3.2. All boring logs have several features in common. Most have a graphic stick log showing the main soil and bedrock materials depicted with different patterns. These patterns can vary between companies and geographic regions. In general, a cross-hatch pattern represents fill, a dot pattern stands for sand, diagonal stripes indicate clay, and a solid fill is used for bedrock. Often graphic symbols are combined in a logical manner. For example, a dot pattern with diagonal stripes would indicate a clayey sand. Graphic logs should be accompanied in all soil reports by a legend of patterns and the associated soil and rock types. The graphic stick logs are commonly annotated with written descriptions of each major soil unit, as shown in the second column in Figure 3.2.

Most boring logs show where soil and rock samples were taken. It is common to show the depth, type, and classification of each sample. In Figure 3.2, two types of samples were taken. The abbreviations SPT and T shown in the sixth column refer to Standard Penetration Test (SPT) sample and thin-walled tube sample, respectively. The SPT test sampler is described in the next section. The thin-walled tube sampler typically consists of a 3-inch- [76-mm-] diameter steel tube with sharpened and tapered end.

Figure 3.1 Typical soil boring rig with hollow stem augers

The thin-walled sampler is hydraulically pushed into the ground and is used for soft or loose soils, because it generally causes less sample disturbance. Soil or rock classification symbols often are listed after the sample type. In the example, all soil is classified as CL and the bedrock is classified as BRCS. Some abbreviations used by engineers are nonstandard, such as BRCS, which is used by local engineers in Colorado for claystone bedrock. Each abbreviation and symbol that appears on boring logs should be defined in the soil report legend. Several common abbreviations are given in Appendix A. The Unified Soil Classification System (USCS) and associated symbols are described later in this chapter.

Many boring logs also have laboratory test results shown next to each sample. In Figure 3.2, the results of moisture content, dry density, and unconfined compressive strength tests are shown. Groundwater levels measured during drilling and several days later should be shown somewhere on the boring log. In Figure 3.2, groundwater level observations are shown in the footer block. In this example, water was found at a

LOG OF EXPLORATORY BORING NO. 3

Page 1 of 1

CLIENT	Last Chance Development, LLC	**ARCHITECT/ENGINEER**	Joe Trainer, P.E.
SITE	Mightyville, OH	**PROJECT**	Proposed Manufactuerering Plant Expansion

GRAPHIC LOG	DESCRIPTION	DEPTH (FT)	USCS SYMBOL	SAMPLES: NUMBER	SAMPLES: TYPE	SAMPLES: SPT - N BLOWS / FT.	TESTS: MOISTURE, %	TESTS: DRY DENSITY PCF	TESTS: UNCONFINED STRENGTH PSF
0.5	6" TOPSOIL								
			CL	1	SPT	4,5,8	17.8		
			CL	2	SPT	2,3,4	20.9	103	2910
	<u>SANDY LEAN CLAY</u> Brown/tan/red, moist to wet, medium stiff to stiff		CL	3	T		11.6	105	
		5							
			CL	4	SPT	5,4,7	18.9	110	
10.0		10							
	<u>CLAYSTONE</u> Dark Brown/Rust, slightly moist, hard								
			BRCS	5	SPT	20,23,37	20.0		
15.0	BOTTOM OF BORING	15							

THE STRATIFICATION LINES REPRESENT THE APPROXIMATE BOUNDARY LINES BETWEEN SOIL AND ROCK TYPES: IN SITU, THE TRANSITION MAY BE GRADUAL.

WATER LEVEL OBSERVATIONS		
WL	10.2' W.D.	9.5' A.B.
WL		
WL	Checked 24 hrs. A.B.	

Mightyville Geotechnical Incorporated

BORING STARTED	4/29/1993		
BORING COMPLETED	4/29/1993		
RIG	CME-55	FOREMAN	DS
APPROVED	HP	JOB#	200158

Figure 3.2 Example log of exploratory boring

depth of 10 feet 2 inches [3.1 m] while drilling (W.D.) and at 9 feet 5 inches [2.9 m] 24 hours after boring (A.B.). Finally, the results of field penetration resistance tests, if any, should be shown next to each drive sample on the log. The SPT blow count is listed in column 7 in Figure 3.2.

3.2 FIELD PENETRATION RESISTANCE

Many exploratory borings are conducted in conjunction with a soil and bedrock sample collection procedure known as a Standard Penetration Test (a.k.a. SPT, see ASTM D1586). The SPT test consists of lowering a split-barrel sampler into a drill hole using sections of drill rod. After reaching the bottom of the drill hole, the sampler is driven into the subsurface by repeated blows with a calibrated hammer. The standard split barrel sampler consists of a 1.5-inch [38 mm] inside diameter (I.D.), usually 24 inches [61 cm] long, thick-walled tube with tapered soil entry tip, as shown in Figure 3.3. The tube is split lengthwise so that it can be separated when the end caps are unscrewed and the soil sample can be recovered. The calibrated hammer weighs 140 pounds [64 kg] and is dropped from a height of 30 inches [76 cm] above the drill rod to deliver a consistent amount of impact energy. The number of blows delivered to the sampler for a given penetration distance is an indication of soil consistency and shear strength.

If possible, the split-barrel sampler is driven a distance of 18 inches [46 cm]. The number of hammer blows required to drive the sampler during the last 12 inches [30 cm] is called the “blow count” (a.k.a. the N-value). Often boring logs list the number of blows for a series of 6-inch [15-cm] intervals, as in Figure 3.2. The blow count is obtained by adding the second and third intervals together. For example, the blow count is 13 (5+8 =13) at the sample interval from 1 to 2 feet in depth in Figure 3.2.

SPT blow counts are accepted indicators of coarse-grain soil density, fine-grain soil stiffness, and bedrock hardness, as shown in Table 3.2. Coarse-grain soils, such as sand and gravel, are commonly described as very loose, loose, medium dense, dense, and very dense based on blow count. Fine-grain soils, such as clay and silt, are commonly described as very soft, soft, medium stiff, stiff, and very stiff based on blow count. N-values shown in Table 3.2 corresponding to these descriptions as well as correlations to unconfined strength were adapted from Peck, Hanson, and Thornburn (1965) with the exception that a definition for fluid soil was added. Field identification rules shown in the table were adapted from ASTM D2488.

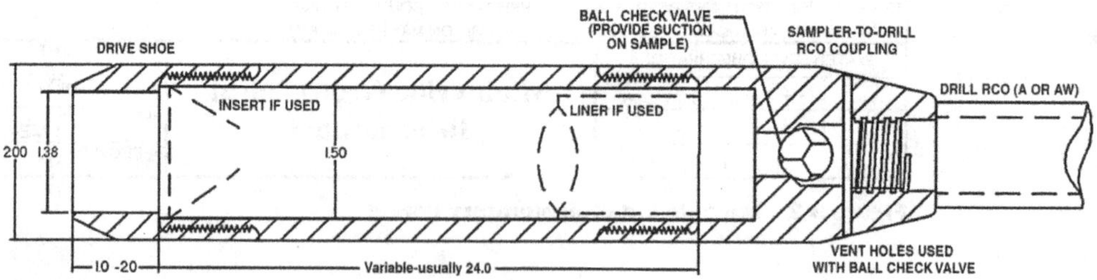

Figure 3.3 Split-barrel sampler (Bowles, 1988)

Table 3.2 Soil and Bedrock Consistency

Coarse-Grain Soils			Fine-Grain Soils					Bedrock	
Density	N-Value N_{55}	Relative Density	Stiffness	N-Value N_{55}	Field Identification	Unconfined Compressive Strength (tsf) [KPa]	Moisture Content	Hardness	N-Value N_{55}
Fluid	WOR – WOH	<10%	Fluid	WOR – WOH	Nontraversable	NIL	>>LL	Fluid	WOR - WOH
Very loose	1–4	10%–15%	Very soft	1–2	Easily penetrated by fist	<0.25 [<24]	>LL	Soft	1-30
Loose	4–10	15%–35%	Soft	2–4	Easily penetrated by thumb	0.25–0.5 [24–48]	LL	Medium	30-50
Medium	10–30	35%–65%	Medium	4–8	Moderate thumb penetration	0.5–1.0 [48–96]	PL-LL	Hard	50-100
Dense	30–50	65%–85%	Stiff	8–15	Difficult thumb penetration	1.0–2.0 [96–192]	PL	Very hard	100-BNC
Very dense	>50	>85%	Very stiff	>15	Indented by thumbnail	>2.0 [>192]	<PL	Competent	BNC

WOR=Weight of rod; WOH=Weight of hammer; LL=Liquid Limit; PL=Plastic Limit; BNC=Bounce

Weathered bedrock or weaker sedimentary rock formations can be described as soft, medium hard, hard, and very hard based on blow count. SPT testing is ineffective in more competent bedrock. Coring and rock quality designation should be used to describe competent bedrock. More information about bedrock sampling, classification, and strength is given in Section 3.4.

In very dense soils or hard bedrock, an excessive number of blows may be required to drive the sampler 12 inches [30 cm]. For practical reasons, sample driving is often stopped after 50 blows and the depth driven is recorded. For purposes of sizing helical piles, the blow count can be determined by estimating the number of additional blows that would be necessary to drive the sampler a full 12 inches (30 cm). Hence, a record of 50/6" would be equivalent to a blow count of 100 blows/ft. A record of 50/3" would be equivalent to a blow count of 200 blows/ft and so on.

In some parts of the United States, variations of the standard penetration resistance test are used. One popular variation in the western United States is the use of a modified California barrel sampler. This sampler has a slightly larger diameter compared to the standard sampler and is used to obtain samples for swell/consolidation testing. Given all the variations typical of natural soils, differences between blow count information obtained using the standard split-barrel sampler and that obtained using a California barrel sampler are small. Reasonable helical pile sizes can be determined based on either type of test results.

It has been suggested that SPT N-values need to be corrected for hammer efficiency. Length of drill rod, hammer type, drop height control, borehole diameter, and other factors can affect hammer efficiency. Bowles (1988) summarizes the various factors developed by others that can be incorporated to correct N-values for hammer efficiency. The efficiency of the penetration system as a whole is defined by the energy ratio, E_r, which is computed by

$$E_r = \frac{\text{Energy Delivered to Sampler}}{\text{Theoretical Input Energy}} \times 100 \tag{3.1}$$

The energy ratio is often indicated in a subscript to the common abbreviation for blow count. For example, N_{55} indicates the SPT blow count at an energy ratio of 55.

Most of the older rope-pulley or cathead systems used in the 1950s and 1960s as well as in some places today are thought to produce an energy ratio of 50 to 60. In North America, modern safety hammers produce an energy ratio of 70 to 80; automatic trip hammers can produce an energy ratio between 80 and 100. Older rope-pulley or cathead systems used in Japan seem to be more efficient than those used in North America. Conversely, safety hammers used in the United Kingdom may have less efficiency than those used in North America (Bowles, 1988).

Sometimes it is beneficial to convert blow counts obtained using a certain energy ratio to those based on a different energy ratio. The product of energy ratio and blow count should be a constant such that (Bowles, 1988)

$$E_{r1} \times N_1 = E_{r2} \times N_2 \tag{3.2}$$

Hence, blow counts can be converted between systems in this way:

$$N_2 = \frac{E_{r1}}{E_{r2}} \times N_1 \tag{3.3}$$

SPT energy conversions are seldom used in helical pile design and installation in North America. However, the helical pile designer needs to be aware that there can be changes in energy ratio across different geographic regions and with different types of equipment. Energy corrections might be considered when pile behavior appears counterintuitive to the experience of the designer.

In many coastal areas where soft soils are present, Cone Penetration Tests (CPTs) may be performed in lieu of SPT tests. In the CPT test, a cone-shaped device with a tubular sleeve is pushed into the subsurface (ASTM D3441). The device measures the pressure exerted on the conical penetrator and the friction along the sleeve. The experienced geotechnical engineer can use CPT measurements directly to estimate soil shear strength, density, and stiffness. They also can be correlated to blow count by

$$N_{55} = \lambda_{CPT} \times q_c \tag{3.4}$$

Where

q_c is CPT cone tip resistance and
λ_{CPT} is the ratio between N_{55} and q_c given in Table 3.3

The correlations given in the table were derived from Bowles' (1988) graphical representation and modified to work with Equation 3.4. Correlations between CPT tip resistance and SPT N-values should be considered approximate and used with caution.

Boring logs with CPT test results generally list cone penetration resistance readings every 3 inches (8 cm). Next to each CPT result may be listed the corresponding correlation to SPT blow count. Since the blow count values were determined from correlations and not from actual tests, their values are approximations of the blow count that would be obtained if a SPT test were started at the depth indicated. Do not be confused by the fact that blow count correlations are given every 3 inches (8 cm) when the actual tests physically can be performed only at 18-inch (46-cm) intervals.

Table 3.3 CPT Correlations

Typical Soil Type	Avg Grain Size (in)	[mm]	λ_{CPT}, Ratio N_{55}/q_c (blow count/psi)	[blow count/kPa]
Clay	0.025	[0.001]	0.069	[0.01]
Silty Clay	0.127	[0.005]	0.046	[0.0066]
Silt	0.25	[0.01]	0.034	[0.005]
Silty Sand	2.5	[0.1]	0.017	[0.0025]
Sand	13	[0.5]	0.011	[0.0016]

When a stable drill hole can be drilled without casing, exploratory borings typically are made using a 4-inch- (10-cm-) diameter solid stem auger that is removed from the hole each time a SPT test is conducted. In locations with soils prone to caving, exploratory borings typically are made using a larger-diameter hollow-stem auger. The sampling device and drill stem can be extended down the central core of a hollow-stem auger without removing it. Another method of drilling is mud rotary whereby drilling fluid is used to maintain borehole stability. SPT results obtained when mud rotary methods are used should be approached with caution, as they can be inaccurate if the drilling fluid is particularly viscous.

Groundwater can affect the results of SPT tests. During drilling in low-permeable soils, it is often possible for the level of groundwater in an exploratory boring to be much lower than that in surrounding soils because insufficient time has been allowed for seepage into the borehole. This phenomenon can result in excess groundwater pressure and lower than normal blow counts. Checking boring logs for groundwater level readings with time can sometimes identify this effect. If groundwater levels appear to be lower during exploratory drilling compared to levels measured at later dates, then there is a possibility that blow counts are artificially low. Some geotechnical engineers will discuss this consequence in their report and will make recommendations as to the true consistency and shear strength of soil revealed in their borings. It may be possible to use smaller helical bearing plates than those indicated by blow counts under excess groundwater pressure. Installation torque provides an important verification of soil strength even in groundwater.

3.3 SOIL CLASSIFICATION

One of the annoyances of geotechnical engineers is the misuse of the terms "soil" and "dirt." "Soil" is defined as the layer of fine material up to and including boulder-size fractions covering the Earth's crust. "Dirt" is defined as misplaced soil. Hence, soil is everything comprising the ground above bedrock. When it lands on hands or pants, it is dirt. Helical piles are affected by soil mechanics. Dirt mechanics will be left for others to discuss.

Soils are mainly classified by particle size as shown in Table 3.4. Another term for particle size distribution is "soil gradation." The main classes of soil, from smallest to largest particle size, are clay, silt, sand, gravel, cobble, and boulders.

Particle size distribution is determined by geotechnical engineers using a laboratory procedure known as sieve analysis (ASTM D6913-04). The test is very simple and consists of placing a known quantity of dry soil on a stack of different sized wire-mesh screens called "sieves." The stack of sieves and the soil is shaken vigorously by machine. Each sieve is removed and weighed to determine the amount of soil contained within the different sieve sizes. Sieves are named for the approximate number of openings per inch [25 mm] of screen. For example, a No. 4 sieve has roughly 4 openings per inch [25 mm]. A No. 50 sieve has roughly 50 openings per inch [25 mm], and so on.

Table 3.4 Particle Sizes

Soil Classification		Particle Diameter in	[mm]	U.S. Standard Sieve Sizes
Fine grain	Clay	Less than 0.00008	[Less than 0.002]	Less than #200
	Silt	0.00008 to 0.003	[0.002 to 0.07]	
Coarse grain	Sand	0.003 to 0.25	[0.07 to 4.8]	#200 to #4
	Gravel	0.25 to 3	[4.8 to 76]	Greater than #4
	Cobble	3 to 12	[76 to 300]	Hand measure
	Boulders	Greater than 12	[Greater than 300]	

Example sieve analysis results are shown in Figure 3.4. In this figure, the percent of the soil sample retained on each sieve is plotted with respect to sieve size. Uniformly graded soils, such as many marine deposits, river and lake deposits, aeolian soils, and beach sands, have a very narrow range of particle sizes and appear in this plot as a near-vertical line. Well-graded soils, such as many floodplain deposits, slope talus, and glacial till, have a broad range of particle sizes and appear as a diagonal line in this plot.

An important sieve size to remember is the No. 200 sieve, which has an opening of 0.003 inches [0.075 mm]. A No. 200 sieve separates what are known as coarse grains (sand, gravel, cobble, and boulders) from what are known as fine grains (clay and silt). The No. 200 sieve size is about at the resolution of the unaided human eye. Hence,

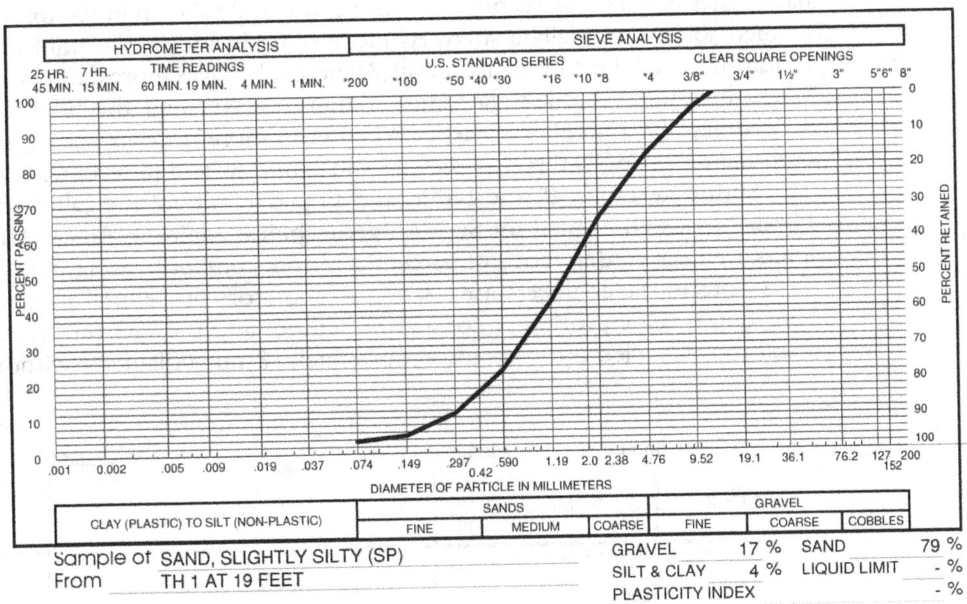

Figure 3.4 Example sieve analysis results

coarse-grain soils can be seen, but fine-grain soils (clay and silt) are too fine to resolve without a magnifying glass or microscope. This is analogous to the threshold resolution for computer printing, which is about 200 dpi (dots per inch). Anything more coarse than this will look grainy. Gradation is also analogous to sandpaper grit size. Sandpaper is available in 40, 60, 80, 100, 120, 200, 240, and 320 grits. "Sandpaper grit size" refers to the particles adhered to the sandpaper and corresponds roughly to sieve size.

Fine-grain soils (clay and silt) are identified in the field by visual examination, texture, and behavior at different moisture contents. Fine-grain soils appear as a uniform paste when moist. Individual grains cannot be resolved by eye and are smooth by touch. Clay soils are very plastic and can be rolled easily into threads when moist. Clay soils are moderate to very strong when dry. Silt soils are slightly plastic, often difficult to roll into threads without crumbling, and, in the absence of chemical cementation, have negligible to low strength when dry. (ASTM D2488-00)

Silt also can be identified based on its behavior under pressure. When moist silt is squeezed in the hand, one might expect water to be released like a sponge. This is not the case. The surface of typical moist silt actually will appear to become dry when squeezed. After hand pressure is removed, the silt will appear moist once again. This phenomenon is called "dilatancy." Some fine sands also exhibit dilatancy. Clay generally does not. Very moist silt also exhibits "pumping" under equipment and foot pressure. Pumping is when the ground appears to compress and rebound in an exaggerated manner under application and removal of load.

For further classification, fine-grained soil behavior can be broken down into four basic states based on increasing moisture content. As shown in Figure 3.5, a dry, fine-grained soil is essentially a solid. As more water is added, the solid goes through a transition called the "Shrinkage Limit," where it behaves more like a pliable semisolid. When more water is added, the semisolid transitions through the "Plastic Limit" where it becomes like a moldable plastic. When even more water is added, the plastic soil eventually transitions through the "Liquid Limit" where it becomes a viscous liquid. The Shrinkage Limit, Plastic Limit, and Liquid Limit are the called the "Atterberg Limits" of a soil. Named for Swedish geotechnical engineer, Albert Atterberg, these limits are measured in terms of percent moisture content by weight.

Geotechnical engineers use a series of laboratory procedures to determine Atterberg Limits. One procedure consists of determining the moisture content at which a soil can be rolled into threads with ⅛ inch (3 mm) diameter without crumbling

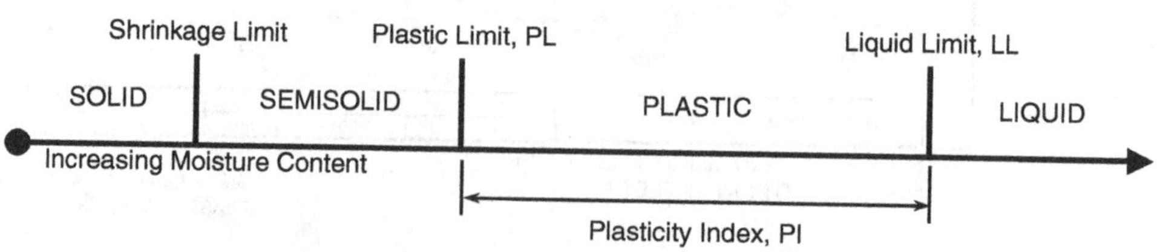

Figure 3.5 Behavior phases of fine-grained soils

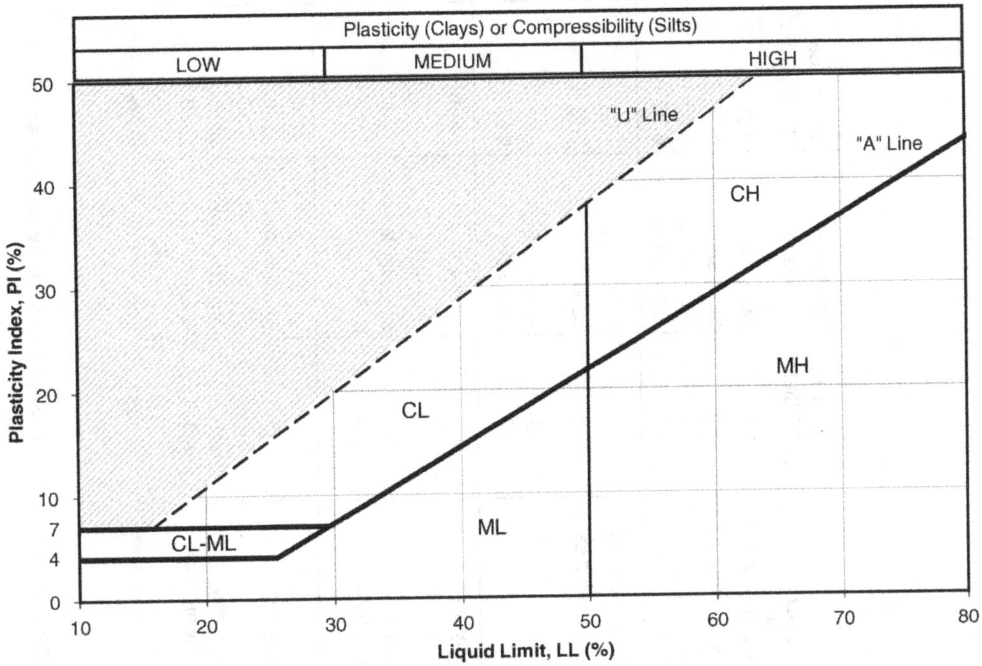

Figure 3.6 Classification based on Atterberg limits

(ASTM D4318-05). Another procedure involves determining the moisture content of the soil flowing a specific distance over a curved surface under a certain number of impacts (ASTM D4318-05). The test procedures are somewhat inexact, but the tests have been used for almost 100 years and are recognized for soil classification.

Roughly speaking, a large difference between Liquid Limit and Plastic Limit indicates clay. This difference between Liquid Limit and Plastic Limit is of sufficient importance that it has been given the name "Plasticity Index." A small Plasticity Index typically indicates silt. More specifically, geotechnical engineers use the chart of Plasticity Index and Liquid Limit shown in Figure 3.6 to classify fine-grained soil. If a sample plots above the A-Line, it is clay. If it plots below the A-Line, it is silt. Soils with a Liquid Limit above 50 percent are termed highly plastic.

The primary soil classification used by geotechnical engineers is the Unified Soil Classification System, or USCS (ASTM D2487-00). A flow chart summarizing the USCS system is shown in Table 3.5. Soils are first classified as coarse grain or fine grain, depending on the fraction of particles passing the No. 200 sieve. Coarse-grain soils are further subdivided into sands and gravels based on percent passing the No. 4 sieve (1/4 inch- [6-mm-] diameter particles). Coarse-grain soils containing less than 12 percent fines are classified as well graded or poorly graded based on the shape of the gradation curve. Fine-grain soils are further subdivided into clays and silts based on Atterberg Limits. Borderline soils receive dual symbols (e.g. SP-SM, GW-GC, etc.).

Table 3.5 Soil Classification

Soil Classification			Passing #200 Sieve	Other Criteria		Description	Similar AASHTO Class[5]	Similar USDA Class[5]	Similar '68 NYCBC Class[5]
Coarse Grain	Gravel	GW	<5%	<50% passing #4 sieve[1]	$C_u>4$ and $1<C_c<3$[3]	Clean, well-graded gravel	A-1	Stony sand	6-65
		GP			not GW	Clean, poorly graded gravel	A-3	Stony sand	6-65
		GM	12%–50%		below "A" line[4]	Silty gravel	A-2	Stony sandy loam	6-65
		GC			above "A" line[4]	Clayey gravel	A-2	Stony sandy clay	6-65
	Sand	SW	<5%	>50% passing #4 sieve[1]	$C_u>4$ and $1<C_c<3$[3]	Clean, well-graded sand	A-1	Sand	7-65
		SP			not SW	Clean, poorly graded sand	A-3	Sand	7-65
		SM	12%–50%		below "A" line[4]	Silty sand	A-2	Sandy loam	8-65
		SC			above "A" line[4]	Clayey sand	A-2	Sandy clay	9-65
Fine Grain	Low Plasticity	ML	>50%	Liquid limit <50%	Below "a" line[4]	Low-plasticity silt	A-4	Silt loam	10-65
		CL			above "A" line[4]	Low-plasticity clay	A-6	Clay, clay loam	9-65
		OL			LL ratio <75%[2]	Low-plasticity organic silt/clay	A-8	N/A	11-65
	High Plasticity	MH		Liquid limit <50%	below "A" line[4]	High-plasticity silt	A-5	Silt, silt loam	10-65
		CH			above "A" line[4]	High-plasticity clay	A-7	Clay, clay loam	9-65
		OH			LL ratio <75%[2]	High-plasticity organic silt/clay	A-8	N/A	11-65

Note: Use dual symbols for all borderline soils (e.g., GP-GC, SP-SM, SC/CL, SM/ML, etc.).

[1] Percent passing based on weight of coarse fraction only.

[2] Ratio of Liquid Limit after oven drying to Liquid Limit of natural soil

[3] Coefficient of Uniformity, Cu = D60/D10, and Coefficient of Curvature, Cc = D302/D10xD60, where D10, D30, and D60 are percent passing #10, #30, and #60 sieve, respectively.

[4] Refers to where Atterburg Limits plot relative to "A" line.

[5] American Association of State Highway Transportation Officials (AASHTO), United States Department of Agriculture (USDA), and 1968 New York City Building Code (NYCBC) use different criteria for classification. The classifications given are examples of overlaps in these various systems and should not be considered absolute.

As a side note, Soil Conservation Service maps and soil surveys are based on the United States Department of Agriculture soil classification system, which is different from that of the USCS. In the agricultural system, the term "loam" is used. For purposes of helical pile design, loam can be taken as synonyms to a silty or clayey sand. Soil Conservation Service maps are available for much of the United States and can be useful when trying to determine subsurface conditions in an area with limited geotechnical data. These maps are based in large part on correlations between soil type and native vegetation as well as site reconnaissance and review of aerial maps. Hence, the information only reflects the upper few feet of soil (less than 5 ft [1.5 m]).

There are several other soil classification systems. The American Association of State Highway Transportation Officials (ASHTO) has a classification system wherein soils are labeled as class A1 through A8. Most state departments of transportation use this system. These classifications generally relate to performance of the soil as a pavement subgrade. The 1968 New York City Building Code (NYCBC) classification system assigns labels from class 1-65 to 11-65 to different soil and bedrock types based in large part on bearing characteristics. The NYCBC system separates fine sand (8-65) from coarse sand (7-65) and has a separate class for hardpan (5-65). Although these two systems are based on different criteria, AASHTO and NYCBC class designations can be roughly correlated to USCS classification as shown in Table 3.5.

In a soil engineering report, soil descriptions typically are written in the body of text as well as on logs of exploratory borings/pits. An excerpt from a soil report is given next.

> In general, the soil boring penetrated 4 feet of dark brown to black, very moist, stiff, sandy clay man-placed fill over light brown, wet, medium dense, slightly silty to clayey sand underlain at depths ranging from 16 to 20 feet by gray-brown, moist, hard, interbedded claystone and sandstone that extends to the depths explored.

When encountering this language, try not to get lost in all the adjectives. Look for the basic soil classifications. The statement above basically says the subsurface consists of 4 feet [1.2 m] of clay fill over sand underlain by bedrock at 16 to 20 feet [4.9 to 6.1 m].

Most soils consist of a mixture of different types of fine and coarse grains and can be difficult to classify. A helical pile installer who pays attention to soil gradation, moisture content, Atterberg Limits, and soil strength and compares his or her own observations with subsurface exploration performed by others will develop an improved sense of soil behavior. He or she will be better able to predict from a set of plans or a soil report just how soil will interact with equipment and helical pile foundations.

3.4 BEDROCK

Many helical piles bear on bedrock. Penetration into bedrock is a common concern. A text on the subject of helical piles would not be complete without a section focused on the basic properties of bedrock and a discussion on how it interacts with helical piles.

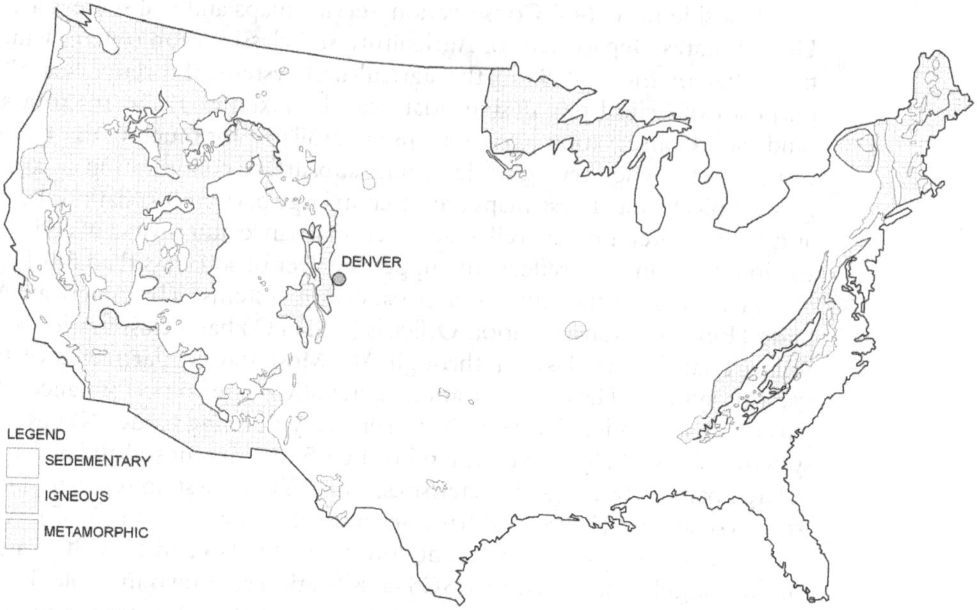

Figure 3.7 Bedrock types in contiguous United States

This section provides some background about bedrock, its strength, bearing capacity, and penetrability.

Bedrock is the continuous rock layer forming the Earth's crust. The depth of bedrock is highly variable; it is exposed at the surface in some places and buried under hundreds to thousands of feet of soil and rock fragments in others. Bedrock is not isolated boulders or cemented layers within the soil; it is the continuous mass of rock, thousands of feet thick, that essentially underlies all soil at some depth.

There are three basic types of bedrock: igneous, metamorphic, and sedimentary. A crude map of the contiguous United States with approximate areas where these types of bedrock occur is presented in Figure 3.7. This map was created based on a number of geologic maps and satellite imagery. It is highly generalized and by no means a substitute for detailed geologic mapping and site-specific exploration. Nonetheless, the figure shows that igneous and metamorphic rocks are found in the mountainous areas of eastern and western U.S. and around the Great Lakes in an area known as the North American shield. Sedimentary rock is found across much of the central and southeastern regions of the United States.

Igneous rock is formed essentially by the cooling of magma either above- or belowground. This rock is often uniform or massive and lacks stratification and planar features. Common igneous rocks include granite, diorite, gabbro, basalt, and tuff. Igneous rock is almost always very hard and cannot be penetrated with a helical pile. Unconfined compressive strengths range from 40 ksi [275 MPa] to 60 ksi [413 MPa], as much as 20 times the strength of ordinary concrete, 3 ksi [20 MPa]. Igneous rock provides exceptional strength for end-bearing applications. The surface of igneous rock can be highly weathered. For example, there may be a layer of decomposed granite (gruss) above the bedrock. Helical piles often can penetrate these weathered materials but not the parent rock.

Metamorphic rock is formed by transformation of other rock types by varying amounts of heat and pressure. Common metamorphic rocks include foliated rocks, such as schist, gneiss, and slate, and massive rocks, such as marble and quartzite, as well as other less common rock types. Metamorphic rock can range from moderate strength to very strong and ordinarily cannot be penetrated by a helical pile. Unconfined compressive strengths range from 15 ksi [103 MPa] to 43 ksi [294 MPa], many times the strength of ordinary concrete. Metamorphic rock typically provides exceptional strength for end-bearing applications. Metamorphic rock can be highly foliated, jointed, and fractured. Foliated metamorphic rocks are more susceptible to weathering. Some of the weak, fractured, or highly weathered metamorphic bedrock can be penetrated a short distance by a helical pile.

Sedimentary rock is a layered or massive rock composed of consolidated rock fragments, mineral grains, or organic materials from deposition, biogenic activity, or precipitation from solution. They are consolidated under less heat and pressure than metamorphic rocks. Common sedimentary rocks include sandstone, mudstone, shale, gypsum, and limestone. Highly plastic, softer shale, and clay-rich mudstones are sometimes called "claystone." Sedimentary rock has a broad range of strengths but is often weaker than other types of bedrock. Despite this, helical piles cannot penetrate most intact, competent sedimentary rocks. Unconfined compressive strength of competent sedimentary bedrock can vary from 6 ksi [41 MPa] to 25 ksi [172 MPa]. This is at least twice the strength of ordinary concrete. Sedimentary bedrock usually provides reasonably high bearing strength for support of helical piles. However, sedimentary rock can be very poor quality, noncemented, and highly weathered. There are documented cases where helical piles have penetrated these materials significant distances. The helical piles with torque measurements shown in Figure 2.4 penetrated claystone between 30 and 54 feet [9 and 16 m]. The helical piles in Figure 2.5 penetrated claystone to a depth of roughly 36 feet [11 m] and farther in a predrilled pilot hole. Highly cemented or competent sedimentary bedrock will cause helical pile refusal. Section 3.5 contains some rules for judging bedrock penetrability.

Table 3.6 lists some index properties of common bedrock types including unit weight, unconfined compressive strength, modulus of elasticity, and Poisson's ratio. The table was compiled from various sources (Bowles, 1988; Das, 1990; Peck, Hanson, and Thornburn, 1965) and the author's experience. Index properties may be used for initial modeling when geotechnical data is deemed inaccurate or lacking. Index properties are not a substitute for good field and laboratory measurements by a geotechnical firm or the opinion of the experienced engineer.

If it is anticipated that helical piles will end bear on bedrock, it is often advisable to use a single helical bearing plate to achieve maximum possible penetration through any weathered material and ensure end bearing on the more competent material. Some engineers have specified that a blunt end pile be used when end bearing on competent rock. They essentially want the pilot point cut off so that the maximum cross section

Table 3.6 Index Properties of Common Bedrock Types

		SPT Blow Count N	Unit Weight (pcf)	g[g/cm^3]	Unconfined Compressive Strength[1] (ksI) q_{peak}	[MPa]	Modulus of Elasticity[1] x10^3 (ksi) E_{peak}	[MPa]	Poisson Ratio N
Sedementary	Shale		140	[2.24]	6	[41]	3	[20]	0.35
	Sandstone		145	[2.32]	20	[137]	6	[41]	0.32
	Mudstone		169	[2.7]			5	[34]	
	Dolomite		160	[2.56]	23	[157]	9	[62]	
	Limestone		165	[2.64]	25	[172]	6	[41]	0.35
Metamorphic	Schist		165	[2.64]	15	[103]	12	[82]	0.2
	Quartzite		166	[2.65]	43	[294]			
	Gneiss		187	[2.99]	28	[196]			
	Marble		169	[2.7]	36	[245]			
	Slate		169	[2.7]	28	[196]			
Igneous	Granite	50/Bounce	166	[2.65]	40	[275]	12	[82]	0.28
	Dolerite		190	[3.04]	50	[343]	16	[110]	0.25
	Gabbro		195	[3.12]	43	[294]	16	[110]	0.25
	Diorite				43	[294]	14	[96]	0.25
	Basalt		178	[2.85]	60	[413]	15	[103]	0.3

RQD %	**Rock Quality**	**Reduction[2] % Factor[2] %**
0–25	very poor	0–7
25–50	poor	7–25
50–75	fair	25–56
75–90	good	56–81
90–100	excellent	81–100

Example:
Granite, RQD = 50
Reduction Factor = 0.5 x 0.5 = 0.25

$q_{peak} = 40$ ksi
$q_u = q_{peak} \times 0.25 = 10$ ksi
$E_{peak} = 12 \times 10^3$ ksi
$E = E_{peak} \times 0.25 = 3 \times 10^3$ ksi

[1] Peak values given for specific rock types. Reduce for rock quality by multiplying values by reduction factor.
[2] Reduction factor = RQD^2

of shaft and helix bears on what they imagine is a smooth, flat, horizontal, planar surface of bedrock. This is absolutely unnecessary. Helical piles have been used on thousands of projects wherein the tapered pilot point and lead helix edge bear directly on competent bedrock without modification.

The shape of the bedrock surface underground can be understood best if one examines local bedrock outcrops whenever present. The surface is often irregular. It can undulate and contain cracks and crevices, even buried ravines. It can be covered by boulders and etched by ancient rivers and creek beds. If one imagines this type of underground landscape, it becomes clear that it is better to use a helical pile with a pilot point when end bearing on bedrock in order to improve penetration through any weathered material and provide the opportunity for the helical pile to engage, or bite, into the bedrock or an overlying boulder some a small amount.

If a single helical bearing plate is used in end bearing on bedrock, the uplift resistance of the helical pile is better judged by the torque immediately prior to hitting the bedrock and the shear strength and bearing capacity of the weathered material above the bedrock. Requirements for large tensile capacity may be a reason for using multiple helical bearing plates in an end-bearing condition or switching to a different type of deep foundation. The requirement for tension often can control the configuration of a helical pile; hence, the designer should avoid overgeneralizing on plans and should be very specific about which piles are required to resist tension.

In many instances, the surface of bedrock does not follow major joint patterns, bedding planes, or ground surface contours. As a rule of thumb, bedrock is often shallowest at higher elevations, because these areas are often more resistant to weathering, whereas lower areas have been experiencing deposition. In addition to the information contained in soil boring logs, surface features such as outcrops, oversteepened slopes, and other evidence can be used to better understand the type, contours, and hardness of bedrock. There is no substitute for a good engineering geologist. He or she understands the geologic origins of the area and can help the project team anticipate underground conditions that will affect helical pile installation and performance.

The properties of bedrock shown in Table 3.6 are for competent, intact rock. Strength, bearing capacity, and other properties of bedrock used in design need to be modified to reflect the presence of fractures, joints, voids, and other structural discontinuities. The competence of bedrock is measured by Rock Quality Designation (*RQD*). *RQD* is determined by summing the lengths of all pieces of intact rock greater than 4 inches (10 cm) in length in a core run and dividing by the total core length as given by Stagg and Zienkiewicz (1968):

$$RQD = \frac{\sum(\text{Length of Intact Rock} > 4"[100\text{mm}])}{\text{Length of Core}} \tag{3.5}$$

Rock quality designation is used to classify bedrock quality as very poor, poor, fair, good, and excellent, as shown in Table 3.6. Design values for strength and elasticity of lower-quality rock may be obtained by multiplying the properties for competent, intact rock by RQD^2 as given by

$$q_u = q_{\text{peak}}(RQD^2) \tag{3.6}$$

$$E = E_{\text{peak}}(RQD^2) \tag{3.7}$$

where q_u and E are the design unconfined compressive strength and modulus of elasticity of rock, and q_{peak} and E_{peak} are the unconfined compressive strength and modulus of elasticity of competent, intact rock.

Values of q_{peak} and E_{peak} should be obtained from laboratory tests or may be approximated from Table 3.6 when tests are unavailable. Examples of these calculations are shown on the right side of Table 3.6. The term "RQD^2" is referred to as the reduction factor.

3.5 SITE SUITABILITY

The overall objective of subsurface exploration is to identify subsurface profiles that are particularly well suited for the use of helical piles and those that prohibit their use. In general, a helical pile can be used in most soil or bedrock conditions. Helical piles are ideal for sites with expansive soils, unknown fill, collapsible soils, soft clays and silts, and most other locations where a deep foundation is being considered. Some project features and subsurface conditions that may limit the use of helical piles are discussed here.

Stable soils and bedrock typically can support shallow foundations, such as spread footings or reinforced mats. Shallow foundations are often the most economical solution for new construction. Examples of stable soils and bedrock include medium dense to very dense sand and gravel, medium stiff to very stiff clays with low expansion/shrinkage potential, and nonexpansive bedrock. Helical piles might be more economical in these subsurface conditions when working inside existing structures, on remote sites, in backyards, or in other areas where it is difficult to obtain or place concrete. Sailer and Soth (2004) reported that helical piles are more economical than footings and concrete stem walls for support of residential additions due to limited access and cost savings associated with less excavation and reduced time for construction.

Typically, a properly manufactured and installed helical pile can penetrate soil and bedrock formations with SPT blow count as high as 100 to 150 blows/ft (50/6" to 50/4"). Helical piles may encounter refusal in materials with higher blow counts. A chart showing the ultimate bearing strength of different soil and bedrock types is shown in Figure 3.8. A typical helical pile can penetrate practically all soil types and some softer bedrock formations. Helical piles generally refuse in bedrock with an ultimate bearing capacity over 70 tsf [7 MPa]. Notably, this roughly correlates to the strength of weak cement grout (1,500 psi [10 MPa]). Helical piles cannot penetrate most igneous and metamorphic bedrock formations or more competent sedimentary bedrock formations.

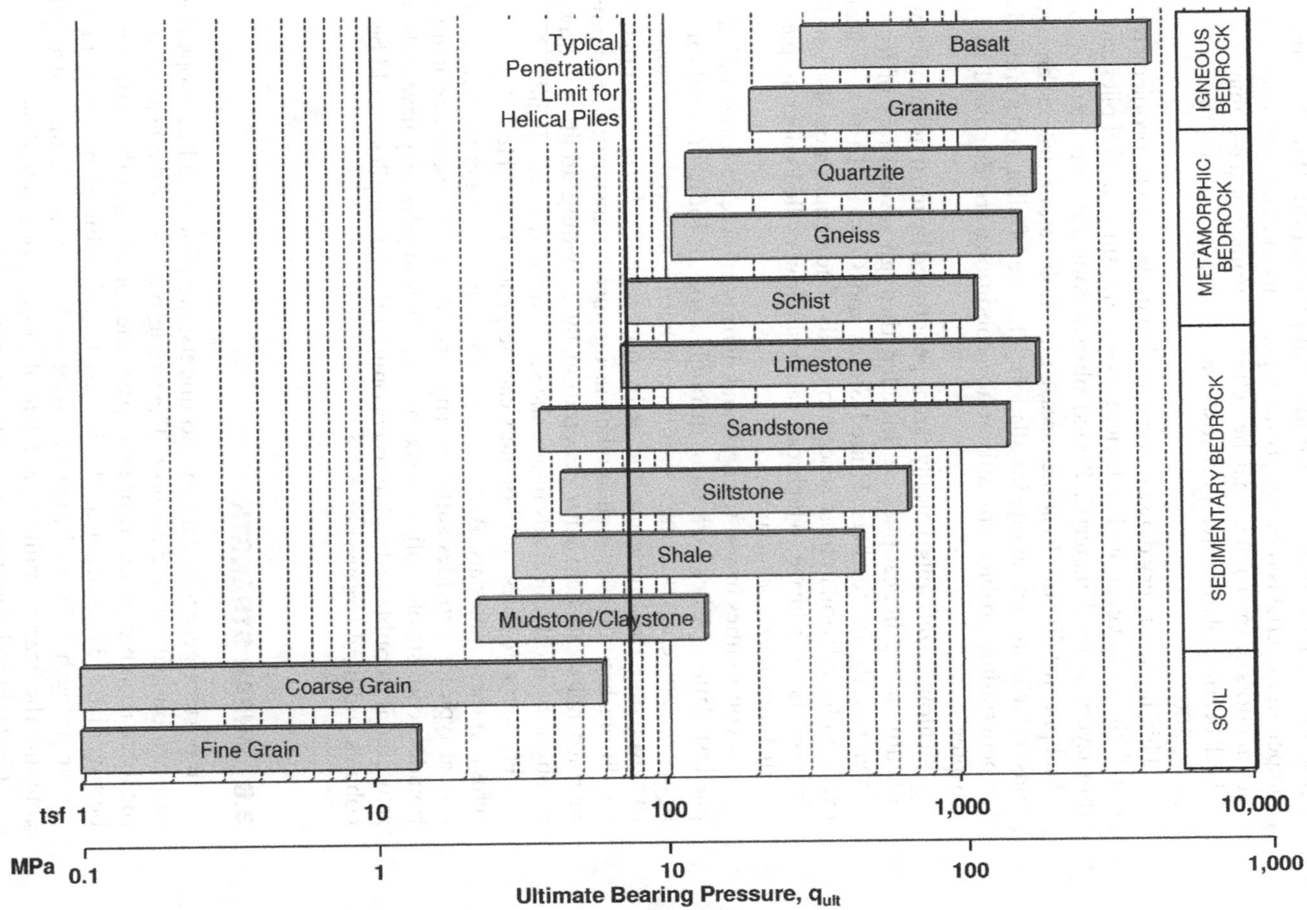

Figure 3.8 Ultimate bearing pressure of soil and rock

To support compressive loads, it is generally sufficient to bear directly on very dense soils or hard bedrock with minimal penetration. Where deeper penetration in competent material is required, other types of foundation and anchor systems, such as rock anchors or micropiles, may be better solutions. Larger-diameter shafts with small helical bearing plates have been used successfully to penetrate very dense oil sands (Nasr, 2007).

Thin layers of very dense or very hard materials can be problematic for helical pile installation. However, with sufficient crowd and time, helical piles can auger through thin layers of hard material. Buried obstructions also can increase the difficulty of helical pile installation. In some cases, a pilot hole can be drilled through the obstructions to facilitate helical pile installation. The use of pilot holes and other techniques for penetrating problematic subsurface conditions are discussed in Chapter 2 and Chapter 9.

Shallow frozen soils typically can be penetrated by most helical piles without undue difficulty. Many helical piles are installed through several feet of frozen soil during winter months in northern United States and Canada. In fact, helical piles are an excellent foundation in frost-susceptible soils. The slender steel shaft and large bearing plate resists frost heave, which can cause shallow-drilled concrete piers to be thrust upward under light structures.

Discontinuities in soil strength and density can create some difficulty with helical pile installation. A helix traveling through a soft or loose material can slip and begin to auger upon encountering a significantly harder material. Increasing crowd typically can overcome this and cause the helix to engage the harder material. When project conditions limit crowd, such as when using portable installation equipment, a second or third helix can help with penetration by providing additional downward thrust through the mechanical advantage the screw formed by the flights.

Frequent cobbles and boulders can be problematic. Helical piles can encounter refusal in these materials. When cobbles and boulders are contained in a dense matrix, refusal typically provides sufficient support. When the matrix-containing cobbles or boulders consists of a soft or weak material, helical piles may have to be removed and relocated if a cobble or boulder is encountered. Helical piles should be avoided when cobbles and boulders are pervasive.

3.6 SHEAR STRENGTH

An important geotechnical topic to understand is how soil develops the strength to support foundations and anchors. The strength of a helical pile in ground depends on the force required to shear, or tear apart, the soil around the perimeter of the helical bearing plates. The resistance of soil or rock to shearing along a planar surface is called "shear strength." As explained in Chapters 4, 5, and 10, shear strength is used to compute the bearing, pullout, and lateral capacity of a helical pile.

Geotechnical engineers measure shear strength using a variety of field and laboratory tests. One of the simplest tests is the direct shear test pictured in Figure 3.9.

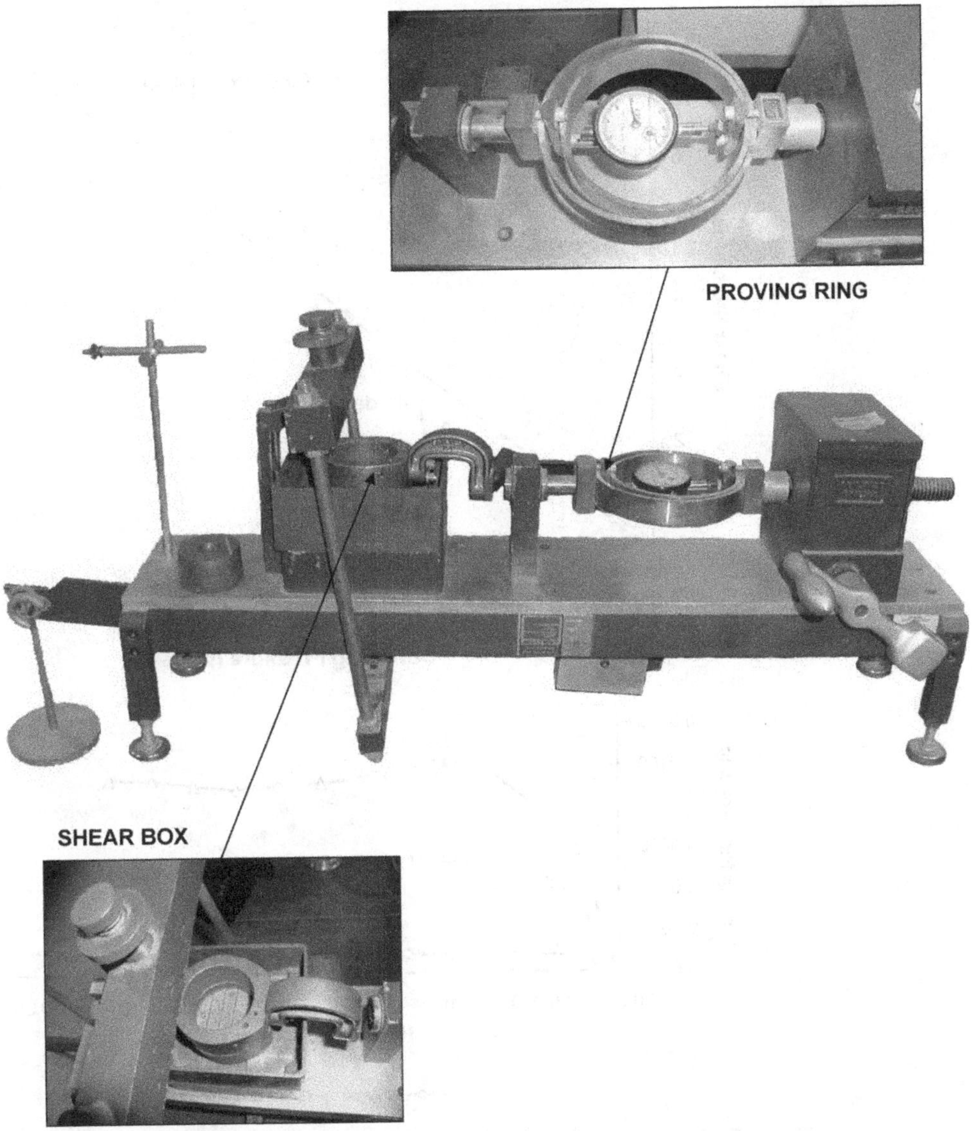

Figure 3.9 Direct shear test

In this test, soil is placed in a metal box separated into two halves. Weight (called the "confining pressure") is placed on top of the soil sample to simulate soil confinement in underground conditions. Then the top half of the shear box is pushed laterally with respect to the bottom half. A proving ring or transducer can be used to measure the lateral force applied to the shear box. Lateral force is increased until eventually the top of the box will move at a constant force (the peak strength has been reached).

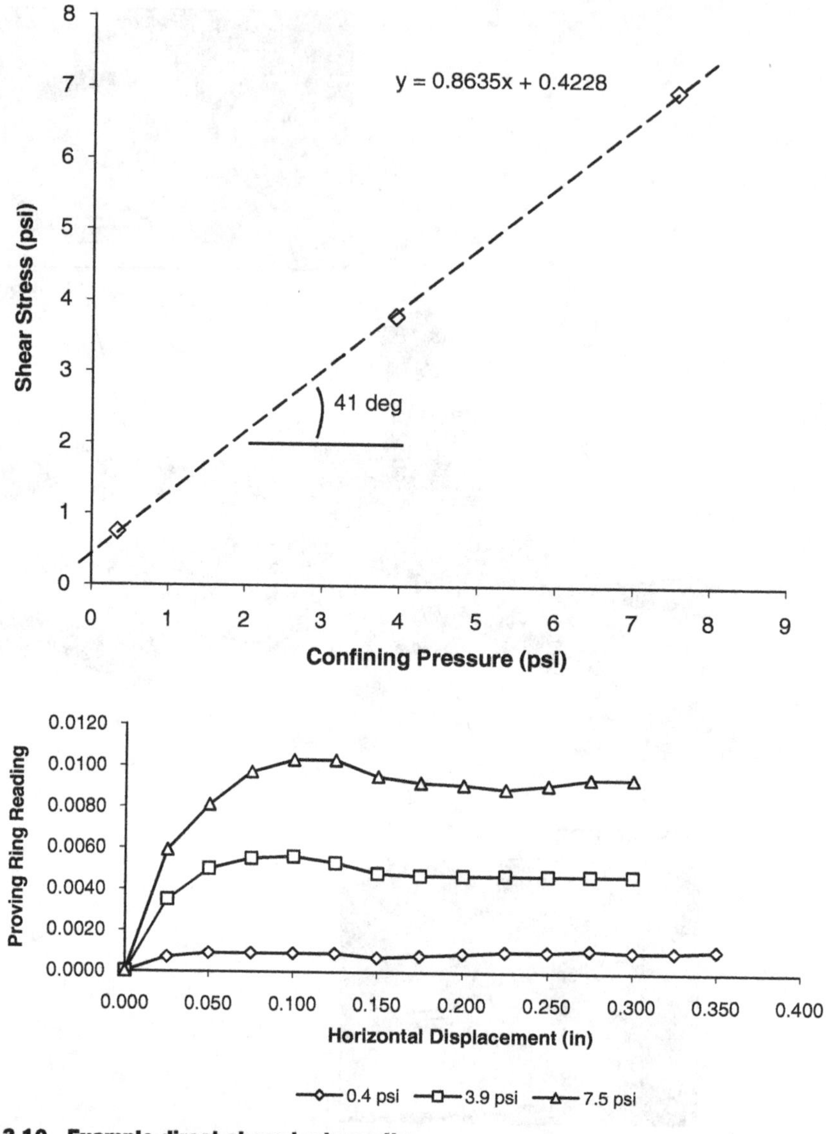

Figure 3.10 Example direct shear test results

The force at this point divided by the cross-sectional area of the sample is the shear strength of the soil at the applied confining pressure.

Example direct shear test results are shown in Figure 3.10. This figure is shown in imperial units only because the magnitude of the measurements has no bearing on the discussion herein. The lower chart contains the force-displacement readings for three direct shear tests run on the same soil at different confining pressures 0.4, 3.9, and 7.5 psi [3,27, and 52 kPa]. Proving ring readings correlate to the lateral forces applied to the shear box during the tests. The shear strength is the maximum proving ring reading obtained at each confining pressure. As can be seen in this example, peak shear strength increases with greater confining pressure. From these data, a plot of shear strength with respect to the confining pressure can be created as shown in the top portion of Figure 3.10; the data generally form a straight line called a "shear strength envelope." The strength of soil without confining stress is the intercept of the shear strength envelope with the y-axis known as the cohesion. The angle that the shear strength envelope makes with respect to the x-axis is termed the angle of internal friction.

The equation for shear strength is given by

$$T = c + \sigma_n' \tan\Phi \tag{3.8}$$

Where

T is shear strength
c is cohesion
σ'_n is effective confining stress, and
Φ is the angle of internal friction.

This equation is exactly analogous to the best-fit line given in Figure 3.10, where y and x represent T and σ'_n, respectively. Cohesion is the intercept given by 0.4228 psi [2.92 kPa] and angle of internal friction is the arctangent of the slope, *arctan(0.8635)*, which equals approximately 41 degrees.

This introduction to shear strength is highly simplified. Pore water pressure, drainage, capillary tension (apparent cohesion), stress paths, and rate of load application can affect shear strength. Shear strength envelopes over large ranges of stress can be a curve rather than a straight line. For the design of helical piles, it is important to use a shear strength envelope derived from tests that best simulate actual loading conditions. Half to two-thirds of the foundation loads in most buildings and utility towers are live loads, or short-duration, transient loads caused by people, wind, snow, or seismic loads. By virtue of these conditions, fairly rapid loading of samples is perhaps best suited for shear strength determination in foundation design. Confining pressures used in shear strength tests should approximate the range of stresses likely to be experienced by the soil at the bearing location.

According to Bowles (1988), undrained shear strength determined from rapid test methods is widely used for fine-grain soils and is conservative where field loading conditions and water content are duplicated by the test method. For purposes of helical pile design, the shear strength of fine-grain soils can be taken as the undrained shear strength, s_u, ($\Phi = 0$ concept) as shown in Equation 3.9.

$$T = s_u \tag{3.9}$$

Common tests for determining undrained shear strength of fine-grain soils include SPT, CPT, and unconfined compressive strength tests (ASTM D2166-00). A correlation between undrained shear strength and blow count for fine-grain soils is given by

Terzaghi and Peck (1967) and shown in Equation 3.10.

$$s_u = \lambda_{SPT} N_{55} \tag{3.10}$$

Where

λ_{SPT} equals 0.065 tsf/blow/ft [6.2 kPa/blow/30 cm].

A correlation between undrained shear strength and CPT tip resistance, q_c, using the empirical correlation given by Lunne et al. (1997) is shown in Equation 3.11.

$$s_u = \frac{(q_c - P'_o)}{N_k} \tag{3.11}$$

Where

P'_o is effective overburden pressure at the depth of the cone, and

N_k is an empirical cone factor that is site dependent. N_k is known to vary between 11 and 19 with an average of 15 (Lunne and Kleven, 1981).

The unconfined compression test consists of placing an undisturbed sample of fine-grain soil in a compression machine and loading it in increments until the peak strength is reached. Undisturbed samples are those whose volume has been unaffected by sampling. If volume is unchanged, effective stress also must be constant. Negative pore water pressures take the place of total overburden stress and hold the sample together. In this way, unconfined tests on undisturbed samples provide a good representation of shear strength for the effective stress present at the depth at which the sample was taken. Undrained shear strength of fine-grain soils can be determined from unconfined compressive strength, q_u, as shown in Equation 3.12.

$$s_u = \frac{q_u}{2} \tag{3.12}$$

Due to inherent higher permeability, coarse-grain soils are almost always loaded and tested in drained conditions (Bowles, 1988). The soil sample used to generate the data shown in Figure 3.11 is a well-graded, normally consolidated, silty sand with gravel. Most coarse grain-soils have little or no cohesion. In helical pile design, the shear strength of coarse-grain soils may be taken as ($c = 0$ concept)

$$\mathrm{T} = \sigma'_n \tan\phi \tag{3.13}$$

Common tests for determining angle of internal friction of coarse-grain soils include SPT, CPT, and the direct shear test described previously. A correlation between angle of internal friction and blow count for fine-grain soils is given by Parry (1977) and shown in Equation 3.14.

$$\Phi = 25 + 28\sqrt{\frac{N_{60}}{\lambda_\phi P'_o}} \tag{3.14}$$

Table 3.7 Index Properties of Common Soils and Weathered Bedrock

		SPT Blow Count N_{55}	Unit Weight γ (pcf)	[g/cm³]	Angle of Friction Φ (deg)	Undrained Strength s_u (psf)	[kPa]	p-y Modulus K (pci)	[kg/cm³]	Stress-Strain Modulus[1] E_s (ksi)	[MPa]	Poisson Ratio μ	Strain at 50% Peak Strength ϵ_{50}
Coarse-grain	Very loose	1-4	70	[1.12]	25	–	–	5	[0.1]	1	[9]	0.35	–
	Loose	4-10	90	[1.44]	29	–	–	25	[0.6]	2	[16]	0.35	–
	Medium	10-30	110	[1.76]	33	–	–	90	[2.4]	3	[23]	0.35	–
	Dense	30-50	120	[1.92]	39	–	–	225	[6.2]	7	[47]	0.35	–
	Very dense	>50	130	[2.08]	45	–	–	500	[13.8]	20	[137]	0.35	–
Fine-grain	Very soft	1-2	80	[1.28]	–	200	[9]	30	[0.8]	1	[7]	0.5	0.06
	Soft	2-4	85	[1.36]	–	400	[19]	100	[2.7]	2	[14]	0.4	0.02
	Medium	4-8	90	[1.44]	–	800	[38]	500	[13.8]	5	[31]	0.3	0.01
	Stiff	8-15	100	[1.6]	–	1,500	[71]	1,000	[27.6]	7	[47]	0.2	0.005
	Very stiff	>15	120	[1.92]	–	3,000	[143]	2,000	[55.3]	10	[71]	0.1	0.003
Weathered bedrock	Soft	<30	120	[1.92]	–	4,000	[191]	2,000	[55.3]	70	[482]	0.25	0.003
	Medium	30-50	130	[2.08]	–	10,000	[478]	3,000	[83]	280	[1931]	0.25	0.002
	Hard	50-100	135	[2.16]	–	20,000	[957]	4,000	[110.7]	520	[3586]	0.25	0.001
	Very hard	>100	140	[2.24]	–	50,000	[2394]	5,000	[138.4]	700	[4828]	0.25	0.001

Where

P_o' is the effective overburden stress at the depth of the SPT test, and
λ_ϕ is a correction factor for imperial units equal to 0.048 blows/ft/psf [1 blow/30 cm/kPa].

CPT tip resistance may be correlated to SPT N-values with caution using Eqation 3.4 in order to estimate angle of internal friction from Equation 3.14.

Table 3.7 contains index properties of common soils and weathered bedrock. It was compiled from various sources (Bowles, 1988; Das, 1990; Peck, Hanson, and Thornburn, 1953) and the author's experience. Many of the properties contained in this table were determined using the relationships contained in this section. Unit weight, angle of friction, and undrained strength are used in Chapter 4 and 5 to compute bearing and pullout capacity of helical piles. P-y modulus and strain at 50 percent peak strength are used in Chapter 10 to compute lateral resistance of helical piles. Stress-strain modulus and Poisson's ratio are used in Chapter 4 to estimate settlement of a helical pile. Here again, index properties may be used for initial modeling when geotechnical data is deemed inaccurate or lacking; however index properties are not a substitute for good field and laboratory measurements by a geotechnical firm or the opinion of the experienced engineer.

Drained strength should be considered in fine-grain and coarse-grain soils when checking long-term stability under sustained loads. An example of when drained shear strength should be considered is the determination of long-term stability of a retaining wall or earth retention project.

Some soils are sensitive to disturbance. Sensitive soils are those that exhibit a lower strength when remolded. A helical pile with multiple helical bearing plates installed in sensitive soils may have reduced strength. The geotechnical report should address soil sensitivity. In some cases, remolded shear strength may provide a better estimate the capacity of helical piles with multiple helices. Load tests are suggested for verification of capacity in sensitive soils.

A similar phenomenon is the reduction in strength of a soil that has experienced considerable strain. An example of this is shown in Figure 3.10. At higher confining pressures, two of the samples showed reduced shear strength at high strain or displacement. The shear strength at high strain is known as the residual strength. In Figure 3.10, the residual strength is about 85 percent of the peak strength. Some clay soils and claystones have residual strength on the order of 10 percent of the peak strength.

Properly installed helical piles should produce minimal disturbance and small strains on the soil between the helical bearing plates and at a small distance away from the shaft. Peak shear strength may be used in most cases to estimate bearing, pullout, and lateral capacity. In some cases, it is desirable to consider the shear strength of soil along the shaft of a helical pile. The soil immediately adjacent a shaft has experienced considerable strain during installation. It may be more appropriate to used residual shear strength values when evaluating the shear strength and adhesion of soil along a helical pile shaft. As discussed in Chapter 4 and 5, it is conservative to neglect adhesion of soil along a helical pile shaft for this and other reasons.

Chapter **4**

Bearing Capacity

In this chapter, methods of calculating the theoretical bearing capacity of a helical pile are presented. The most common methods are based on traditional geotechnical engineering limit state analysis, such as individual bearing and cylindrical shear methods. Another method, called the Laboratoire Central des Ponts et Chaussees (LCPC) method, utilizes cone penetrometer data to estimate bearing capacity. This chapter pertains to conventional helical piles as defined in Chapter 1. Calculating the capacity of grouted helical pile systems is addressed in Chapter 16.

Also discussed herein are buckling and down drag. Buckling is the loss of lateral stability of a helical pile shaft under axial loads in soft or loose soils. Down drag occurs when soil consolidation exerts negative skin friction on the shaft of a helical pile. Both of these factors can affect the bearing capacity of a helical foundation in certain circumstances.

In practice, the methods described in this chapter are used mainly to size a helical pile, that is, to determine the number and diameter of helical bearing plates and the strength of shaft required to support the intended design loads based on the subsurface conditions at a particular site. When used in conjunction with installation torque measurements and occasional load tests, determination of helical pile capacity can be accomplished with a high degree of confidence.

4.1 HELIX SPACING

There are two methods for determining bearing capacity based on theoretical soil mechanics: individual bearing and cylindrical shear. If the spacing between helical bearing plates is very large, as in Figure 4.1a, then each helix will act independently. The bearing capacity of the helical pile in this case is the sum of the individual capacities of all the helical bearing plates. This is called the "individual bearing" method.

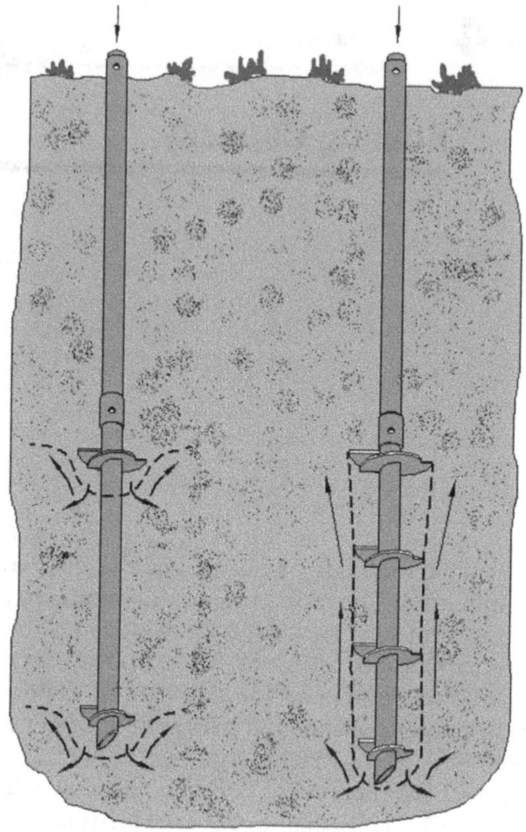

(a) (b)

Figure 4.1 Modes of helical pile failure

If the spacing between helical bearing plates is small, as in Figure 4.1b, then the helical bearing plates will act as a group. The bearing capacity of the helical pile in this case is the combination of bearing of the bottom helical bearing plate and side shear along the cylinder of soil encased between the helical bearing plates. This is called the "cylindrical shear" method.

The closeness of helical bearing plates is a relative term that depends on the geometry of the helical pile and surrounding soil conditions; it is not generally known in advance whether the helical bearing plates are close together or far apart. Capacity should be determined using both methods; the least value obtained is the correct or limiting state. The process of calculating all possible modes of failure to find the minimum value is called limit state analysis.

There have been a number of studies to determine the spacing where the transition between individual bearing and cylindrical shear occurs (Bassett, 1978; Narasimha

Rao and Prasad, 1993a; Narasimha Rao, Prasad, and Shetty, 1991; Narasimha Rao, Prasad, and Veeresh, 1993). In these studies, laboratory experiments were conducted on model helical piles and underreemed ground anchors in clay-filled test cylinders. The helix or underreem spacing to diameter ratio was varied between approximately 1 and 5 for the model piles. Effects of helix or underreem spacing on cylindrical shear capacity were analyzed. Experimental results by Narasimha Rao and fellow researchers indicated that the model piles exhibited individual bearing failure rather then cylindrical shear at a helix spacing to diameter ratio of 1.5. Whereas, experimental results from Bassett suggest the transition from cylindrical shear to individual bearing occurs at an undereem spacing to diameter ratio of 2.1 to 3.4. The main difference between these studies is probably scaling phenomena as described in the next paragraph and remolded soil consistency.

The ideal spacing of helical bearing plates is that where results from individual bearing and cylindrical shear methods are equal. Spacing farther than this results in an unnecessarily long shaft. Spacing less than this can be a waste of helical bearing plates. The spacing where the two methods converge varies with soil density, strength, and consistency as well as groundwater conditions, depth below ground, and the diameter of the helical pile. The foregoing experiments were conducted on small laboratory models. The maximum helix diameter was approximately 6 inches [152 mm] in the experiments led by Narasimha Rao. Since cylindrical shear increases with the radius of the helical bearing plates and individual bearing capacity with the square of the helical bearing plates, it is believed that the optimal helix spacing to helix diameter ratio increases for larger diameter helical piles. For shaft sizes from 1.5 inches square [38 mm] to 3.5 inches [89 mm] diameter in most soil types, the optimal spacing is commonly taken as two to three times the average diameter of the helical bearing plates (Seider, 2004). As an aside, the spacing should be in increments of the pitch so that the helical bearing plates track the same path during installation.

4.2 INDIVIDUAL BEARING METHOD

The assumed failure mechanism in the individual bearing method consists of each helical bearing plate displacing the soil or bedrock in a characteristic deep bearing capacity failure mode. A free body diagram showing an idealized distribution of forces on a helical pile in the individual bearing method is shown on the right side of Figure 4.2. A uniform pressure distribution is assumed on the underside of each helical bearing plate. Adhesion stresses are assumed along the length of the shaft. An axial load is applied to the shaft at the pile butt. Ultimate bearing capacity, P_u, of the pile is the sum of individual bearing capacities of n helical bearing plates plus adhesion along the shaft, given by

$$P_u = \sum_n q_{ult} A_n + \alpha H(\pi d) \tag{4.1}$$

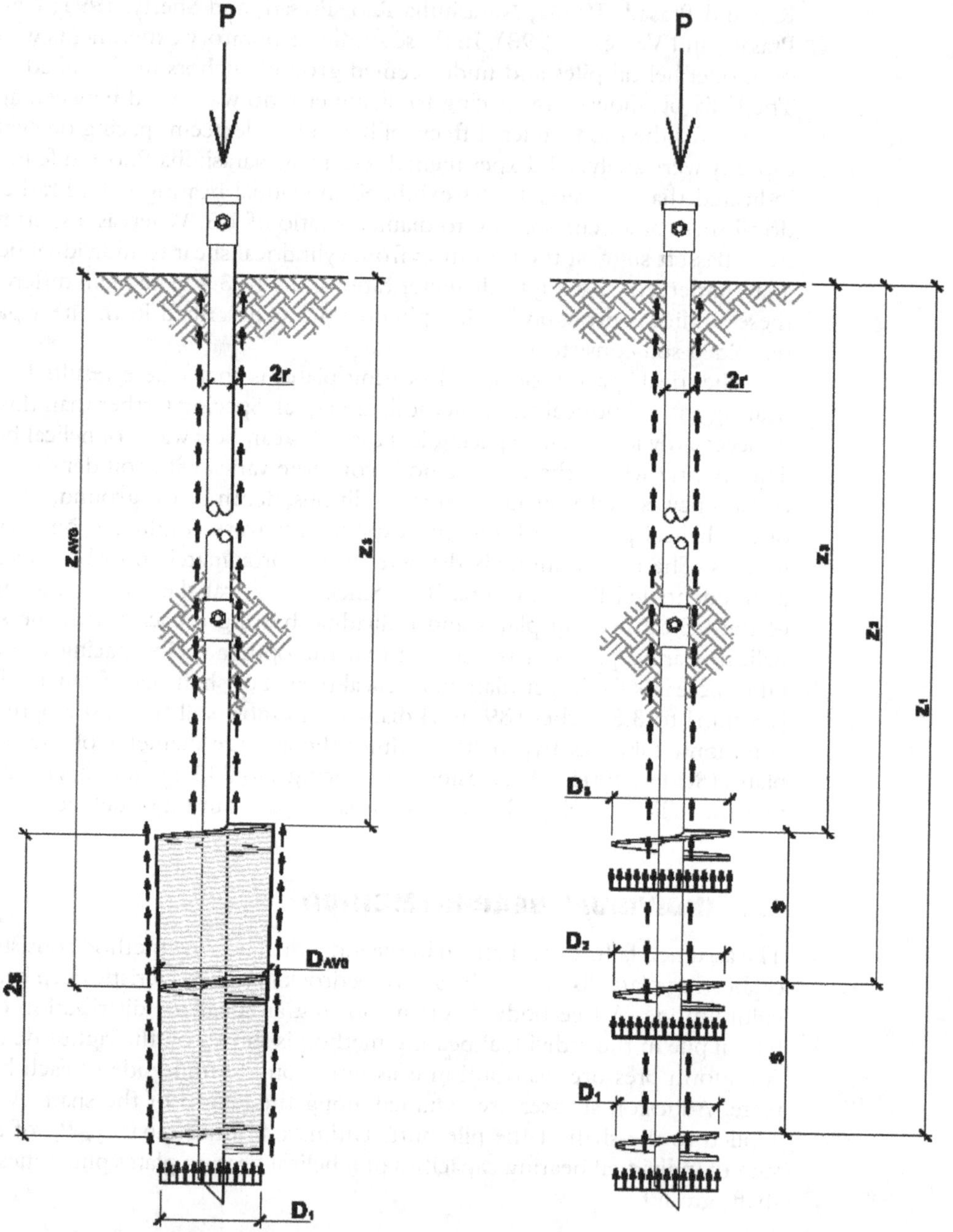

Figure 4.2 Individual bearing and cylindrical shear methods

Where

q_{ult} is the ultimate bearing pressure
A_n is the area of the nth helical bearing plate
α is adhesion between the soil and the shaft
H is the length of the helical pile shaft above the top helix, and
d is diameter of a circle circumscribed around the shaft.

Ultimate bearing pressure sometimes is provided in the geotechnical engineering report for a project. When it must be computed, the ultimate bearing pressure of soil may be determined using the familiar bearing capacity equation for circular bearing elements given by Terzaghi (1943):

$$q_{\text{ult}} = 1.3cN_c + q'N_q + 0.3\gamma BN_\gamma \tag{4.2}$$

Where

c is cohesion
q' is effective overburden stress at the bearing depth
γ is soil unit weight
B is the width of the bearing element, and
N_c, N_q, and N_γ are bearing capacity factors

The bearing capacity factors originally derived by Terzaghi are shown in Figure 4.3. The computation of bearing capacity factors is mathematically tedious so they are usually shown graphically. All three bearing capacity factors increase exponentially with the angle of internal friction. Very high bearing pressures are computed at large friction angles.

Meyerhof (1951) modified Terzaghi's formula to include factors for shape of the bearing element and depth. The modified equation took the form

$$q_{\text{ult}} = cN_c s_c d_c + q'N_q s_q d_q + 0.5\gamma BN_\gamma s_\gamma d_\gamma \tag{4.3}$$

Where

s_c, s_q, and s_γ are shape factors and
d_c, d_q, and d_γ are depth factors.

Meyerhof also redefined the bearing capacity factors, N_c, N_q, and N_γ. The redefined factors are given by Equations 4.4, 4.5 and 4.6 and represented by the gray lines in Figure 4.3. As can be seen, the factors derived by Meyerhof are slightly less than the original Terzaghi bearing capacity factors.

$$N_q = e^{\pi \tan \Phi} \tan^2\left(45 + \frac{\Phi}{2}\right) \tag{4.4}$$

$$N_c = (N_q - 1)\cot \Phi \tag{4.5}$$

$$N_\gamma = (N_q - 1)\tan(1.4\Phi) \tag{4.6}$$

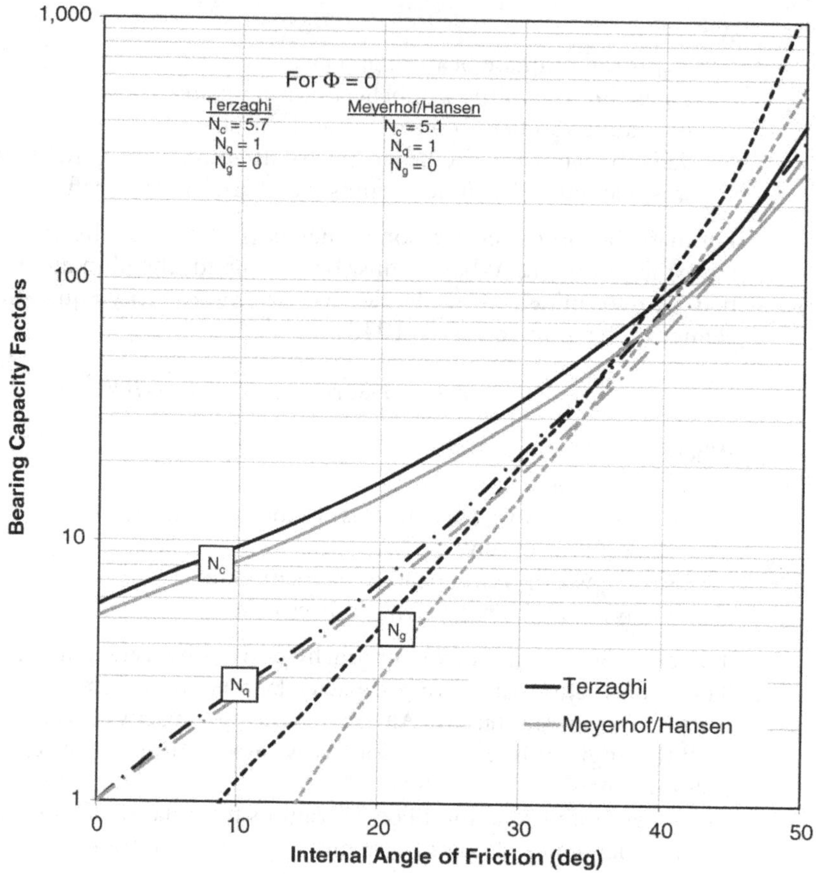

Figure 4.3 Bearing capacity factors (adopted from Terzaghi, 1943, and Meyerhof, 1951)

Shape and depth factors were refined by Hansen (1970) and Vesic (1973). They used Meyerhof's bearing capacity factors with the shape and depth factors given in Equations 4.7 to 4.13.

$$s_c = 1 + \frac{N_q}{N_c}\frac{B}{L} \tag{4.7}$$

$$s_q = 1 + \frac{B}{L}\tan\Phi \tag{4.8}$$

$$s_\gamma = 1 - 0.4\frac{B}{L} \tag{4.9}$$

$$d_c = 1 + 0.4K \tag{4.10}$$

$$d_q = 1 + 2K \tan \Phi(1 - \sin \Phi)^2 \tag{4.11}$$

$$d_\gamma = 1 \tag{4.12}$$

$$K = \arctan\left(\frac{H}{B}\right) \tag{4.13}$$

Where

L is the length of the foundation element,
K is a scaling parameter, and
Φ is the angle of internal friction of the soil.

Vesic also had some additional factors for inclination of the bearing element and for sloping ground surface. Inclination and ground slope factors are important for shallow foundations but are usually an unnecessary consideration for deep foundations.

A number of simplifications are possible when the forgoing bearing capacity theory is applied to helical piles. For a helical pile, *B* and *L* equal the diameter of the helix, *D*, and the ratio of B/L is equal to 1. Typically the depth of the helical bearing plates is much greater than their diameter, such that the ratio H/B becomes very large. The scaling parameter, *K*, approaches a value $\pi/2$ for large values of H/B. If *K* and B/L are constant, the shape and depth factors vary only with angle of internal friction. Thus, shape and depth factors can be grouped together with bearing capacity factors and plotted with respect to angle of internal friction such that $N'_c = N_c s_c d_c$, $N'_q = N_q s_q d_q$, and $N'_\gamma = N_\gamma s_\gamma d_\gamma$. The combined bearing, shape, and depth factors for helical piles are shown in Figure 4.4. In comparison to the corresponding factors in Figure 4.3, the combined factor N'_c is slightly higher; N'_q is relatively unchanged, and N'_γ is slightly lower.

To account for the weight of the pile or, in the case of helical piles, the weight of soil over the helical bearing plates, the overburden pressure, q', is typically subtracted from the second term in Equation 4.3. Helical bearing plate diameter, *D*, may be substituted for the foundation element width, *B*. The simplified bearing pressure equation for helical piles can be written as

$$q_{ult} = cN'_c + q'\left(N'_q - 1\right) + 0.5\gamma D N'_\gamma \tag{4.14}$$

For fine-grain soil where $\Phi = 0$, Hansen and Vesic equations yield a N'_c equal to 10. However, Skempton (1951) showed both theoretically and experimentally that N'_c approaches a constant value of 9 for deep foundations. Most practitioners used Skemption's result for the $\Phi = 0$ condition. Under this condition, the second and third terms in Equation 4.14 go to zero because $N'_q = 1$ and $N'_\gamma = 0$ per Figure 4.4. For helical piles, the cohesion, *c*, can be taken as the undrained shear strength, s_u, as discussed in Chapter 3. Hence, the ultimate bearing pressure for fine-grain soils based on Skempton is simply given by

$$q_{\text{ult}} = 9 s_u \tag{4.15}$$

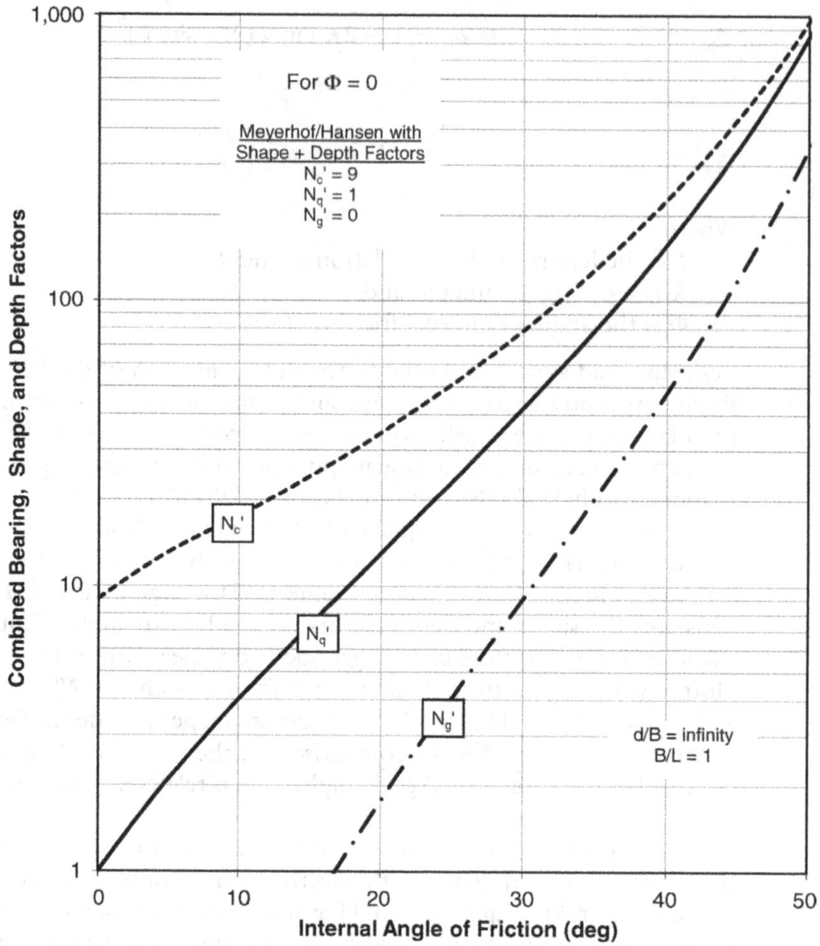

Figure 4.4 Combined bearing, shape, and depth factors

If Eq. 4.15 is combined with Eq. 3.10 from Chapter 3, then ultimate bearing pressure of fine-grain soils can be written in terms of standard penetration test blow count. However, Equation 3.10 is based on an energy ratio of 55, which is for older rope-pulley cathead systems. A correction can be made to relate ultimate bearing pressure to SPT blow counts with an energy ratio of 70, which is more indicative of modern equipment. The resulting relationship is given by

$$q_{\text{ult}} = 11\lambda_{\text{SPT}} N_{70} \tag{4.16}$$

To determine the capacity of a helical pile in fine-grain soil using the individual bearing capacity method, the ultimate bearing pressure from either Equation 4.15 or 4.16 may be substituted into Equation 4.1. To be conservative, the adhesion along the shaft can be ignored. Shaft adhesion is discussed further later in the chapter.

The accuracy of using Equations 4.16 and 4.1 was examined by comparing predicted axial capacity based on standard penetration test blow count with measured capacity in 47 full-scale load tests performed in fine-grain soils on helical piles with round shafts ranging from 2.875 inches [73 mm] to 10.75 inches [273 mm] diameter and between one and five helical bearing plates ranging from 8 inches [203 mm] to 30 inches [762 mm] diameter. The load tests were from various references, contributions to the book by industry, and the private files of CTL|Thompson, Inc. These load tests and others are summarized in Appendix C. The results of the comparison are shown in Figure 4.5. On average, measured axial capacity is 1.03 times the predicted capacity, and the standard deviation of the data is 0.47, indicating reasonably good correlation. It should be noted that use of a factor of safety of 2.0 with Equation 4.13 would result in 98 percent of the measured capacities being above the predicted capacities.

Another way to view the accuracy of Equation 4.16 is to plot measured bearing pressure against Standard Penetration Test blow count as shown in Figure 4.6. As can be seen, Equation 4.16 fits the helical pile test data in fine-grain soils fairly well. Ultimate bearing pressures in the 47 tests vary from approximately 2 tsf [192 kPa] to almost 50 tsf [4,788 kPa].

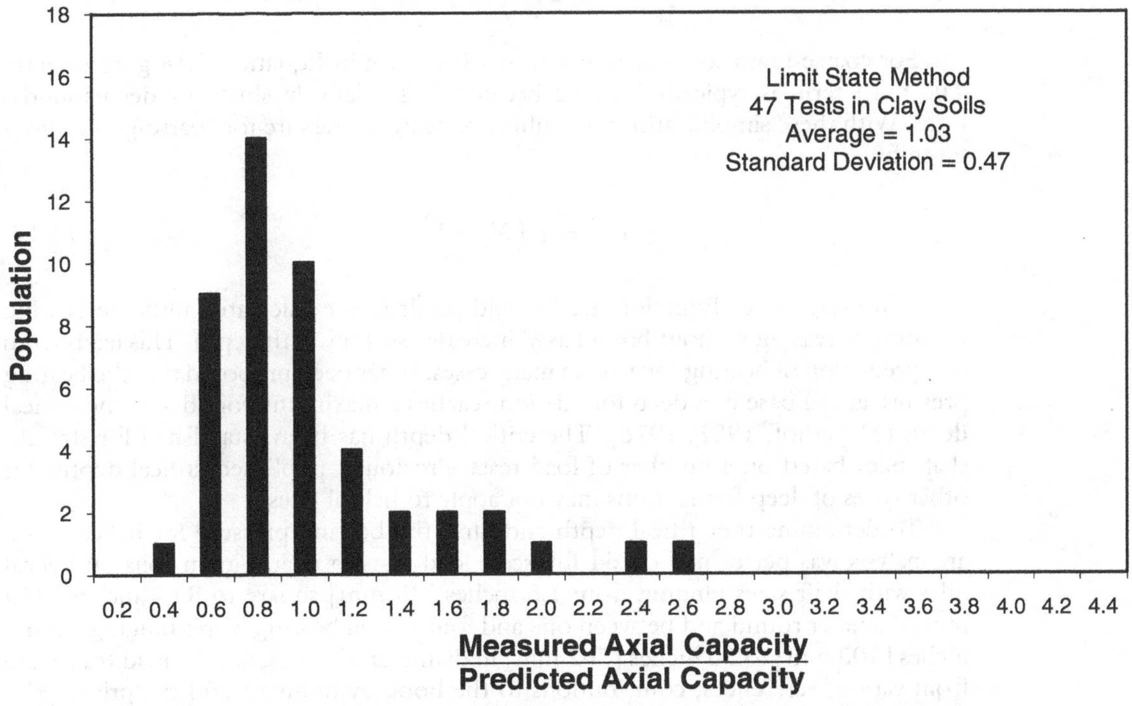

Figure 4.5 Comparison of measured and predicted capacity in clay using individual bearing method

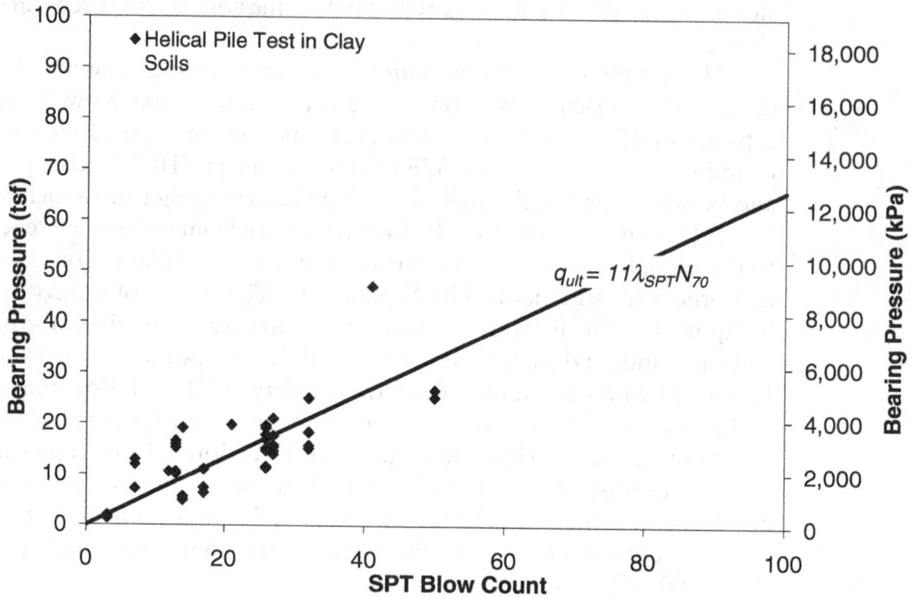

Figure 4.6 Correlation between bearing pressure and blow count in clay

For coarse-grain soil, where $c = 0$, the first term in Equation 4.14 goes to zero. The third term is typically ignored because it is relatively small for deep foundations. With these simplifications, the ultimate bearing pressure for coarse-grain soils is given by

$$q_{\text{ult}} = q'\left(N_q' - 1\right) \tag{4.17}$$

However, use of Equation 4.17 would result in the calculated ultimate bearing pressure increasing without bound as q' increases steadily with depth. This leads to an overprediction of bearing capacity in many cases. It has been proposed that the bearing pressure at the base of a deep foundation reaches a maximum bound at some critical depth (Meyerhof, 1951, 1976). The critical depth has been established for straight shaft piles based on a number of load tests. Previously published critical depths for other types of deep foundations may not apply to helical piles.

To determine the critical depth and limit the bearing pressure for helical piles, an analysis was performed on 54 full-scale load tests in coarse-grain soils on helical piles with shaft sizes ranging from 1.5 inches [38 mm] square to 8.62 inches [219 mm] diameter round and between one and four helical bearing plates ranging from 8 inches [203 mm] to 30 inches [762 mm] in diameter. Once again, the load tests were from various references, contributions to the book by industry, and the private files of CTL|Thompson, Inc.; these tests and others are summarized in Appendix C. As expected, Equation 4.17 significantly overestimates the measured capacity for all cases because the bearing pressure increases without bound. Critical depths recommended for straight shaft piles (Meyerhof, 1951, 1976), which are on the order of 4 to 10 times the square root of passive earth pressure, also result in significant overestimation of the measured capacity when used in conjunction with Equation 4.17. From a regression analysis, it was determined that the best fit to the load test data for helical piles is obtained by setting the critical depth equal to two times the average diameter of the helical bearing plates.

Using the critical depth of two times the diameter, the bearing capacity factor N_q' was back calculated for the available load test data. A soil unit weight equal to 120 psf [1.9 g/cm^3] was assumed for all tests. Results are shown in Figure 4.7 along with various published relationships for determining N_q and N_q'. These relationships include

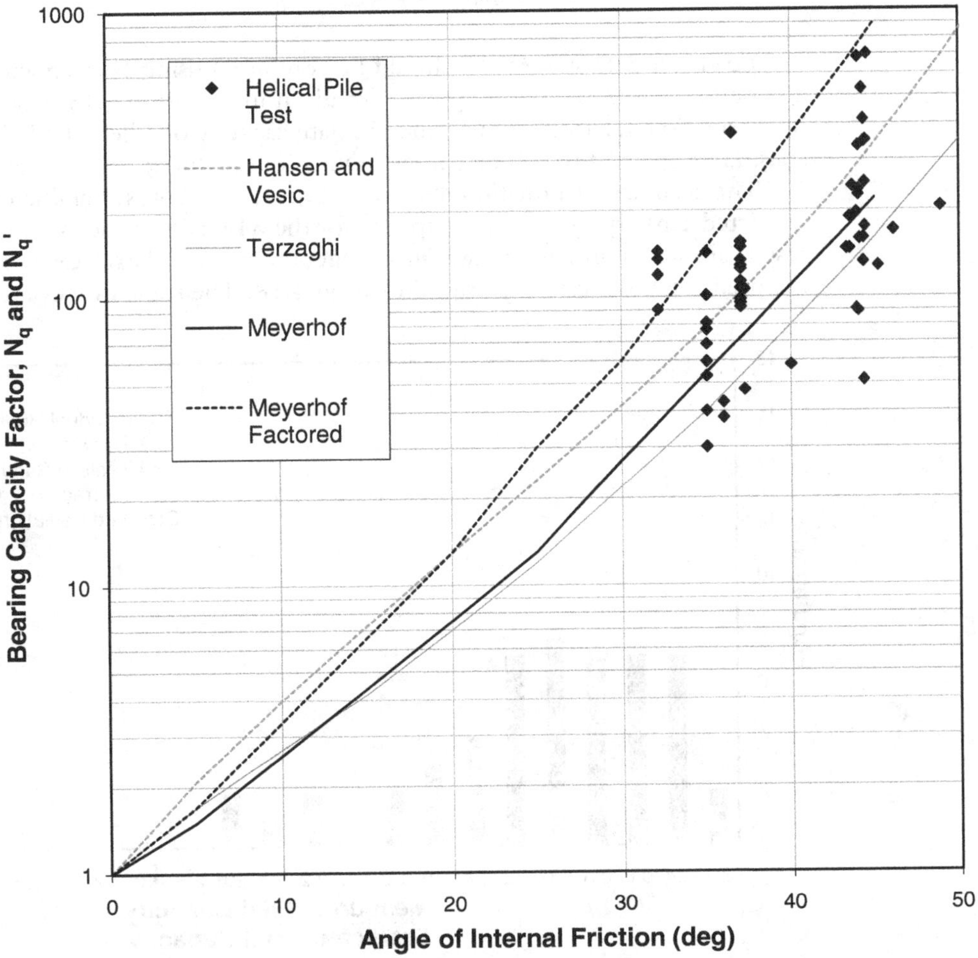

Figure 4.7 Bearing capacity factor Nq from helical pile load tests in sand

the bearing capacity factor, both uncorrected and corrected for shape and depth, given by Meyerhof (1976) for deep foundations as shown by the solid and dashed black lines in the figure, respectively. The original bearing capacity factor by Terzaghi (1943) is shown by the solid gray line, and the combined bearing capacity factor using the shape and depth factors from Hansen and Vesic is shown by the dashed gray line. The relationship that best intersects the average of the helical pile load test data is the combined bearing capacity factor, N'_q, using Hansen and Vesic shape and depth factors.

In summary, the ultimate bearing pressure for helical piles in coarse-grain soils may be computed using traditional bearing capacity theory by replacing the effective overburden stress, q', in Equation 4.17 with the product of soil unit weight, γ, and two times the average helix diameter, D_{AVG}, as shown in Equation 4.18.

$$q_{\text{ult}} = 2D_{\text{AVG}}\gamma\left(N'_q - 1\right) \tag{4.18}$$

Load test data suggest N'_q should be computed using Hansen and Vesic's shape and depth factors. The bearing pressure given in Equation 4.18 may be substituted into Equation 4.1 to determine the ultimate capacity of a helical pile in coarse-grain soils. Here again, it would be conservative to omit adhesion along the shaft.

The accuracy of Equation 4.18 with Equation 4.1 was examined by comparing predicted capacity to measured capacity for the 54 load tests described above and the results are shown in Figure 4.8. On average, the measured axial capacity is 1.16 times the predicted axial capacity using Equation 4.18. The standard deviation of the data

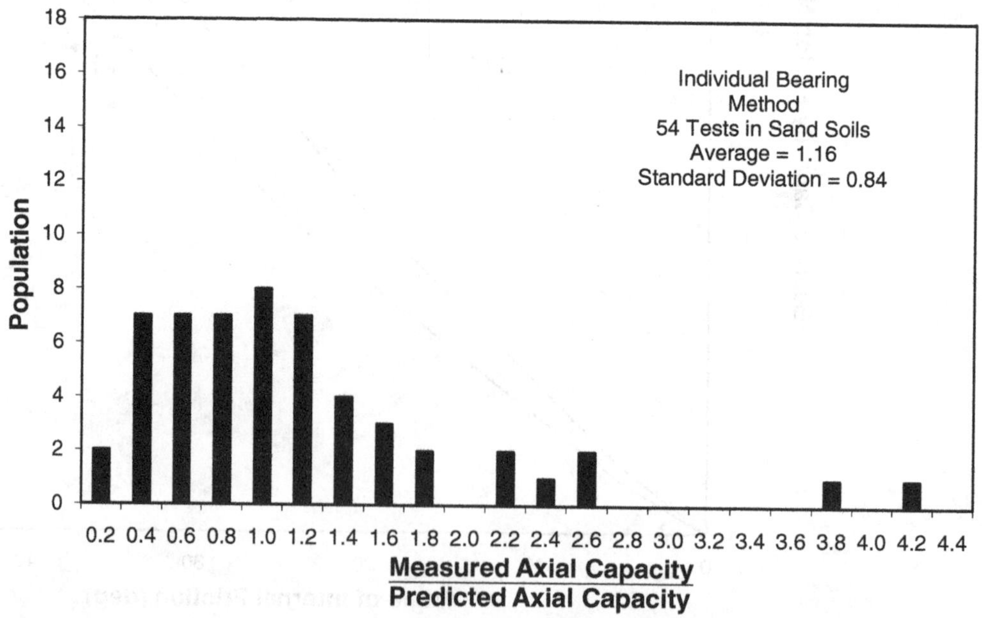

Figure 4.8 Comparison of measured and predicted capacity in sand using individual bearing method

is 0.84 indicating mediocre correlation. It is noted that use of a safety factor of 2.0 with Equation 4.18 would result in measured capacity exceeding predicted capacity 83 percent of the time.

Parry (1977) proposed computing ultimate bearing pressure of coarse-grain soils with a simple correlation to standard penetration test blow count given by

$$q_{\text{ult}} = 5\lambda_{\text{SPT}} N_{55} \tag{4.19}$$

Where

$\lambda_{SPT} = 0.065$ tsf/blow count [6.2 kPa/blow count] and

N_{55} is the standard penetration test blow count at an energy ratio of 55.

This relationship is compared with the results from the 54 load tests on helical piles (described above) in Figure 4.9. A correction was made for the hammer energy in order to change the energy ratio from 55 to 70; the resulting relationship is shown by the dashed line. The measured axial capacity in each of the 54 helical pile tests was divided by the total area of the helical bearing plates to obtain measured ultimate bearing pressure given on the y-axis. Where blow counts were unavailable, estimates were calculated from Equation 3.14 using published values of angle of internal friction and sample depth. As can be seen, the correlation by Parry (1977), which was originally derived for shallow foundations, does not correlate well for helical piles even with the energy correction.

A new relationship is suggested for estimating ultimate bearing capacity directly from standard penetration blow count for helical piles in coarse-grain soils given by

$$q_{\text{ult}} = 12\lambda_{\text{SPT}} N_{70} \tag{4.20}$$

The author has used this relationship as a rule of thumb with success for many years. It seems to fit the data for helical pile tests in coarse-grain soils shown in Figure 4.9 better than the relationship by Parry. As can be seen in the figure, the measured bearing pressures range from approximately 5 tsf [479 kPa] to almost 100 tsf [9,576 kPa] for the helical pile load tests in coarse-grain soil.

The accuracy of Equation 4.20 with Equation 4.1 was examined by comparing predicted capacity to measured capacity for the 54 load tests described earlier and the results are shown in Figure 4.10. Compared with Figure 4.8, using blow count to determine bearing pressure in coarse-grain soils appears to provide more consistent results than using bearing capacity factors. On average, the measured axial capacity from field tests is 1.34 times the predicted axial capacity using Equation 4.20. The standard deviation of the data is 0.82, indicating about the same correlation as was obtained from Equation 4.18. It is noted, however, that use of a safety factor of 2.0 with Equation 4.20 would result in measured capacity exceeding predicted capacity 91 percent of the time, which is better than that obtained with Equation 4.18.

Besides bearing in fine-grain and coarse-grain soil, a helical pile might also bear directly on bedrock. The ultimate bearing pressure of highly weathered rock can be determined by correlations with unconfined compressive strength using Equation 4.15

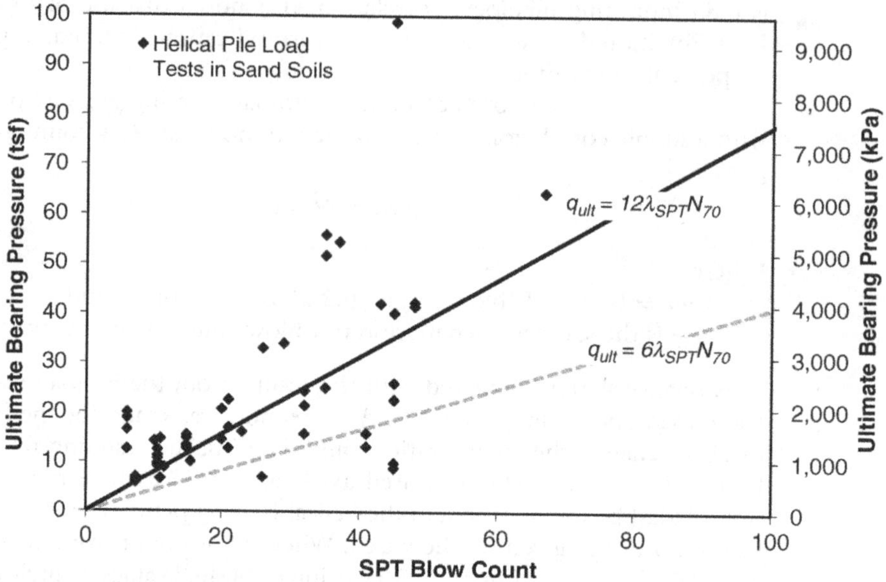

Figure 4.9 Correlation between bearing pressure and blow count in sand

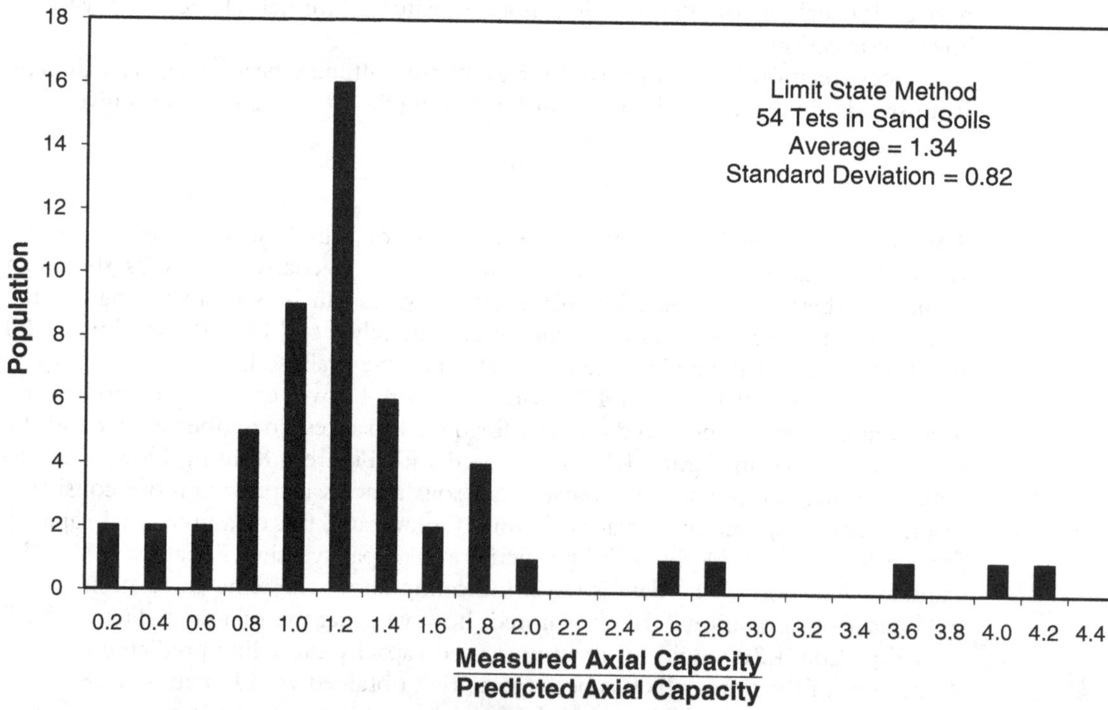

Figure 4.10 Comparison of measured and predicted capacity in sand using individual bearing method with SPT correlations

and Equation 3.12 from Chapter 3. The ultimate bearing pressure of more competent rock can be determined from RQD values using Equations 3.5 and 3.6 along with ultimate unconfined compressive strength as discussed in Chapter 3. The structural capacity of helical pile shafts often governs the overall axial capacity when end bearing on competent rock. Structural capacity of shafts is discussed later in this chapter.

Highly weathered rock can behave more like a soil than like competent rock. It makes sense that the ultimate bearing pressure of highly weathered rock could be estimated from Standard Penetration Test blow count. The correlation in Equation 4.21 is proposed based on many years of experience.

$$q_{\text{ult}} = 13\lambda_{\text{SPT}}N_{70} \qquad (4.21)$$

All parameters have been defined previously. Ultimate bearing pressure determined using Equation 4.21 is compared with the results of 23 load tests performed on helical piles in weathered rock in Figure 4.11. The load tests included helical piles with diameters ranging from 2.875 inches [73 mm] to 8.6 inches [218 mm]. Each helical pile had one helical bearing plate ranging from 8 inches [203 mm] to 16 inches [406 mm] diameter. These and other load tests are given in more detail in Appendix C. The relationship given by Equation 4.21 seems to fit the data shown in Figure 4.11 within reason, given the inherent variability of weathered rock. As can be seen in the figure, the measured bearing pressure in weathered rock ranges from approximately 5 tsf [479 kPa] to almost 180 tsf [18,000 kPa] for the available helical pile load tests.

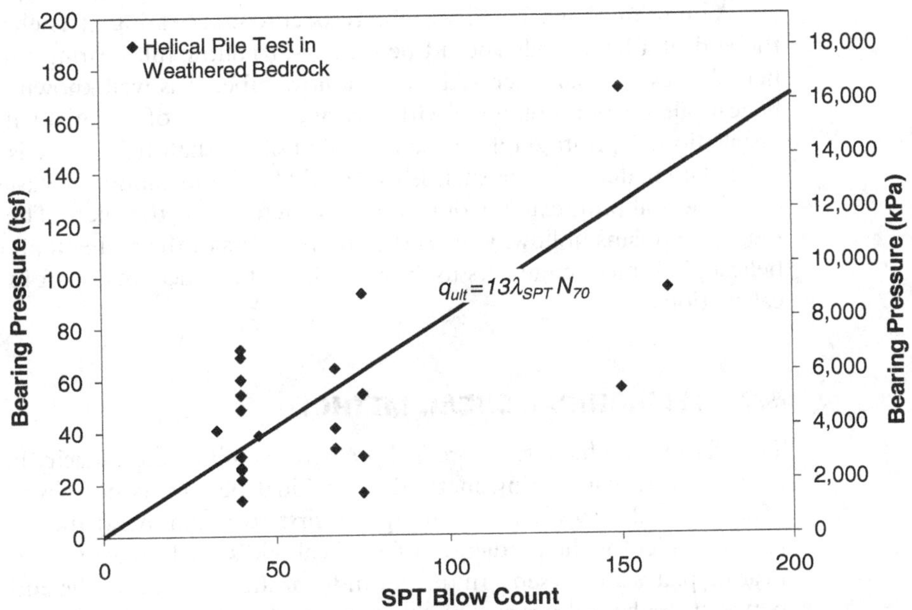

Figure 4.11 Correlation between bearing pressure and blow count in weathered bedrock

Figure 4.12 Plugged end of a typical helical pile

The rock types in these tests included claystone, sandstone, and shale. Penetration into the rock varied from a few inches [centimeters] to several feet [meters].

A question that often arises with respect to end bearing of a helical pile is whether the end of a helical pile should be plugged to allow full bearing on the shaft. Many helical piles are manufactured using hollow tube. It is well known that the end of a helical pile becomes plugged with a very dense layer of material within a few feet of installation. A photograph of the end of a hollow shaft helical pile is shown in Figure 4.12. The material in the shaft is extremely hard and cannot be easily pried from the end. The soil plug extends only a few diameters into the shaft. The rest of the shaft typically remains hollow. Due to the compactness of the material plugging the end of helical piles, most engineers use the full area of a helical pile shaft in bearing capacity calculations.

4.3 CYLINDRICAL SHEAR METHOD

Thus far in this chapter, theoretical predictions of bearing capacity have been focused on the individual bearing method. Individual bearing is not always the limit state. Mooney et al. (1985) were among the first to recommend the use of a cylindrical failure model for the prediction of a helical pile's axial capacity when multiple helical bearing plates are present. In the "cylindrical shear" method, the entire volume of soil between the helical bearing plates is assumed to be mobilized. A free body diagram showing the forces on a helical pile for the cylindrical shear method is shown in the left side of Figure 4.2. Uniform pressure is distributed under the lead helix, and shear stresses surround the soil encapsulated between the helical bearing plates. Adhesion stresses act along the length of the helical pile shaft located above the top helix. An axial force is applied to the pile butt.

Ultimate bearing capacity of a helical pile based on the cylindrical shear method is found by taking the sum of shear stress along the cylinder, adhesion along the shaft, and bearing capacity of the bottom helix given by

$$P_u = q_{\text{ult}} A_1 + T(n-1)s\pi D_{\text{AVG}} + \alpha H(\pi d) \tag{4.22}$$

Where

A_1 is the area of the bottom helix
T is soil shear strength
H is the length of shaft above the top helix
d is the diameter of the pile shaft, and
the term $(n - 1)s$ is the length of soil between the helices. All other parameters have been defined previously.

Shear strength, T, may be calculated using the methods described in Chapter 3. In fine-grain soil, T may be taken as the undrained shear strength, s_u, based on Equation 3.9. As discussed previously, s_u may be determined from SPT blow count. Predicted capacity based on the cylindrical shear method was compared with the results of 32 load tests on helical piles with shafts ranging from 2.875-inch [73-mm] to 10.75-inch [273-mm] diameter with between 2 and 5 helical bearing plates having 8-inch [203-mm] to 30-inch [762-mm] diameters. These and other load tests are detailed in Appendix C. Results of this comparison are shown in Figure 4.13. To be conservative, adhesion along the shaft was ignored. On average, measured capacities based on the cylindrical shear method were 0.82 times the predicted capacity, suggesting that the method is slightly unconservative. The standard deviation of the data is 0.26, indicating much better correlation than the individual bearing method shown in Figure 4.5. In this case, a factor of safety of 2.0 would result in 94 percent of the measured capacities being above the predicted capacities.

As discussed in Chapter 3, the shear strength of coarse-grain soil is represented by Equation 3.13. The difficult part of determining shear strength in coarse-grain soil is defining the effective confining stress, σ'_n. In undisturbed ground, the effective lateral confining stress at a depth z in the ground is given by

$$\sigma'_n = K_o P'_o \tag{4.23}$$

Where

K_o is the at-rest coefficient of lateral earth pressure and
P'_o is the effective overburden stress at depth z.

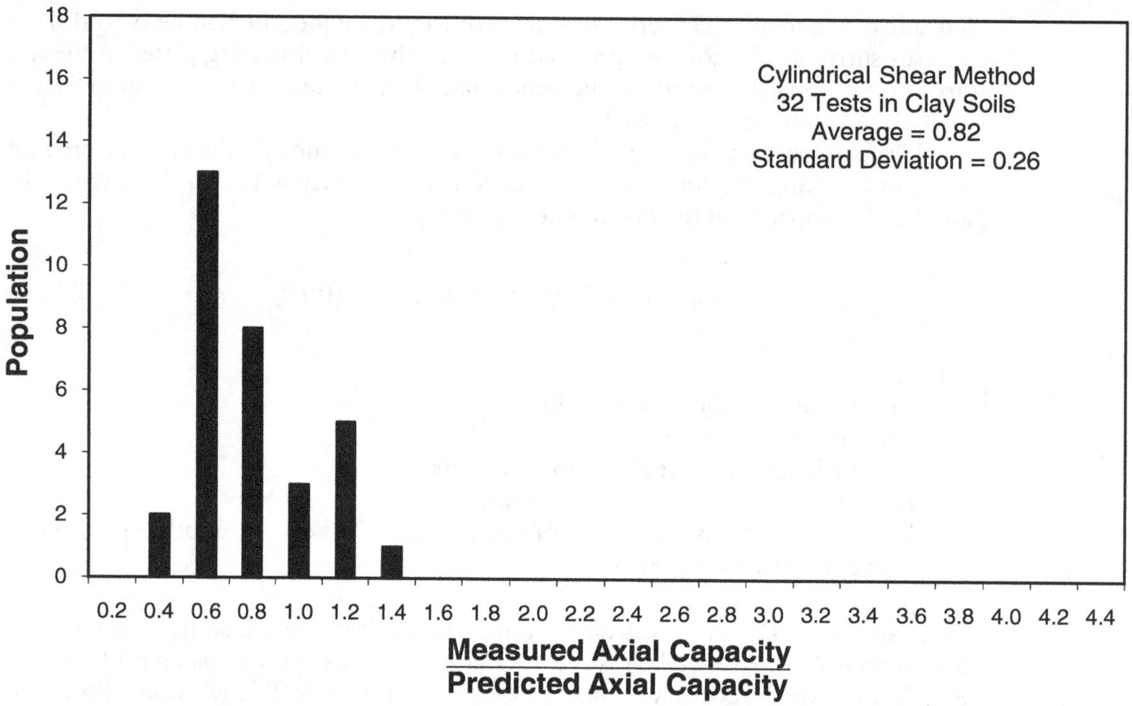

Figure 4.13 Comparison of measured and predicted capacity in clay coils using cylindrical shear method

Effective overburden stress may be computed from

$$P'_o = \gamma z - \gamma_w h_w \tag{4.24}$$

Where

γ_w is the unit weight of water

γ is the unit weight of soil (technically it should be the moist unit weight above the water table and saturated unit weight below the water table), and

h_w is the height of water above depth z

An ideally installed helical pile will thread into the soil with minimal disturbance. However, the lateral displacement of soil during insertion of the pile can cause an increase in lateral stresses immediately adjacent the pile. Mitsch and Clemence (1985) showed the magnitude of lateral stress around helical piles is proportional to the initial relative density of the soil. They computed inter-helix shear stress based on several load tests in sand and recommended lateral earth pressure coefficients shown in Figure 4.14. The original values suggested by Mitsch and Clemence are shown in the table at the right of the figure and by the data points in the chart. A best-fit regression analysis was performed to find an equation for the relationship between the lateral earth pressure,

Angle of Internal Friction, Φ	25	30	35	40	45
Coefficient of Lateral Pressure, K_h	0.7	0.9	1.5	2.35	3.2

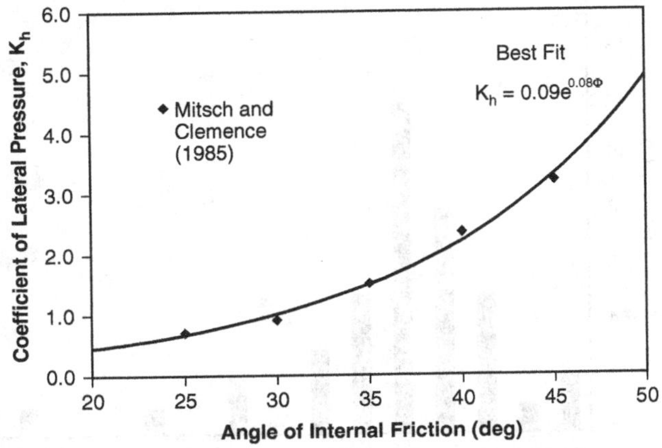

Figure 4.14 Lateral earth pressure coefficients for cylindrical shear method (Adapted from Mitsch and Clemence, 1985)

K_h, and internal angle of friction, Φ, using Mitsch and Clemence recommended values. The resulting equation is:

$$K_h = 0.09e^{0.08\Phi} \tag{4.25}$$

The relationships given in Equations. 4.23, 4.24 and 4.25 can be substituted into Equation 3.13 to find the shear stress acting on the cylinder of soil located between helical bearing plates, as given by

$$T = \left(0.09e^{0.08\Phi}\right)(\gamma z - \gamma_w h_w) \tan \Phi \tag{4.26}$$

This equation and Equation 4.22 form the crux of the cylindrical shear method for finding the theoretical ultimate capacity of a helical pile with multiple helical bearing plates in coarse-grain soil. To be conservative, adhesion along the shaft above the top helix may be neglected. The accuracy of this method was examined by comparing predicted capacity with the results of 42 load tests on helical piles with shafts ranging from 1.5 inches [38 mm] square to 8.62 inches [219 mm] in diameter with between two and four helical bearing plates having 8-inch [203 mm] to 30-inch [762-mm] diameters. These and other load tests are detailed in Appendix C. Results of this comparison are shown in Figure 4.15. On average, the measured capacity in the load tests was 1.07 times the cylindrical shear method predicted capacities. The standard deviation of the data is 0.58, indicating better correlation than the individual bearing method shown in Figure 4.8. Ultimate bearing pressure was determined based on blow count for this comparison. Use of a factor of safety of 2.0 with this method would result in 93 percent of the measured capacities being above the predicted capacities.

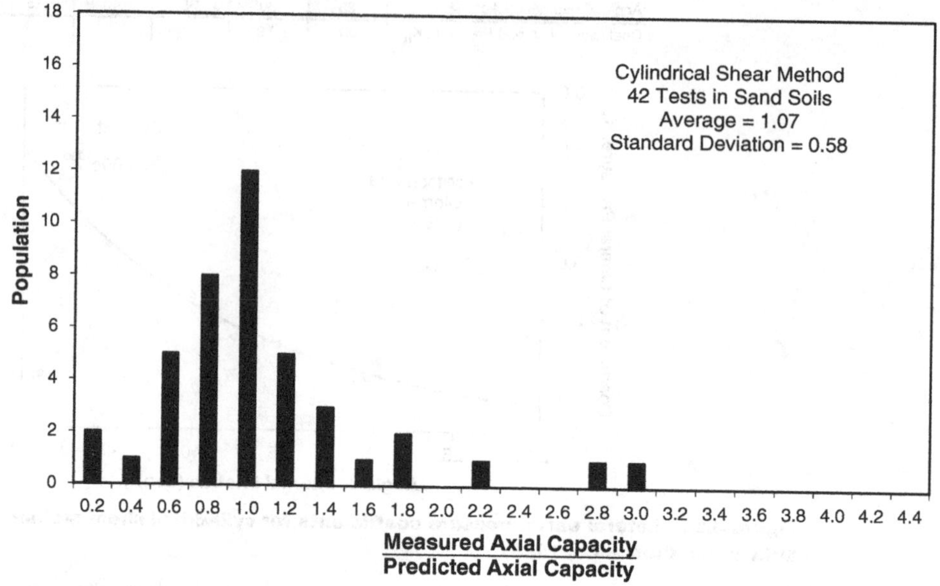

Figure 4.15 Comparison of measured and predicted capacity in sand soils using cylindrical shear method

4.4 LIMIT STATE ANALYSIS

In limit state analysis, the capacity should be determined using both the individual bearing and cylindrical shear methods, and the least value calculated should be reported as the predicted capacity. To evaluate the effectiveness of limit state analysis, the minimum value computed using the foregoing methods was compared with load tests described previously. Results are shown in Figure 4.16 for fine-grain soils and in Figure 4.17 for coarse-grain soils. Both graphs show good correlation between measured and predicted capacities, lending credence to this type of analysis and giving confidence to the theoretical prediction of the capacity of a helical pile.

All of the load test data for fine-grain soils, coarse-grain soils, and weathered bedrock were combined and compared with predicted capacity based on limit state analysis in Figure 4.18. A normal distribution curve is shown by the solid curve in the figure. The mean value of the data is 0.97, indicating practically 1:1 correlation. The standard deviation of 0.51 and the visually narrow distribution of the data further reinforce the premise that the capacity of helical piles can be theoretically determined with a reasonable degree of confidence. In the preparation of this figure, ultimate bearing pressure was estimated based on standard penetration test blow count. The predicted capacity was found by computing the capacity using both individual bearing and cylindrical shear methods and taking the limit state. A total of 112 load tests on helical piles were analyzed to produce this figure.

Narasimha Rao, Prasad, and Veeresh (1993) suggested correcting the individual bearing method to take into account helix spacing. The correction is supposed to

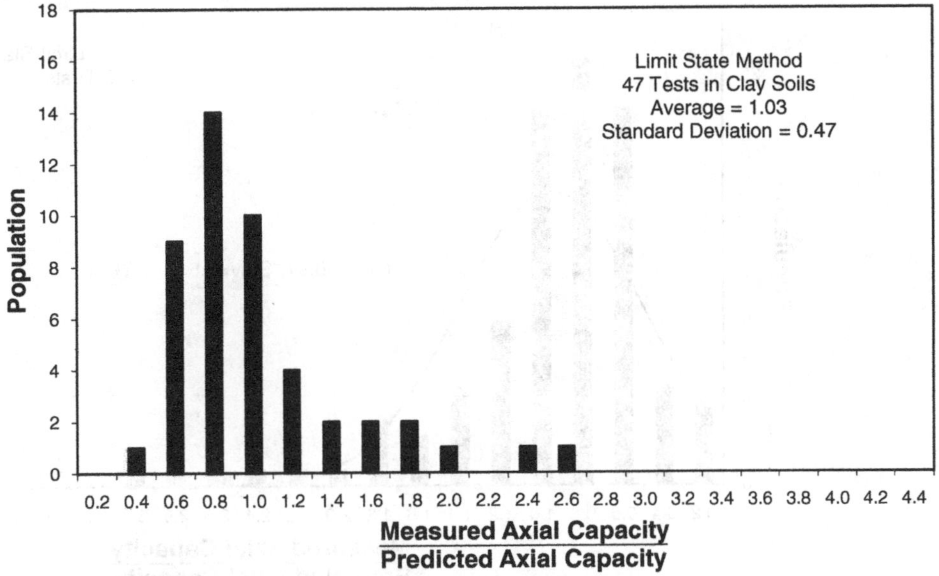

Figure 4.16 Comparison of measured and predicted capacity in clay soils using limit state analysis

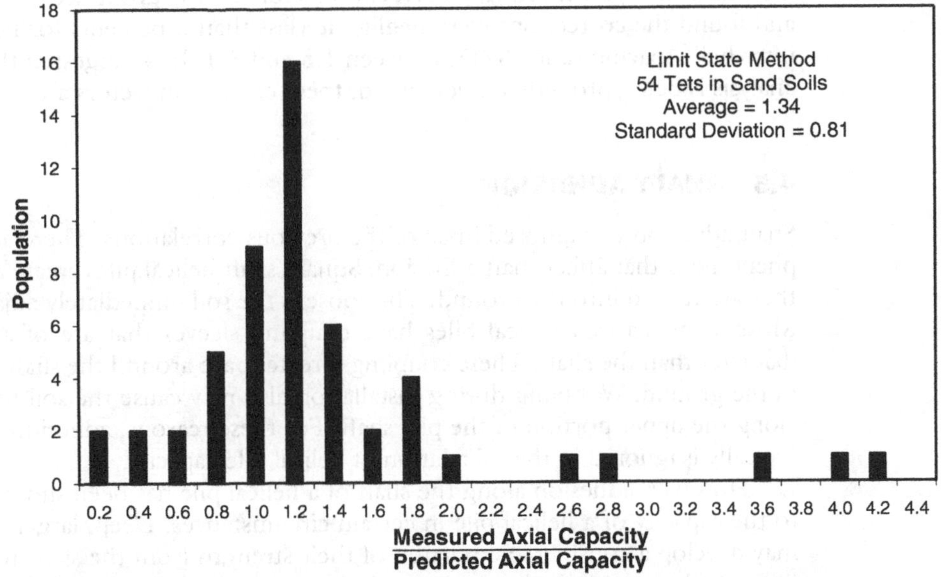

Figure 4.17 Comparison of measured and predicted capacity in sand soils using limit state analysis

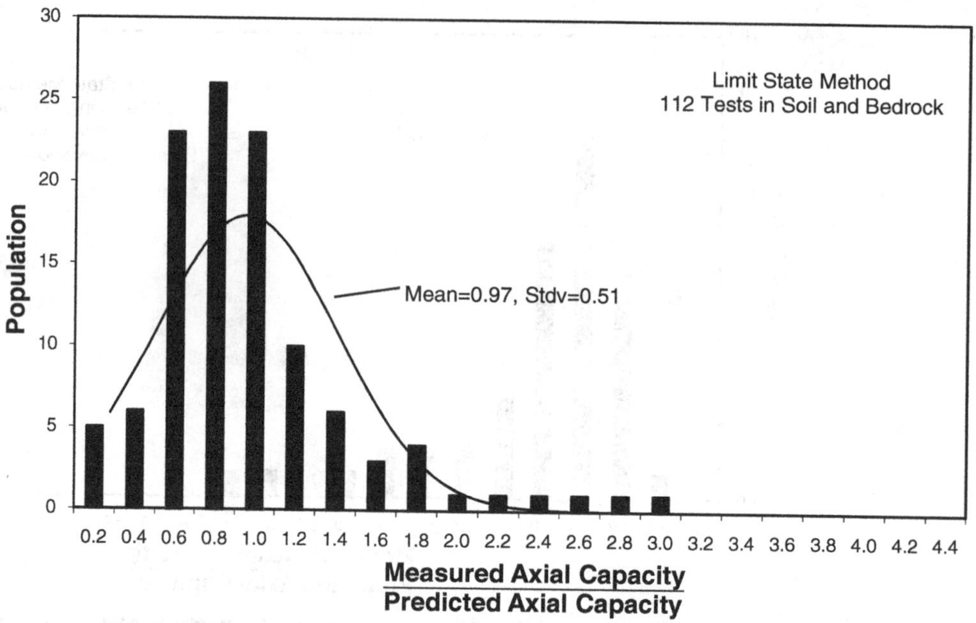

Figure 4.18 Comparison of measured and predicted capacity for all available load lests

better model a transition zone between fully individual bearing and fully cylindrical shear. Tappenden (2004) compared results computed using the individual bearing method with and without the correction factors by the group led by Narasimha Rao and found the corrections were negligible (less than 2 percent) for helical piles with inter-helix spacing ratio (S/D) between 1.5 and 3.4. It is suggested that if limit state analysis is being properly implemented, these correction factors are unnecessary.

4.5 SHAFT ADHESION

Shaft adhesion was ignored in all of the previous correlations. There are a number of phenomena that affect shaft adhesion. Square shaft helical piles create a round hole as they are turned into the ground. This loosens the soil immediately adjacent the shaft. Most manufactured helical piles have coupling sleeves that are of a slightly larger diameter than the shaft. These couplings create space around the shaft upon insertion in the ground. Wobbling during installation also may cause the soil to separate from along the upper portion of the pile shaft. For these reasons, adhesion along the shaft typically is ignored in the calculation of helical pile capacity.

However, adhesion along the shaft of a helical pile has been shown to contribute to the capacity of a helical pile in certain circumstances. Deep, larger-diameter shafts may develop a considerable portion of their strength from the shaft to soil interface. Smooth shafts with flush couplings also may generate some capacity from adhesion. Ghaly and Clemence (1998) showed that significant adhesion can even occur around square shafts in clean coarse-grain soils. If shaft adhesion is accounted for, it may be approximated by

$$\alpha = 2/3T \quad (4.27)$$

Where

T is the shear strength of the soil from Equation 3.9 (fine grain) or Equation 4.26 (coarse grain).

The factor of 2/3 is used to account for the reduced friction of soil on bare or galvanized steel. Different reduction factors may be appropriate for paint, epoxy coating, or other surface treatments. A factor of 1.0 may be used to calculate adhesion of grouted helical pile shafts.

Since soils immediately adjacent helical pile shafts experience high strains from shearing the soil during installation, residual shear strength parameters may be more appropriate for determining adhesion. Particularly sensitive soils with residual shear strength that is much lower than peak strength would have minimal adhesion. For this reason, the adhesion between soil and driven pile shafts has been observed to decrease with increasing overconsolidation ratio (Meyerhof, 1976).

According to Narasimha Rao, Prasad, and Veeresh (1993), adhesion can be used for helical piles but it should be limited. Their research suggests limiting adhesion to an effective shaft length, H_{eff}, equal to the length of the shaft above the helical bearing plates, H, minus 1.4 to 2.3 times the diameter of the uppermost helical bearing plate, D_T, as shown in Figure 4.19. H_{eff} is less than the total shaft length as a result of void forming above the top helix and subsequent interference with the adhesion along the shaft. Based on full-scale load tests conducted on instrumented multi-helix

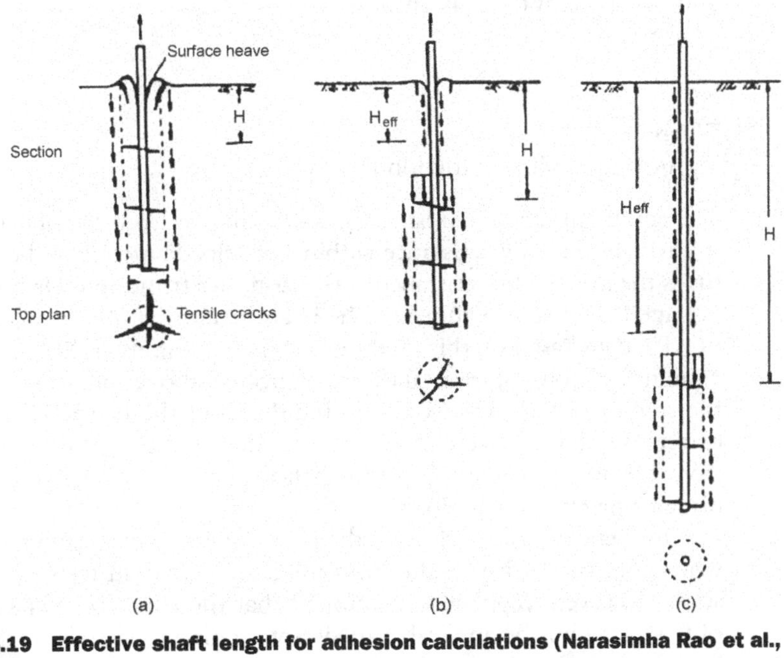

Figure 4.19 Effective shaft length for adhesion calculations (Narasimha Rao et al., 1993c)

piles installed in coarse-grain and fine-grain soils, Zhang (1999) concluded that H_{eff} is approximately equal to the available shaft length minus the diameter of the upper helix.

4.6 LCPC METHOD

The LCPC method (named for the Laboratoire Central des Ponts et Chaussees in Paris, France) is based on 197 static load tests on 48 test sites with several different pile types including drilled piles, driven piles, and cast screw piles. The method is described in detail by Bustamante and Gianeselli (1982) and is said to be useful for complex layered sites with soft or loose soils interdisbursed with partial cemented layers, cobbles, or other compact materials where prediction of bearing capacity factors and soil shear strength is highly random.

The LCPC method is a way to estimate ultimate soil bearing pressure and side shear strength using cone penetrometer test (CPT) data. The resulting values can be used in the individual bearing and cylindrical shear methods to find helical pile capacity. According to the LCPC method, ultimate soil bearing pressure, q_{ult}, is given by

$$q_{ult} = q_{ca}k_c \tag{4.28}$$

Where

q_{ca} is the equivalent cone tip resistance at the depth of the helix and
k_c is the penetrometer bearing capacity factor. Tappenden (2007) used k_c of 0.45 for calculations on helical piles.

Side shear strength, T, is given by

$$T = q_s \tag{4.29}$$

Where

q_s is the unit skin friction from the CPT test.

The equivalent cone tip resistance is determined from a profile of CPT data by taking the average cone tip resistance within 1.5 helix diameters of the pile tip. Data over 1.3 times the average tip resistance in this zone are truncated. Data less than 0.7 times the average tip resistance above the helix are truncated. An example showing truncation of CPT data based on this method is shown in Figure 4.20. As can be seen, the lower extremes of cone tip resistance are truncated above the bottom of the pile, which in the case of helical piles would be the depth of the helical bearing plate. The higher extremes of cone tip resistance are truncated over the zone of consideration spanning from 1.5 times the diameter of the helical bearing plate above the pile tip to that same distance below the pile tip.

Tappenden (2004) showed the LCPC method correlates well (within 30 percent) when compared with results from full-scale field load tests on helical piles for many sites. However, Tappenden cautioned that the penetrometer bearing capacity factor of 0.45 may not be applicable to sites with glacial till soils, such as that encountered

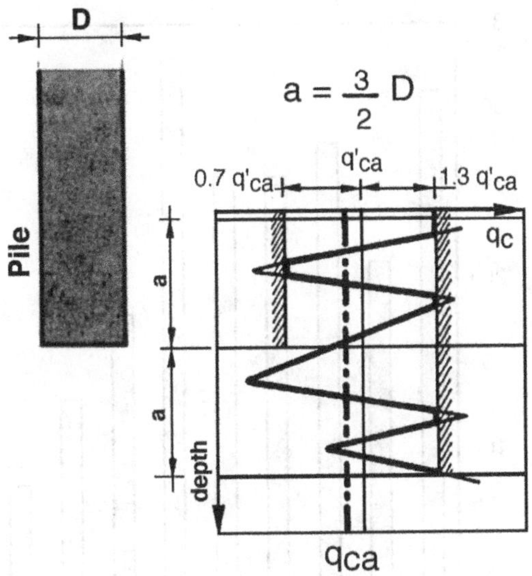

Figure 4.20 LCPC method of CPT data truncation

in the vicinity of Ft. McMurray, Alberta, or for highly overconsolidated clay, such as that found in Edmonton, Alberta. The LCPC method significantly overestimated capacity in these soils by up to 374 percent. It is thought that till soils affect CPT results and that disturbance during helical pile installation may have reduced capacity in highly overconsolidated clay. A lower penetrometer bearing capacity factor should be developed for these conditions.

4.7 PILE DEFLECTION

Individual bearing, cylindrical shear, and LCPC methods of limit state analysis deal only with strength. Helical piles designed using a factor of safety of at least 2.0 generally exhibit acceptable deflections in competent bearing material. Based on many years of experience, settlement of a properly designed and installed helical pile in good bearing material is typically on the order of 1/2 to 1 inch [13 to 25 mm] or less under design loads. Good bearing material is defined as any soil with a blow count higher than 20.

The net deflections measured in 120 load tests on helical piles are shown in Figure 4.21. Net deflection is defined as the total deflection of the pile head minus the elastic elongation or shortening of the shaft. Calculation of net deflection is discussed further in Section 7.4. The average net deflection at load test capacity is 1.1 inches [28 mm]. A majority of the tests exhibiting large deflections were in soft soils. If load test capacity were divided by a factor of safety of 2.0, then the net deflection under allowable load would be less than half the net deflection at the test load due to the nonlinear behavior of most load tests on helical piles. The net deflection at allowable load would be less than 1 inch [25 mm] for almost all tests. For a majority of tests, the net deflection at allowable load would fall between 0.5 and 1 inch [13 to 25 mm], thereby justifying the statements made above based on experience.

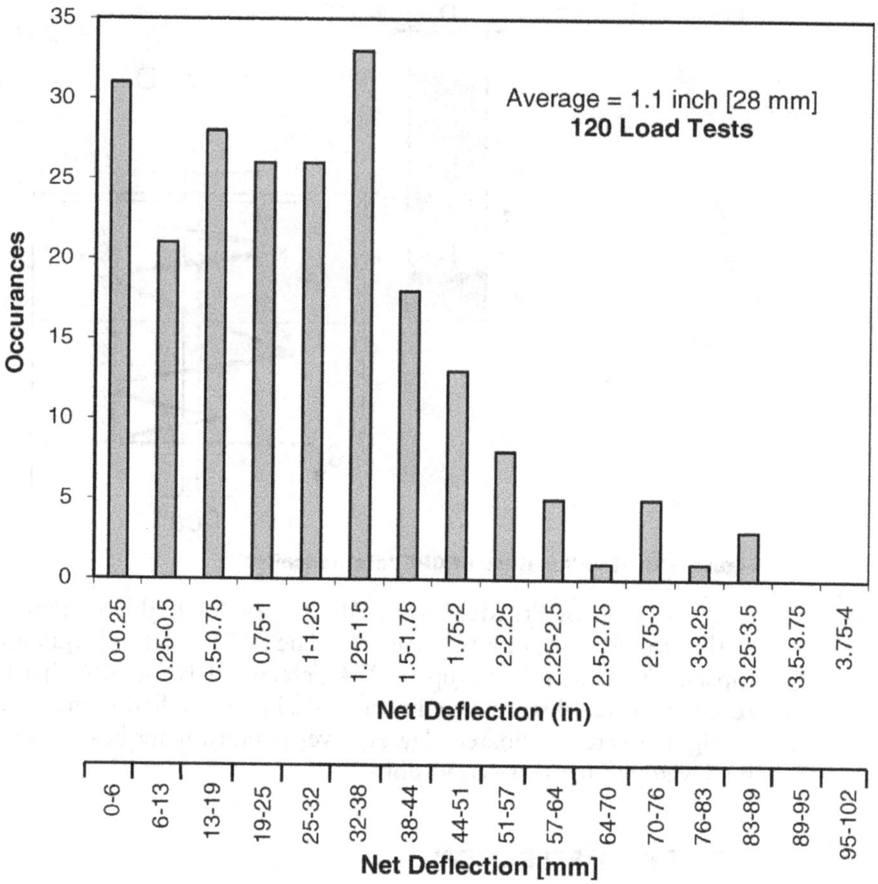

Figure 4.21 Helical pile load test net deflections

The load tests from which these data were taken are described in more detail in Appendix C. All load tests were included in Figure 4.21 where pile length and pile diameter were known. These tests included both tension and compression. Elastic lengthening and shortening were estimated assuming 0.25-inch [6 mm] wall thickness for round shafts. A number of load test interpretation methods were used to determine the load test capacity. A discussion of axial load testing and various interpretation methods is contained in Chapter 7.

Of the data contained in Figure 4.21, the average SPT blow count at the pile tip depth was available for 86 tests. Net deflection measurements are plotted against SPT blow count in Figure 4.22. Tests on helical piles with shafts between 2.875 inches to 3.5 inches [73 to 89 mm] are represented in the figure by solid black diamonds. Tests on helical piles with larger diameter shafts are represented by the symbol "x." In general, net deflection decreases with increasing blow count, as might be expected. The two extreme outliers at a blow count of about 40 are pullout tests in highly

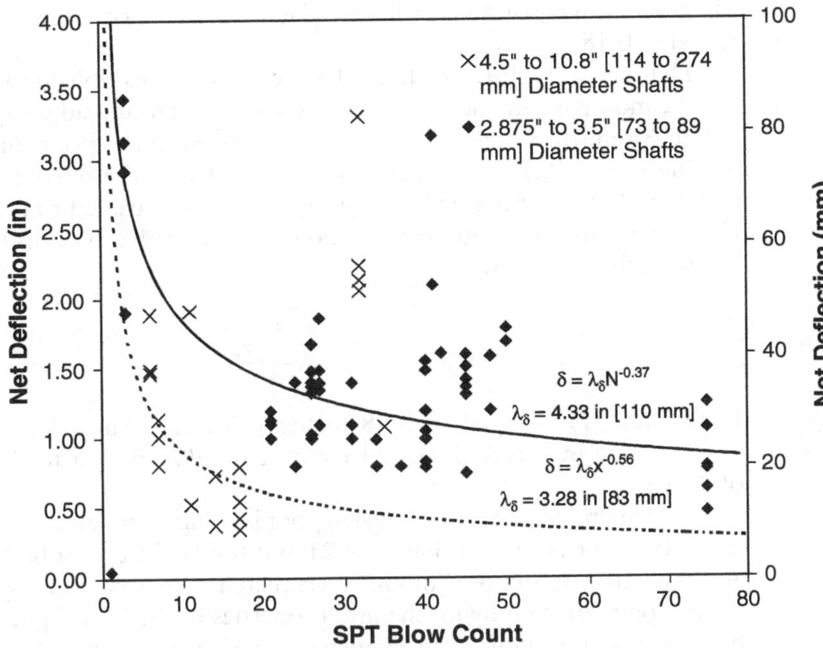

Figure 4.22 Helical pile load test deflections

variable glacial till. The till may have affected blow counts. Helical piles with larger-diameter shafts seem to exhibit less net deflection than helical piles with smaller shafts. The reason for this is unknown except that many of the load tests on helical piles with larger diameters were conducted to verify load capacity and not necessarily taken to plunging or ultimate capacity.

A best-fit power function regression analysis was performed to find an equation to model the data for 2.875-inch- to 3.5-inch- [73- to 89-mm-] diameter shafts. A power function was selected to match the obvious boundary conditions that net deflection should approach infinity as blow count approaches zero for fluid soils and that net deflection should become negligible at extremely high blow counts. The equation for the best-fit power function shown by the solid curve in Figure 4.21 is given by

$$\delta = \frac{\lambda_\delta}{{N_{70}}^{0.37}} \tag{4.30}$$

Where

δ is deflection

λ_δ is a fitting constant equal to 4.33 inches [110 mm], and

N_{70} is SPT blow count.

The hammer energy used for the geotechnical exploration of all these load tests is unknown. An energy ratio of 70 was assumed since most of the tests are fairly recent.

The regression equation matches the data with a mediocre least square deviation (R^2-*value*) of 0.48.

Helical piles with 4.5-inch to 10.8-inch diameter exhibit considerably more scatter. For the most part, net deflection is very small. The four load tests with 2 to 3.5 inches [51 to 89 mm] of deflection had one to two 30-inch- [762-mm-] diameter helical bearing plates. Helix thickness was not provided in the reference. It is possible that punching flexure of the helical bearing plates may have led to the higher deflections. If the four outliers are ignored, the power function that best fits the load test data for larger shafts is given by

$$\delta = \frac{\lambda_\delta}{N_{70}{}^{0.56}} \tag{4.31}$$

In this case, λ_δ is equal to 3.28 inches [83 mm]. The equation is shown by the dashed curve in Figure 4.21 and matches the data with a mediocre deviation (R^2) of 0.51.

Settlement becomes a concern in poor bearing materials and under sensitive structures. It can be seen from Figure 4.21 and the deviations of best-fit Equations. 4.30 and 4.31 that deflection cannot be estimated very accurately by blow count alone. Deflection is affected by mechanical properties of the helical pile, such as helix pitch, bearing plate thickness, shaft diameter, and shaft thickness. Deflection is affected by the soil conditions immediately around and several feet (meters) away from the pile tip, such as soil layering, effective stress, overconsolidation ratio, adhesion on the shaft, density, stiffness, elasticity, lateral earth pressure coefficient, and perhaps most important the coefficient of consolidation.

Several studies have shown that settlement of a helical pile can be estimated effectively using commercially available finite element or discrete element software. An example of how a helical pile can be modeled is shown in Figure 4.23. The triangular pattern represents the mesh of elements. The intersections of the mesh lines are called nodes. A helical pile with two helical bearing plates is drawn in the center of the mesh as represented by the bold lines. The helical pile is approximately 21 feet [6.5 m] deep. The design load is being applied to the pile at the ground surface. Deflections are exaggerated by 200 times. The maximum deflection is 0.22 inch [5.5 mm].

Finite element and discrete element programs enable the engineer to find a solution to complex mathematical problems without closed form solutions. These methods are also called numerical modeling. Numerical modeling is actually quite simple. Effective stress at each node and the resulting strain of each element is computed through multiple iterations until the underlying relationships between stress and strain and the boundary conditions are satisfied for every node and every element. The accuracy of these estimates depends only on the reliability of the input properties of soil and the boundary conditions used in the model.

The boundary conditions used in the example shown in Figure 4.23 include fixing the nodes at the bottom of the model in both horizontal and vertical directions and fixing the nodes at the sides of the model in the horizontal direction only. Some boundary conditions need to be input into the model, or it will be unable to find a

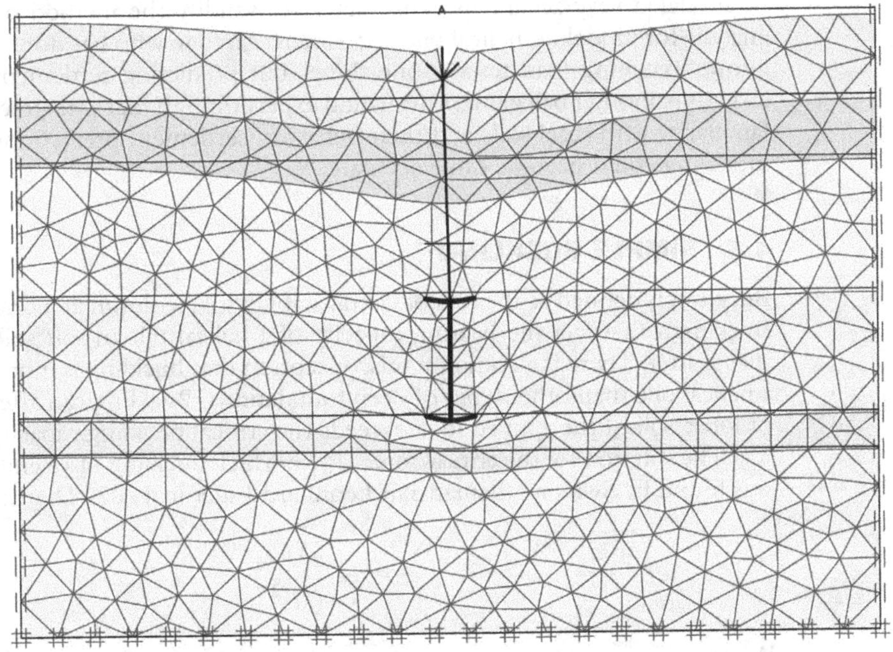

Figure 4.23 Finite element model of a helical pile in layered soils (Nasr, 2007)

solution. As long as the bottom and side elements are located far away from the helical pile, these conditions should have little effect on the results.

Less obvious boundary conditions are the fixity of individual nodes. The fact that the portion of helical pile shaft located above the top helix crosses finite elements without intersecting any nodes means that that portion of shaft is decoupled from the soil (zero adhesion). The portion of shaft located between the helical bearing plates crosses every element at a node so this section of shaft can be fixed to the soil. Affixing the soil nodes between the helical bearing plates to the pile shaft enables the pile to capture the soil and drag it down with the helical bearing plates, as in the case of the cylindrical shear method. The shaft can be decoupled from the soil by releasing the nodes from the shaft to model the individual bearing method.

The subsurface shown in the model is broken into several layers. Each soil element must have a defined stress-strain modulus and Poisson ratio. Index values for several common soil and bedrock types are shown in Tables 3.6 and 3.7 in Chapter 3. Stress-strain modulus and Poisson ratio have not been studied as extensively as other soil properties, such as density and shear strength. Determination of the correct soil and rock properties to input into a numerical model is difficult. Perhaps the best way to determine these properties is through analysis of a load test performed at the site in question. Once the appropriate soil and rock properties are back-calculated from modeling the load test, numerical modeling can be used to evaluate other helical pile configurations.

Nasr (2007) designed the numerical model shown in Figure 4.23 and compared it with two load tests. According to the designer, the results of the load tests could be modeled very accurately. This fact gives validity the method of using numerical analysis for modeling helical pile settlement. Anyone who has done a lot of modeling knows matching load tests is one thing, but predicting settlement from numerical analysis is another ball game. One would expect though that with practice and increased familiarity with local soils and their behavior, settlement predictions using numerical modeling could become quite accurate.

4.8 SIMPLE BUCKLING

Buckling is defined as the loss of lateral stability of a column at a certain critical force. Buckling can affect the bearing capacity of a helical pile in soft soils, or any pile for that matter. There is a common misconception that buckling is more of a problem for a helical pile than other types of deep foundations. Buckling can affect any pile in soft or loose soils if the applied load exceeds the critical buckling capacity.

The solution for the critical force, P_u, at which buckling of a slender column occurs was found by Swiss mathematician Leonhard Euler in 1757. The Euler formula is

$$P_u = \frac{\pi^2 EI}{(kl)^2} \tag{4.32}$$

Where

E is modulus of elasticity,
I is the area moment of inertia,
k is the effective length factor, and
l is the unsupported length of the column.

Examination of Equation 4.32 reveals that buckling does not depend on the strength of materials. The critical load varies directly with the stiffness of the shaft (EI) and inversely with the square of the effective length (kl).

It is important to note that buckling is a condition of elastic stability. It was found from the solution of a complex harmonic equation defining the stability of a column under load. It was not found by simple static analysis or equilibrium between stress and strain. A column may be strong enough to carry a certain load, but it is of little use if it is unstable. It is for this reason that simple stress-strain based modeling may be inappropriate for an unbraced column or a column in air or fluid soils.

According to Tomlinson (1986), piles embedded wholly in the ground need not be considered as columns for the purposes of structural design. This is echoed in the *International Building Code* (IBC) (2006) Section 1808.2.9.1, which states that any soil other than fluid soil shall be deemed to afford sufficient lateral support to prevent buckling of braced piles. Section 1808.2.5 explains that all piles should be laterally braced. Lateral bracing may be accomplished by the intersection of two walls or grade beams from different directions. A minimum of two piles must be used under pile caps braced in only one direction by a wall or grade beam. A minimum of three piles must be used to provide bracing under an isolated pile cap. Piles in a continuous pile cap under a wall or other linear structure must be staggered a minimum of 1 foot [30 cm] from the centerline of the pile cap except in light residential construction. Further details regarding foundation layout are given in Chapter 12.

Tomlinson (1986) further explains that where a pile projects aboveground, the portion aboveground must be considered a column. Tomlinson goes on to state that the *European Code of Practice of Foundations* recommends that in good ground, the lower point of counterflexure can be taken as 5 feet [1.5 m]. When the top stratum is in soft clay or silt, this point may be taken as half the depth of penetration into this stratum but not necessarily more than 10 feet [3.1 m]. IBC (2006) Section 1808.2.9.2 again echoes this by stating that any pile standing unbraced in air, water, or fluid soils should be checked for buckling using a depth of fixity of 5 feet (1.5 m) in firm soils and 10 feet [3.1 m] in soft soils plus the length of the shaft in air, water, or fluid soils. ICC-Evaluation Services, Inc. (2007) defines soft soils for helical piles as any soil with a SPT blow count greater than zero and less than five. Some designers believe that all helical piles should be checked for buckling regardless of the bracing at the top of the pile. Their argument is that the method of installation may cause a small part of the upper portion of the shaft to be unsupported.

Examples showing the unbraced length of helical pile shafts under two different conditions are shown in Figure 4.24. On the left side of the figure, the pile butt is laterally braced by a concrete pile cap. Presumably the concrete pile cap itself is braced by a system of walls, grade beams, or multiple piles under the same cap. The pile butt is laterally braced but not rotationally braced because it extends only a short distance into the plain concrete cap. The appropriate end condition for the pile is a pinned head connection. This is fairly typical of most pile cap systems. There are no fluid soils present under the pile cap in this example. The depth to fixity and hence, the unsupported length, l, is either 5 feet [1.5 m] or 10 feet [3.1 m], depending on subsurface conditions. The effective length factor, k, may be obtained from AISC (2001) Table C-C2.1 for the pinned-fixed condition (shape "b," design $k = 0.8$).

The example on the right side of Figure 4.24 consists of a helical pile shaft that extends some distance into the bottom of a concrete pile cap. Reinforcement is placed around the pile butt to provide rotational as well as lateral restraint. The appropriate end condition for the pile is a fixed head connection. In this example, there is a layer of fluid soils (N = WOH) located below the pile cap. The unsupported length of the pile shaft is the thickness of this layer plus the depth of fixity. This time, the effective length factor, k, may be obtained from AISC (2001) Table C-C2.1 for the fixed-fixed condition (shape "a," design $k = 0.65$).

The original Euler buckling equation given in Equation 4.32 is not used in practice because modern developments have shown that there is a transition zone between a compact column and a truly slender column such that experimental results do not always match Equation 4.32. Buckling is better determined according to AISC (2001), which states that the nominal (LRFD) compressive strength for flexural buckling, P_n, is given by

$$P_n = A_g F_{cr} \tag{4.33}$$

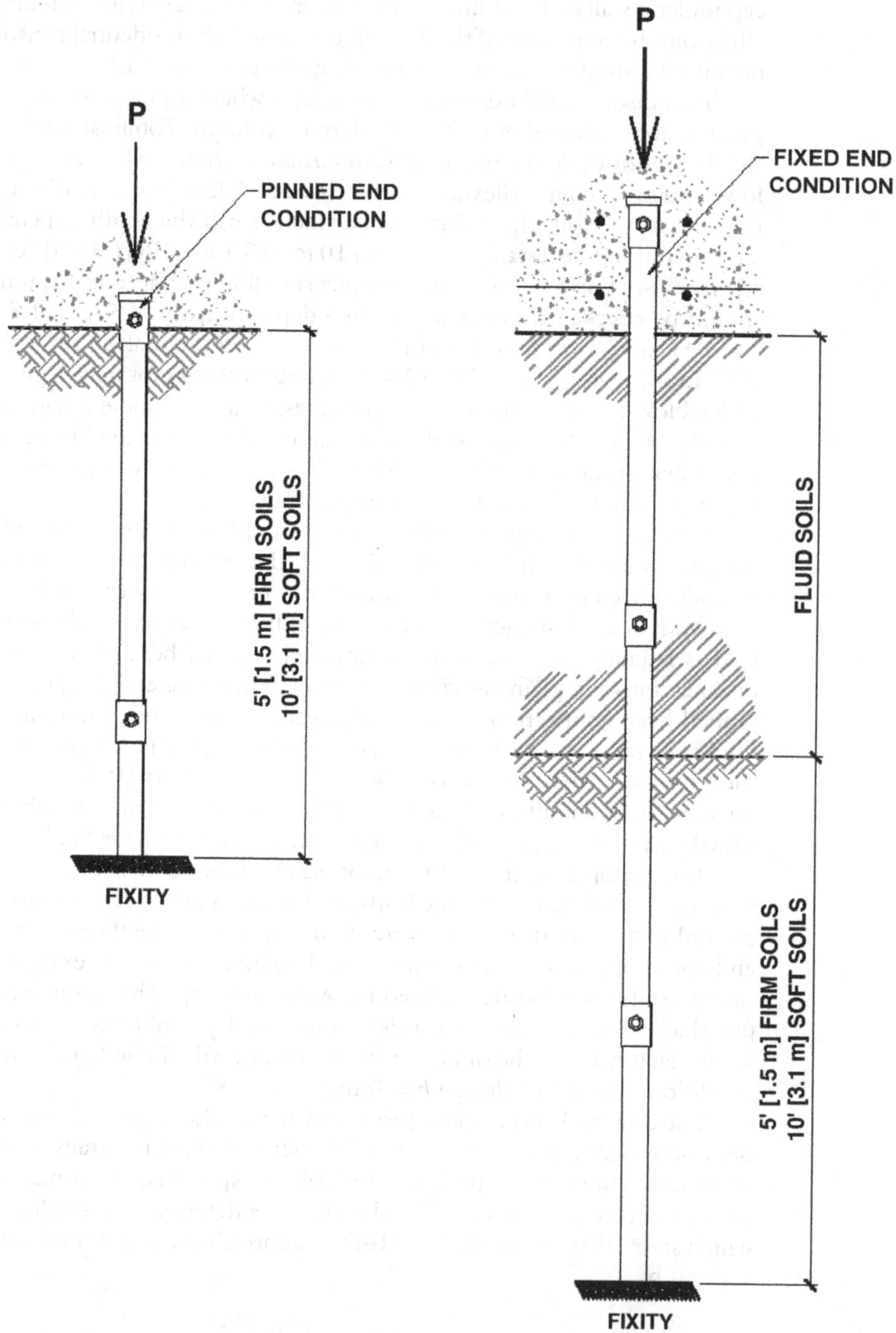

Figure 4.24 Unbraced length examples for helical pile shafts

Where

A_g is the gross cross-sectional area of the pile shaft and

F_{cr} is the critical buckling stress that depends on the column slenderness parameter λ_c, in this way:

for $\lambda_c \leq 1.5$

$$F_{cr} = \left(0.658\lambda_c^2\right) F_y \tag{4.34}$$

for $\lambda_c > 1.5$

$$F_{cr} = \left[\frac{0.877}{\lambda_c^2}\right] F_y \tag{4.35}$$

In Equations 4.34 and 4.35, the column slenderness parameter is given by

$$\lambda_c = \frac{kl}{r\pi}\sqrt{\frac{F_y}{E}} \tag{4.36}$$

Where

F_y is the yield strength of the steel comprising the helical pile shaft, and

r is the radius of gyration of the helical pile shaft about the axis of buckling. All other parameters have been defined previously.

The radius of gyration of a solid square steel bar is given by (AISC, 2001)

$$r = \frac{d}{\sqrt{3}} \tag{4.37}$$

Where

d is the side dimension of the square bar (e.g., for 1.5 inch × 1.5 inch square,

d =1.5 inches [38 mm]).

The radius of gyration of a round tube is given by (AISC, 2001)

$$r = \frac{\sqrt{d^2 - d_1^2}}{2} \tag{4.38}$$

Where

d is the outside diameter of the shaft, and

d_1 is the inside diameter of the shaft.

Using the simple buckling analysis prescribed by the IBC, buckling equations given in AISC, a resistance factor of 0.85, and a load factor of 1.5, the allowable buckling strengths of various helical pile shafts were computed for various conditions of pile butt fixity in soft and firm soils. Table 4.1 shows results for common round shafts, Table 4.2 shows results for common square-shaft helical piles, and Table 4.3 shows results

Table 4.1 Allowable Buckling Capacity of Common Round Shaft Helical Piles

		HSS1.75 × 0.11	HSS2.88 × 0.2	HSS2.88 × 0.27	HSS3 × 0.12	HSS3 × 0.25	HSS3.5 × 0.21	HSS3.5 × 0.3
		65 ksi [448 MPa] Min. Yield Strength Steel						
		Allowable Capacity (kip [kN])						
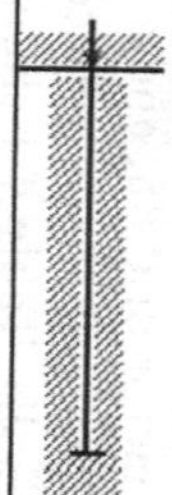 K=0	Firm soil (L = 60 in [1.5 m])	21 [93]	63 [280]	83 [369]	40 [177]	80 [355]	82 [364]	111 [493]
	Soft soil (L = 120 in [3.0 m])	21 [93]	63 [280]	83 [369]	40 [177]	80 [355]	82 [364]	111 [493]
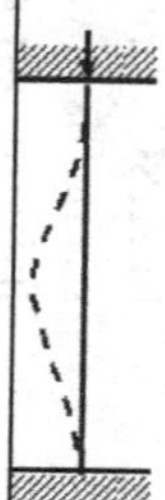 K=0.65	Firm soil (L = 60 in [1.5 m])	14 [62]	54 [240]	70 [311]	35 [155]	68 [302]	74 [329]	99 [440]
	Soft soil (L = 120 in [3.0 m])	4 [19]	33 [146]	42 [186]	23 [102]	43 [191]	54 [240]	71 [315]

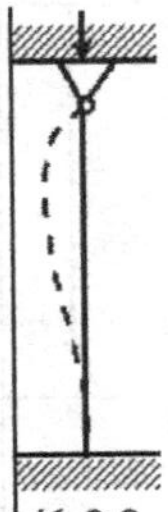	Firm soil (L = 60 in [1.5 m])	11	[48]	49	[217]	64	[284]	32	[142]	63	[280]	70	[311]	94	[418]
K=0.8	Soft soil (L = 120 in [3.0 m])	3	[13]	24	[106]	30	[133]	17	[75]	32	[142]	43	[191]	56	[249]
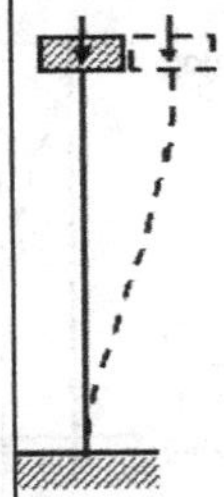	Firm soil (L = 60 in [1.5 m])	5	[22]	36	[160]	47	[209]	25	[111]	47	[209]	57	[253]	76	[338]
K=1.2	Soft soil (L = 120 in [3.0 m])	1	[4]	11	[48]	13	[57]	8	[35]	14	[62]	21	[93]	27	[120]
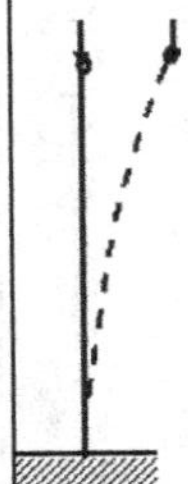	Firm soil (L = 60 in [1.5 m])	2	[8]	14	[62]	14	[62]	10	[44]	18	[80]	27	[120]	35	[155]
K=2.1	Soft soil (L = 120 in [3.0 m])	0	[1]	3	[13]	4	[17]	3	[13]	5	[22]	7	[31]	9	[40]

Table 4.2 Allowable Buckling Capacity of Common Square Shaft Helical Piles

	RCS1.5 × 1.5	RCS1.75 × 1.75	RCS1.25 × 1.25	RCS1.5 × 1.5	RCS1.75 × 1.75	RCS2 × 2	RCS2.5 × 2.5
	50 ksi [345 MPa] Min. Yield Str Steel		75 ksi [517 MPa] Min. Yield Strength Steel				
	Allowable Capacity (kip [kN])						
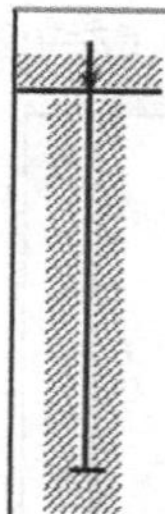 Firm soil (L = 60 in [1.5 m])	64 [283]	87 [385]	66 [295]	96 [425]	130 [578]	170 [756]	266 [1181]
Soft soil (L = 120 in [3.0 m]) K=0	64 [283]	87 [385]	66 [295]	96 [425]	130 [578]	170 [756]	266 [1181]
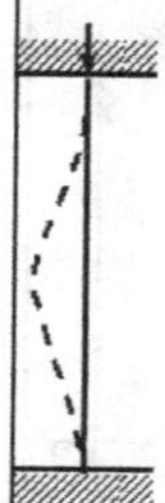 Firm soil (L = 60 in [1.5 m])	35 [156]	56 [249]	19 [84]	39 [174]	68 [301]	103 [458]	193 [857]
Soft soil (L = 120 in [3.0 m]) K=0.65	10 [43]	18 [81]	5 [21]	10 [43]	18 [81]	31 [138]	76 [338]
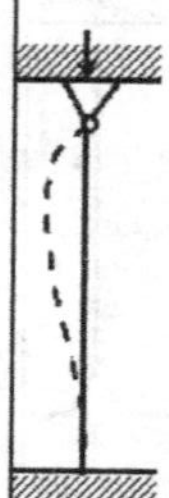 Firm soil (L = 60 in [1.5 m])	26 [115]	45 [199]	13 [55]	26 [115]	48 [214]	80 [354]	164 [727]
Soft soil (L = 120 in [3.0 m]) K=0.8	7 [28]	12 [53]	3 [13]	7 [28]	12 [53]	21 [91]	50 [223]

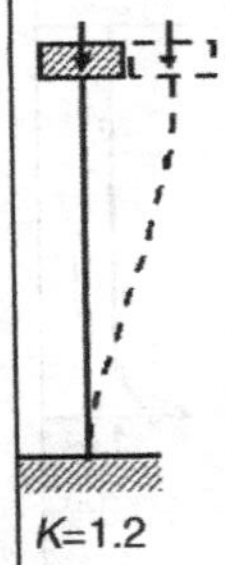	Firm soil (L = 60 in [1.5 m])	12	[51]	21	[95]	6	[24]	12	[51]	21	[95]	37	[162]	89	[397]
	Soft soil (L = 120 in [3.0 m])	3	[12]	5	[23]	1	[6]	3	[12]	5	[23]	9	[40]	22	[99]
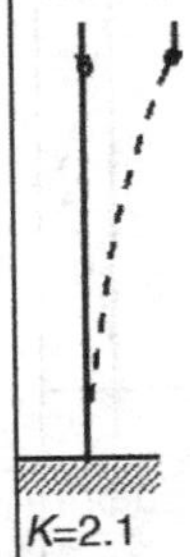	Firm soil (L = 60 in [1.5 m])	4	[16]	7	[31]	2	[8]	4	[16]	7	[31]	12	[53]	29	[129]
	Soft soil (L = 120 in [3.0 m])	1	[4]	2	[7]	0	[2]	1	[4]	2	[7]	3	[13]	7	[32]

Table 4.3 Allowable Buckling Capacity of Larger Round Shaft Helical Piles

		HSS4 × 0.22		HSS4.5 × 0.23		HSS6.63 × 0.28		HSS6.63 × 0.43		HSS8.63 × 0.32		HSS8.63 × 0.5		HSS10.8 × 0.5	
		50 ksi [345 MPa] Min. Yield Strength Steel													
		Allowable Capacity (kip [kN])													
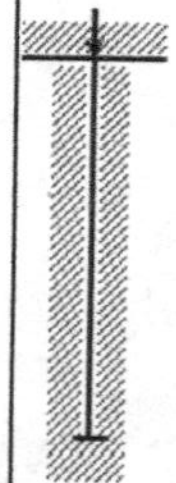 K=0	Firm soil (L = 60 in [1.5 m])	76	[337]	90	[400]	158	[703]	238	[1060]	238	[1059]	362	[1609]	458	[2039]
	Soft soil (L = 120 in [3.0 m])	76	[337]	90	[400]	158	[703]	238	[1060]	238	[1059]	362	[1609]	458	[2039]
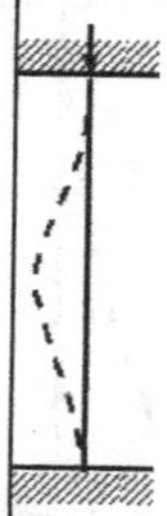 K=0.65	Firm soil (L = 60 in [1.5 m])	71	[317]	86	[380]	155	[688]	233	[1035]	235	[1045]	357	[1588]	455	[2022]
	Soft soil (L = 120 in [3.0 m])	59	[263]	74	[329]	145	[644]	217	[966]	226	[1006]	343	[1525]	443	[1972]
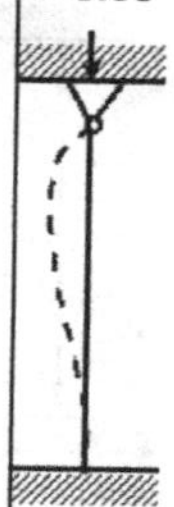 K=0.8	Firm soil (L = 60 in [1.5 m])	69	[307]	84	[371]	153	[680]	230	[1023]	234	[1038]	355	[1577]	453	[2013]
	Soft soil (L = 120 in [3.0 m])	52	[231]	67	[297]	138	[616]	207	[921]	220	[979]	334	[1483]	436	[1938]

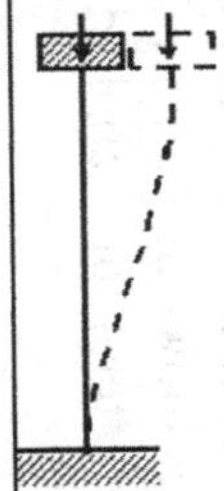 K=1.2	Firm soil (L = 60 in [1.5 m])	61	[273]	76	[338]	147	[653]	220	[980]	228	[1013]	346	[1537]	446	[1981]
	Soft soil (L = 120 in [3.0 m])	32	[144]	46	[205]	117	[521]	174	[774]	200	[888]	301	[1340]	409	[1819]
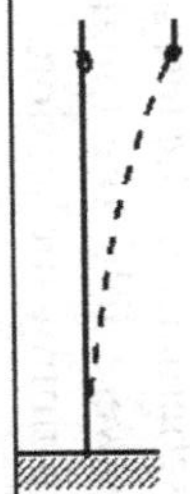 K=2.1	Firm soil (L = 60 in [1.5 m])	40	[176]	54	[240]	126	[559]	187	[833]	208	[926]	315	[1399]	420	[1868]
	Soft soil (L = 120 in [3.0 m])	11	[47]	16	[72]	63	[280]	91	[404]	139	[618]	207	[919]	323	[1437]

for larger-diameter round shaft helical piles. Boundary conditions and soil types are shown on the left side of the tables. Shaft sizes are shown along the top. The symbol RCS stands for round corner square-bar. Yield strength of steel was varied from 50 ksi to 75 ksi [345 to 517 MPa] based on the most common availability of shafts in industry. In all cases, the flexural strength and rigidity of shaft couplings was taken as equal to or greater than that the shaft.

As can be seen in the tables, the buckling strength for all shafts falls off sharply in soft soils without lateral or rotational bracing. This may in part explain why common practice does not require buckling analysis for fully braced piles and why the IBC recommends that all piles be laterally braced. Buckling appears to limit capacity of square shaft piles more quickly under unbraced conditions than round shaft helical piles. A method of grouting around square shaft helical piles has been invented that helps reduce the risk of buckling. This method is discussed in Chapter 16.

Tables 4.1 through 4.3 may be used as a handy reference for initial selection of pile shafts. However, helical piles by different manufacturers vary in steel grade, available shaft diameters, and wall thickness. Not all shaft couplings are as rigid as the shaft. Once the specific helical pile is selected for the project, the designer should perform buckling analysis using the actual properties for the manufactured helical piles being considered. A benefit of the tables is to provide a baseline from which the designer can compare his or her calculations. It is further cautioned that many in industry believe this method of predicting helical pile allowable buckling strength is overly conservative. Few cases of helical pile shaft buckling have been documented.

4.9 ADVANCED BUCKLING

Software for underground pile buckling computation is not readily available to the practicing engineer. This is partly because buckling is a stability criterion that is not easily modeled using conventional methods of stress-strain analysis as discussed earlier. Some buckling studies have been attempted using LPILE software (Hoyt et al., 1995; Perko, 2003). LPILE is one of the most widely used software packages for lateral pile analysis. This software package incorporates a nonlinear discrete element p-y method of analysis to determine lateral pile deflections under various boundary conditions. It is difficult to apply LPILE to determine buckling capacity because the designer needs to introduce some perturbation to destabilize the pile in a lateral direction and to force it to begin buckling.

The buckling of a 1.5-inch × 1.5-inch [38 × 38-mm] square shaft helical pile used for underpinning was studied by Hoyt et al. (1995). The researchers conducted a laboratory investigation whereby a typical helical pile underpinning bracket was attached to a solid block. Load was gradually applied to a short section of helical pile shaft through the bracket. The shaft was supported on a roller assembly and was restrained laterally by a load cell. A schematic of the test setup is shown in Figure 4.25. Through this study, the researchers were able to estimate the lateral force applied to the top of a helical pile in an underpinning application.

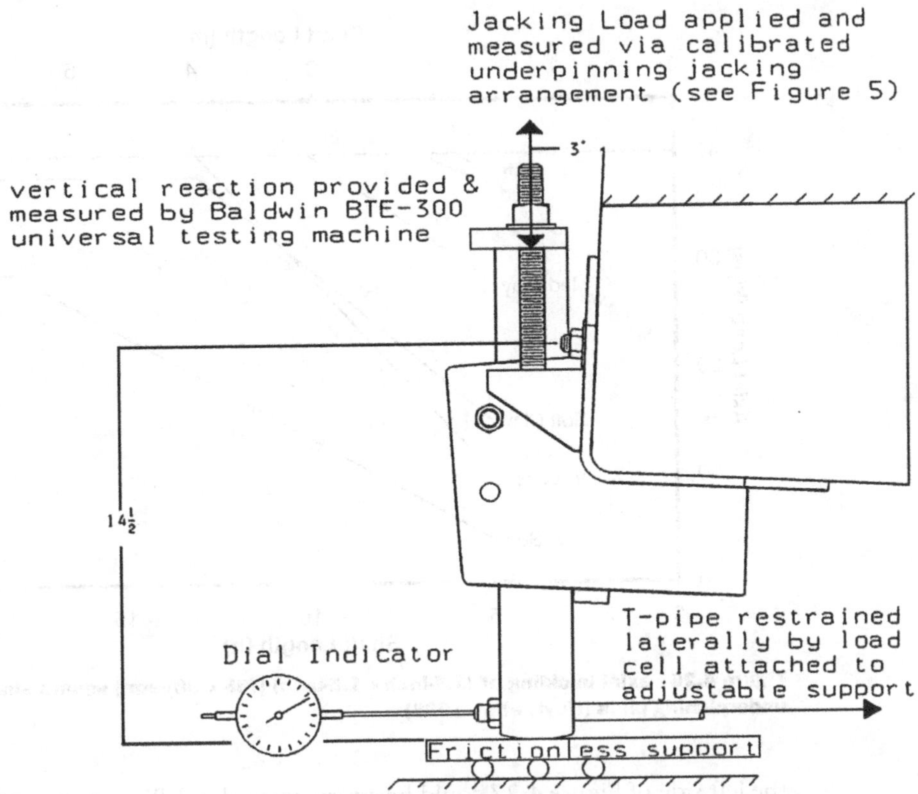

Figure 4.25 Laboratory test setup of Underpinning bracket (Hoyt, et al. 1995)

Hoyt et al. (1995) then used LPILE software to simulate underground buckling in different soil conditions. Lateral forces determined in their laboratory investigation were applied as boundary conditions to the top of the pile. The results obtained from LPILE were found to be in agreement with full-scale field tests. Hoyt's LPILE results, summarized in Figure 4.26, indicate that the buckling capacity decreases with shaft length. This is counterintuitive to conventional Euler theory wherein the buckling capacity of slender columns generally decreases with increasing unsupported shaft length. A close examination of their results shows that the helical piles being modeled were failing due to overturning caused by bracket eccentricities rather than pure buckling. The main conclusion of their study was that buckling of deeply embedded square shaft helical piles with underpinning brackets occurs at less than the allowable mechanical capacity of the subject helical pile (40 kips [178 kN]) only in soft to very soft clay. This is shown by the solid horizontal lines located below dashed line in the figure.

Perko (2003) used LPILE in an attempt to model buckling of helical piles for new foundations. Perko claimed that a pile with pinned end conditions as shown on

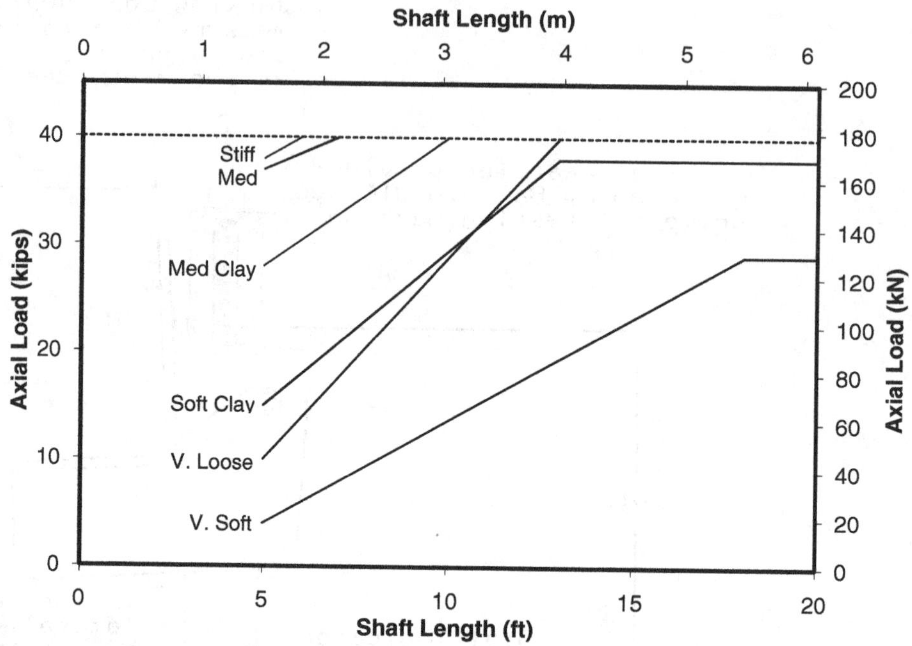

Figure 4.26 axial buckling of (1.5-inch× 1.5-inch) [(38× 38)-mm] square shaft underpinning piles (Hoyt, et al. 1995)

the left side of Figure 4.27 could be approximated in LPILE using a pile twice as long with fixed-slope, free-translation-boundary conditions at the top, as shown on the right side of Figure 4.27. Although buckling for the pinned-end conditions cannot be determined readily using LPILE, buckling with fixed-slope and free-translation conditions could be easily modeled. Since the elastic curve for the buckled portion of each of these conditions has the same effective length, Perko suggested that the two configurations should yield approximately the same critical buckling load, and hence LPILE could be used to model traditional buckling.

Using this method, Perko modeled a variety of commonly available helical pile shafts including 1.5-inch × 1.5-inch [38 × 38-mm] and 1.75-inch × 1.75-inch [44 × 44 mm] square shafts, 2.5-inch and 3.0-inch [64 and 76 mm] nominal diameter, schedule 80 pipe shafts, and 3.0-inch- [76 mm] O.D., 0.12-inch- and 0.25-inch- [3- and 6-mm-] thick wall, high-strength structural tube shafts. Lateral load on the helical pile shafts was set equal to that caused by a departure from plumbness equal to 1.5 percent of the length. This gave the perturbation necessary to initiate buckling. The soil conditions incorporated in the model were similar to those shown in Table 3.7.

Perko determined that buckling is a critical constraint on the design capacity of helical piles only in very soft to soft clays and very loose to loose sands. Buckling capacity was in excess of manufacturer's recommended maximum allowable axial capacity

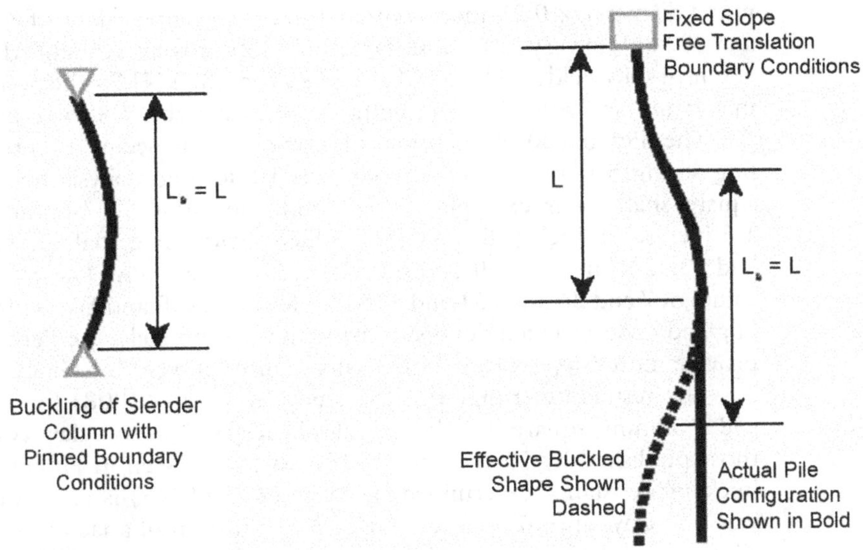

Figure 4.27 LPILE model for buckling analysis (Perko, 2003)

of the helical piles in the other soil conditions. The results of buckling calculations are shown in Table 4.4. Allowable buckling capacity was determined from ultimate buckling capacity using a factor of safety of 1.5. The results in the table represent the maximum recommended axial design capacity for these helical pile shafts in the soil conditions shown. Buckling failure does not exclude the use of helical piles in weak soils. Rather, Perko suggested that the design axial capacity be lowered to the limits shown in the table in order to avoid buckling-type failure.

Table 4.4 can be compared with results given for pinned end conditions in Tables 4.1 and 4.2. The two methods yield surprisingly similar results for heavier round shafts. For example, Perko (2003) found an allowable buckling capacity for the 3-inch- [76-

Table 4.4 Allowable Buckling Capacity based on LPILE (Perko 2003)

		1.5″ × 1.5″ Square Bar	1.75″ × 1.75″ Square Bar	3.0″ *O.D.* 0.12 Wall HSST	2.5″ *Nom.* Schd. 80 Pipe	3.0″ *O.D.* 0.25 Wall HSST	3″ *Nom.* Schd. 80 Pipe
Sand	Very loose	23 Kips [102 kN]	28 Kips [125 kN]	38 Kips [169 kN]	51 Kips [227 kN]	64 Kips [285 kN]	79 Kips [351 kN]
	Loose	28 Kips [125 kN]	41 Kips [182 kN]	55 Kips [245 kN]	75 Kips [334 kN]	81 Kips [360 kN]	115 Kips [512 kN]
Clay	Very soft	15 Kips [67 kN]	21 Kips [93 kN]	28 Kips [125 kN]	34 Kips [151 kN]	38 Kips [169 kN]	50 Kips [222 kN]
	Soft	28 Kips [125 kN]	38 Kips [169 kN]	50 Kips [222 kN]	63 Kips [280 kN]	68 Kips [302 kN]	89 Kips [396 kN]

mm-] diameter × 0.25-inch- [6-mm-] thick helical pile shaft of 38 and 68 kips [169 and 302 kN] in very soft and soft clay soils, respectively. Straight AISC buckling in soft and firm soils yielded 32 and 63 kips [142 and 280 kN] for the pinned head condition in soft and firm soil. The two methods are within about 8 to 16 percent.

The Perko (2003) method of using LPILE seems to predict higher buckling strength than traditional code based buckling analysis for smaller round and square shafts. For example, Perko found an allowable buckling capacity for the 1.5-inch × 1.5-inch [38 × 38 mm] square helical pile shaft of 15 and 28 kips [67 and 125 kN] in very soft and soft clay soils, respectively. These are compared to the values of 7 and 26 kips [31 and 116 kN] found in soft and firm soils using AISC buckling and code specified depths to fixity. In very soft soils, the Perko method seems to provide more support to slender shafts so as to prevent buckling.

The results determined in the study by Perko (2003) for 1.5-inch × 1.5-inch [38 × 38 mm] square shaft helical piles in very soft to soft clays correspond well with those published by Hoyt et al. (1995). As can be seen in Figure 4.25, the ultimate buckling resistance determined by Hoyt et al. (1995) is approximately 28 kips for very soft clays and 37 kips for soft clay. Application of a factor of safety of 1.5 yields allowable buckling capacities of 19 and 25 kips, respectively. These values for the same shaft in similar soil conditions shown in Table 4.4 are 15 and 28 kips, respectively. The results determined in the Perko study for 1.5-inch × 1.5-inch [38 × 38 mm] square shaft helical piles in very loose to loose sand soils are less than those determined by Hoyt et al. (1995). One reason for this difference is that the angle of internal friction for very loose sand used by Perko was 25 degrees instead of 28 degrees as assumed by Hoyt.

Interestingly, the length of shaft affected by buckling in all soil conditions investigated by Perko (2003) varied from approximately 7 to 12 feet [2.1 to 3.7 m]. Provided that at least this length of helical pile shaft was surrounded by weak soils, the buckling capacity was independent of any additional length bounded by weak soils. This somewhat contradicts traditional Euler buckling theory but is supported by past experience (Tomlinson, 1986). Perko concluded that buckling capacity depends only on the consistency of near surface soils and is unaffected by the presence of weak soils below about 7 to 12 feet in depth [2.1 to 3.7 m]. The length of helical pile shafts used in the Perko study was 30 feet [10 m]. The length of shaft over which buckling occurred was determined by the depth where pile shaft deflections became insignificant in the LPILE models.

The free movement of helical pile couplings was not taken into account in either of the aforementioned buckling studies. Hoyt et al. (1995) recommended that additional studies should be performed to determine the effect of couplings on buckling capacity. Since then there have been significant strides by manufacturers to make helical pile couplings more rigid. ICC-Evaluation Services, Inc. (2007) recommends that coupling rigidity tests be performed by manufacturers to evaluate the free departure from straightness. The test consists of applying a lateral load equal to 0.4 percent of the desired allowable axial load to the end of an unsupported length of pile shaft with maximum number of possible couplings and recording the deflection. The excess deflection due to the couplings is treated as an eccentricity in overall pile compression capacity calculations.

The discussion in Sections 4.8 and 4.9 regards buckling of the upper portion of shaft. When a helical pile extends through a layer of unstable or fluid soil located at some distance below the ground surface, it may be prudent to design the length of shaft in the fluid soil as a column following procedures given in the Chicago Building Code for caissons. According to this code, the unsupported length of shaft shall be taken as the thickness of the layer plus 4 times the shaft diameter. The coefficient of buckling for the column shall be taken as 1.0 to indicate pinned-pinned conditions (Concrete Reinforcing Steel Institute, 1996).

4.10 DOWN DRAG

Down drag is the phenomenon where soft soils surrounding a pile consolidate and produce a downward force on the pile shaft, tending to cause settlement. Consolidation of soils can be triggered by changes in groundwater, placement of fill, or other surcharge loads placed on the ground surface. In addition, low permeability soils that were deposited relatively quickly by humans or by natural geologic processes may be underconsolidated and are not in equilibrium with their own weight. Underconsolidated soils can experience consolidation over long periods of time.

Consolidation differs from settlement in that it involves the drainage of pore water over time in reaction to applied pressures. Consolidation is normally associated with fine-grain soils that have low permeability. Settlement is associated with coarse-grain soils. Settlement is an immediate change in volume under applied loads due to a mechanical restructuring of the soil. The term "settlement" also is used when referring to the downward movement of a pile. Piles do not consolidate; that would imply that they are shrinking because pore water is moving out of them. One of the more common conditions that result in down drag is shown in Figure 4.28. In this example, a layer of compacted fill has been placed above soft or fluid soils. If the soft or fluid soils are saturated and have low permeability, the pore water trapped within the soil will temporarily increase in pressure and support the weight of the new fill. With time, these pore pressures will dissipate as pore water fluid drains from the area under pressure. The loss of pore fluid allows the soil to restructure and decrease in volume. Consolidation of soft or fluid soils will cause settlement or downward movement of the fill. A net result is that adhesion between the fill material and the helical pile shaft as well as adhesion between the soft soils and the helical pile shaft will cause a net downward force on the pile shaft. The negative adhesion stresses as a result of down drag are shown along the shaft in Figure 4.28. This in turn places more force on the helical bearing plates that are trying to resist downward movement of the pile.

Down drag affects all pile types. Helical piles provide better resistance to down drag than any other pile type due to their slender shaft and larger bearing elements.

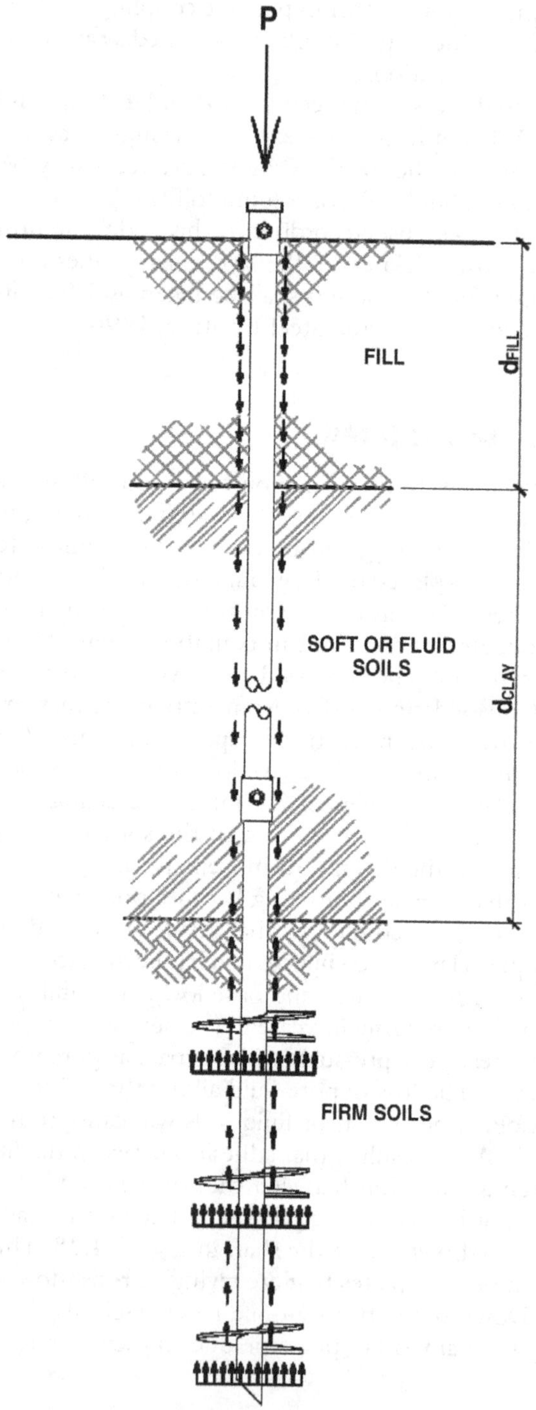

Figure 4.28 Example load conditions resulting in down drag

All the factors that reduce adhesion along helical pile shafts listed previously in this chapter are beneficial when it comes to down drag. Shaft adhesion may be estimated using Equation 4.28. The adhesion stress will be different in the fill from those in the soft or fluid material. In fact, adhesion in the soft or fluid material often can be ignored. Down drag can be taken into account in the determination of pile capacity by inserting a negative value for adhesion in the formulas for individual bearing, Equation 4.1, or cylindrical shear, Equation 4.22.

Even with the added benefits of helical piles, down drag can be significant and must be taken into account in the design and sizing of helical piles. The author has computed reductions in helical pile capacity as high as 15 to 33 percent or more in some cases. For example, this would mean that a helical pile bearing in firm soils that is capable of ordinarily supporting an ultimate capacity of 45 tons [400 kN] might be reduced to only supporting an ultimate load of 30 tons [267 kN] at the ground surface due to down drag.

Grouting of a helical pile shaft can increase down drag forces by enlarging the circumference of the shaft and increasing the adhesion between the soil and the shaft. This is not necessarily an issue as long as the designer takes the increased down drag into account. In one case, the author computed that the capacity of a helical pile with a grouted shaft was reduced by 100 percent by down drag. In other words, the predicted negative adhesion on the shaft was equal to or higher than the bearing capacity of the helical bearing plates. This happened because the developer raised a site by placing a thick layer of fill over an area that had soft soils. Helical piles with grouted shafts were used to support about a dozen homes. The helical piles did not extend to sufficient depth. In fact, they were stopped within the soft soils. Some of the homes exhibited as much as 6 inches [15 cm] of downward movement at the time of the author's involvement, which was only about two years after the homes were built.

As stated previously, helical piles are one of the better foundations in areas where down drag is likely to occur. There have been many cases where down drag caused conventional timber piles or augered cast-in-place piles actually to pull down on a structure. Helical piles were used successfully to underpin some of these structures so that the failed piling systems could be cut off. As long as the designer is cognizant of the effects of down drag and takes these forces into account, helical piles with and without grout can be applied to address down drag situations quite successfully.

Chapter 5

Pullout Capacity

Almost all of the theoretical methods explained in Chapter 4 also apply to the determination of pullout capacity of deeply embedded helical anchors. Limitations and special considerations for theoretical determination of pullout resistance are described in this chapter. Methods for determining the minimum embedment to promote a deep mode of behavior are presented. The chapter includes discussions on the effects of groundwater, group efficiency, structural capacity in tension, and cyclic loading.

5.1 THEORETICAL CAPACITY

The pullout capacity of helical anchors can be determined following the same procedures described in Chapter 4 for determination of bearing capacity, provided the anchors are embedded to sufficient depth to ensure a deep mode of behavior (A.B. Chance, 1993b; Ghaly and Clemence 1998). The minimum embedment depth required to ensure a deep mode of behavior is discussed in the next section. Adams and Klym (1972) established that the soil resistance mobilized in uplift above the top helix is similar to the bearing resistance mobilized beneath a deep foundation. Depending on the spacing of helical bearing plates and subsurface conditions, helical anchors can exhibit individual bearing and cylindrical shear modes of failure (Narasimha Rao et al., 1989).

Model helical anchors pulled out of the ground with inter-helix spacing ratios of 1.5, 2.3, and 4.6 are shown in Figure 5.1. Inter-helix spacing ratio is defined as the spacing between helical bearing plates divided by their average diameter (s/D_{AVG}). The model anchor with more closely spaced helical bearing plates ($s/D_{AVG}=1.5$) displayed on the left side of Figure 5.1 clearly shows cylindrical shear behavior as evidenced by the continuous cylinder of soil surrounding all of the bearing plates. The model anchor with greater helix spacing ($s/D_{AVG}=4.6$) displayed on the right side of Figure 5.1 has

Figure 5.1 Model helical anchors with different inter-helix spacing ratios (Narasimha Rao et al. 1989)

cone-shaped soil deposits above each helix, which is more indicative of individual bearing behavior.

Schematic diagrams of the stresses on helical anchors in cylindrical shear and individual bearing are shown in Figure 5.2. These diagrams are very similar to those shown in Figure 4.2 for compression. The cylindrical shear method pictured on the left side of the figure has a uniform pressure distribution above the top helix and shear stresses surrounding the soil encapsulated between the helical bearing plates. Adhesion stresses act along the length of the helical pile shaft located above the top helix. The individual bearing method pictured on the right side of the figure shows a uniform pressure distribution on the upper side of each helical bearing plate. Adhesion stresses are shown along the entire length of the shaft. In both methods, an upward axial load is applied to the shaft at the ground surface.

According to the cylindrical shear method, ultimate pullout capacity, P_u, of a helical anchor is found by taking the sum of shear stress along the cylinder, adhesion

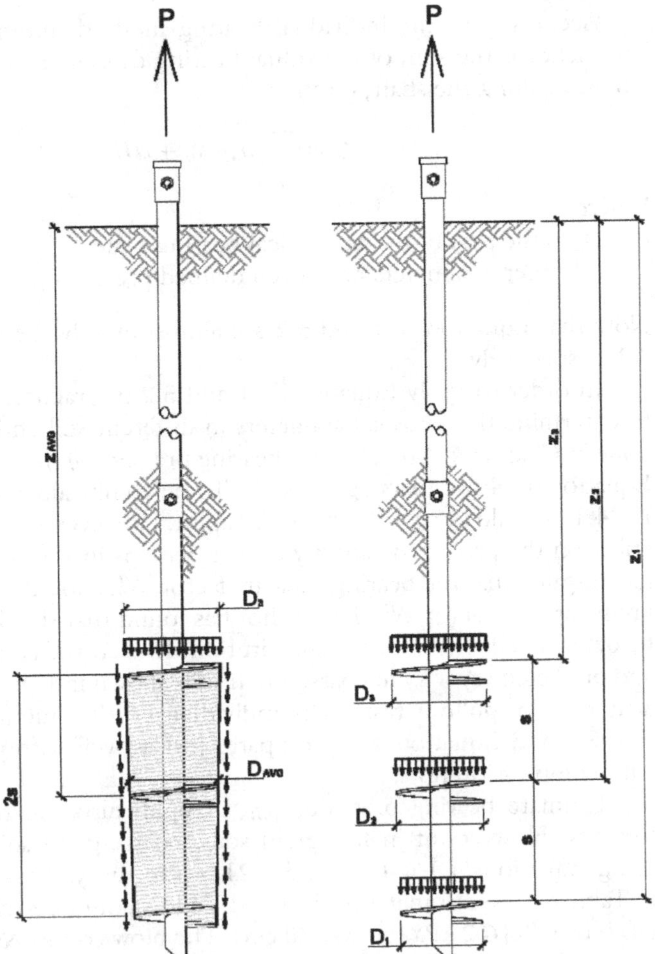

Figure 5.2 Cylindrical shear and individual bearing methods for helical anchors

along the shaft, and bearing on the upper helix given by

$$P_u = q_{ult} A_T + T(n-1) s\pi D_{AVG} + \alpha H_{eff} (\pi d) \tag{5.1}$$

Where

q_{ult} is the ultimate bearing pressure
A_T is the area of the upper helix
T is soil shear strength, α is adhesion between the soil and the shaft
H_{eff} is the length of shaft above the top helix
d is the diameter of the pile shaft, and
$(n-1)s$ is the length of soil between the helices

According to the individual bearing method, ultimate pullout capacity, P_u, of the anchor is the sum of individual bearing capacities of *n* helical bearing plates plus adhesion along the shaft, given by

$$P_u = \sum_n q_{ult} A_n + \alpha H_{eff} (\pi d) \tag{5.2}$$

Where

A_n is the area of the *n*th helical bearing plate.

All other parameters have been defined previously.

Note that Equations 5.1 and 5.2 are almost exactly the same as Equations 4.22 and 4.1, respectively.

In order to apply Equations 5.1 and 5.2 in practice, it is useful to examine how to determine the various parameters in different soil and bedrock conditions as was done in Chapter 4. The ultimate bearing pressure, q_{ult}, in uplift can be estimated from Equation 4.14 for coarse-grain soils. To be technically correct, the term $(N_q' - 1)$ in Eq. 4.14 should be replaced by N_q' for uplift. However, this replacement has a negligible effect on the predicted capacity except perhaps in very loose soils. Some researchers have argued that the bearing capacity factor, N_q', should be replaced with an empirical uplift capacity factor, N_u'. The author has found that suitable estimates of capacity can be determined without such substitution. In fact, the comparison between measured and predicted capacity in coarse-grain soils shown in Figure 4.8 of Chapter 4 includes a number of pullout tests. The individual bearing method computed using Equation 5.1 and Equation 4.14 compares just as well with pullout tests as it compares with compression tests.

Ultimate bearing pressure, q_{ult}, in uplift also can be estimated from penetration test blow count in fine-grain soils, coarse-grain soils, and weathered bedrock using Equations 4.16, 4.20, and 4.21, respectively. These relationships are repeated in Table 5.1. Recall that λ_{SPT} is the SPT blow count correlation factor equal to 0.065 tsf/blows/ft [6.2 kPa/blows/30 cm]. The blow count, N_{70}, used with these relationships can be determined through engineering judgment or by applying probabilistic soil mechanics. Corrections for energy ratio should be made for the particular hammer in use.

The relationships between bearing pressure and blow count given in Figures. 4.9, 4.6, and 4.11 for coarse-grain soils, fine-grain soils, and bedrock are redisplayed in Figures 5.3, 5.4, and 5.5. In the new figures, the symbol types differentiate pullout tests

Table 5.1 Correlations Between Ultimate Bearing Pressure and SPT Blow Count

Primary Soil Condition	Relationship	Approximate Bearing Pressure (tsf/blows/ft)	Approximate Bearing Pressure [kPa/blows/30 cm]
Fine-grain soil	$q_{ult} = 11\lambda_{SPT}N_{70}$	$0.72N_{70}$	$68N_{70}$
Coarse-grain soil	$q_{ult} = 12\lambda_{SPT}N_{70}$	$0.78N_{70}$	$74N_{70}$
Weathered bedrock	$q_{ult} = 13\lambda_{SPT}N_{70}$	$0.85N_{70}$	$81N_{70}$

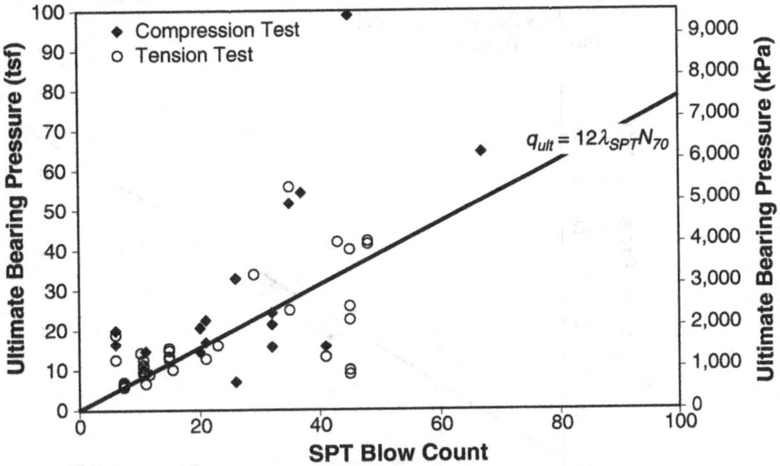

Figure 5.3 Correlation between bearing pressure and blow count in sand soils

from compression tests. Open circle symbols represent pullout tests. Solid diamond symbols represent compression tests. As can be seen in the figures, the ultimate bearing pressure correlates just as well for pullout capacity as it does for bearing capacity. There are no apparent trends shown in any of the figures that would indicate a need for different SPT correlations for pullout capacity versus compression capacity.

When determining helical anchor capacity using the cylindrical shear method, shear strength, T, may be calculated using the methods described in Chapter 3. In coarse-grain soil, T is determined using Equation 4.26. In fine-grain soil, T may be

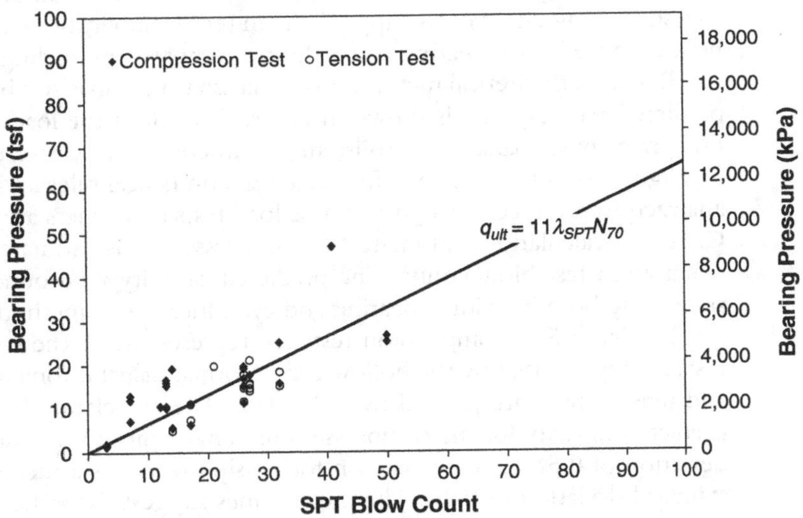

Figure 5.4 Correlation between bearing pressure and blow count in clay soils

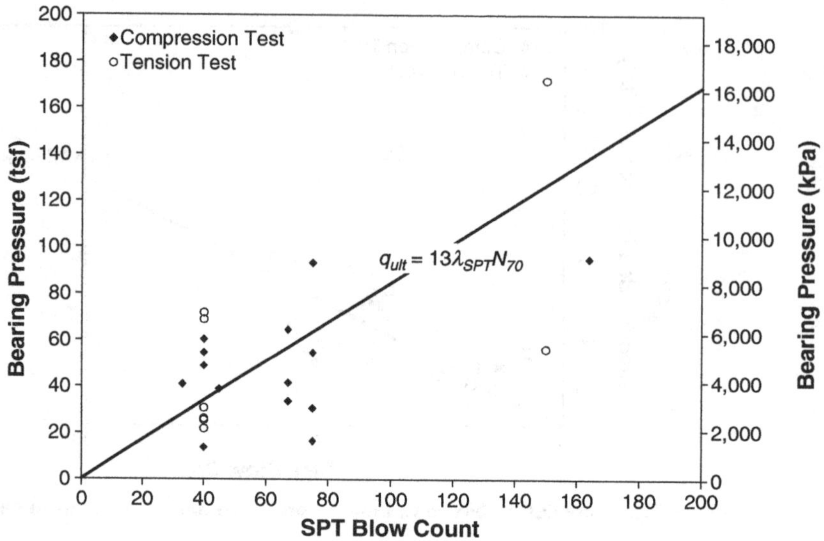

Figure 5.5 Correlation between bearing pressure and blow count in weathered bedrock

taken as the undrained shear strength, s_u, based on Equation 3.9. Undrained strength, s_u may be determined from laboratory unconfined compression strength tests or SPT blow count measurements.

Adhesion along helical anchor shafts may be ignored to be conservative. If it must be taken into account, then the methods described in Chapter 4 can be used to determine the adhesion and the length of shaft over which it acts, H_{eff}. Zhang (1999) showed that H_{eff} is the same regardless of loading direction or soil type. This is due to a shadowing effect of the upper helix in tensile applications and the formation of a hollow above helical bearing plates during compression loading.

Based on theoretical methods just discussed, a comparison between measured and predicted axial capacity is shown in Figure 5.6. All of the load test data available for fine-grain soils, coarse-grain soils, and weathered bedrock were combined to create this figure. In total, 112 load full-scale tests on helical piles and helical anchors were analyzed to produce this figure. These load tests and others are detailed in Appendix C. In the calculations, ultimate bearing pressure was estimated based on standard penetration test blow count. The predicted capacity was found by taking the limit state using both individual bearing and cylindrical shear methods.

In Figure 5.6, compression tests are represented by the solid bars, and tension tests are represented by the hollow bars. Normal distribution curves for compression and tension tests are plotted over the data. The correlation between measured and predicted capacity for the compression tests has a mean value of 1.06 and a standard deviation of 0.58. The correlation for tension tests has a mean value of 0.87 and a standard deviation of 0.40. The mean values suggest theoretical limit state methods overestimate pullout capacity by 13 percent and underestimate bearing capacity by 6 percent on average. The standard deviations of the data indicate that theoretical

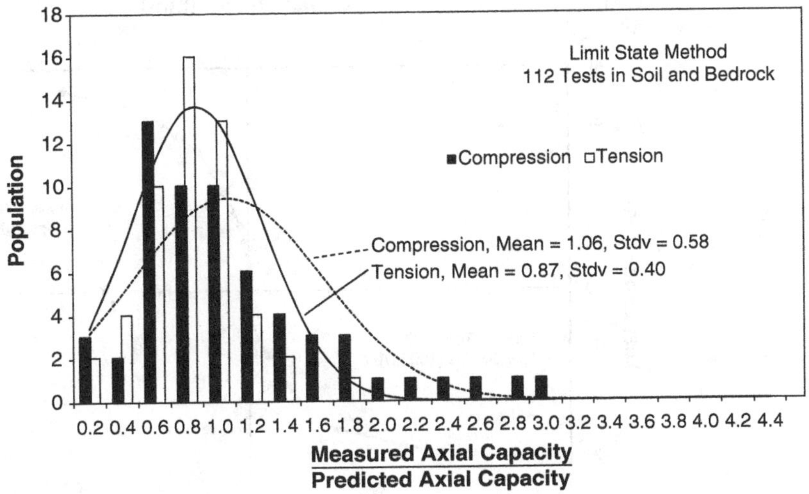

Figure 5.6 Comparison between measured and predicted axial capacity for all available load tests in soil and rock

determinations of capacity are more precise for pullout than compression. Given the variability in SPT data on most sites and the other factors that can affect helical pile and helical anchor capacity, both sets of data support the premise that capacity can be determined theoretically with a reasonable degree of confidence.

To evaluate the relationship between bearing and pullout capacity more directly, compression and tension load tests were performed on the same helical pile. A graph with the results of both tests is provided in Figure 5.7. The helical pile had a 3-inch-[76-mm-] diameter shaft and two helical bearing plates with 8-inch and 12-inch [203- and 305-mm] diameters. It was installed into glacial till to a depth of approximately 15 feet [4.6 m]. The glacial till exhibited highly variable blow counts. The compression test was run on the helical pile first, then the load frame was reset and the tension test was performed on the same pile. The ultimate capacity in bearing was 36 kips [160 kN] and in tension was 30 kips [130 kN]. The final installation torque was only 3,700 foot-pounds [5,000 N-m], indicating pile installation was halted in a fairly soft or loose material. The capacity of the pile in compression was 20 percent more than in tension. Interestingly, this is almost exactly the difference between the mean values shown in Figure 5.6 (21.8 percent). For all practical purposes, the measured capacities in compression and tension are fairly close.

In theory, bearing and pullout capacity of a deeply embedded helical pile in a uniform soil or bedrock should be similar. In practice, the slight difference in capacity may be explained by disturbance of the soil or bedrock above the helical bearing plates during installation. If the designer wants to be conservative, pullout capacity determinations may be multiplied by a reduction factor of 0.87 given the results shown in Figure 5.6. Hence, the adjusted theoretical pullout capacity of a pile, P_{ut}, may be

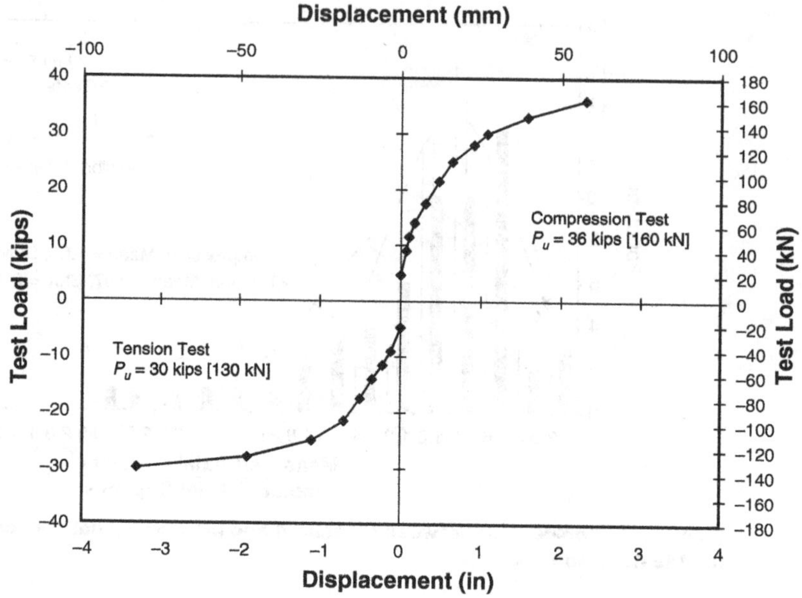

Figure 5.7 Compression and tension tests performed on the same pile

taken as

$$P_{ut} = \lambda_t P_u \tag{5.3}$$

Where

P_u is the capacity determined by limit state analysis and
λ_t is a factor intended to account for disturbance. The disturbance factor, λ_t, has an average value of 0.87 based on the data from Figure 5.6.

The magnitude of λ_t would be expected to vary with subsurface conditions such as overconsolidation ratio. The factor also would be expected to vary with installation equipment and possibly techniques such as the amount of crowd applied, installation speed, and different installation operators. Weathered bedrock and highly overconsolidated soils may be more sensitive to disturbance during helical pile installation. As a consequence, the author has experienced decreased pullout capacity in highly overconsolidated soils and weathered bedrock even though the data in Figure 5.5 do not exhibit any strong variance. Caution should be exercised when working in these soil conditions. It may be necessary to conduct load tests to verify the pullout capacity of helical anchors in highly overconsolidated soils and weathered bedrock.

5.2 MINIMUM EMBEDMENT

One of the most important topics governing the performance of helical anchors is minimum embedment. All of the foregoing theoretical capacity calculations are based on a deep mode of behavior. If a helical anchor is too shallow, the weight of soil above

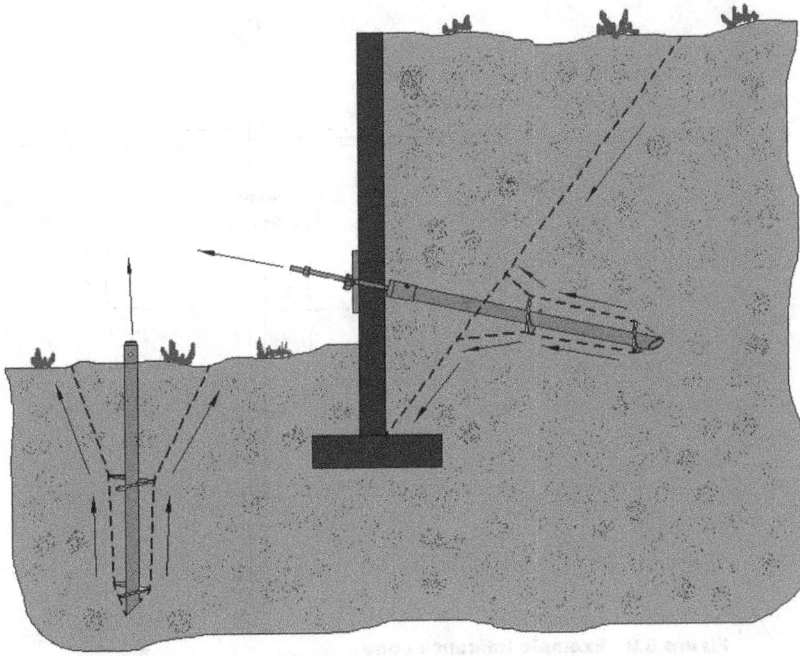

Figure 5.8 Insufficient embedment length (Courtesy of Magnum Piering, Inc.)

it will be insufficient to provide the required pullout pressure, and the anchor may fail in a shallow mode as shown in Figure 5.8. Shallow failure can occur for piles with bearing plates located too close to the ground surface or for anchors with bearing plates located too close to the active soil wedge. A shallow mode of failure consists of shearing the soil around the helical bearing plates and literally lifting up the cone of soil above the top helix. For piles, this is manifest by mounding of the ground surface around the pile. For earth retention systems, the active wedge moves downward and the helical anchors dislodge from the soil behind the active wedge. In order to ensure proper performance, helical anchors must be embedded to sufficient depth below the ground surface and behind the active zone to avoid shallow failure.

The required minimum depth of embedment is that depth where the weight of a cone of soil above the shallowest helix is sufficient to provide the necessary pullout pressure. A diagram showing a theoretical 45-degree influence cone is shown in Figure 5.9. A simple equation that can be used to estimate of the weight of the influence cone is given in the figure. This estimate is based on the known volume of a cone. It is conservative because it assumes a vertex at the shallowest helix.

A more rigorous estimate of the weight of an influence cone can be determined by integrating discs of soil with differential thickness dz from the shallowest helix to the ground surface. The weight of these disks is simply the product of surface area, thickness, and soil unit weight. The diameter of an arbitrary disk located at a distance z above the shallowest helix is found from simple trigonometry, $D_T+2z\tan\Theta$, where D_T is the diameter of the top helix and Θ is the angle of the influence cone from vertical.

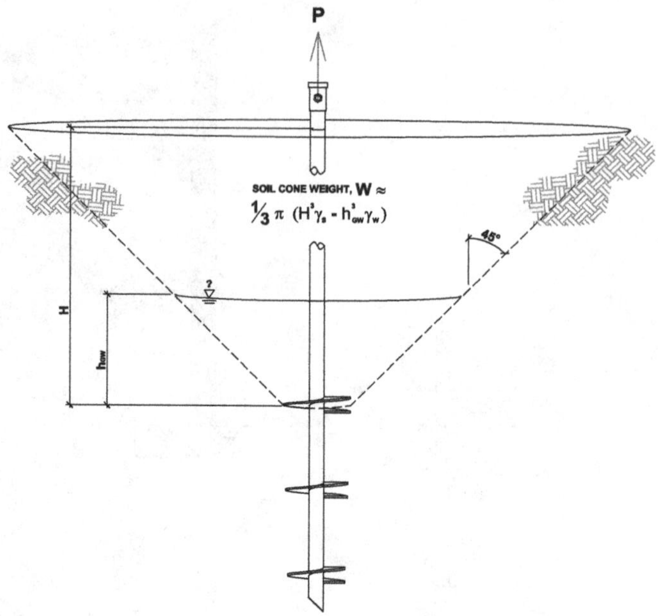

Figure 5.9 Example influence cone

The total weight, W, of the entire influence cone is given by the integral

$$W = \frac{\pi}{4}\gamma' \int_0^H (D_T + 2z \tan\Theta)^2 dz \tag{5.4}$$

Where

H is the depth to the shallowest helix
z is distance above the shallowest helix, and
γ' is effective soil unit weight.

It is standard in industry to reference the depth of embedment in terms of the relative embedment ratio, N_T, defined as depth to the shallowest helix, H, divided by its diameter, D_T. Integration of Equation 5.4 and replacement of H with $N_T D_T$ yields

$$W = \frac{\pi}{4}\gamma' D_{T3} \left(N_T + 2N_{T2} \tan\Theta + \frac{4}{3} N_{T3} \tan^2\Theta \right) \tag{5.5}$$

This result can be used to determine the weight of an influence cone that is either entirely above ground water or entirely submerged, since it allows for the use of only one effective unit weight of soil. Weight of a partially submerged influence cone can be determined by integrating from the top helix to the groundwater table using the buoyant unit weight and integrating from the groundwater table to the ground surface using the moist unit weight of soil. The integral results in a long and cumbersome equation so it will not be given here. The designer also may use simple geometry to compute the volume of the cone of soil above and below the water.

It is useful to compare the equations governing required embedment with those for capacity. Theoretical ultimate pullout capacity of a helical anchor is given by Equations 5.1 and 5.2. When shaft adhesion is ignored and only the shallowest helix is considered, ultimate pullout capacity simplifies for both equations to

$$P_u = q_{ult}\left(\frac{\pi}{4}D_{T^2}\right) \tag{5.6}$$

The pressure, q_{ult}, can be estimated from SPT blow count using the formulae given in Table 5.1.

The relative embedment ratio required to provide enough weight to hold down a helical anchor can be determined by setting the weight of the influence cone, Equation 5.5, equal to the theoretical ultimate pullout capacity for the shallowest helix, Equation 5.6. The weight of the influence cone acting on the second and deeper helical bearing plates would be much larger than that acting on the shallowest helix so they do not need to be considered in the computation of minimum embedment.

Sample results obtained from equating Equation 5.5 with Equation 5.6 are shown in Figure 5.10. In the preparation of this figure, the relationship between SPT blow count and pullout pressure for coarse-grain soil was used. Unit weights were varied with SPT blow count and are shown along the top of the figure. These unit weights were derived according to Table 3.7. The angle of the influence cone from vertical was assumed to be 45 degrees.

As can be seen in the figure, the required embedment ratio varies from approximately 4 for loose soil to 7 for dense soil. At first glance, this would seem counterintuitive. Dense soil should create a heavier cone of influence than loose soil so one might think that the influence cone could be made shallower. However, a helical anchor in dense soil can theoretically support much more pullout force. In fact, the theoretical pullout capacity increases at a greater rate than the weight of the influence cone. Therefore, the correct interpretation of Figure 5.10 is that the required embedment must be increased for higher blow counts in order to enable higher pullout capacities.

Ghaly and Hanna (1992) and Ghaly, Hanna, and Hanna (1991a) tested miniature helical anchors with different geometries in a sand-filled testing tank equipped with stress transducers. It was determined by physical observations and transducer measurements that the angle of the pullout "cone" from vertical is more closely given by 2/3 times the angle of internal friction of the soil ($\Theta = 2\Phi/3$) based on their tests. Meyerhof and Adams (1968) said that the angle of the influence cone varies from $\Theta = \Phi/2$ to $\Theta = \Phi/4$. Clemence and Veesaert (1977) used $\Theta = \Phi/4$. The resistance provided by the cone of soil in these previous studies includes frictional stresses along the sides of the cone. Ignoring the side friction and calculating the weight based on $\Theta = 45$ degrees

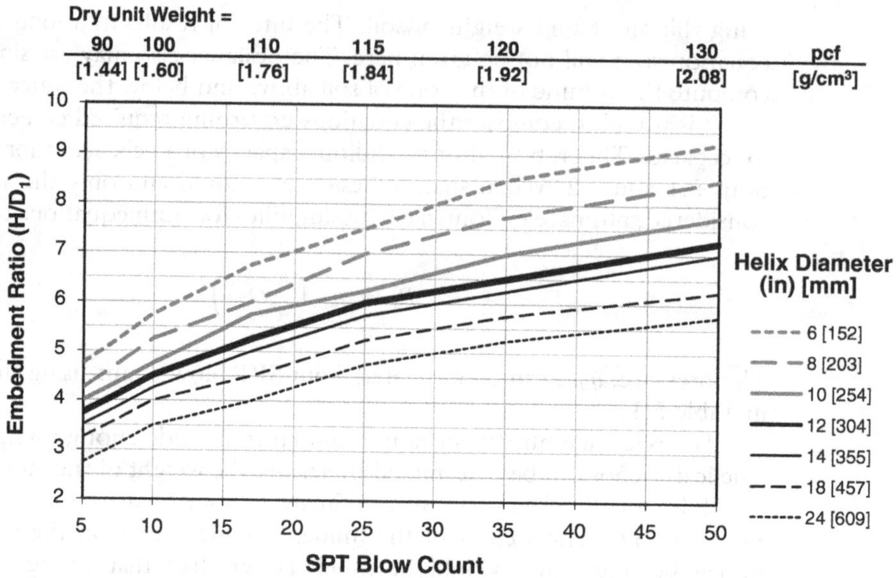

Figure 5.10 Required minimum helical anchor embedment

is simpler and seems to provide suitable estimates based on the author's experience. Using a 45-degree cone is also common in various other disciplines of material science.

In their tests with miniature anchors, Ghaly and Hanna (1992) and Ghaly, Hanna, and Hanna (1991a) determined that the size of the influence zone surrounding the bearing plates of a helical anchor is a function of the relative embedment ratio and the density of the surrounding sand. Minimum embedment ratios based on their results are shown in Table 5.2. In particular, a transition between significant and minimal strain occurred at H/D_1 ratios of 7, 9, and 11 for loose, medium, and dense sand, respectively. These results confirm that relative embedment ratio should increase with soil density. The laboratory test results for miniature anchors also generally match the calculations based on a 45-degree influence cone for a small helix diameter.

According to Rao et al. (1993a), piles with $H/D_T < 2$ are defined as shallow. Transition piles have relative embedment ratios such that $2 < H/D_T < 4$, and deep helical piles have $H/D_T > 4$. Mitsch and Clemence (1985b) showed helical piles follow Meyerhof and Adams's (1968) theory that piles with $H/D_T < 5$ exhibit shallow

Table 5.2 Minimum Helical Anchor Embedment (Ghaly and Hanna, 1992)

Soil Condition	Relative Embedment Ratio (H/D_T)
Fine-grain	5
Coarse-grain (loose)	7
Coarse-grain (medium)	9
Coarse-grain (dense)	11

behavior and piles with $H/D_T > 5$ respond according to deep failure condition. Again referencing Figure 5.10, calculations based on a 45-degree influence cone match these major conclusions fairly well. The calculated minimum embedment ratio varies from 2.5 to 5 in loose soils and from 4.5 to 7.5 in medium-dense soils. Typical sizes of helical anchor bearing plates group around relative embedment ratios similar to the conclusions of these previous investigations.

Ghaly and Clemence (1998) showed theoretically and experimentally that the pullout capacity of helical anchors installed at an inclination angle is greater than that of vertical anchors. This difference was explained by the development of a larger zone of soil mobilization. However, in the case of retaining walls, it is anticipated that this effect is canceled out by infringement with the active soil wedge. Additional study is required. At present, it is recommended that the increased strength due to inclination angle be ignored in order to be conservative and the pullout capacity of inclined and horizontal anchors be estimated following the same procedures as vertical anchors.

As discussed by Perko (1999), it is believed that the transition between shallow and deep behavior for helical anchors in fine-grain soils occurs at smaller embedment ratios compared to those in coarse-grain soils. The results shown in Figure 5.10 are based on the SPT blow count correlation for coarse-grain soil. The SPT correlation for fine-grain soil shown in Table 5.1 is only slightly less than this. Lower bearing pressure requires less embedment, which means Figure 5.10 would be conservative with respect to fine grain soils. Hence, Figure 5.10 can be used to estimate required relative embedment ratios in fine-grain soils in addition to coarse-grain soils. It is prudent to be somewhat conservative in fine-grain soils with respect to long-term loads on structures since blow count measurements can vary with changes in moisture content. This is discussed further in the next section.

5.3 EFFECT OF GROUNDWATER

One of the factors that can affect the pullout capacity of helical anchors, or any anchor for that matter, is groundwater. There have been catastrophic failures of helical anchor guyed structures caused by rising groundwater. Buoyant forces reduce the effective unit weight of soil by approximately half. Effective unit weight is the saturated unit weight minus the unit weight of water. The 45-degree influence cone method can be extended easily to account for changing groundwater conditions and to revise the required relative embedment ratio. Results shown in Figure 5.11 were found by equating influence cone weight with required pullout pressure as described in Section 5.2, except the effective soil unit weight was adjusted to account for ground water at the ground surface. These results suggest that the relative embedment ratio needs to be increased by approximately 20 percent to account for buoyant forces.

In addition to reducing the effective weight of the influence zone over a helical anchor, groundwater also can reduce the effective stress in soil around helical bearing plates and cause a decrease in the ultimate bearing pressure in both compression and tension applications. Bearing pressure obtained from SPT blow count correlations are

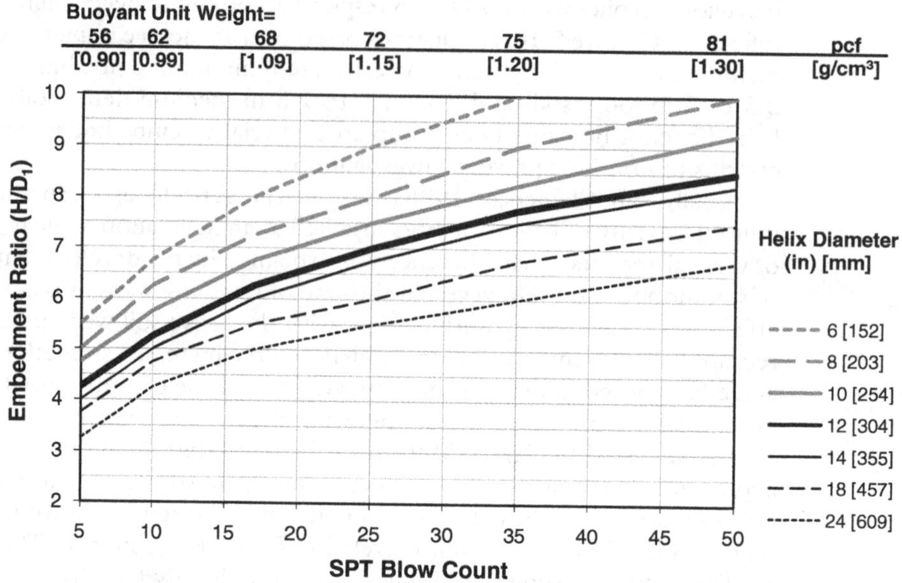

Figure 5.11 Required minimum helical anchor embedment in submerged soil

representative of total stresses at the time of exploration. If groundwater conditions fluctuate significantly, the correlations provided by blow counts can be in error.

An example of the effect of groundwater on helical anchor capacity is shown in Figure 5.12. The two anchors in this study are essentially identical except that anchor V4 was installed to a depth approximately 5 feet [1.5 m] greater than anchor V1. Both anchors consisted of a 1.75-inch [44 mm] square shaft with 8-inch and 10-inch [203 mm and 254 mm] diameter helical bearing plates. Both anchors were installed with similar final installation torque. Groundwater was at a depth of approximately 13 feet [4 m] during installation of anchor V1. Immediately prior to testing, groundwater rose to a depth of approximately 5 feet [1.5 m] below ground surface. The SPT blow count of the silty sand is unknown. However, the load tests suggest medium density. According to Figure 5.11, both anchors were installed past the required relative embedment ratio of 6 to 8 for medium-dense soil. The low capacity of helical anchor V1 cannot be explained using the 45-degree cone of influence model alone. The low capacity of anchor V1 was likely caused by a reduction in effective stress due to groundwater and also may be due in part to the clay layer located just above the shallowest helix.

In areas where groundwater is anticipated to fluctuate significantly compared to the depth of the helical anchor, an effective stress analysis is more appropriate for determination of ultimate bearing pressure, especially if SPT blow counts are obtained during a dry period. In coarse-grain soils, ultimate bearing pressure may be calculated using the effective soil unit weight in Equation 4.18 and taking hydrostatic pressure into account in Equation 4.24. In fine-grain soils, it may be necessary to perform

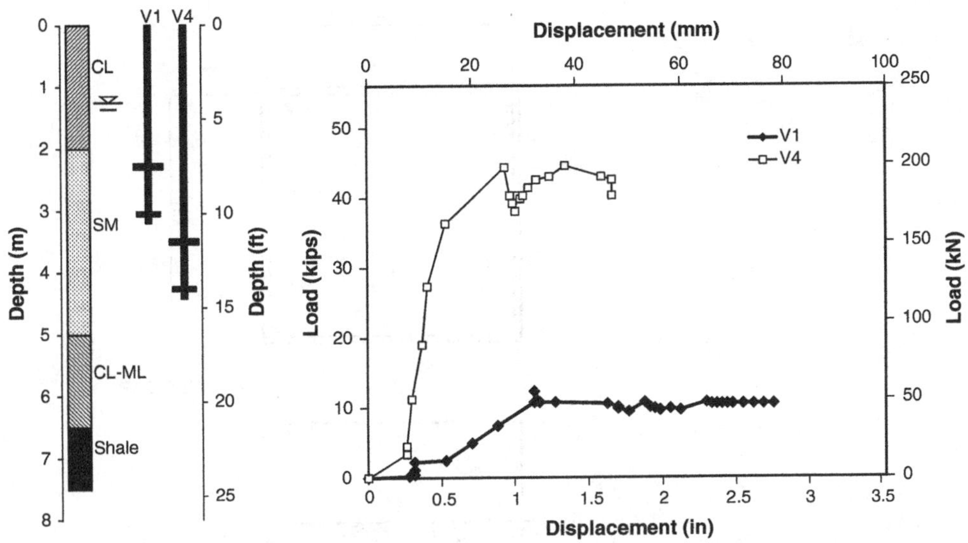

Figure 5.12 Effect of Groundwater on helical anchor load tests (Adapted from Victor and Cerato, 2008)

laboratory tests to obtain the effective stress envelope unless SPT tests can be conducted during conditions of high groundwater. Cerato and Victor (2008; in press) show in a graph of predicted versus measured capacity that the lower capacity of anchor V1 could be predicted fairly accurately using effective stresses.

5.4 GROUP EFFICIENCY

Helical anchors and helical piles should be designed with a minimum spacing to avoid group effects. A group effect occurs when several anchors or piles are installed at close spacing such that the capacity of the group is less than the sum of individual capacities. Most of the laboratory and field research regarding minimum helical anchor spacing has been done by private companies and is proprietary. Report AC358 of ICC-Evaluation Services, Inc. (2007) states that a minimum on-center spacing of 4 helical bearing plate diameters should be used to avoid group effects.

On-center spacing should be measured at the midpoint of helical bearing plates as shown in Figures 5.13 and 5.14. Pile caps may be kept small by splaying out helical piles at a batter angle while keeping pile butts closely spaced. An example of this is shown in Figure 5.14. As can be seen in the figures, the bearing area and side shear surface area are approximately the same for both the vertical piles and the battered piles.

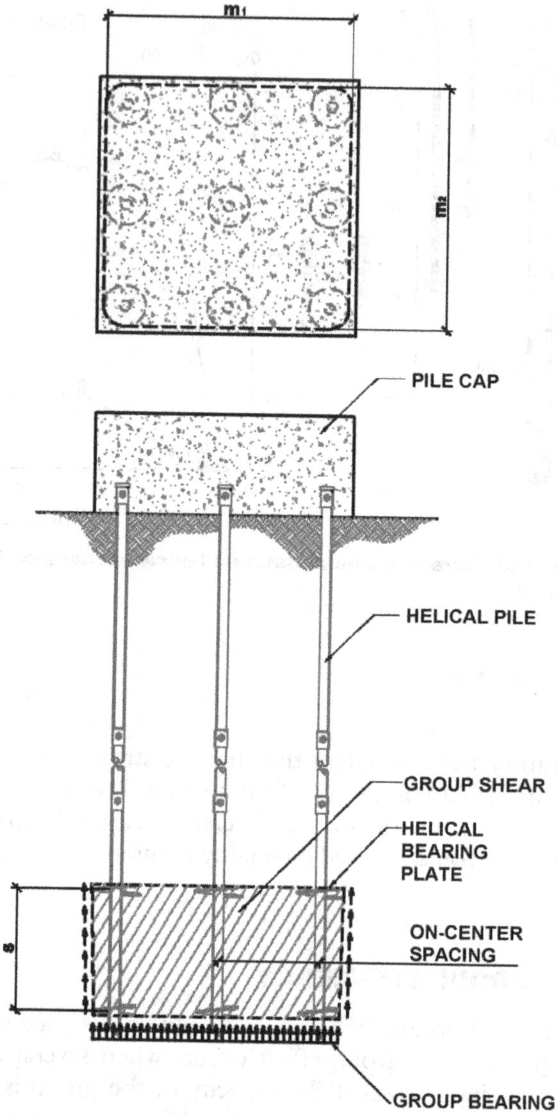

Figure 5.13 Group effect for vertical piles

Spacing closer than 4 helical bearing plate diameters does not necessarily indicate group efficiency will be reduced; rather, it indicates that a group analysis should be performed. To investigate the capacity of a closely spaced group of helical piles or helical anchors, a theoretical envelope is drawn about the helical bearing plates, as shown in Figures 5.13 and 5.14. The ultimate capacity of the group, P_{ug}, is determined using a method similar to determining cylindrical shear. The bearing capacity of the entire group is summed together with the shear stress on the surface of the envelope

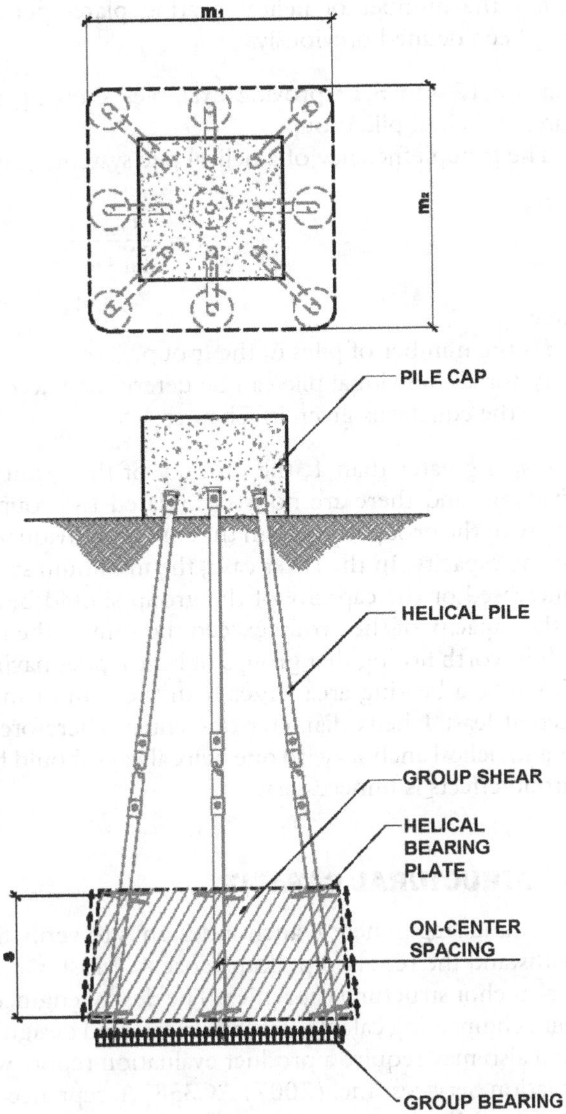

Figure 5.14 Group effect for battered piles

encasing all of the helical bearing plates given by

$$P_{ug} = q_{ult}\,(m_1)(m_2) + 2\mathrm{T}s(n-1)(m_1+m_2) \tag{5.7}$$

Where

m_1 and m_2 are the width and breadth of the pile group in plan view,
s is the spacing of the helical bearing plates along the length of the shaft, and

n is the number of helical bearing plates per pile. All other parameters have been defined previously.

Figures 5.13 and 5.14 provide examples showing the dimensions of the envelope encircling helical pile groups.

The group efficiency of a helical pile system, η, is defined as

$$\eta = \frac{P_{ug}}{\sum_i P_u} \tag{5.8}$$

Where

i is the number of piles in the group.

P_u for an individual pile can be determined according to limit state theory using the equations given in Chapter 4.

If η is greater than 1, the capacity of the group is greater than the sum of the individuals, and there are no issues related to group effects. If η is less than 1, the capacity of the group is less than the sum of individual capacities and group effects will limit the capacity. In the latter case, the minimum spacing between piles may need to be increased or the capacity of the group should be downgraded to P_{ug}. In no case can the capacity of the group exceed the sum of the individuals.

It is worth noting that groups of helical piles having only one helical bearing plate always have a bearing area larger than the sum of the individuals, provided they are spaced at least 1 helix diameter on-center. Therefore, the group efficiency of helical piles and helical anchors with one helix always should be greater than 1, and an analysis of group effects is unnecessary.

5.5 STRUCTURAL CAPACITY

A remaining step in helical anchor design is to verify that the anchor itself is sufficient to withstand the required pullout capacity. Most manufacturers provide a schedule of helical anchor structural capacities. The design engineer can rely on these capacities or require engineering calculations by a registered design professional. The local building official also may require a product evaluation report written in accordance with ICC-Evaluation Services, Inc. (2007) AC358, Acceptance Criteria for Helical Foundation Systems and Devices.

The structural capacity of a helical anchor should include an evaluation of both shaft and helix capacity. The evaluation of shaft capacity should include calculations of the gross yielding of the shaft and fracture of any couplings. Analysis of couplings must include rupture of any welds, block shear of the shaft and collar sleeve, shearing of any pins, and bolt hole bearing strength. All of these calculations can be performed in accordance with methods prescribed in the *Manual of Steel Construction* (AISC, 2001 or more recent edition) using either LRFD or ASD. Evaluation of helix capacity can be done using a plate-punching analysis or numerical modeling software. The results of these calculations can be used to check the welds between the helix and the shaft in combined shear and flexure.

The structural capacity of helical pile shafts in compression was discussed partly in the buckling sections of Chapter 4. In addition to checking buckling capacity on a project-by-project basis, the structural capacity of a helical pile in compression must include analysis of the couplings and helical bearing plates as described earlier.

In lieu of calculations, the shaft, couplings, and helical bearing plates may be tested in laboratory. An example of a laboratory test to determine the punching capacity of a helical bearing plated mounted to a short section of shaft is shown in Figure 5.15. The helix may be placed over a 6-pin mandrel or a spiral fixture. Load is applied to the shaft such that weld rupture and helix punching flexure are tested simultaneously. ICC-Evaluation Services, Inc. (2007) states that the allowable strength of helical anchor

Figure 5.15 Helix punching flexure test (Courtesy of Magnum Piering, Inc.)

devices tested in laboratory may be taken as 0.5 times the measured ultimate strength or 0.6 times the measured yield strength, whichever is less.

Long-term corrosion should be taken into account when checking the structural capacity of a helical anchor or helical pile. One can do this by reducing the thickness of all helical pile components by a sacrificial thickness to account for corrosion loss over the design life span. Engineering properties such as gross area, moment of inertia, and section modulus are computed using the corroded sections. Sacrificial thickness depends on the protective coating on the helical pile and project subsurface conditions, such as resistivity, pH, moisture content, dissolved oxygen, and various contaminants. Corrosion is discussed in Chapter 11.

An important factor to consider in the structural design of helical piles is the torsional resistance. In order to achieve a specified axial capacity in-ground, it will be necessary to apply a certain amount of torque to the shaft during installation. Correlations between torque and capacity are discussed in Chapter 6. Ordinarily, corrosion does not need to be accounted for in torsion calculations because the shaft will experience torsion only during installation. One exception is helical piles used for sign poles or other structures that place permanent dead or live torsional loading on the foundation.

Helical anchors are typically manufactured of high-strength carbon steel with yield strength in the range of 50 to 70 ksi [345 to 483 MPa]. Unless an accredited evaluation report is provided, manufacturers' mill certificates should be checked for steel yield strengths and other mechanical properties. To ensure quality, helical pile manufacturers should have an accredited quality control plan in place, such as ISO 9001.

5.6 CYCLIC LOADING

Cyclic loading of helical anchors can occur when they are used as a cable guy to resist wind loads on structures such as communications towers, power poles, or wind turbines. A number of studies have been conducted to evaluate the resistance of helical anchors, dead anchors, and piles subject to cyclic loading. Previous research (Dejong et al., 2003; Dejong et al. 2006; Hanna et al., 1978; Trofimenkov and Maruipolshii, 1965) has shown that cyclic loading has a degrading effect on the strength of both coarse-grain and fine-grain soils and on anchor performance. However, some repeated load applications have caused an increase in static capacity by stiffening the soil-anchor system (Clemence and Smithling, 1984). It was suggested by Victor and Cerato (2008) that these contradictory results may be explained by differences in the amount of disturbance caused by installation.

All of this is of little help to the design engineer dealing with cyclic loading conditions. Ideally, helical anchors should cause minimal disturbance during installation. Ghaly and Clemence (1998) showed upward creep is almost 100 percent recoverable if the cyclic load is less than 25 percent of the ultimate static resistance. Given this result, it may be advisable to design helical anchors such that cyclic loads are kept below 25 percent of the ultimate capacity of the anchor.

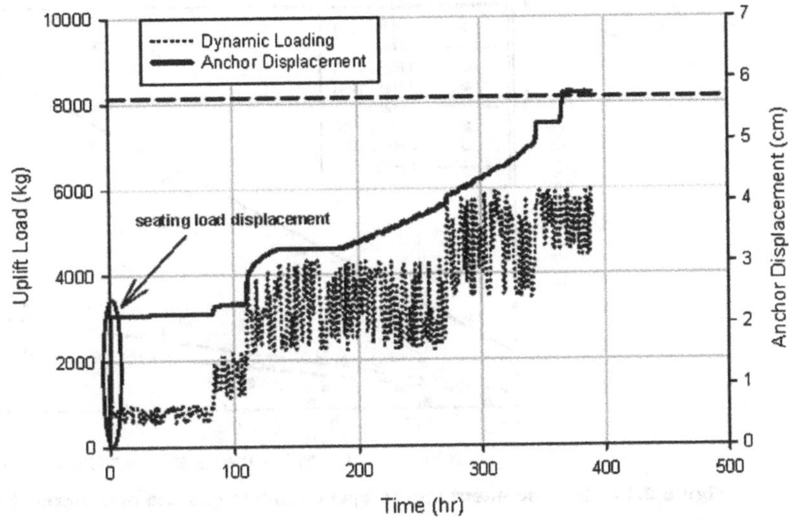

Figure 5.16 Cyclic load testing of a helical anchor (Cerato and Victor, 2008)

An example of a cyclic load test is shown in Figure 5.16. In this study, a seating load was applied to the anchor at the beginning of the test. The test consisted of applying a cyclic load with a frequency of approximately 3 Hz to a vertical anchor. The dashed horizontal line in the figure represents the design pullout capacity. The ultimate pullout capacity of this anchor based on limit state methods is approximately 26 kips [116 kN]. As can be seen in the results, cyclic loading at roughly 5 to 15 percent of the predicted ultimate capacity resulted in minimal displacement beyond that caused by the seating load. Higher cyclic loads caused some additional displacement of the anchor. The anchor continued to perform well with average cyclic load at roughly 25 percent and peak loads exceeding 35 percent of the predicted ultimate load until 200 hours (over 2 million cycles) at which time the anchor began to creep. This test confirms the conclusion of Ghaly and Clemence (1998) that service loads on anchors subject to cyclic loading conditions should be maintained below 25 percent of the ultimate pullout capacity.

The ultimate pullout capacity of the anchor shown in Figure 5.16 was determined from a static load test after cyclic loading. The measured ultimate capacity was on the order of 44 kips [196 kN or 20,000 kg]. This is more than three times the ultimate capacity that would be predicted based on conventional limit state theory. Cerato and Victor (2008) suggest that the increased capacity may have been from compaction of the soil during cyclic loading. This result confirms the conclusions by Clemence and Smithling (1984) that cyclic loading sometimes results in a stiffening of the soil-anchor system.

Cerato and Victor (2008) conducted a number of tests similar to that shown in Figure 5.16 using anchors with two, three, and four helical bearing plates. Their results are shown in Figure 5.17. The cyclic load test conducted on the helical anchor

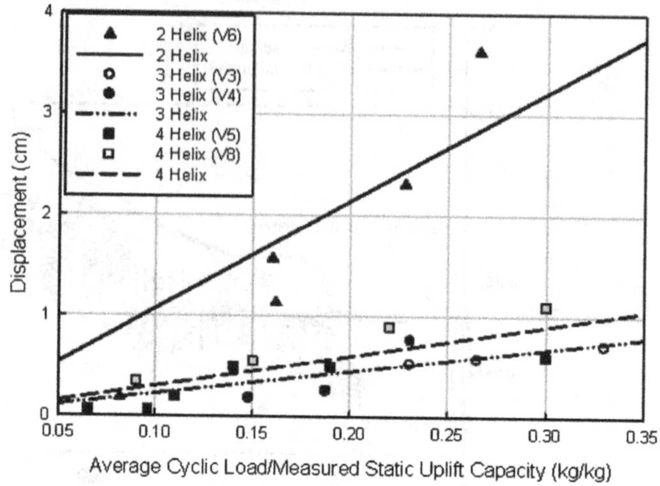

Figure 5.17 Displacement due to cyclic loading (Cerato and Victor, 2008; in press)

having only two helical bearing plates, shown in Figure 5.16, exhibited considerably higher displacements and was more prone to creep than the anchors with three and four helical bearing plates. This figure is based on ultimate pullout capacity measured in load tests after cyclic loading. If the data are adjusted for the ultimate capacity predicted before cyclic loading, then the two-helix anchor results would fall into place with the other data.

Chapter 6

Capacity-to-Torque Ratio

It is generally accepted that installation torque can be used to verify the axial capacity of a helical pile in both tension and compression applications. Over the last 40 years, this unique feature has helped helical piles gain recognition and popularity. This chapter provides an overview of the origin, reliability, pitfalls, and applicability of correlations between torque and capacity.

The process of a helical bearing plate cutting through soil or weathered bedrock in a circular motion is analogous to a plate penetrometer test. Common sense dictates that the torque required to advance a helix plate would be indicative of soil consistency and strength. It is reasonable that the installation torque should provide an indication of maximum bearing and pullout pressure. Most engineers agree that helical pile capacity should always be verified in the field through installation torque measurements.

6.1 EARLY EMPIRICAL WORK

The relationship between helical pile capacity and installation torque has been used as a general rule of thumb in practice since the 1960s. However, the data were kept proprietary. Data first appeared in presentations and reports to various public agencies by Cole (1978) and by Gill and Udwari (1980). A relationship between capacity and torque was first published in professional literature by Hoyt and Clemence (1989), whose work is considered to be a milestone in the helical pile industry. In their benchmark publication from the *Proceedings of the 12th International Conference on Soil Mechanics and Foundation Engineering* in Rio de Janeiro, the elegant expression given in Equation 6.1 was presented; it relates final installation torque, T, with ultimate axial capacity, P_u.

$$P_u = K_t T \tag{6.1}$$

The parameter K_t has become known as the capacity-to-torque ratio and has units of ft^{-1} [m^{-1}]. Hoyt and Clemence suggested that K_t is a constant that depends primarily on shaft diameter. In their work, K_t was assumed to be independent of the number and size of helical bearing plates and also the subsurface conditions. The equation relating torque to capacity was based on empirical data and experience.

In order to test the relationship between torque and capacity, Hoyt and Clemence (1989) analyzed 91 pullout load tests on helical anchors at 24 different sites with various soil types (fine grain and coarse grain). The helical anchors in the study had shaft sizes varying from 1.5 inch [38 mm] square to 3.5 inch [89 mm] diameter. Each anchor had from 2 to 14 helical bearing plates with diameters varying from 6 inches to 20 inches [152 to 508 mm]. The data were carefully selected to cover a broad range of circumstances such that no more than two tests were selected for the same anchor type at the same site and similar depth.

Hoyt and Clemence (1989) compared the results of these load tests with predicted capacity using the cylindrical shear, individual bearing, and torque correlation methods. Results of these three comparisons are shown in Figures 6.1, 6.2, and 6.3. In the figures, the ratio of measured capacity to predicted capacity is plotted along the x-axis. The number of occurrences is plotted on the y-axis. The mean value, median value, and standard deviation of the correlation data are shown in the top right corner of each graph.

In all three correlations, the median value is close to unity, indicating that cylindrical shear, individual bearing, and torque correlation methods are all reasonable approximations on average. However, the standard deviation suggests that calculations based on either method alone are not very precise. The distribution for all three correlations is skewed toward conservatism such that the measured capacity is greater than the predicted capacity about 60 to 67 percent of the time. This also means that

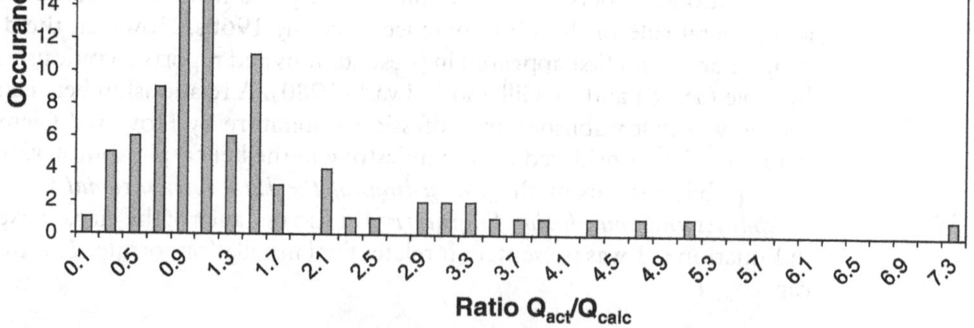

Figure 6.1 Cylindrical shear comparison (Hoyt and Clemence, 1989)

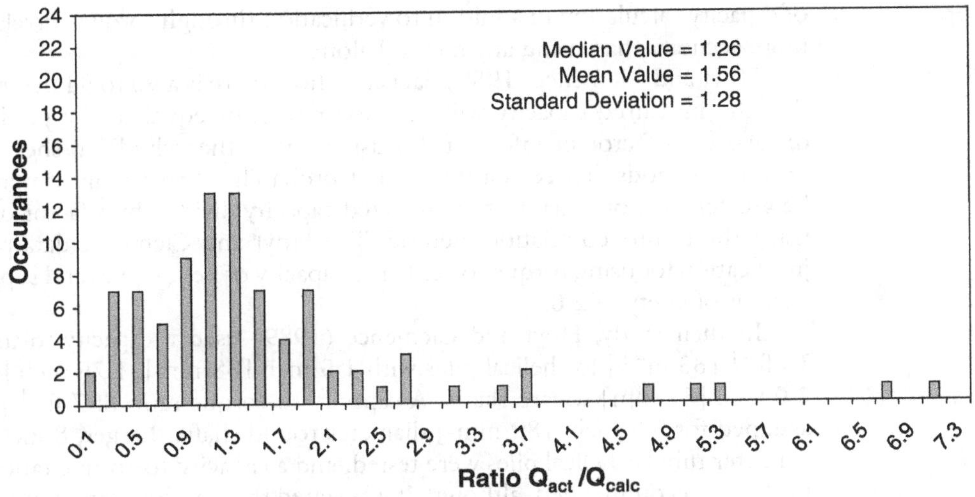

Figure 6.2 Individual bearing comparison (Hoyt and Clemence, 1989)

the measured capacity will be less than the predicted capacity 33 to 40 percent of the time.

From the figures, it is apparent that the distribution is narrower and the standard deviation is less for the torque correlation method compared to the other two methods. This finding indicates that torque correlations are more precise and has led some to conclude that the torque correlation method is better than any other method. It will be shown later in this chapter that the combination of traditional limit state methods

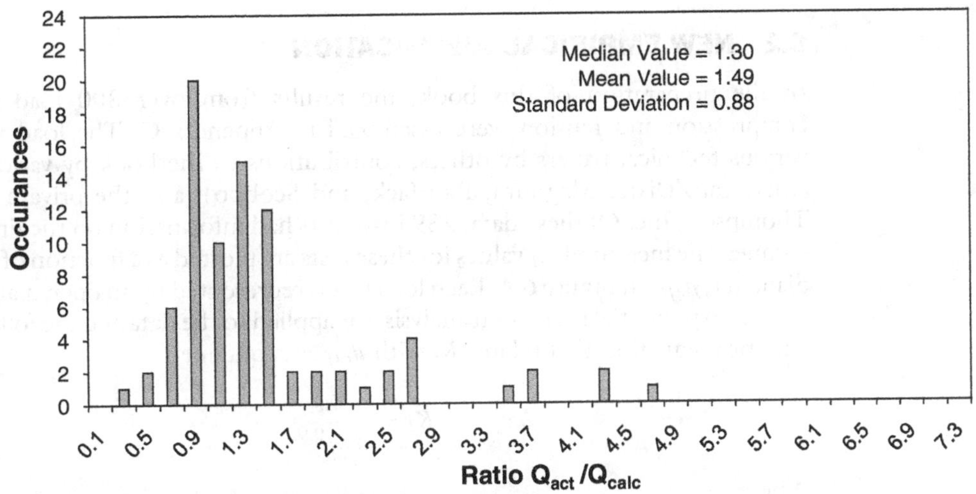

Figure 6.3 Torque comparison (Hoyt and Clemence, 1989)

of capacity calculation in addition to verification through torque correlations is much more accurate than using any method alone.

Hoyt and Clemence (1989) data also show there is a 90 to 94 percent probability that the measured capacity will be greater than or equal to the predicted capacity divided by a factor of safety of 2.0 using either the cylindrical shear or individual bearing methods. There is a 99 percent probability that the measured capacity will be greater than or equal to the predicted capacity divided by a factor of safety of 2.0 using the torque correlation method. The Hoyt and Clemence data provide strong justification for using torque to verify the capacity of helical piles and support for using a factor of safety of 2.0.

In their study, Hoyt and Clemence (1989) used a capacity-to-torque ratio of 10 ft^{-1} [33 m^{-1}] for helical piles with 1.5-inch [38-mm], 1.75-inch [44-mm], and 2.0-inch [51-mm] square shafts. A capacity-to-torque ratio of 7 ft^{-1} [23 m^{-1}] was assumed for 3.5-inch- [89-mm-] diameter round shafts. Larger, 8-inch- [203-mm-] diameter tubular helical piles were tested, and a capacity-to-torque ratio of 3 ft^{-1} [10 m^{-1}] was recommended, although it was stated that the correlation was not as strong.

It is important to note that the tests analyzed by Hoyt and Clemence (1989) were in tension. It was shown in a subsequent study that the tensile capacity of helical piles measured on three sites was 16 to 33 percent less than the measured compression capacity. The difference was attributed to the fact that the lead helix rests on relatively undisturbed soil in compression applications, but in tension the trailing helix bears on the soil affected by the installation of both the lead and trailing helices (Hargrave and Thornsten, 1992). For practical purposes, it has been common practice to use the same capacity-to-torque ratio in both tension and compression and to ignore the slight increase in the capacity for compression applications. This is essentially what was done in AC358 (ICC-ES, 2007).

6.2 NEW EMPIRICAL JUSTIFICATION

In the preparation of this book, the results from over 300 load tests in both compression and tension were assembled in Appendix C. The load tests are from various technical papers by others, contributions to the book by various companies (MacLean/Dixie, Magnum, RamJack, and Scobbo), and the private files of CTL| Thompson, Inc. Of these data, 239 load tests had information on the final installation torque. The measured K_t values for these tests are plotted as a function of effective shaft diameter, d_{eff}, in Figure 6.4. Each load test is represented by an open diamond symbol.

An exponential regression analysis was applied to the data and the following best-fit empirical equation that relates K_t with d_{eff} was obtained:

$$K_t = \frac{\lambda_k}{d_{eff}{}^{0.92}} \tag{6.2}$$

Where

λ_k is a fitting factor equal to 22 $in^{0.92}/ft$ [1433 $mm^{0.92}/m$].

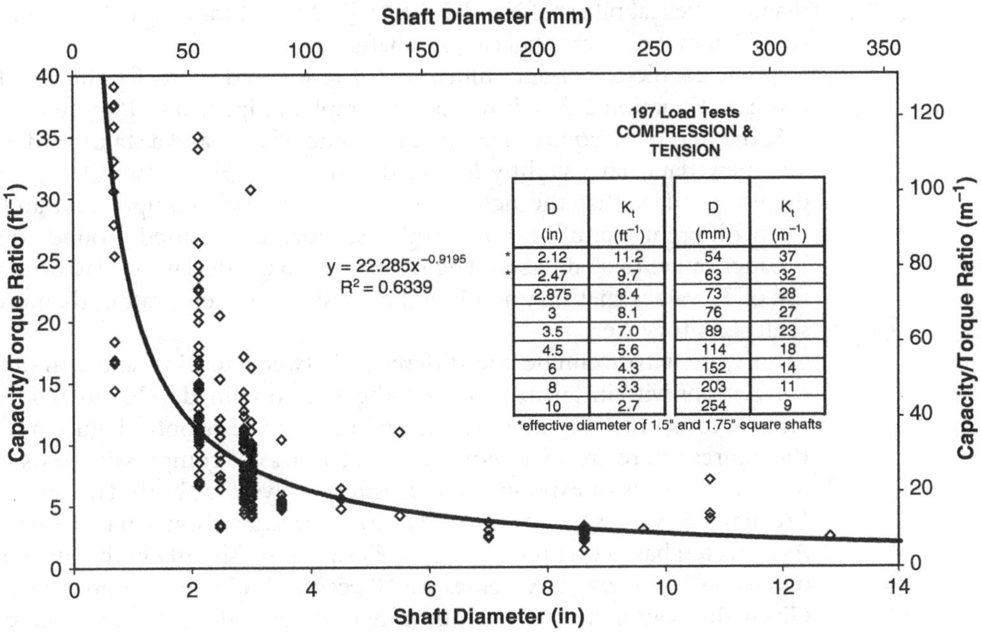

D (in)	K_t (ft^{-1})	D (mm)	K_t (m^{-1})
*2.12	11.2	54	37
*2.47	9.7	63	32
2.875	8.4	73	28
3	8.1	76	27
3.5	7.0	89	23
4.5	5.6	114	18
6	4.3	152	14
8	3.3	203	11
10	2.7	254	9

Figure 6.4 Empirical capacity-to-torque ratio

The unusual units of λ_k are an outcome of the regression and allow d_{eff} to be input in units of inches [mm] and outputs K_t in units of ft^{-1} [m^{-1}]. In the construction of Figure 6.4, effective shaft diameter was defined as the diameter of the hole that would be created by the pile shaft rotating 360 degrees in ground. Hence, d_{eff} simply equals the shaft diameter for a round shaft. For a square shaft, d_{eff} is the diameter of a circle circumscribed around the shaft or the diagonal distance between opposite corners of the square shaft. The piles used in these tests varied from laboratory models with d_{eff} less than 1 inch [25 mm] to full-scale piles with d_{eff} over 12 inches [305 mm].

The K_t values obtained from the regression analysis for several standard shaft sizes are shown in the table within the graph. The new values match the previous work done by Hoyt and Clemence (1989) and Hoyt (2007) very well. The regression analysis indicates K_t values of 11.1, 9.7, 7.0, and 3.3 ft^{-1} [37,32, 23, and 11 m^{-1}] for 1.5-inch-square, 1.75-inch-square, 3.5-inch-diameter, and 8-inch- diameter [38-mm-square, 44-mm-square, 89-mm-diameter, and 203-mm-diameter] shafts, respectively. This compares well with K_t values of 10, 10, 7, and 3 ft^{-1} [33,33, 23, and 10 m^{-1}] assumed by Hoyt and Clemence (1989) for similar shaft sizes.

The K_t values obtained from regression analysis also correspond fairly well to the work done by Tappenden (2006) on larger-size shafts. The regression analysis in Figure 6.4 indicates K_t values of 5.6 ft^{-1} [18 m^{-1}] for 4.5-inch- [114 mm] diameter shafts and between 4.3 and 2.7 ft^{-1} [14 and 9 m^{-1}] for shafts between 6 inches and 10 inches [152 and 254 mm] diameter. From analysis of the results from 29 load tests, Tappenden found K_t values of 5.2 ft^{-1} [17 m^{-1}] for 4.5-inch- [114-mm-]

diameter helical pile shafts and 2.8 ft^{-1} [9.2 m^{-1}] for 5.5-inch- to 10.75-inch- [140- to 273-mm-] diameter helical pile shafts.

The coefficient of determination, or R-squared value, for the best-fit relationship given by Equation 6.2 is shown on the graph in Figure 6.4. If plotted similar to Figure 6.3, the new data would have a median value of 1.01 and a standard deviation of 0.51. The new data have slightly less scatter than the Hoyt and Clemence data. This is despite the fact that the new data is based on both compression and tension tests, many different manufacturers' products, more sites, varied ground conditions, and a broader range of shaft sizes. The smaller standard deviation may be explained by the use of K_t values that vary as a function of shaft diameter rather than grouping several shaft sizes together.

In order to examine the difference between tension and compression capacity correlations with installation torque, the data contained in Figure 6.4 is separated into tension tests only (Figure 6.5) and compression tests only (Figure 6.6). As shown in the figures, there are 98 pullout tests and 141 axial compression tests included in the data. The results of exponential regression analysis for both data sets are very similar. Predicted K_t values for common shaft sizes are again shown in tables within the graphs. As expected based on previous work, K_t values are slightly higher in compression than in tension. On average, K_t values are 10 percent higher in compression than in tension. Given the scatter in the data and inherent variability of subsurface conditions, the practice of generally assuming the same capacity-to-torque ratios for both compression and tension tests seems reasonable. If the designer desires more rigor, slightly lower capacity-to-torque ratios can be used in tension.

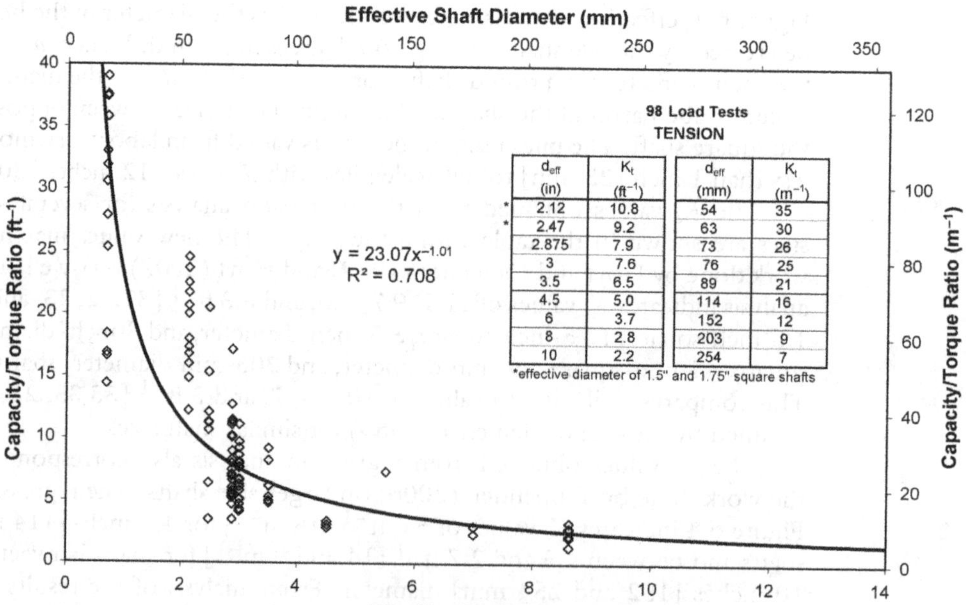

Figure 6.5 Empirical capacity-to-torque ratio (tension only)

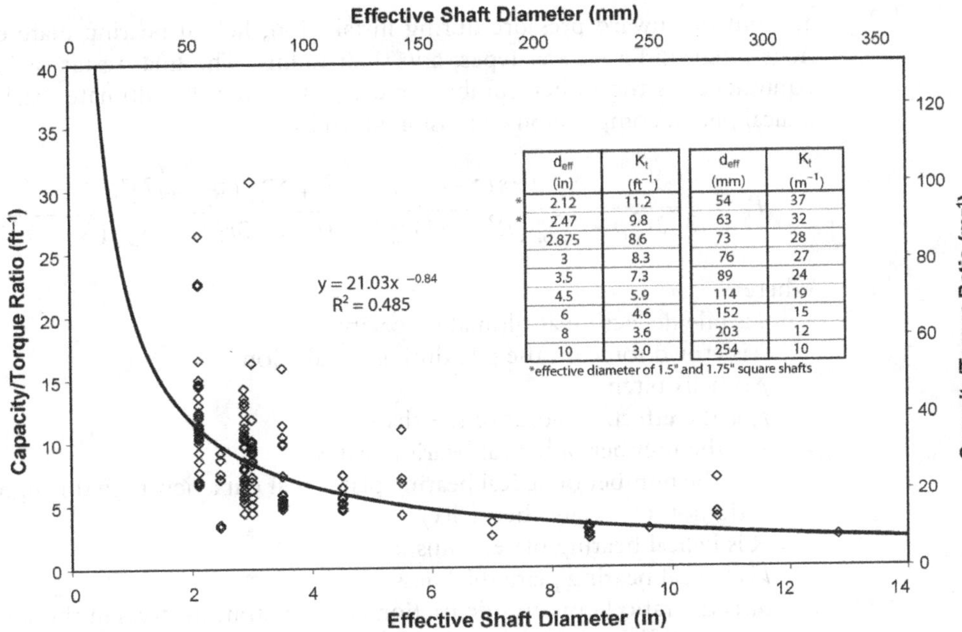

Figure 6.6 Empirical capacity-to-torque ratio (compression only)

The load tests analyzed by Hoyt and Clemence (1989) in the derivation of K_t values were all taken to ultimate bearing capacity (plunging failure) rather than some allowable deflection. The load limit for a majority of the load tests in Appendix C were obtained using the modified Davisson offset method (discussed in Chapter 7), which limits net deflection to 10 percent of the average helix diameter. Both load test interpretation methods show good correlation with installation torque.

6.3 ENERGY MODEL

Based on all of the empirical evidence, it is clear that torque measurements taken during installation of a helical pile indicate soil shear strength and consistency at the depth through which the helical bearing plates are passing. It would be beneficial to theoretically derive bearing and pullout capacity from installation torque in order to further justify the relationship between torque and capacity as well as to explore factors that might affect K_t. Unfortunately, due to the complex interaction of the helical bearing plates with the soil, it is difficult theoretically to derive bearing pressure and shear strength from torque measurements.

In order to avoid this difficulty, a model was proposed by Perko (2000) wherein the capacity of a helical pile was directly related to the installation torque by energy equivalence. Perko calculated the energy required to penetrate a unit volume of soil during axial loading and equated that to the energy expended to penetrate a unit volume of soil during rotation of a helical pile into the ground. This method took into account downward pressure during installation, helical bearing plate configuration, helix pitch, different soil types, and shaft radius. The final result of Perko's energy equivalency is the rather cumbersome equation for the ultimate axial capacity of a helical pile in compression or tension given by

$$P_u = \frac{12\delta\,(2\pi T + Fp)\left[r_s^2 + \sum_n \left(R_n^2 - r_s^2\right)\right]}{3\left[2r_s^3 p + \sum_m \left(R_m^2 - r_s^2\right) t_m^2\right] + 16\pi\alpha_s\left[3r_s^3\lambda_s + \sum_n \left(R_n^3 - r_s^3\right) t_n\right]} \tag{6.3}$$

Where

δ is pile deflection at ultimate capacity
F is crowd force on the pile during installation
p is helix pitch
r_s is the effective radius of the shaft
n is the number of helical bearing plates
m is the number of helical bearing plates that cut a new path through the soil (i.e., do not follow another helix)
R is helical bearing plate radius
t is helical bearing plate thickness
α_s is the ratio between side friction and penetration stress in the soil, and
λ_s is the effective shaft length subject to friction.

The model compares well with empirical work by Hoyt and Clemence (1989). The model indicates that K_t is largely independent of crowd force during installation, final installation torque, number of helical bearing plates, and helix pitch. The model indicates a strong dependence of K_t on effective shaft diameter. Average K_t values of 11 ft^{-1} [37 m^{-1}] and 7 ft^{-1} [23 m^{-1}] are predicted by the model for three common sizes of square-shaft helical piles and a 3.5-inch- [89-mm-] diameter round shaft helical pile, respectively. These values compare well with the values of 10 ft^{-1} [33 m^{-1}] and 7 ft^{-1} [23 m^{-1}] from Hoyt and Clemence (1989).

In the original work, Perko (2000) also compared predictions based on the model to measured K_t values from a number of laboratory and full-scale investigations. Model predictions generally matched laboratory and field measurements within the range of error exhibited by the repeatability of those measurements. One of the more interesting facets of the model is its ability to explain higher K_t ratios measured in laboratory tests compared to field tests. The Perko (2000) work is perhaps the first to explain the interesting behavior of helical piles that K_t does not correlate between small-scale laboratory and full-scale field tests.

In the preparation of this book, the Perko (2000) model was compared with other previously published field and laboratory data listed in Appendix C. In the calculations, the effective shaft length, λ_s, was set equal to the number of helical bearing plates times the pitch, np. This value was used because at a minimum the shaft located between the leading and trailing ends of the helix are clearing a path through the soil. Beyond this, it is anticipated that shaft friction becomes significantly less due to wobbling and high strains. The ratio of side friction and penetration stress, α_s, was taken as 0.6, which

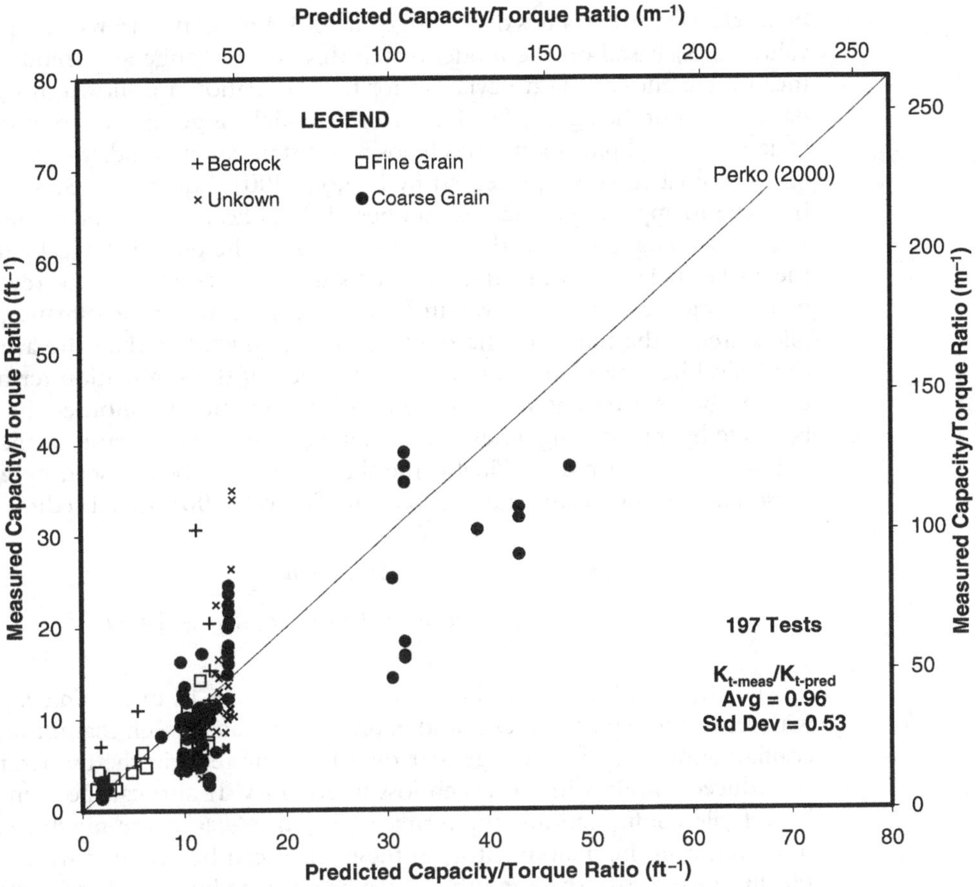

Figure 6.7 Energy model for capacity-to-torque ratio

indicates an angle of friction between the soil and pile steel of 30 degrees. There may be some dependence of α_s on soil consistency, but this dependence is expected to be small because the friction coefficient between steel and soil is largely independent of density (Das, 1990). Based on field observations that helical bearing plates seldom follow the exact same path due to slippage, the number of helical bearing plates cutting a new path through the soil, m, was set equal to the total number of helical bearing plates, n. To be consistent with one of the more commonly recognized load test interpretation methods, the deflection at ultimate capacity, δ, was set equal to 10 percent times the average helix diameter or 1 inch [25 mm], whichever is less.

The resulting model predictions are compared with 197 measured values in Figure 6.7. The diagonal line in the figure represents a 1:1 correlation between predicted and measured K_t. As described in the legend on the top of the figure, subsurface conditions in these tests included fine-grain and coarse-grain soils as well as weathered bedrock. Measured values of K_t ranged from 1.4 to 39 ft^{-1} [5 to 128 m^{-1}]. Predicted values of K_t based on the model match this general range and trend fairly well. The mean value and standard deviation for the correlation are shown in the lower right-hand corner of the figure. The fact that the model has good correspondence for a large variety of helical pile geometries helps to substantiate its validity.

The final equation presented by Perko (2000), Equation 6.3, is somewhat cumbersome to apply in practice. A number of simplifications can be made to the model. First, it is recognized that the crowd force times the pitch is typically much less than the final installation torque times 2π. This allows the equation to be rewritten in terms of K_t using the definition given in Equation 6.1. Also, for most manufactured helical pile systems, the radius of the shaft squared is much less than the average radius of the helical bearing plates squared. Hence, each of the summation terms in Equation 6.3 can be rewritten as n times the average helix radius. Another simplification can be made by introducing a new parameter, θ, defined as the ratio of the average helix radius to the shaft radius. Finally, the shaft radius can be replaced by $d_{eff}/2$. If all of these simplifications are implemented, the Perko (2000) model reduces to

$$K_t = \frac{6\delta\pi\left[1+n\theta^2\right]}{\left(\frac{3}{4}p+6\pi\lambda\alpha_s+2\pi\alpha_s n\theta^3 t\right)d_{eff}+\frac{3}{4}n\theta^2 t^2} \tag{6.4}$$

Further simplification of Eqation 6.4 can be made by examining each of the terms with respect to typical values found in practice. It can be seen that for most helical pile configurations, $n\theta^2$ is much greater than 1, so the term in the numerator can simply be reduced to $n\theta^2$ without much loss in accuracy. It also can be seen that for most helical pile configurations, the terms $3/4p$ and $3/4n\theta^2 t^2$ are much smaller than the other terms in the denominator, so these terms can be dropped with minimal effect. Finally, the effective shaft length, λ_s, can be replaced by np as discussed above. Hence, Eqation 6.4 can be re-written as

$$K_t = \frac{3\delta\theta^2}{\left(3p+\theta^3 t\right)\alpha_s}\cdot\frac{1}{d_{eff}} \tag{6.5}$$

The first term in Equation 6.5 depends only on the pitch and thickness of helical bearing plates, the ratio of helical bearing plate radius to shaft radius, the ratio of soil friction to penetration stress, and final pile head deflection. This term should be fairly constant for a given helical pile system and project site. It is independent of the number and size of helical bearing plates as long as the ratio of average helix radius to shaft radius is held constant. If the first term in Equation 6.5 is grouped into a new parameter, H, defined as the helical pile–soil interaction coefficient, the Perko (2000) model reduces to simply

$$K_t = \frac{\mathrm{H}}{d_{eff}} \tag{6.6}$$

where

$$H = \frac{3\delta\theta^2}{(3p + \theta^3 t)\,\alpha_s} \tag{6.7}$$

The result given in Equation 6.6 is very similar to the empirical equation given in Equation 6.2. Depending on the pile geometry, H can vary between about 1.0 to 4.0. For large helical piles with 3/4-inch- [19-mm-] thick helical bearing plates, H is on the order of 1.4. Small laboratory models can have H varying from 1.9 to 4 depending on how they are constructed and the final deflection. For most standard-size helical piles with typical helical bearing plate radius to shaft radius ratios and standard helix thickness, H has an average value of 2.4. If this value is input in Equation 6.6 and a conversion is made so that the effective diameter can be input in inches [mm] and K_t output in ft^{-1} [m^{-1}], the theoretical model is very close to the empirical equation determined from regression analysis.

The simplified theoretical model based on energy, Equation 6.6, was superimposed over the data shown in Figure 6.4 using a value of 2.4 for H. The result is shown in Figure 6.8 wherein the simplified energy model is represented by the upper gray line. As anticipated from the similarity between Equation 6.6 and Equation 6.2, the model matches the data fairly well. At the extreme values of shaft diameter, both theoretical and empirical equations suggest very high capacity-to-torque ratios as the friction along the shaft becomes negligible. Both equations also suggest that the capacity-to-torque ratio decreases with the inverse of effective shaft diameter.

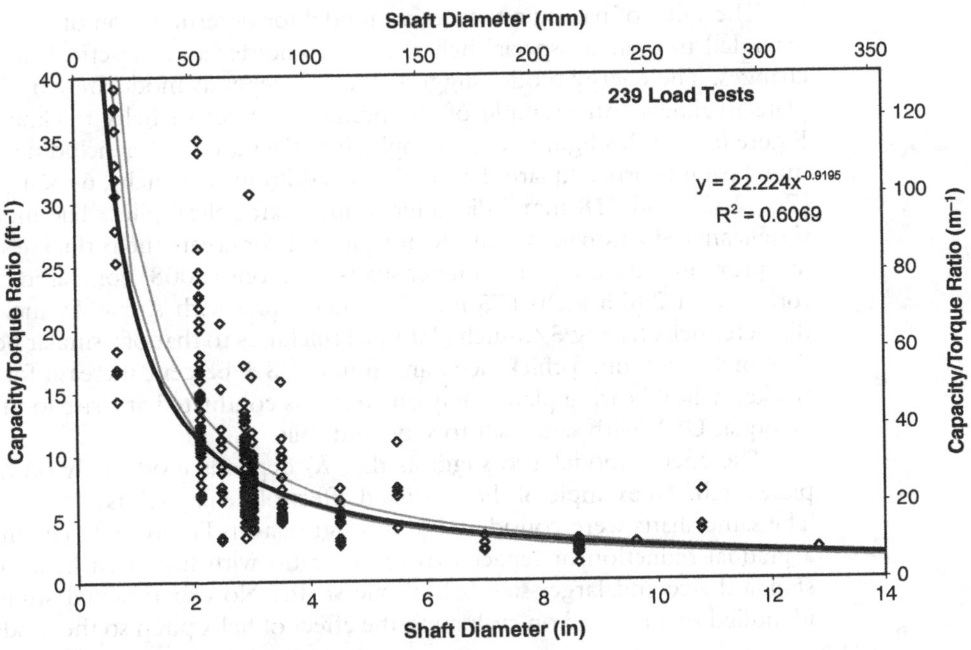

Figure 6.8 Theoretical capacity-to-torque ratio

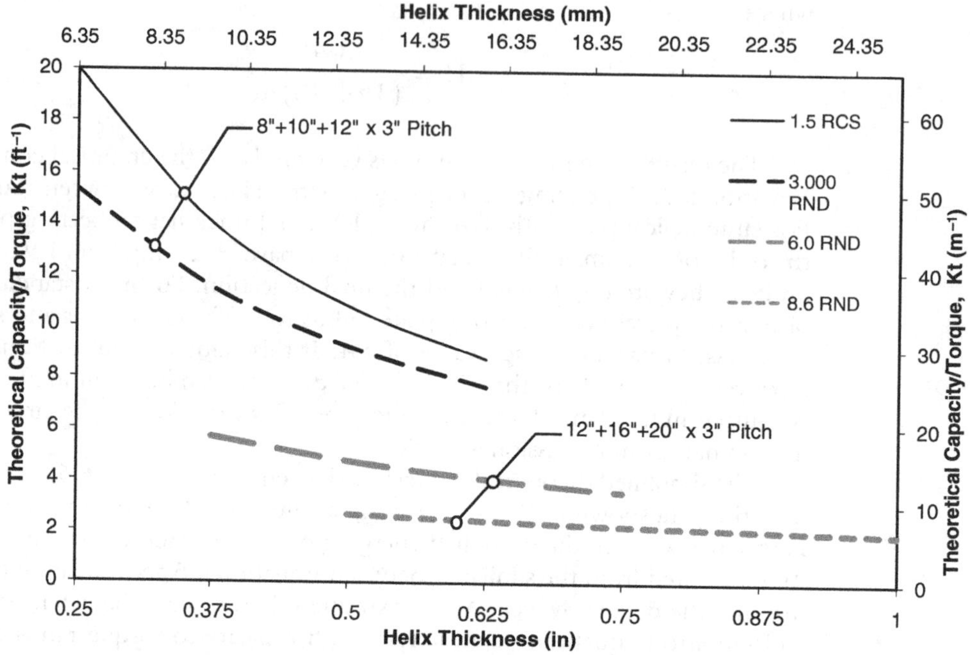

Figure 6.9 Theoretical effect of helical bearing plate thickness

The value of having a theoretical model for determination of K_t is that it can be extended to evaluate special helical pile geometries and the effect of various design changes. The energy model suggests that K_t depends moderately on helical bearing plate thickness. An example of the predicted effect of helix thickness is shown in Figure 6.9. In this figure, four example shaft sizes are shown including 1.5-inch- [38-mm-] round corner square shaft (RCS) in addition to 3-inch-, 6-inch-, and 8.6-inch- [76-, 152-, and 218-mm-] diameter round shaft helical piles. The model suggests a significant reduction in capacity-to-torque ratio for greater helix thickness. The effect is less pronounced for larger-diameter shafts. Slemons (2008) compared the installation torque of a 2.875-inch- [73-mm-] diameter pile with single 14-inch- [356-mm-] diameter helix having 3/8-inch [10-mm] thickness to that of a similar helical pile with 3/4-inch- [19-mm-] thick helix and found a 3.5 percent increase for the pile with thicker helical bearing plate. Only one test was conducted at a site located in Atlanta, Georgia, USA, with sandy silt to silty sand soils.

The energy model also suggests that K_t depends moderately on helical bearing plate pitch. An example of the predicted effect of helix pitch is shown in Figure 6.10. The same shafts were considered in this figure as in Figure 6.9. The model suggests a gradual reduction in capacity-to-torque ratio with increased helix pitch for both standard-size and larger-size helical pile shafts. No experimental studies have been identified that focused on evaluating the effect of helix pitch so the model predictions

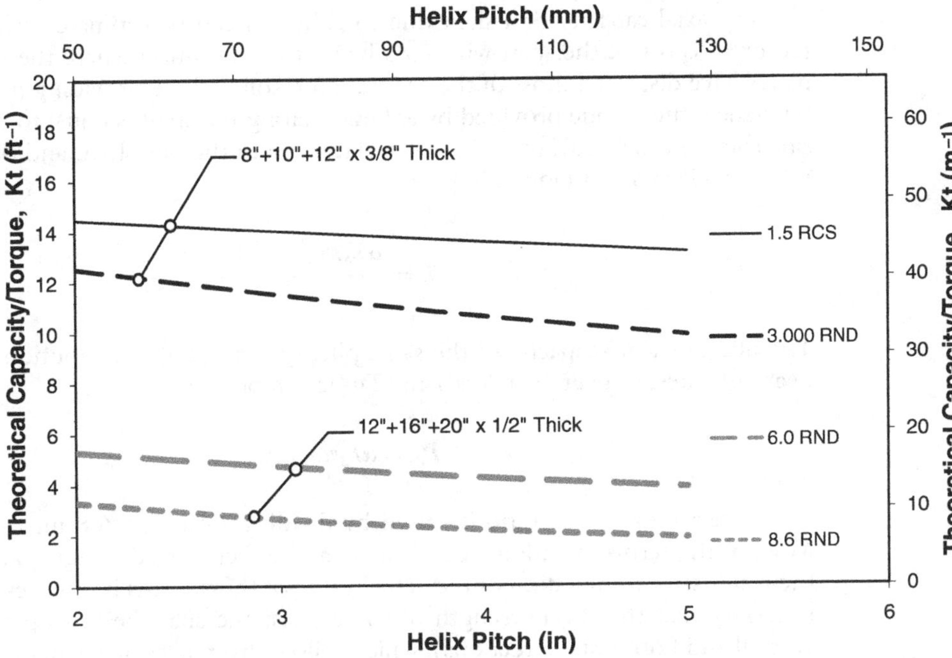

Figure 6.10 Theoretical effect of helix pitch

should be approached with caution. One might expect that although the capacity-to-torque ratio is only nominally affected by increasing helix pitch, it may be much more difficult and require more crowd to advance helical piles with a high degree of helix pitch into the ground properly without augering.

6.4 SIMPLE SHAFT FRICTION MODEL

During the installation of helical piles, helical bearing plates sometimes fail to track properly in ground. If insufficient crowd is being applied, if a strength transition is encountered, or if an obstruction is met, helical bearing plates can slip and rotate without forward advancement. When this occurs, torque measurements may not indicate the strength and consistency of the ground at the depth of the helical bearing plates. Rather, the bearing plates may be spinning in a disturbed zone created after several revolutions. At that point, the torque typically decreases to an amount that depends only on shaft friction. The lower bound limit of the capacity of a helical pile is often of interest.

In the Perko (2000) energy model, adhesion along the shaft was considered in the calculation of energy required for installation but was ignored in the calculation of the energy required to penetrate the soil under axial loads. As a consequence, the results from the energy model are incorrect as helical bearing plate diameter approaches the shaft diameter.

The axial capacity of a shaft due to adhesion can be estimated from the torque required to rotate the shaft without a helix. In a condition where the helical bearing plates have displaced most of the surrounding soil and are providing minimal torsion resistance, the torque provided by adhesion along the shaft is equal to the product of side shear on the shaft, the effective surface area of the pile shaft, and the shaft radius given by (Perko and Doner, in press):

$$T = \frac{\alpha \lambda_s \pi d_{eff}^2}{2} \tag{6.8}$$

The ultimate axial capacity of the same pile, ignoring any contribution of the helical bearing plates, is given by (Perko and Doner, in press):

$$P_u = \alpha \lambda_s \pi d_{eff} \tag{6.9}$$

Capacity-to-torque ratio is found by dividing Equation 6.8 into Equation 6.9. Most of the terms including adhesion, α, and effective shaft length, λ_s, drop out of the result so determination of the actual value of these terms is unnecessary. It is only necessary that the shear strength of soil against the shaft be isotropic (the same in vertical and horizontal directions), which follows from basic mechanics. The resulting simple expression after cancellation of terms is given by Perko and Doner (in press):

$$K_t = \frac{2}{d_{eff}} \tag{6.10}$$

The expression given in Equation 6.10 is similar to Equations 6.2 and 6.6, which were determined from empirical regression analysis and theoretical energy equivalency. In order to compare this expression with other models and actual load tests, K_t values obtained from Equation 6.10 are represented by the lower gray line in Figure 6.8. As can be seen in the figure, the simple shaft friction model falls just below the empirical regression line for shaft diameters corresponding to load tests on full-scale helical piles and just above the empirical regression for laboratory models.

For common 1.5-inch ×1.5-inch- (38 ×38 mm-) square shaft helical piles, the theoretical capacity-to-torque ratio based on the energy model is 13.6 [45 m^{-1}], the best-fit exponential regression of empirical load test data is 11.2 [37 m^{-1}], and simple shaft friction is 11.3 [37 m^{-1}]. For common 2.875-inch- [73-mm-] diameter shaft helical piles, the theoretical capacity-to-torque ratio based on the energy method is 10.0 [33 m^{-1}], the best-fit exponential regression is 8.4 [28 m^{-1}], and simple shaft friction is 8.3 [27 m^{-1}]. The simple shaft friction model may be considered a lower bound for verification of helical pile capacity in the field for a helical pile that is spinning without advancement.

The simple shaft friction model also may be useful for the verification of the capacity of any helical piles wherein the majority of their capacity is from shaft friction.

Some examples of this include larger-diameter helical pile shafts with very small helical bearing plates, helical piles with extremely long shafts, and helical piles in layered coarse- and fine-grain soil strata.

6.5 OTHER THEORETICAL MODELS

Ghaly and Hanna (1991a) presented a unique relationship between torque and capacity for a single helix in sand using the results of tests on laboratory models. The relationship considered area of the helix, overburden stress, and helix pitch. However, it did not take shaft diameter into account, which is known to be one of the largest factors affecting the relationship between torque and capacity.

Tappenden (2004) tested Ghaly and Hanna's relationship by comparing it to test results on full-scale field load tests on large-diameter helical piles and found that the correlation is not very good. Ghaly and Hanna's relationship overestimated capacities by 5 to 10 times. Tappenden stated that Ghaly and Hanna's relationship is not recommended for use with full-scale helical piles.

G. L. Bowen (in press) presented a method for determination of capacity-to-torque ratios based on a static analysis with non-dimensional parameters. It was shown that the capacity-to-torque ratio varies with a number of parameters but the inverse of the shaft radius had the largest effect. Predictions based on the static model compare remarkably well with the empirical capacity-to-torque ratios given in Section 6.2. Another unique finding from the analysis was that the crowd required to install a helical pile increases with the square of the ratio of pitch to shaft radius.

6.6 PRECAUTIONS

A number of factors affect torque measurements. This section contains certain precautions that the pile designer and installation contractor should be aware of. In addition, helical pile manufacturers' literature should be consulted to verify the K_t values that apply to their products and any special instructions for torque measurement.

Augering, defined as rotation of the helix where forward advancement is stalled, generally causes a significant decrease in torque. Empirical correlations between torque and capacity and those based on the energy model may be conservative when augering occurs. Augering is most common when the helix advances from a softer stratum into a harder stratum. Augering is also common with untrue helical bearing plates. Traditional limit state theory can be used to determine helical pile capacity along with field verifications based on the simple shaft friction model if augering cannot be avoided.

Most of the empirical work that has been published to date is limited to true helix-shaped bearing plates. An example of a true helix is shown in Figure 6.11. A true helix is mounted perpendicular to the shaft at all points and the leading and trailing edges are parallel. ICC-ES (2007) uses the term "conforming" to define the conditions that are most represented in past literature. A conforming helical pile meets

Figure 6.11 True helix shape

all of the criteria shown in Table 6.1. Helical pile products manufactured outside of these tolerances are not excluded but should be approached with caution with regard to using previously published capacity-to-torque ratios. In other words, capacity-to-torque ratios should not be blindly applied to all manufactured products without first evaluating the product geometry and installation conditions.

In their evaluation of 91 tension load tests on helical anchors, Hoyt and Clemence (1989) recommended that the final installation torque should be averaged over a distance equal to three times the average helix diameter. In compression applications, the final installation torque is often taken as the final torque reading at the termination depth. The installation speed during all past research has generally been between 10 and 30 revolutions per minute. It is unknown how slower or faster rates of installation affect the capacity-to-torque ratio.

Most researchers agree that the torque correlation does not apply when the bottom helix rests on hard material in a refusal condition and when the top helix is less than 5 diameters below the surface (Pack, 2000). When a helical pile is in an end-bearing condition, load tests and limit state calculations may be substituted for installation torque correlations. Capacity may be limited by mechanical strength of the helical pile and its components. In shallow applications, the minimum embedment must be obtained in addition to achieving the minimum required torque, as discussed in Chapter 5.

The installation of a helical pile can cause a temporary increase in pore water pressure immediately around the shaft and helical bearing plates in sensitive saturated soils. This phenomenon is discussed in more detail in Chapter 7, Section 7.1. In these cases, there may be some pile freeze that occurs over time. Pile freeze is generally

Table 6.1 Torque Correlation Conformance Criteria (modified from ICC-ES, 2007)

	Criteria
1	Square shafts with dimensions between 1.5 inches by 1.5 inches [38 mm by 38 mm] and 1.75 inches by 1.75 inches [44 mm by 44 mm], or round shafts with outside diameters between 2.875 inches [73 mm] and 3.5 inches [89 mm].
2	True helix-shaped bearing plates that are normal with the shaft such that the leading and trailing edges are within 1/4 inch [6 mm] of parallel.
3	Allowable helical pile capacity is less than 30 tons [67 kN].
4	Helical plate diameters between 8 inches [203 mm] and 14 inches [356 mm] with thickness between 3/8 inch [10 mm] and 1/2 inch [13 mm].
5	Helical plates and shafts are smooth and absent of irregularities that extend more than 1/16 inch [1.6 mm] from the surface, excluding connecting hardware and fittings.
6	Helical plates spaced along the shaft between 2.4 to 3.6 times the helix diameter.
7	Helical pitch is 3 inches ± 1/4 inch [76 mm ± 6 mm].
8	All helical plates have the same pitch.
9	Helical plates are arranged such that they theoretically track the same path as the leading helix.
10	For shafts with multiple helices, the smallest-diameter helix shall be mounted to the leading end of the shaft with progressively larger diameter helices above.
11	Helical pile shaft advancement equals or exceeds 85% of the helix pitch at time of final torque measurement.
12	Helical piles shall be installed at a rate less than 25 revolutions per minute.
13	Helical plates have generally circular geometry.

defined as an increase in strength with dissipation of pore pressures over time. If pore pressures are expected, the capacity-to-torque ratio may be artificially low. It may be permissible in these circumstances to wait a set time period before obtaining a final installation torque measurement by re-setting up on a pile several days later and exerting torque until movement occurs. The author is unaware if this has been attempted, so the technique should be used with caution and in conjunction with load tests.

Some engineers believe that the capacity-to-torque ratio is a function of soil type. As shown in the previous section, the capacity-to-torque ratio does exhibit significant scatter. Some of this scatter may be due to different soil types and other subsurface conditions, such as groundwater and soil disturbance. Other factors may include minor effects of the size and number of helical bearing plates, crowd applied during installation, and helix shape among others. The values of K_t obtained from exponential regression analysis represent average values across a myriad of different subsurface conditions and helical pile geometries. Variations should be expected. Using an average value of K_t means that load tests will exceed expectations in half the instances and fall short of expectations in the other instances. Given the high degree of variability of the ground, there are no certainties in geotechnical engineering or capacity determination of any pile system. However, if multiple methods of capacity determination are implemented with appropriate factors of safety, there can be a high degree of certainty, as discussed in Chapter 8.

6.7 EXPLORATION WITH HELICAL PILE

SPT blow count and CPT tip resistance may be crudely estimated from torque measurements. This is accomplished by first estimating the predicted ultimate capacity of the helical pile by multiplying K_t by the measured installation torque at a particular depth. Then either the limit state or the LCPC method described in Chapter 4, as appropriate, can be used to back-calculate SPT blow count or CPT tip resistance based on the configuration of the helical pile acting as a "soil probe". In Chapter 8, helical pile sizing charts are provided that can be used to quickly estimate SPT blow count when combined with correlations between capacity and torque.

Using a helical pile to explore the subsurface can be beneficial on projects with limited geotechnical information but should be approached with caution. Torque measurements should not be used as a substitute for more traditional geotechnical exploration, rather they may be considered a supplement. Obviously, the soil type cannot be determined by installing a helical pile. Capacity-to-torque ratios exhibit a significant amount of scatter and may vary with changing subsurface conditions. It might be more accurate to install a helical pile in the location of an exploratory boring with Standard Penetration Test blow count or Cone Penetration Test tip resistance information to calibrate the helical pile "soil probe" for the particular project site.

Accumulation of torque readings with depth during the normal course of helical pile installation can add to the geotechnical information on a project. An example would be the use of shallow torque readings to verify the extents of removal and replacement of soft materials for support of floor systems, which is often difficult to determine with widely spaced exploratory borings. Another example would be the identification of previously undetected soft or weak lens or changes in bedrock profile. As familiarity with installation torque increases, the data produced with a helical pile becomes more and more useful in examining soil density and consistency. Exploration using a helical pile is an accepted method of practice in the Municipality of Anchorage (Bowen, 2009).

Chapter 7

Axial Load Testing

Some engineers recommend load testing at sites with a particularly large number of helical piles to verify the axial capacity and the capacity-to-torque ratio at specific sites. Load tests are specified under certain conditions in some building codes and by some municipalities. Load tests also should be considered when deflection is a critical concern.

Procedures for axial compression and axial tension load testing are discussed in this chapter. Several methods exist for interpreting the capacity of a deep foundation based on a measured load-deflection curve. A number of these load test interpretation methods are described. Advantages and disadvantages of each of these methods are discussed relative to capacity determination for helical piles.

7.1 COMPRESSION

Helical pile compression load tests are typically conducted in accordance with ASTM D1143-07. In this method, a load frame is constructed over the test pile. An example load frame is shown in Figure 7.1. Reaction piles, normally consisting of a pair or set of four helical anchors, are installed a minimum clear distance away from the test pile of five times the maximum diameter of the largest reaction anchor or test pile but no less than 8 feet [2.5 m]. For helical piles, maximum diameter should be taken as the largest helical bearing plate. The loading procedure depends on the subsurface conditions, the type of project, and the designer's preference. A number of different loading procedures covered in ASTM D1143-07 are discussed in this section.

ASTM D1143-07 requires that a primary and a secondary method of measurement be incorporated to monitor axial movement of the test pile. The primary method typically consists of two dial-gage extensometers. The secondary method often consists of a ruler or engineer's scale affixed to the pile shaft, pile cap, or ram housing.

Figure 7.1 Example compression load frame (Courtesy of Magnum Piering, Inc.)

Movement of the pile is detected through the use of an optical level or survey transit trained on the scale. A wire, mirror, and scale system also may be considered as a secondary method of measurement.

The photograph in Figure 7.2 shows a typical setup of equipment over the top of a helical pile. Twin dial-gage extensometers are located at opposite sides of the pile cap. The dial gages are affixed to reference beams supported at a significant distance away from the top of the pile. A hydraulic ram is centered over the pile. An engineer's scale is affixed to the side of the ram. An electronic load cell is positioned between the ram and the main reaction beam. The reaction beam is centered over the ram and the helical pile. Hydraulic pressure also may be used to measure applied loads provided the ram, pressure gauge, and hydraulic pump are calibrated as a unit.

Due to the slender nature of helical pile shafts, alignment of the pile, hydraulic ram, and reaction beam is critical to the proper measurement of axial compressive capacity. Misalignment can exert combined axial and flexural loads on the pile shaft. Eccentricity of several inches can cause failure of the shaft to occur in combined flexure and buckline at much less load than the ultimate compression capacity of the pile.

The theoretical reduction in capacity of a helical pile shaft under eccentric loads is shown in Figure 7.3. The capacity of several common helical pile shafts ranging from 2.5 inches to 10.5 inches [64 to 267 mm] were analyzed using AISC equations for combined shaft flexure and buckling. The unsupported length of the shafts was assumed to be 5 feet, which is indicative of firm soils, as discussed in Section 4.8. The effective length factor was set equal to 0.8 for fixed-pinned boundary conditions.

Figure 7.2 Example instrumentation and load application equipment

Relative capacity is shown in the y-axis of the figure and defined as the ratio of combined flexure and buckling capacity at a certain eccentricity to the capacity of the shaft under a concentric load. Eccentricity is the distance between the point of load application and the central axis of the shaft. Relative eccentricity is shown on the x-axis in the figure and is defined as the ratio of load eccentricity to the diameter of the shaft. When the results are normalized in this way, all of the shaft sizes evaluated approach a uniform curve.

If the total eccentricity or misalignment of the load frame, hydraulic cylinder, and pile shaft is on the order of 2 inches [51 mm], then the predicted reduction in capacity of a 3-inch [76 mm] diameter shaft is roughly 0.4 times the capacity that would have been measured under concentric loads. Clearly, alignment of the load frame and jacking assembly is critical in the load testing of helical piles. In design applications, eccentricity is prevented from being transmitted to the shaft through properly designed lateral bracing, pile caps, or bracket assemblies.

Another factor that can affect helical pile capacity measurements in full-scale load testing is the plumbness or inclination of the helical pile shaft with respect to the load frame. If the pile shaft is out of plumb when the test load is applied, lateral forces exerted on the pile shaft tend to cause buckling at a load significantly less than the true axial capacity. An example of this phenomenon is demonstrated by the two load tests pictured in Figure 7.4. Both helical piles had exactly the same configuration in virtually the same soil conditions. One pile, represented by the open symbols in the figure, encountered an obstruction during installation and was forced out of plumb by

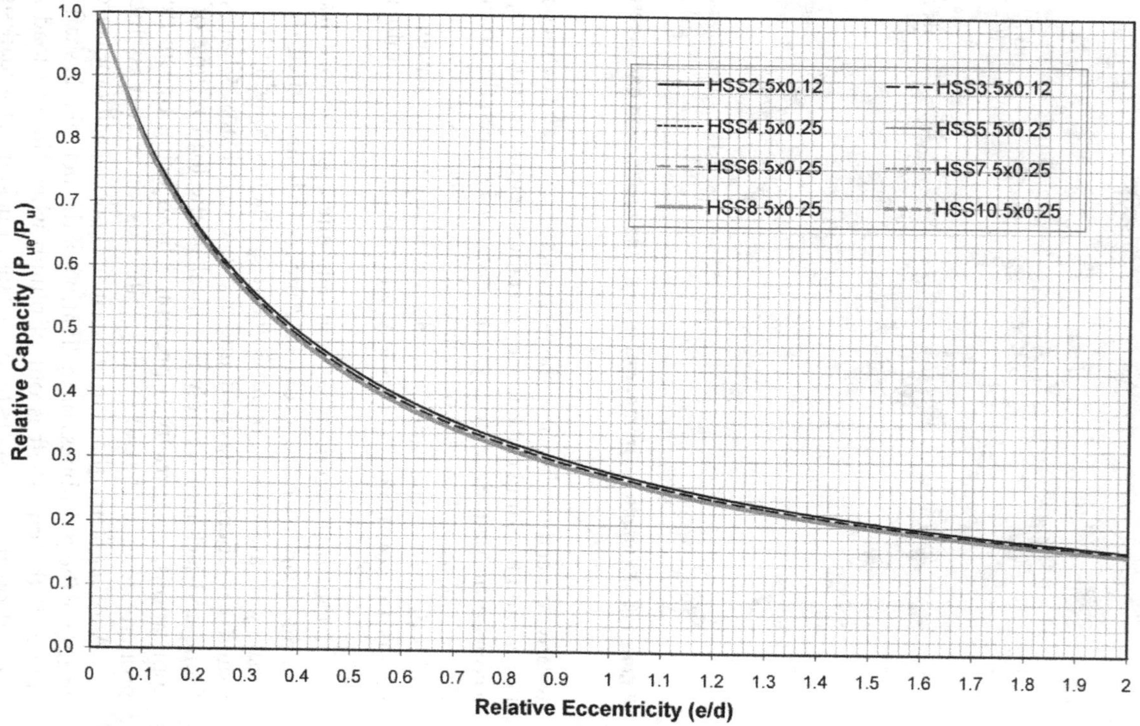

Figure 7.3 Effect of eccentricity on shaft capacity

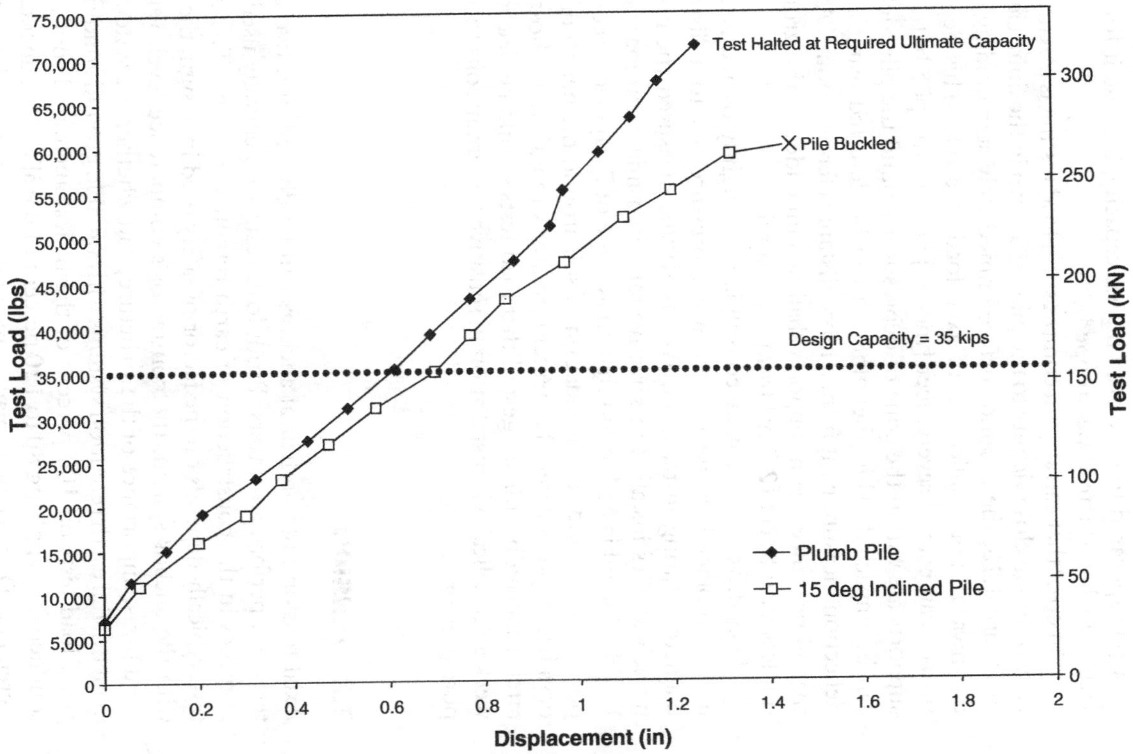

Figure 7.4 Effect of accidental shaft Inclination on load test results

approximately 15 degrees. The pile shaft buckled at a test load of 60 kips [267 kN]. A new pile, represented by the closed symbols in the figure, was installed with improved plumbness and tested a few feet away. The load test on the new pile successfully reached the required ultimate capacity of 70 kips [311 kN] without complications. In fact, the new pile appeared to have even higher capacity, because it had not exhibited signs of yielding when the test was stopped.

An error that can arise in compression load testing is placing supports for the reference beams in close vicinity to the pile. As shown in the finite element model contained in Figure 4.23, the ground surface immediately surrounding a pile can be deflected downward upon application of axial load to a pile. This depression of the ground surface can extend several feet [meters] around the pile butt. If the reference beam supports are within the cone of depression around the pile, deflection measurements can be impacted. This phenomenon can be detected using a secondary method of deflection measurement from a more distant reference point. ASTM D1143-07 states that all reference beam supports shall be embedded in the ground a clear distance of not less than 8 feet [2.5 m] from the test pile.

Piles driven in normally consolidated to lightly overconsolidated soils can result in a temporary increase in pore water pressure out to a distance of 20 to 30 pile diameters during installation. Weech (1996) measured the pore pressure generated by helical piles installed in sensitive saturated fine-grain marine soil in southwestern British Columbia and found that an average of 7 days were required for dissipation of pore pressures and reconsolidation of soil around the helical piles. Piles tested after 7 days had approximately 17 percent greater strength then those tested only 19 hours after installation. It is suggested that load tests should be delayed by a minimum of 1 week after helical pile installation in sensitive fine-grain soils to allow for dissipation of pore pressures.

7.2 TENSION

Axial tension tests on helical anchors can be divided into two categories: proof load tests and performance tests. Proof load tests are generally used for quality assurance purposes in the construction of earth retention systems. The number of proof load tests typically depends on the factor of safety used in design, the designer's experience with the anchor system, the contractor's competence level, variability of subsurface conditions, importance of the structure, and whether the system is temporary or permanent. Tall earth retention systems with variable ground conditions may benefit from proof load tests on 100 percent of the helical anchors. Proof load tests typically are conducted by applying 80 to 120 percent of the allowable anchor load in as few as one increment. Displacement may or may not be measured. The proof load is held until anchor movement stops based on hydraulic ram bleed, at which time the load may be removed completely or reduced to the post-tensioning load and locked off.

Performance tests are more rigorous and generally are used to verify anchor performance against design assumptions. Performance tests may be specified on all types of projects including temporary and permanent shoring, tension membrane structures, tall buildings, tie-back retaining walls, and cable guy systems. With the exception of smaller residential projects, performance tests typically are conducted on one or two helical anchors at the beginning of a project and each time there is a significant change in soil conditions. On small projects with only a few helical anchors, it may be more economical to design with a higher factor of safety and omit anchor load tests; often the cost of the test can exceed the cost of installing a few extra anchors. The choice of whether to require performance tests is ultimately up to the designer, owner, and local building authority.

Performance tests on helical piles or helical anchors can be conducted in general accordance with ASTM D3689. Federal Highway Administration (FHWA), Deep Foundation Institute (DFI), and the association of American Drilled Shaft Contractors (ADSC) anchor load test methods also are common. Performance tests consist of applying either 150 or 200 percent of the anchor design load in 10 to 25 percent increments. Anchor displacement is measured using dial-gage extensiometers and a secondary method such as an optical transit or wire, scale, and mirror system. Each load increment is held for a set period of time or until movement halts. The final load increment is maintained for a minimum hold period depending on which standard method is being followed. Some performance test methods require the anchor be unloaded in uniform decrements. Other test methods require repeated loading and unloading of the anchor. Repetitive loading may be considered for structures subject to alternating live loads, such as shear walls. Long hold periods also may be considered if creep is suspected. The type of performance test should be selected by the designer to evaluate those factors that most affect the overall function and serviceability of the structure. Several loading procedures addressed in ASTM D3689 are discussed in Section 7.3.

The setup for performance tests varies with the type of project. A photograph showing the setup for a performance test on a helical anchor installed to support the membrane roof of an outdoor amphitheater is shown in Figure 7.5. The anchors on this project were required to be installed at a very shallow angle. An excavation around the helical anchor shaft had to be made to accommodate the test frame. The near-surface soils were very soft, so a considerable amount of dunnage was required to prevent settlement of the test frame. The dunnage shown here consists of timber beams. Any material of suitable strength and stiffness may be used as dunnage. Steel beams are used to span between the two footing pads formed by the dunnage. The hydraulic jack is placed over the thread bar attached to the helical anchor and secured with a cap plate and nut. After successful testing, heavily reinforced concrete pile caps were cast over the helical anchors. Some pile caps contained up to five anchors. Performance tests were required on four anchors with an equal number of test anchors located on opposite sides of the project.

A photograph showing the setup for a tension test on a vertical helical pile is shown in Figure 7.6. In this test, movement is being measured with two dial-gage extensiometers attached to the reference beam located just to the side of the hydraulic ram. The force applied to the pile is being measured using a calibrated hollow-core

Figure 7.5 Performance test on a helical anchor (Courtesy of Magnum Piering, Inc.)

hydraulic jack by relating pressure to load. Like the static axial compressive load test, ASTM D3689 requires that the hydraulic jack, pump, and pressure gauge be calibrated as a unit. A hollow-core load cell also can be used to measure test loads. A secondary method of displacement measurement is being accomplished using the laser level and

Figure 7.6 Helical pile tension test

target rod (not shown). It is important in any anchor load test that the load be applied parallel with the primary axis of the helical anchor shaft. Otherwise, combined flexure and tension can cause the thread bar to fracture at a load much less than the maximum test load.

The first course of dunnage should be placed as far from the shaft as practicable. Any part of the test frame or dunnage should not be in contact with the helical pile shaft. The helical pile shaft should be free to move upward. ASTM D3689 requires the clear distance between the test pile and the dunnage shall be at least five times the pile butt diameter. For a typical helical pile, this is on the order of 8 to 30 inches [0.2 to 0.8 m]. The ASTM also says this distance shall be not less than 8 feet (2.5 m). However, it goes on to state that judgment should be used with regard to how far away from the test pile the reaction supports have to be. The 8-foot criterion typically is waved for helical piles that are deeply embedded. It is known by practitioners that the reaction supports at the ground surface have no effect on the tension capacity of a helical pile with a depth to helix diameter ratio greater than 5.

Dunnage, as shown in Figures 7.5 and 7.6, is often used to support the jacking apparatus and to spread out the loads on the ground surface. The amount of dunnage required is a function of the soil bearing capacity. Two common ways of arranging dunnage are shown in Figure 7.7. The total area of dunnage required to distribute loads properly and avoid settlement may be estimated from Table 7.1. The table is separated into two major soil types. For the purposes of sizing the required dunnage pad, soft soil is defined as soil with ultimate bearing pressure of approximately 2,000 pounds per square foot (psf) [100 kPa] such as topsoil, wet clay, disturbed sand, and uncompacted fill. Firm soil is defined as soil with ultimate bearing pressure of approximately 10,000 psf [500 kPa] such as stiff clay, dense sand and gravel, and well-compacted fill. Required dunnage area for intermediate or extreme soils can be interpolated or extrapolated. Table 7.1 has columns labeled "A" and "B." These columns give the required minimum dimensions for the two configurations shown in Figure 7.7. Table 7.1 was prepared using the ultimate bearing pressure and does not include a factor of safety. Settlement of the testing system on the order of 1 inch [25 mm] or more may occur using these dimensions but should not affect results.

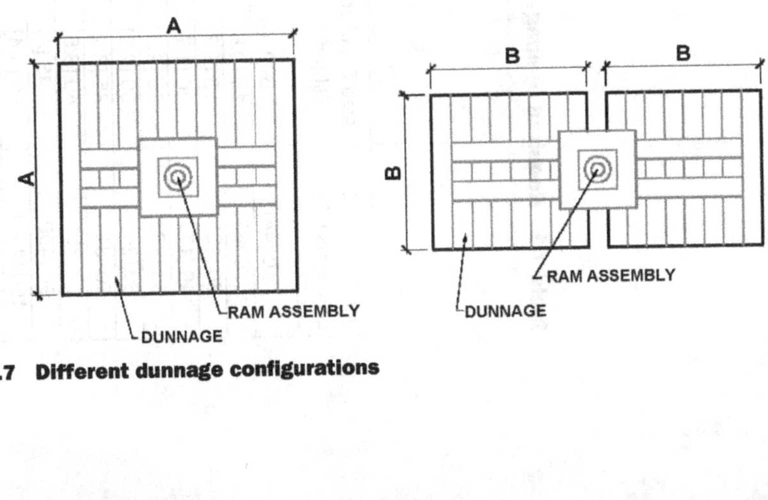

Figure 7.7 Different dunnage configurations

Table 7.1 Required Dunnage Area for Tension Tests

	Soft Soils (q_u = 2,000 psf [100 kPa]) *top soil, wet clay, disturbed sand, uncompacted fill*			**Firm Soils** (q_u = 10,000 psf [500 kPa]) *hard dry clay, dense sand and gravel, well compacted fill*		
Maximum Test Load (kips [kN])	Area Required (sf [m^2])	A (ft [m])	B (ft [m])	Area Required (sf [m^2])	A (ft [m])	B (ft [m])
10 [44]	5 [0.4]	2.2 [0.6]	1.5 [0.3]	1 [0]	1 [0.2]	0.7 [0.2]
20 [88]	10 [0.8]	3.1 [0.9]	2.2 [0.4]	2 [0.1]	1.4 [0.4]	1 [0.2]
30 [133]	15 [1.3]	3.8 [1.1]	2.7 [0.5]	3 [0.2]	1.7 [0.5]	1.2 [0.3]
40 [177]	20 [1.7]	4.4 [1.3]	3.1 [0.6]	4 [0.3]	2 [0.5]	1.4 [0.4]
50 [222]	25 [2.2]	5 [1.4]	3.5 [0.7]	5 [0.4]	2.2 [0.6]	1.5 [0.4]
60 [266]	30 [2.6]	5.4 [1.6]	3.8 [0.8]	6 [0.5]	2.4 [0.7]	1.7 [0.5]
70 [311]	35 [3.1]	5.9 [1.7]	4.1 [0.8]	7 [0.6]	2.6 [0.7]	1.8 [0.5]
80 [355]	40 [3.5]	6.3 [1.8]	4.4 [0.9]	8 [0.7]	2.8 [0.8]	2 [0.5]
90 [400]	45 [4]	6.7 [2]	4.7 [1]	9 [0.8]	3 [0.8]	2.1 [0.6]
100 [444]	50 [4.4]	7 [2.1]	5 [1]	10 [0.8]	3.1 [0.9]	2.2 [0.6]
110 [489]	55 [4.8]	7.4 [2.2]	5.2 [1.1]	11 [0.9]	3.3 [0.9]	2.3 [0.6]
120 [533]	60 [5.3]	7.7 [2.3]	5.4 [1.1]	12 [1]	3.4 [1]	2.4 [0.7]

If an anchor testing system experiences significant differential movement such that loads are no longer concentric or deflection measurements become skewed, the test should be halted and re-set. To reset a test, tensile load should be removed and any permanent deflection of the anchor should be recorded. The dunnage pad should be re-built to offer more support and all loading and measurement fixtures should be repositioned. Any permanent deflection of the anchor during the initial loading sequence should be added to the deflection measured during the second sequence.

7.3 LOADING PROCEDURES

ASTM D1143 and D3689 contain several different loading procedures for static axial compressive and tensile load testing of piles. The loading procedures contained in both standards are identical with the exception of the frequency of load, time, and displacement readings. The various procedures contained in both ASTM documents include maintained cyclic, quick, excess load, constant time intervals, constant rate of penetration, and constant settlement increments. General rules for load increments, hold times, and readings for three of these procedures are shown in Tables 7.2, 7.3, and 7.4. These three methods and a few others are described in more detail below.

The maintained load test procedure (Table 7.2) involves loading the pile in 25 percent increments to 200 percent and holding the test load for a minimum of 12 hrs. After the hold period, the pile is unloaded in 25 percent decrements and a final reading is taken 12 hrs after removing all loads. Time, deflection, and load readings are taken before and after applying each load increment/decrement and at regular intervals during hold times. The standard load test procedure should be considered for helical piles and helical anchors when pile head movements associated with settlement or pullout need to be verified or when long-term creep is suspected.

The cyclic load test (Table 7.3) involves loading and unloading the pile to 50, 100, 150, and 200 percent of the design load. Initial loading increments and final unloading decrements are similar to the maintained test procedure. Cyclic loading and unloading intervals have 20-minute minimum hold times. The test involves 32 total increments/decrements and takes a minimum of 43 hours to complete without counting setup and removal time. Naturally, the cyclic load test should be considered when helical piles or helical anchors are expected to support fluctuating live loads and when a detailed assessment of their deflection characteristics is required.

The quick load test (Table 7.4) involves loading the pile in 5 percent increments until plunging failure or until the capacity of the load frame is reached, whichever occurs first. Each load increment is held for 4 minutes. Readings are taken at 0.5, 1, 2, and 4 minutes after each load increment. After the final test load is reached, load is removed in 5 to 10 equal decrements with rebound readings taken at the same intervals as previously described. The quick test procedure is preferred by contractors, because the load increments can be completed in a few hours compared to several days for the maintained and cyclic load test procedures. The quick load test is a good method to

Table 7.2 Maintained Load Test Procedures (Adapted from ASTM D1143 and ASTM D3689)

	Load[1]	Hold Time	Readings[2]
Loading	15%[3]	NA	Zero
	25%		
	50%	2 hr unless d < 0.01 in/hr [0.25 mm/hr] then stop	Compression: 5, 10, and 20 min, then every 20 min Tension: 2, 4, 8, 15, 45, 60, 80, 100, and 120 min
	75%		
	100%		
	125%		
	150%		
	175%		
Test load	200%	12 hr unless d > 0.01 in/hr [0.25 mm/hr] then 24 hr	Compression: 5, 10, and 20 mins, then every 20 min for 2 hr, every 1 hr for 10 hr, then every 2 hr Tension: same intervals as for loading sequence
Unloading	175%	1 hr	Compression: every 20 min Tension: every 30 min
	150%		
	125%		
	100%		
	75%		
	50%		
	25%		
	0%	12 hr	Compression: after 12 hr Tension: 1, 2, and 12 hr

1 Percent of design load
2 Time, load, and movement before and after application of each load increment/decrement
3 Not specifically referenced in ASTM

verify helical pile or helical anchor capacity when accurate long-term deflections are less of a concern.

The excess load test procedure involves loading and unloading the pile following the maintained load test procedure. Then the pile is reloaded in increments of 50 percent of the design load to the previous maximum test load and thence in increments of 10 percent of the design load until pile failure occurs. This procedure is applied when it is important to evaluate the maximum capacity of the pile rather than just 200 percent of the design load.

The constant time interval test procedure is the same as the maintained load test procedure except loading and unloading of the pile is done in intervals equal to 20 percent of the design load with 1-hour hold times. In some cases, the constant time interval test procedure can shorten the total test time. Another reason to use this method is to simplify the overall test procedure.

Table 7.3 Cyclic Load Test Procedures (Adapted from ASTM D1143 and ASTM D3689)

<table>
<tr><th></th><th>Load[1]</th><th>Hold Time</th><th>Readings[2]</th></tr>
<tr><td rowspan="23">Cyclic loading</td><td>15%[3]</td><td>N/A</td><td>Zero</td></tr>
<tr><td>25%</td><td>Same as maintained procedure</td><td rowspan="32">Same as maintained procedure loading sequence</td></tr>
<tr><td>50%</td><td>1 hr</td></tr>
<tr><td>25%</td><td rowspan="3">20 min</td></tr>
<tr><td>0%</td></tr>
<tr><td>50%</td></tr>
<tr><td>75%</td><td>Same as maintained procedure</td></tr>
<tr><td>100%</td><td>1 hr</td></tr>
<tr><td>75%</td><td rowspan="5">20 min</td></tr>
<tr><td>50%</td></tr>
<tr><td>0%</td></tr>
<tr><td>50%</td></tr>
<tr><td>100%</td></tr>
<tr><td>125%</td><td>Same as maintained procedure</td></tr>
<tr><td>150%</td><td>1 hr</td></tr>
<tr><td>125%</td><td rowspan="7">20 min</td></tr>
<tr><td>100%</td></tr>
<tr><td>50%</td></tr>
<tr><td>0%</td></tr>
<tr><td>50%</td></tr>
<tr><td>100%</td></tr>
<tr><td>150%</td></tr>
<tr><td>175%</td><td rowspan="11">Same as maintained procedure</td></tr>
<tr><td>Test load</td><td>200%</td></tr>
<tr><td rowspan="9">Unloading</td><td>175%</td></tr>
<tr><td>150%</td></tr>
<tr><td>125%</td></tr>
<tr><td>100%</td></tr>
<tr><td>75%</td></tr>
<tr><td>50%</td></tr>
<tr><td>25%</td></tr>
<tr><td>0%</td></tr>
</table>

[1] Percent of design load
[2] Time, load, and movement before and after application of each load increment/decrement
[3] Not specifically referenced in ASTM

Table 7.4 Quick Load Test Procedures (Adapted from ASTM D1143 and ASTM D3689)

	Load[1]	Hold Time	Readings[2]
Loading	5%	4 min minimum, 15 min maximum	Zero
	10%		0.5, 1, 2, 4 mins, then 8 and 15 min if required
	15%		
	20%		
	25%		
	30%		
	35%		
	40%		
	45%		
	50%		
	55%		
	60%		
	65%		
	70%		
	75%		
	80%		
	85%		
	...etc.[3]		
Unload	5 to 10 equal decrements	4 min minimum, 15 min maximum	0.5, 1, 2, 4 mins, then 8 and 15 min if required

1 Percent of anticipated failure load.
2 Time, load, and movement.
3 Continue test until plunging failure occurs or capacity of load frame is reached, whichever happens first.

The constant rate of penetration test procedure is performed such that the rate of pile penetration is held constant. The rate depends on the soil type. The pile is loaded until total pile penetration is at least 15 percent of the average pile diameter. The constant settlement increment test procedure involves applying test load increments such that movement of the pile head is approximately 1 percent of the average pile diameter between each increment. Pile loading is continued until total pile head settlement equals about 15 percent of the average pile diameter. For helical piles, pile diameter shall be taken as the average diameter of the helical bearing plates.

Helical piles generally react quickly to applied loads. The most frequently applied load test procedure for helical piles is the quick load test. Rather than continue increasing loads until plunging failure, the quick load test often is stopped at 200 percent of the design load. Some engineers require that the 200 percent test load increment be held for a minimum of 30 minutes, or until the movement rate decreases to a rate less than 0.01 inch/hour [0.25 mm/hr]. Although the ASTM procedure does not specifically address the issue, it is common to apply a setting load of 10 to 15 percent of the design load to a helical pile or helical anchor prior to taking initial readings.

7.4 INTERPRETATION OF RESULTS

There are several methods for interpreting the capacity of a pile obtained from an axial compression or tension load test. The basic definition of ultimate capacity is the highest load that can be applied to a pile or anchor until deflection continues without application of additional loads (e.g., plunging resistance). This definition is purely strength based and does not limit pile head deflection. Many structures are sensitive to movement and require limitations to total movement. The effective stiffness of the foundation also may be of interest. For this reason, the capacity of piles often is limited based on deflection.

One method of interpreting pile load test data is simply to define the capacity as the load at a predefined amount of deflection. A historic method is to limit pile movement to 10 percent of the pile diameter. Maximum deflection limits between 3/4 inch [19 mm] and 1.5 inch [38 mm] have been published in some local building codes. Some professionals limit the deflection at the design load rather than the ultimate load. The maximum deflection at the design load depends on the sensitivity of the structure to movement, the desired rigidity of the foundation, and local experience. Typical values range from 3/8 inch [10 mm] to 1 inch [25 mm]. Limiting pile capacity based on a minimum factor of safety relative to the ultimate capacity at failure and including a criterion for maximum tolerable total pile head movement at the design load satisfies the demands of most structures very well.

Pile capacity determined from a predefined maximum deflection can depend mainly on the structural properties and elasticity of the pile and can have less to do with pile-soil behavior. In addition, load test equipment utilized in pile load testing is limited in the amount of load that can be applied to the pile and often cannot reach the full ultimate capacity of the pile. This makes it difficult to compare load test capacities from different sites and to know the true factor of safety against bearing or pullout failure. As a result, practitioners and academics have introduced a number of pile capacity interpretation methods. Some methods were established to decouple the effect of pile shaft stiffness from soil resistance, others separate side shear from end bearing, and still others were created to try to better understand pile-soil behavior.

Fleming, Weltman, Randolph, and Elson (1985) state that the adhesion along a pile shaft is mobilized in very small deformations, typically less than 0.2 inch [5 mm].

End bearing resistance is not fully mobilized until large settlements occur, up to 20 percent of the base diameter in coarse-grain soils and 10 percent of the base diameter in fine-grain soils. Due to their unique shape with slender shafts and large bearing elements, ordinary (nongrouted) helical piles behave more akin to end bearing piles. As such, the method of load test interpretation used with helical piles should be one that allows for full mobilization of end bearing resistance. Otherwise, the true capacity of helical piles will be underestimated.

Many practitioners in the helical pile industry have been advocating the use of a modified Davisson offset method for helical piles. According to this method, the ultimate capacity of a helical pile is defined as the load causing a net deflection equal to 0.1 times the average helical bearing plate diameter (ICC-ES, 2007). Net deflection is defined as the total deflection at the pile head minus elastic shortening or lengthening of the shaft. The elastic change in shaft length may be computed from the well-known equation

$$\delta = \frac{Pz}{A_g E} \tag{7.1}$$

Where

- δ is deflection
- P is the load applied to the pile
- z is the length of pile shaft
- A_g is the gross cross-sectional area of the pile shaft, and
- E is the modulus of elasticity of the shaft steel

Elastic shortening/lengthening also may be determined from the rebound of the pile head upon removal of the axial load. Determining elastic shortening/lengthening in this way takes into account the elastic movement of the helical bearing plates and soil rebound.

An example of the modified Davisson method is contained in Figure 7.8. The data represented by diamond symbols in the figure were obtained from an actual load test on a helical pile with 3-inch- [76-mm-] diameter shaft and three helical bearing plates with 12-inch [305-mm] average diameter. The pile bottomed in glacial till of varying consistency. The inclined dashed lines in the figure were drawn using Equation 7.1. The second inclined dashed line intersects the x-axis at a value corresponding to 0.1 times the average helical bearing plate diameter. The intersection of this line with the load-displacement curve represents the load limit from the modified Davisson offset method.

Many common helical piles have an average helix diameter of 12 inches [305 mm] and typically exhibit about 3/8-inch [10-mm] elastic shortening/lengthening under test loads equal to 200% of the design load. This means that net deflections are on the order of 0.8 inch [20 mm] at test loads and 0.4 inch [10 mm] at design loads under the modified Davisson criterion, which is reasonable. The modified Davisson criteria may be unreasonable for helical piles with larger-diameter helical bearing plates. In

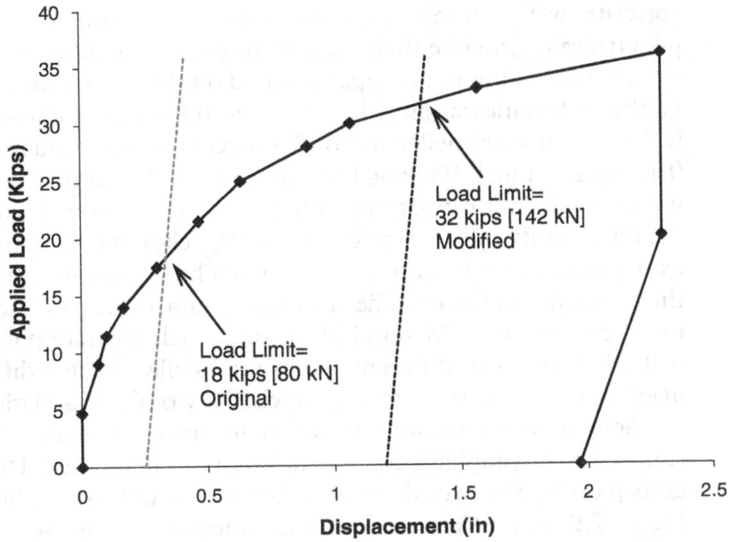

Note: 1 kip= 4.45 kN, 1 in= 25.4 mm

Figure 7.8 Load limit based on Davisson (1972) modified Davisson (ICC-ES, 2007) methods

these cases, an addition constraint involving the allowable deflection under design loads should be considered.

It is interesting that the slope of the dashed lines in Figure 7.8 are slightly steeper than the rebound curve for the example load test. The additional movement of the pile head during unloading may be due to in part to soil rebound. It also may be due to release of elastic flexure of helical bearing plates. The elastic behavior of helical bearing plates in soil is not discussed in literature and requires further study.

7.5 OTHER INTERPRETATIONS

Although the modified Davisson method is the most commonly used method of interpretation for helical pile load tests, there are many methods of which the helical pile designer needs to be aware. For example, there are several methods for finding the point where shaft adhesion is fully mobilized. The original Davisson method is one such method and tries to estimate this point by compensating for pile stiffness (Fellenius, 2001b). The method consists of drawing a line with a slope equal to the elastic lengthening/shortening of the pile given by Equation 7.1 offset by a value of 0.15 inch [4 mm] (Davisson, 1972). The point at which this offset line intersects the load-deflection curve is taken as the load limit. An example of this method is shown by the first inclined dashed line in Figure 7.8. The original Davisson offset method significantly underestimates the ultimate capacity of helical piles. The method is typically not appropriate for helical piles or other end bearing elements because they require much greater deflection to mobilize their full strength. The original Davisson method correlates well with wave equation analysis for dynamic testing of driven piles. Driven piles typically generate their capacity through shaft adhesion.

Another definition of capacity based on shaft adhesion is that given by the Federal Highway Administration (FHWA-HI-96-033), which states that the capacity of a pile is the load at a net deflection of 0.8 percent of the average diameter of the pile plus 0.15 inch [4 mm]. If applied to the load test data shown in Figure 7.8, this method would result in a capacity of only 9 kips. This method would again underestimate the full capacity of helical piles. However, helical pile designers should be aware of its existence and may need to plan for much higher conservatism in their calculations if this method is enforced. The structural engineer and architect need to be aware that methods such as FHWA and Davisson, which separate pile stiffness from allowable deflection, result in different values total deflection for different pile types and are inappropriate criteria for specifying elasticity of the foundation.

Sometimes trends are difficult to discern when analyzing load test data. For this reason, several graphical constructs have been developed. DeBeer (1967/1968) suggests plotting the load-deflection data in a double logarithmic diagram as shown in Figure 7.9. These data are from the same load test as used in the example described previously. DeBeer states that if the yield load of the pile was reached in the test, two line approximations will appear. The intersection of these lines is taken as the yield load. For the example helical pile test data, the DeBeer yield load falls between the values obtained using the original Davisson and modified Davisson approaches.

Other graphical methods are the Fuller-Hoy and the Hansen 90 percent methods. In the Fuller-Hoy method, ultimate load is obtained by plotting a line tangent

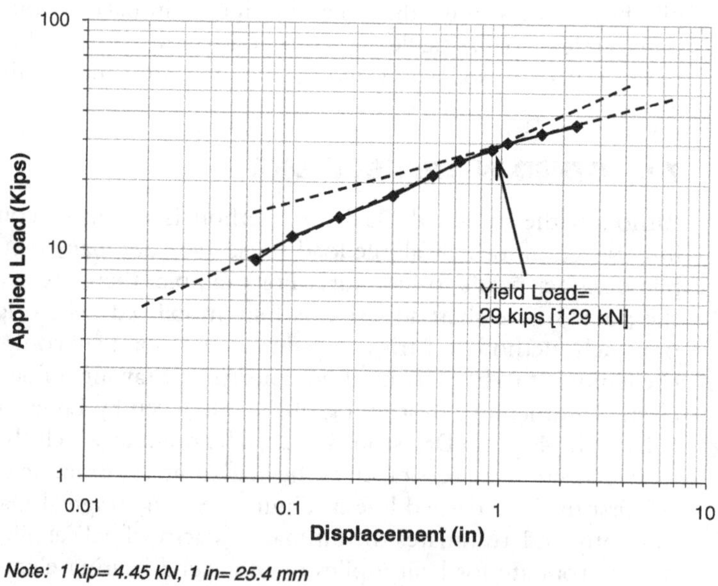

Figure 7.9 Yield load based on DeBeer (1967/1968) method

to the load curve at a slope of 0.14 mm/kN. Intersection of the load curve with the line is considered as ultimate load of the pile (Duzceer and Saglamer, 2002). According to the Hansen 90 percent method (Brinch-Hansen, 1963), the load limit is where the deflection at that load is two times greater than the deflection at 90 percent of that load. The ultimate capacity is found graphically through trial and error.

Still other interpretations of load test data are based on hyperbolic, parabolic, and other polyregression analysis. The Brinch-Hansen method known as the Hansen 80 percent criterion is an example. According to this method, the failure load is the load at which pile head movement is four times that obtained at 80 percent of that load. Normally, the Hansen criterion agrees well with the intuitively perceived "plunging failure" of the pile (Fellenius, 2001b). The Hansen 80 percent load can be found graphically or mathematically. To find the Hansen 80 percent load mathematically, the load-deflection data is plotted such that the square root of pile deflection divided by the load is on the y-axis and the corresponding deflection values are plotted on the x-axis as shown in Figure 7.10. Here again, the load test data are the same as previous plots. As can be seen, the load test data exhibits some variation at the beginning of the test and then approaches a straight-line approximation. A linear regression analysis is performed on the straight-line portion. The ultimate capacity, P_u, is determined from

$$P_u = \frac{1}{2\sqrt{C_1 C_2}} \tag{7.2}$$

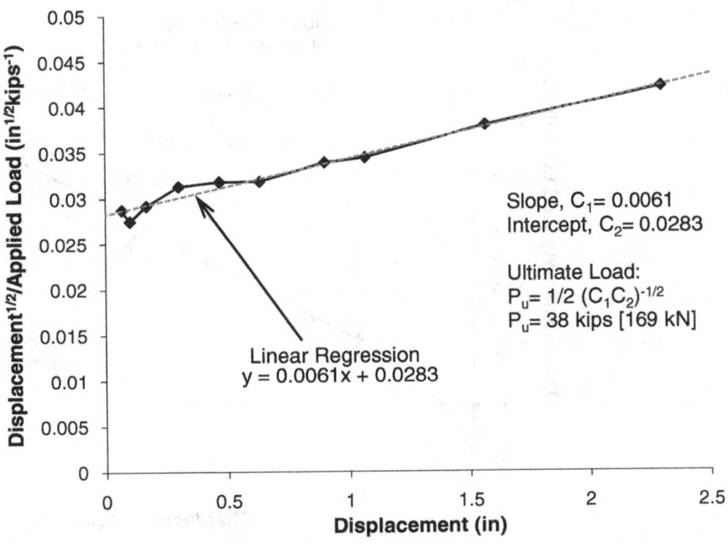

Note: 1 kip= 4.45 kN, 1 in= 25.4 mm

Figure 7.10 Ultimate capacity based on Hansen (1963) method

Where
C_1 is the slope of the linear regression, and
C_2 is the y-intercept.

When the analysis is applied to the sample helical pile load test data in Figure 7.10, the resulting ultimate capacity is slightly higher than the load limit found using the modified Davisson method. The Hansen method also can be used to model the load-deflection curve. If the load test date were to plot as a straight line throughout the test, the ideal load curve would be represented by

$$P = \frac{\sqrt{\delta}}{C_1\delta + C_2} \tag{7.3}$$

Where
P is the axial load on the pile and
δ is pile head deflection.

Fellenius (2001b) explained that Chin proposed another method of pile load test analysis based on the work of Kondner, although the original references for the work were not given. This method, termed the Chin-Kondner method, consists of plotting the measured pile head displacements divided by the corresponding loads on the y-axis and the pile loads on the x-axis, as shown in Figure 7.11. In a typical load test, the values will fall along a straight line after some initial variation. The inverse slope of this

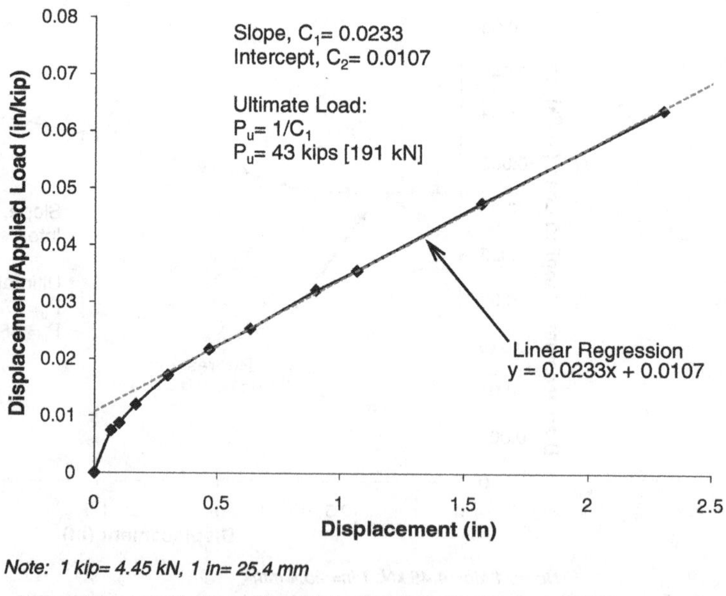

Figure 7.11 Ultimate capacity based on Chin-Kondner (Fellenius, 2001b) method

line is the Chin-Kondner extrapolation of the ultimate load given by

$$P_u = \frac{1}{C_1} \tag{7.4}$$

Where

C_1 is the slope of the linear regression.

Applying this analysis to the example load test data shown in the figure results in an ultimate capacity of 43 kips [191 kN]. This result is higher than any of the previous methods.

According to Fellenius (2001b), if a weakness developed in the pile during the load test, the Chin-Kondner method would show a change in slope of the straight-line approximation so there is merit in plotting the readings in this way as the test progresses. Fellenius also states that the straight-line approximation in the Chin-Kondner method typically materializes after the original Davisson offset load limit has been reached, so the test needs to be run out to large deflections. Otherwise, the linear approximation could be lost, and a false value could be obtained. Like the Hansen method, the Chin-Kondner method also can be used to model the load-deflection curve. The ideal Chin-Kondner load-deflection curve is given by

$$P = \frac{\delta}{C_1\delta + C_2} \tag{7.5}$$

Decourt (1999) proposed a method similar to but opposite the Chin-Kondner method. To apply the method, each load increment is divided by the corresponding pile head displacement and plotted on the y-axis and the pile load is plotted on the x-axis of a graph. An example of this method is shown in Figure 7.12. The example load test data are the same as previously discussed. Here again, the load test data are supposed to approximate a straight line at larger values of displacement. A linear regression analysis is performed on the straight-line portion. The ultimate capacity of the pile is given by

$$P_u = -\frac{C_2}{C_1} \tag{7.6}$$

Where

C_1 and C_2 have been defined previously.

The Decourt method suggests an ultimate capacity very similar to the Chin-Kondner method for the example load test data. The ideal Decourt load-deflection curve is given by

$$P = \frac{C_2\delta}{1 - C_1\delta} \tag{7.7}$$

The relationships between P and δ in Equations 7.3, 7.5, and 7.7 can be used in more sophisticated structural modeling software to approximate the deflection of

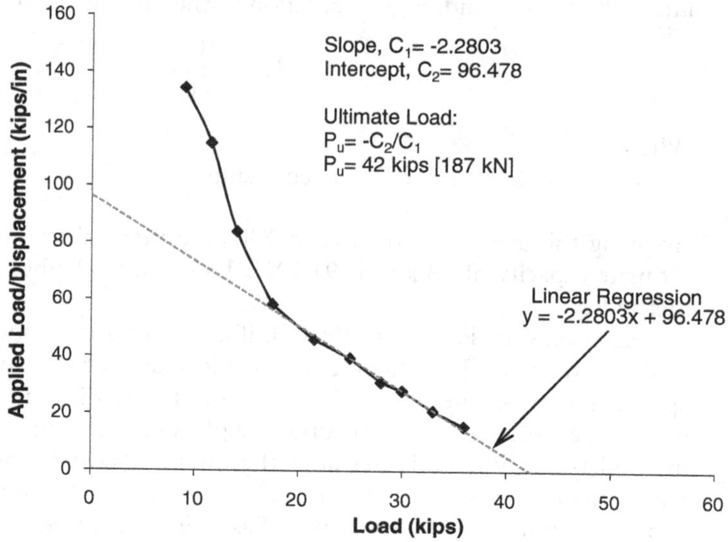

Note: 1 kip= 4.45 kN, 1 in= 25.4 mm

Figure 7.12 Ultimate capacity based on Decourt (1999) method

pile supports. The modeling parameters C_1 and C_2 can be estimated initially based on previous load test results on piles similar to those anticipated for the project and in similar subsurface conditions. The modeling parameters can be confirmed in a field pile test program.

The ideal load-deflection curves based on Hansen, Chin-Kondner, and Decourt methods for the example load test data are shown in Figure 7.13. As can be seen, all three methods provide a very close approximation of the actual helical pile load test data. Pile designers and structural engineers should not be misled by the closeness of the approximations in this example. Often helical pile load test data can exhibit a variety of shapes. Some helical pile tests exhibit a bilinear load-displacement curve. End bearing helical piles show a very linear initial load curve followed by a sharp peak and subsequent decrease in capacity. The shape of the load-deflection curve depends on properties of the pile and surrounding subsurface conditions. The ideal curves presented here may not match other load tests as closely. It should also be observed that the lower half of the load-deflection curve is nearly linear below the allowable load if a factor of safety of 2.0 is used. As is discussed in Chapter 12, the initial load deflection of helical piles can be modeled more simply using a linear spring analogy.

The ultimate capacities determined for the example helical pile load test data using the Chin-Kondner and Decourt methods are higher than the maximum test load of 36 kips [160 kN]. Fellenius (2001b) warns that extrapolation of results is unadvisable. The maximum test load should be used as the ultimate capacity in these cases. One exception may be proof load tests where the ultimate capacity has been verified to a higher load in a number of other tests.

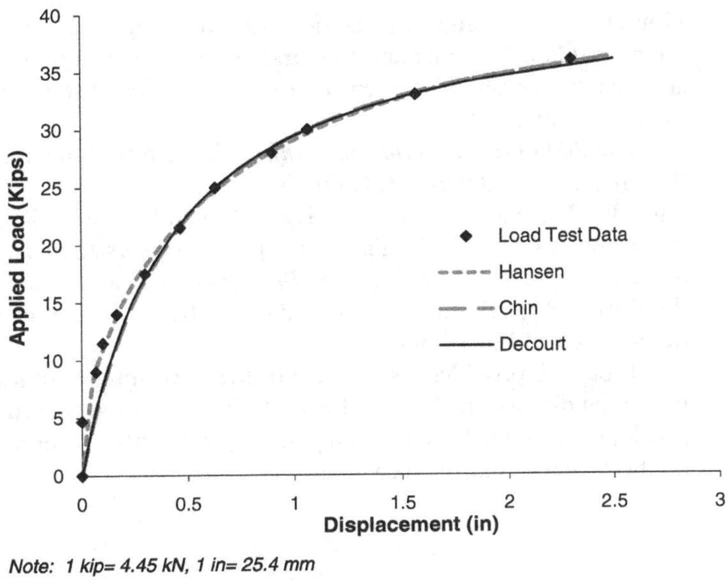

Figure 7.13 Various methods for pile deflection modeling

Duzceer and Saglamer (2002) compared 12 different methods of interpreting the capacities of piles from 25 static load tests. Methods of interpretation included the original Davisson, Brinch-Hansen 80 percent, Chin-Kondner, Shen-Niu, DeBeer, Housel, Mazurkiewicz, Fuller-Hoy, Brinch-Hansen 90 percent, Corps of Engineers, Butler & Hoy, and Tangent Intersection methods. Pile types included cased augered piles and driven precast and tubular steel piles. All piles derived their strength from side friction. Soil types varied but primarily consisted of silty clay and occasional silty to clayey sands. Duzceer and Saglamer compared the capacity interpretations with predicted capacity based on the FHWA method and found the standard Davisson, Shen-Niu, DeBeer,

Table 7.5 Load test interpretation methods

Category	Description	Load Test Interpretation Methods
Strength	Minimum factor of safety	Ultimate strength
Serviceability	Deflection limits	2007 NYC Building Code
End bearing	Net deflection 10% of pile base diameter	Modified Davisson (ICC-ES, 2007)
Shaft Adhesion	Net deflection 0.2 inch [4 mm]	FHWA (1996); original Davisson (1972)
Graphical	Various graphical constructs	DeBeer (1967/1968); Fuller-Hoy; Hansen 90% (Brinch-Hansen, 1963)
Extrapolation	Hyperbolic, parabolic, and other polyregression analysis	Chin-Kondner Decourt (1999); Hansen 80% (Brinch-Hansen, 1963)

Housel, Corps of Engineers, Butler-Hoy and Tangent Intersection methods were conservative. Chin-Kondner and Mazurkiewicz methods were unconservative. Fuller-Hoy and Brinch Hansen 90 percent methods were closest to predicted capacities based on the FHWA method.

The *2003 International Building Code* permitted the use of the original Davisson, Brinch-Hansen 90 percent, Chin-Konder, and other methods approved by the building official. Since Duzceer and Saglamer (2002) showed the Chin-Konder method to be unconservative, the Pile Driving Contractors Association proposed an amendment for the *2006 International Building Code*, and Chin-Konder was replaced by the Butler-Hoy criterion. Commentary on the code change stated that extrapolation methods should be avoided.

Table 7.5 provides a summary of load test interpretation methods based on the foregoing discussion. The modified Davisson method is used most frequently to interpret load tests on helical piles. Depending on specific requirements of the project, other methods may be considered.

Chapter 8

Reliability and Sizing

The ultimate capacity determined by limit state analysis (Chapters 4 and 5) and torque correlations (Chapter 6) must be divided by a factor of safety to obtain the working or allowable capacity. This chapter contains a brief discussion of factors of safety used in foundation design with recommendations for helical piles.

Once it is understood how to compute the allowable capacity of a helical pile, it is possible to size helical bearing plates to suit particular design loads. A simple graphical method for helical pile sizing is presented. Several commercially available software packages for helical pile sizing are discussed. One of the challenges of sizing helical piles appropriately is the selection of geotechnical criteria. Some simple probabilistic soil mechanics techniques are presented that can be employed for this purpose. Despite a thorough geotechnical investigation and the best statistical analysis of the data, ground conditions can vary. It is often necessary to conduct a field test program that can be used for final helical pile sizing and selection. Examples of field adjustments for final pile sizing are discussed.

The chapter concludes with a discussion of the reliability of satisfactory helical pile performance. It is shown that combining limit state methods for helical pile sizing and capacity-to-torque relationships for field verification assures reasonable reliability from a geotechnical standpoint.

8.1 FACTOR OF SAFETY

Thus far, this book has focused on the determination of ultimate capacity of a helical pile. In practice, the ultimate capacity must be divided by an appropriate factor of safety to obtain the allowable capacity to be used in design. The allowable capacity (a.k.a. working capacity), P_a, of a helical pile is computed simply from

$$P_a = \frac{P_u}{F_S} \tag{8.1}$$

Where

P_u is ultimate capacity based on theoretical calculations, installation torque correlations, or load tests, and

F_S is the factor of safety

A factor of safety of 3.0 is commonly used in bearing capacity calculations for footing foundations, drilled shafts, and augered cast-in-place piles. A larger factor of safety is required where direct observation or measurement of the bearing stratum at each bearing element is limited. However, when foundation installation includes an indirect measurement of soil strength at the foundation depth, a smaller factor of safety is permissible. A traditional example of this is pile driving, where a much lower factor of safety is often allowed.

Load tests are one way to improve bearing capacity predictions and allow for a lower factor of safety. According to Fellenius (2001b), practice has developed toward using a range of safety factors depending on the load test program. Where pile design is based on load tests conducted on piles that are not necessarily the same type, size or length as those which will be used for a project, a high safety factor, usually 2.5, is used to account for the unknowns. When load tests are performed to verify final pile design, such that tests are conducted on piles intended for the project by the actual installation contractor, a factor of safety of 2.0 is common. Fellenius (2001b) goes on to say that lower factors of safety are warranted when frequent proof tests are incorporated into quality control and on sites where limited variability is confirmed by detailed site investigation and quality assurance observations.

The installation torque of helical piles provides an indication of soil strength at the depth of the helices as discussed in Chapter 6. Typically, a factor of safety of 2.0 is used in helical pile design when capacity is verified through torque correlations. A factor of safety as low as 1.5 may be used when a significant percentage of helical piles are load tested. For example, some earth retention projects may involve proof tests on a majority of helical anchors. A larger factor of safety may be appropriate when installation torque is not utilized for capacity verification, when load tests are omitted, or for nonconforming helical piles (see ICC-ES, 2007 and Chapter 6). Nonconforming helical piles are those wherein the capacity-to-torque ratio has not been proven.

The use of a standard factor of safety of 2.0 for helical piles is justified through statistics. The comparison of measured and predicted capacity for 112 load tests shown previously in Figure 4.18 may be approximated by a normal distribution. The standard deviation of the normal distribution is 0.51. This indicates that if a safety factor of 2.0 is used with theoretical predictions of helical pile capacity, there is an 84 percent probability that the actual capacity measured in the field will exceed the theoretical prediction. Combining theoretical predictions with correlations between capacity and installation torque can improve the probability of the piles exceeding the required capacity to very high certainty, as discussed in the last section of this chapter.

8.2 HELIX SIZING

The primary reason to learn how to calculate the theoretical capacity of a helical pile is so that one can properly size the helical bearing plates for the anticipated loads and subsurface conditions. All of the tools developed in previous chapters allow this ability. Proper sizing of a helical pile may take multiple iterations until the appropriate number and size of helical bearing plates is computed. Several techniques available in industry are to aid the designer in the selection of the proper lead configuration. Design charts are provided herein for preliminary sizing of helical bearing plates. Two proprietary software programs are described in the next section.

Provided helical piles are manufactured with sufficient helix spacing such that the limit state is defined by the individual bearing method, the capacity of a helical pile can be predicted using Equation 4.1. Ultimate bearing pressure can be estimated from blow count measurements using Equations 4.15, 4.20, and 4.21 for fine-grain soils, coarse-grain soils, and weathered bedrock, respectively. Using these techniques, the ultimate capacities of several common helical bearing plate configurations are shown in Figures 8.1, 8.2, and 8.3. The primary soil or rock category for which the charts

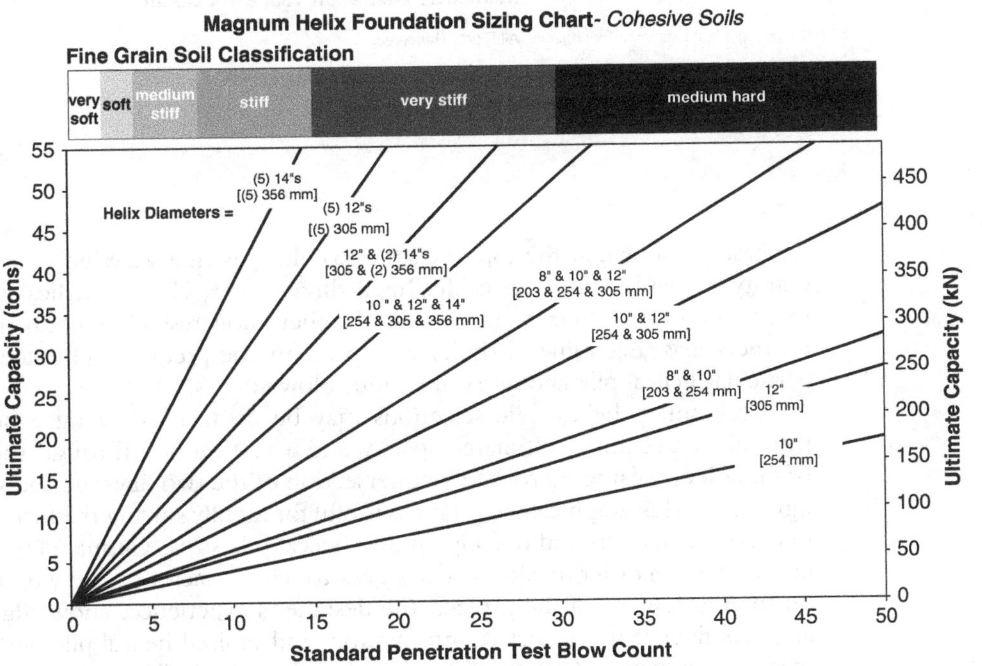

Figure 8.1 Helical pile configurations in fine-grain soils

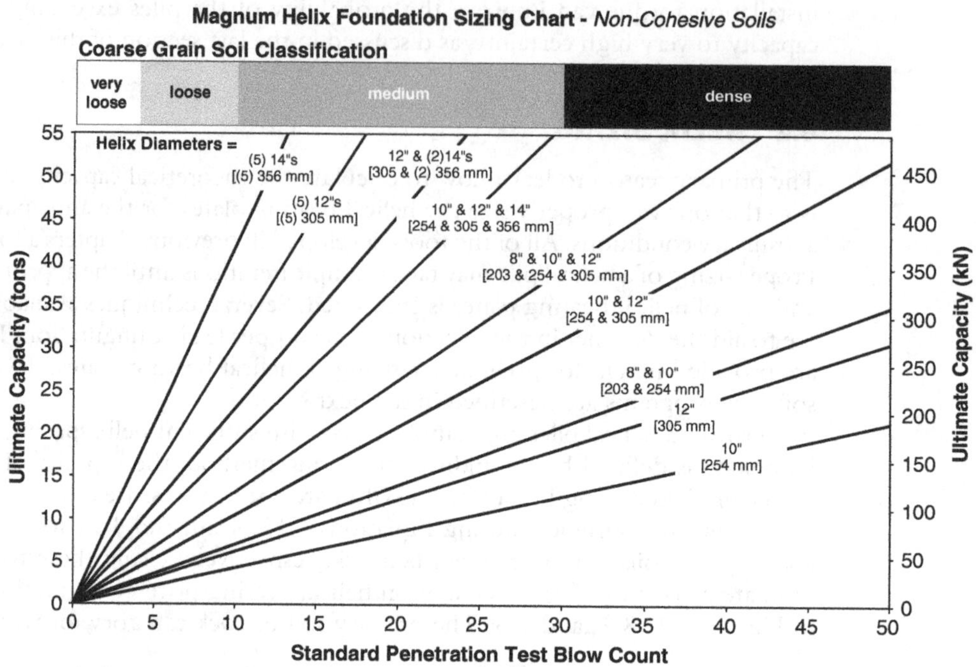

Figure 8.2 Helical pile configurations in coarse-grain soils

are based is shown at the top of each chart along with a scale indicating consistency, density, or hardness as applicable. In all three charts, ultimate helical pile capacity is represented on the y-axis and standard penetration test blow count is represented on the x-axis. Each line in the chart represents the predicted ultimate capacity of a particular helical pile geometry at various blow counts.

Preliminary helical pile selections may be made by drawing a horizontal line through the required ultimate capacity and a vertical line through the blow count of the soil or weathered rock. The intersection of the two lines indicates the pile configuration. This graphical technique is useful for rapidly solving the formulae provided in previous chapters and quickly finding helical pile sizes. Ultimate capacity is shown in the charts in order to allow the designer to select a factor of safety to use depending on the importance of the project, the designer's experience, knowledge of the local area, confidence in subsurface information, and desired helical pile performance and reliability. A factor of safety of 2.0 is typical for helical piles when capacity is verified through torque correlations.

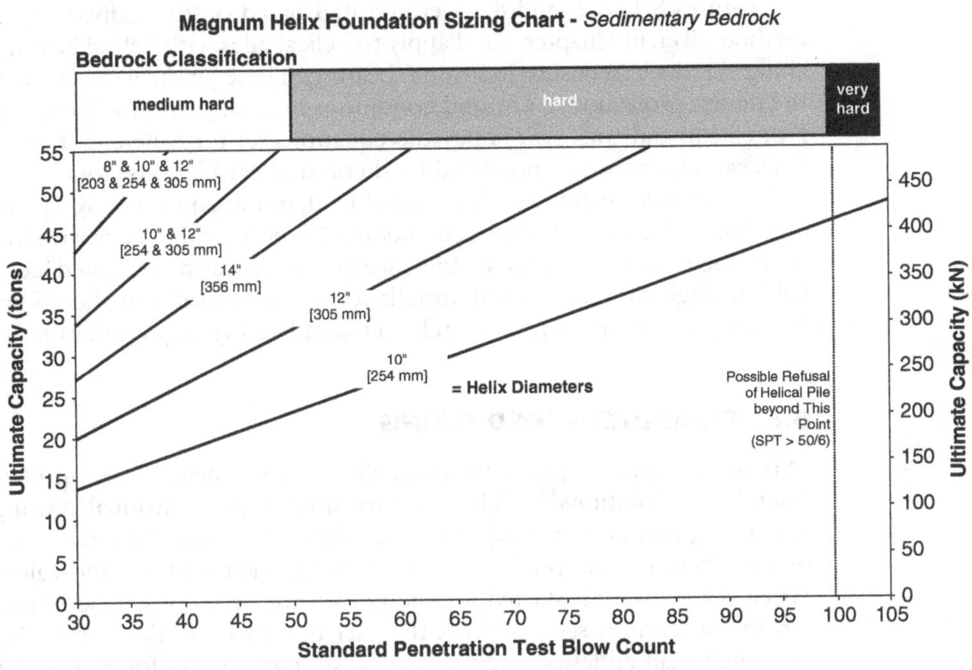

Figure 8.3 Helical pile configurations in weathered bedrock

Example 8a

Problem: Estimate the size of a helical pile required to support project design loads.

Given: Helical piles are required to support a design load of 15 tons [133 kN] with a factor of safety of 2.0. A project soil report indicates a possible bearing stratum is coarse-grain soil (sand) with a SPT blow count of 30.

Answer: The appropriate chart to use in coarse-grain soils is Figure 8.2. Begin by drawing a vertical line in the chart through a blow count of 30, then draw a horizontal line through an ultimate capacity of 2(15)=30 tons. The vertical and horizontal lines intersect at a pile configuration with 10-inch- and 12-inch- [254- and 304-mm-] diameter helical bearing plates. This is the recommended size. Torque should be used to verify capacity in the field.

Figures 8.1, 8.2, and 8.3 were created based on the individual bearing capacity method given in Chapter 4 and apply to helical piles with helical bearing plates spaced along the shaft to ensure individual bearing. These charts were created as a guide to the design professional. Ground conditions may vary with location, depth, and time. Not all soil and rock types obey the equations set forth herein. The performance of a helical pile cannot be predicted by theoretical calculations alone. The final size of a helical pile used in construction should be determined by a design professional based on laboratory and field measurements and his or her experience with local conditions. As is discussed in Chapter 6, the capacity of helical piles should be verified in the field through correlations with installation torque. Load tests should be performed as necessary to validate torque correlations and bearing or pullout capacity.

8.3 COMPUTER-AIDED SIZING

One of the software programs available for sizing helical bearing plates is Ramjack Foundation Solutions™. The software utilizes the individual bearing method for calculating ultimate axial capacity. The software is organized in a series of modules: project/options, soil profile, geometric data, anchor input, and calculation results. When the software is started, the design wizard begins with a job information input screen. An options screen allows the user to customize the analysis by including or excluding shaft adhesion and mechanical strength checks for Ram Jack products.

Soil data is input in a tabular format as shown in Figure 8.4. The software allows up to 10 soil stratums. Each stratum is defined by cohesion, adhesion coefficient, angle of internal friction, coefficient of external friction, and unit weight.

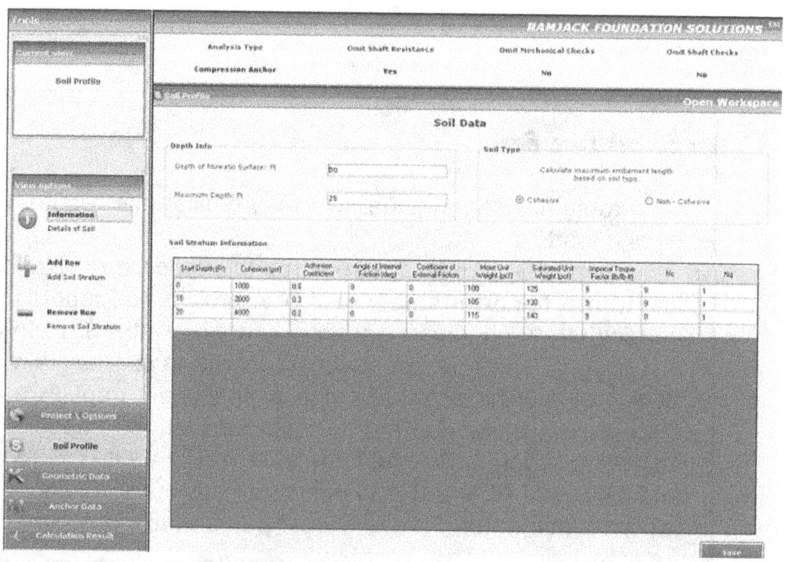

Figure 8.4 Ramjack Foundation Solutions™ soil profile input window

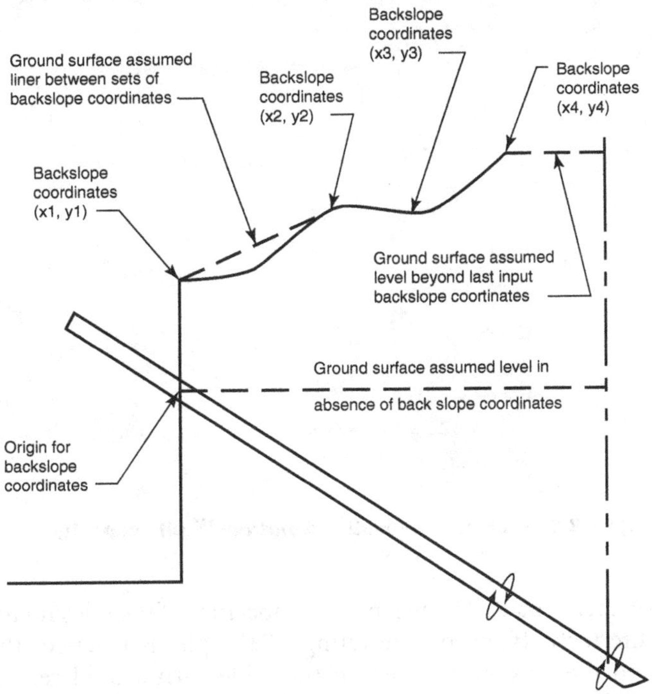

Soil strata start depths must be input relative to the highest input point on the backslope (x4, y4 in illustration; anchor insection point in absence of input backslope coordinates).

Figure 8.5 Ramjack Foundation Solutions™ geometric reference image

Capacity-to-torque ratios and bearing capacity factors for deep local shear are built in as defaults but may be overridden. After a soil profile is constructed, the next step is to enter geometric data. The reference image for the geometric data is shown in Figure 8.5. The software allows the user to enter in the inclination of the pile and the back slope of the overburden if the pile head is not at the surface.

The anchor input screen allows the user to choose the helical pile diameter and helix configuration. Up to six helical bearing plates can be placed on the pile. The software automatically places the helical bearing plates from smallest to largest starting from the tip of the pile. The helical bearing plates are spaced apart three times the diameter of the smallest helix. This finishes the design wizard phase.

A screen shot of the calculation results is shown in Figure 8.6. The software provides a graph of installation torque and theoretical capacity. Capacity is based on Terzaghi's bearing capacity equation and the individual bearing method. Installation torque is estimated by dividing the capacity by the input or default capacity-to-torque ratio. Results are not displayed until the minimum embedment depth has been reached. The program calculates both tension and compression but does not calculate buckling

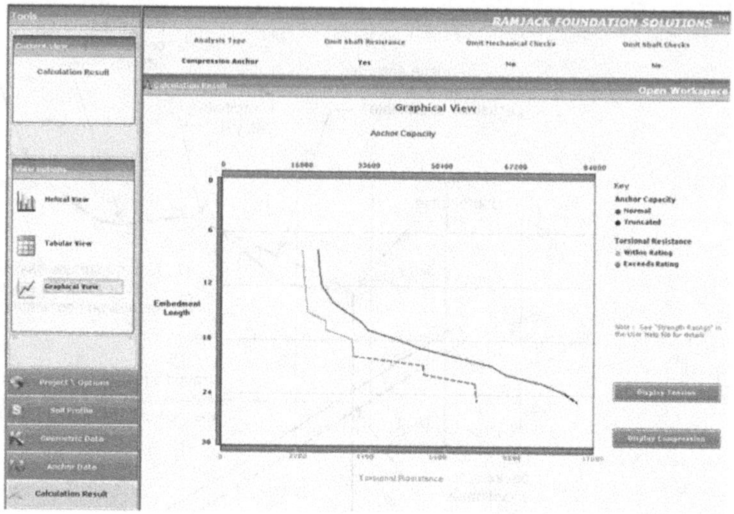

Figure 8.6 Ramjack Foundation Solutions™ pile capacity and installation torque results

or lateral loads. The mechanical capacities of the pile and helical plates are programmed internally. If the torque rating of the pile is reached, the program generates a flag and the torsion numbers in the table turn a bold red. If the compression capacity of the pile or plates is exceeded, the numbers under ultimate capacity turn a bold black.

Another software program available for sizing helical bearing plates is HeliCAP® Helical Capacity Design Software by the Hubbell/Chance company. This software also uses the individual bearing method to estimate capacity. An application and some of the features of this software are demonstrated in the following example.

To begin an analysis in HeliCAP®, the user is required to enter a soil profile made up of three possible soil types. Clay soil utilizes cohesion only. Sand soil is applicable for a soil with angle of internal friction only. A mixed soil type may be used for soil with both friction and cohesion. The program calculates cohesion and friction automatically for the first two of these types. The user must enter both cohesion and a nonzero friction angle for mixed soil type. The software does not list bedrock as a possible soil stratum, so the user must elect to use one of the other soil types to simulate a soft bedrock, such as a claystone, siltstone, or weak sandstone. The user has the ability to override any of the automatically computed values of friction, cohesion, unit weight, or the bearing capacity factors.

A soil type must be entered for a depth of zero feet. From there, sample information is entered. The software assumes the soil type and properties entered continue until the next sample is input. In the example shown in Figure 8.7, a mixed fill material is shown at a depth of 0 feet and 4 feet. A blow count of 12 blows per foot, 300 pounds per square foot (psf) [14 kPa] cohesion, 25 degrees angle of internal friction, and a unit weight of 90 pounds per cubic foot (pcf) are entered manually for these

Soil Profile (click on profile icon in toolbar to close)									
	Depth (ft)	SoilType	N	Cohesion (psf)	Angle of internal friction degrees	Nq-sand bearing capacity factor	In-situ Unit weight (pcf)	Nc-clay bearing capacity factor	Bond Value (psf)
	0	3-Mixed F	12	300	25	7.01	90	9	0
	4	3-Mixed F	12	300	25	7.01	90	9	0
	9	Sand	15	0	31.6	16.14	105		0
	19	Clay	6	750	0	0	92	9	0
	24	Clay	7	875	0	0	94	9	0
	29	Clay	75	9375	0	0	125	9	0
▶	34	Clay	100	12500	0	0	125	9	0
*									

Figure 8.7 HeliCAP® soil profile input window

two depths. The unit weight is total moist or saturated weight. The software accounts for groundwater to calculate effective unit weight.

In the example, the next three samples of clay and sand are entered with blow count values. The software estimates cohesion and friction respectively for these samples. The software also assigns bearing capacity factors for each sample. The last two samples in the example simulate samples of claystone bedrock with blow counts of 50/8" (75 blows/ft) and 50/6" (100 blows/ft), respectively. The software automatically calculates cohesion and the other properties. Note that the software estimated a value of 140 pcf for the unit weight. This was overridden by entering a value of 125 pcf for both samples.

The next step in setting up a HeliCAP® model is to toggle the soil profile input window off by clicking on the profile button. The soil profile resulting from the samples just entered is shown in Figure 8.8. The software automatically shows a basic pile in this view. The properties of the pile can be modified by clicking on the series button and using the pop-up screen. The software allows analysis of only Chance square shaft (SS) and round shaft (RS) products. For the example, a SS175 is selected with 8- and 10-inch-diameter, ⅜-inch-thick, grade 80 helical bearing plates. There is an option on the pop-up screen to calculate bearing, friction, or both. The friction method is used to calculate the capacity of piles with grouted shafts. This is discussed in Chapter 15. For this example, the button for bearing only is selected.

The HeliCAP® software allows the user to set the installation angle, datum depth, and pile length. In the example, the default angle of 90 degrees (vertical) is shown. The datum depth is used to adjust the starting point for top of the helical pile. For many jobs, the tops of the helical piles are located below existing grade to allow for various types of below-grade construction. In the example, the datum depth is set to 3 feet to account for an anticipated below-frost-grade beam. The pile length is set to 35 feet and the "calculate now" button is pressed.

The results are shown in Figure 8.9. Estimated pile ultimate capacity is shown in the middle graph. As can be seen, the software predicts an ultimate capacity of approximately 90 kips at a depth of roughly 36 feet below the ground surface. At that

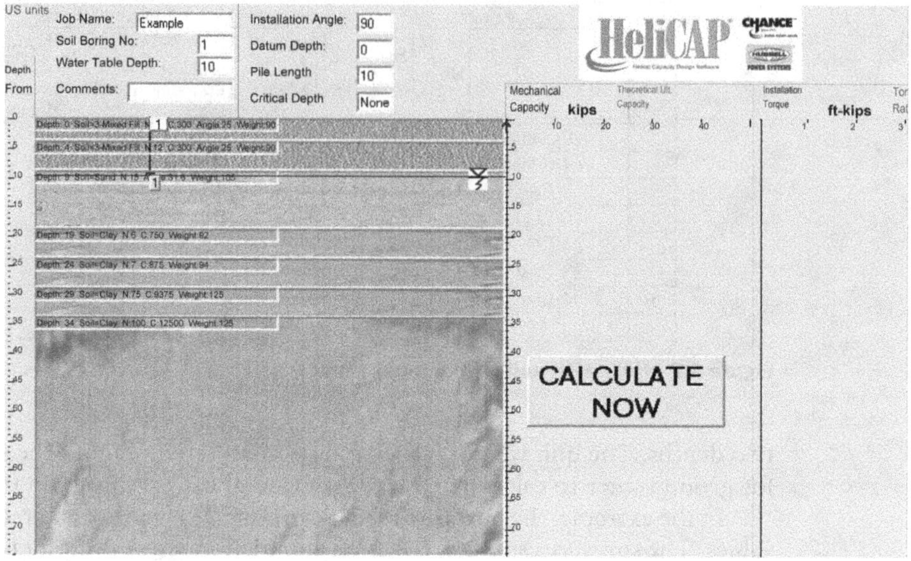

Figure 8.8 HeliCAP® graphic soil profile

point, the software shows that the maximum mechanical strength of the product is exceeded as shown by the gray line. Anticipated installation torque at various depths is shown in the graph to the right. Through a method of trial and error, the designer can use the software to select an appropriate-size helical pile.

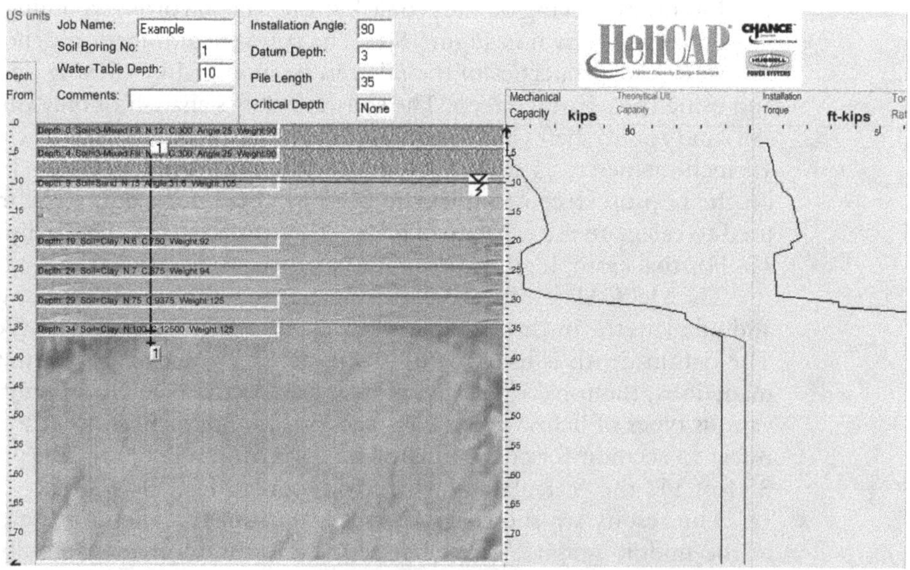

Figure 8.9 HeliCAP® helical pile capacity and installation torque results

8.4 STATISTICS

One of the challenging aspects in selecting helical pile configurations or working in geotechnical design in general is the selection of appropriate soil properties. Many have advocated the use of statistics for evaluation of geotechnical test results. Often there is sufficient Standard Penetration Test (SPT) or Cone Penetration Test (CPT) data such that statistical techniques make sense.

An example showing the SPT test results for a particular bearing stratum at a project site is shown in Figure 8.10. The number of tests for each blow count value is shown on the y-axis. The total population of data shown in the example is 109 tests. For analysis of a large amount of data such as this, it is useful to compute the mean, standard deviation, and coefficient of variation. A spreadsheet may be used to compute these values. Many definitions of mean value have been introduced in statistics, including arithmetic mean, geometric mean, generalized mean, f-mean, harmonic mean, and various weighted means. The most often used definition in geotechnical engineering is the arithmetic mean, G_M, which is simply the average of the data given by

$$G_M = \frac{\sum y_n}{n} \tag{8.2}$$

Where

n is the total population of SPT test samples and
y_n is SPT blow count for samples 1 to n.

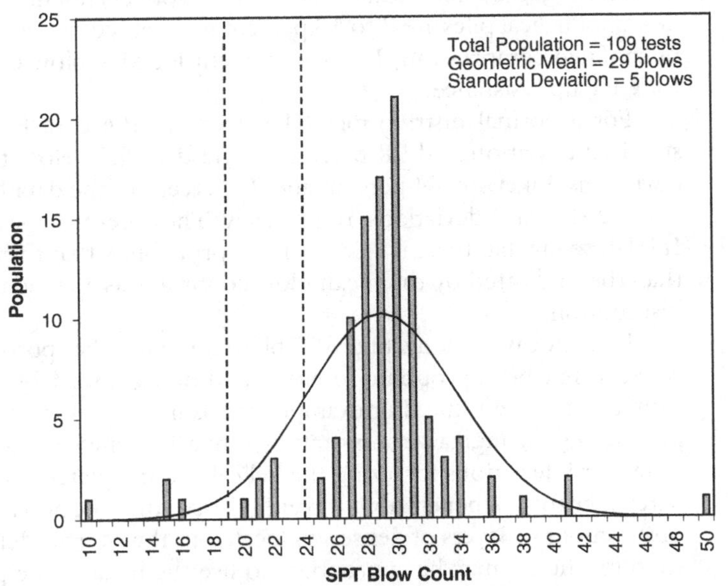

Figure 8.10 Statistical analysis of sample SPT test results

In a normal distribution, the algebraic mean is the exact middle of the data where 50 percent of the measurements fall above the mean and 50 percent of the measurements fall below the mean.

Standard deviation, σ_M, can be calculated from

$$\sigma_M = \sqrt{\frac{n\sum y_n^2 - (\sum y_n)^2}{n(n-1)}} \tag{8.3}$$

Where

all parameters have been defined previously.

The standard deviation generally defines the distribution of data about the mean. A high standard deviation indicates wide scatter. A low standard deviation indicates a narrow grouping.

The coefficient of variation, COV, is the standard deviation divided by the expected value of a variable, which for practical purposes can be taken as the average (Duncan and Wright, 2005). The COV is typically reported as a percentage.

$$COV = \frac{\sigma_M}{G_M} \times 100\% \tag{8.4}$$

The coefficient of variation is a convenient measure of scatter because it is dimensionless. Duncan and Wright (2005) present a table of published coefficients of variation for various geotechnical tests. Some of the more pertinent COV values with respect to design of helical piles are 2 to 13 percent for drained direct shear, 13 to 40 percent for unconfined compression, 15 to 45 percent for SPT blow count, and 5 to 37 percent for CPT tip resistance.

For a normal distribution, 84 percent of the data fall below the mean *plus* 1 standard deviation and 98 percent of the data fall below the mean *plus* 2 standard deviations. Likewise, 84 percent and 98 percent of the data fall above the mean *minus* 1 and 2 standard deviations, respectively. These percentages also indicate probability. It is fair to say that there is an 84 percent probability that the blow count will be higher than that indicated by the mean blow count minus 1 standard deviation for a normal distribution.

In many cases, the average SPT blow count may be appropriate for design. In other cases, it may be appropriate to either add or subtract 1 or more standard deviations from the average value. For example, if it is important to minimize the probability of penetrating through a certain stratum, it may be beneficial to use the mean value minus 2 standard deviations for sizing the helical bearing plates. In this way, there is only a 2 percent chance of penetrating through the stratum. If, however, there are a number of obstructions or layers of dense soil located in the overburden above a desired bearing stratum, then it may be appropriate to use the mean value plus 1 standard deviation for sizing the helical bearing plates. This may result in deeper penetration into the bearing stratum but improve the likelihood of achieving the requisite depth without encountering refusal prematurely.

Example 8b

Problem: Assuming a normal distribution of SPT data, find the blow count values for which there are 84 percent and 98 percent probabilities that the soil will have a blow count at least equal to or higher than these values.

Given: The SPT measurements shown in Figure 8.10.

Answer: According to Figure 8.10, the mean blow count is 29 blows and the standard deviation is 5 blows. These values are calculated most efficiently using a spreadsheet. In a normal distribution, there is an 84 percent probability of a value greater than the mean minus 1 standard deviation. For the subject data, the blow count with 84 percent probability is

$$29 - 5 = 24 \text{ blows} \tag{8b.1}$$

There is a 98 percent probability of a value greater than the mean minus 2 standard deviations. For the subject data, the blow count with 98 percent probability is

$$29 - 5 - 5 = 19 \text{ blows} \tag{8b.2}$$

It is important to note that the preceding discussion regarded normal distributions of data. A normal distribution is simply a mathematical model that statisticians use to approximate a group of data. The normal distribution is also known as the Gaussian distribution, because German mathematician Johann Carl Friedrich Gauss used it in many of his analyses. A normal distribution curve is plotted over the data in Figure 8.10. One of the main features of this distribution is that it is symmetric about the mean and has a bell shape. Normal distributions are not always a good representation of a set of data. There are many other distribution models including binomial, chi-squared, exponential, gamma, and hypergeometric. The probabilities just quoted for adding and subtracting standard deviations are inaccurate for nonnormal distributions. The designer needs to be aware of this especially when accurate probabilities are requested.

For much simpler sets of data consisting of only a few points, normal distributions and standard deviations are of little value. The mean value still may be computed by taking the average. If it is desired to estimate a blow count wherein a certain percent of the data fall above this value, it can simply be done by hand using a sheet of paper and a pencil. First, write down all the blow count values in the bearing stratum of interest in sequential order from lowest to highest as in the example contained in Table 8.1. Next, add up the number of tests (e.g. total sample population). Multiply the total population by 85 percent, or whatever percent probability desired. This is the number

Table 8.1 Example Determination of 85th Percentile Value for Small Data Sets

Step 1 Reorder from lowest to highest.	Step 2 Count number of SPT tests and multiply by desired probability.	Step 3 Starting from the bottom, cross off the number of tests found in Step 2 and circle the next highest.
18	Population = 12 tests	18
18		(18)
19		~~19~~
19		~~19~~
20		~~20~~
22	12 × 85% = 10.2 tests	~~22~~
23		~~23~~
25		~~25~~
25		~~25~~
27		~~27~~
30		~~30~~
32		~~32~~

of tests in the data set that will fall above the blow count value being sought. Starting from the bottom of the list, cross off that number of tests and circle the highest remaining value. This is the blow count wherein 85 percent of the data fall above this value. Computing probabilities in this way is simple yet effective when dealing with a small population of SPT data, especially when the data form a non-normal distribution. The same procedure could be used when evaluating CPT, unconfined compression, direct shear, or other field and laboratory tests.

When there exist even fewer data, the "Three-Sigma Rule" can be applied to estimate standard deviation (Dai and Wang, 1992). This rule of thumb, as originally presented, uses the fact that 99.73 percent of all values of a normally distributed parameter fall within 3 standard deviations of the average. Therefore, standard deviation can be estimated for a normal distribution by taking the difference between the highest and lowest conceived values and dividing by 6. However, there is a tendency in geotechnical engineering to estimate a range of values between conceived extremes that is too small (Duncan, 2000). Duncan and Wright (2005) suggest standard deviation may be estimated from

$$\sigma_M = \frac{V_{HC} - V_{LC}}{4} \tag{8.5}$$

Where

V_{HC} is the highest conceivable value and
V_{LC} is the lowest conceivable value.

The Three-Sigma Rule can be used to quickly evaluate the reasonableness of a statistical analysis and applies to many other distributions besides normal distributions.

As a word of caution, the accuracy of probability estimates depends on how well the data represent the ground conditions. Reliability increases with the number of test samples. The simplicity of the hand method just presented and of the Three-Sigma Rule should not be used as an excuse for inadequate geotechnical investigation. When encountering a substandard geotechnical investigation, one should obtain and analyze additional data when possible.

8.5 FIELD ADJUSTMENTS

The required ultimate capacity of a pile can be divided by the capacity-to-torque ratio to estimate the minimum installation torque. When prescriptive specifications are required for a project, it is necessary for a helical pile designer to show the minimum installation torque in the project plans and specifications. On projects with performance-based specifications, the design or ultimate capacity can be specified and the installation contractor and the design professional can be made responsible for proper installation torque determinations.

Example 8c

Problem: Find the minimum installation torque required for a helical pile.

Given: Required ultimate capacity of 50 kips [222 kN] and a manufacturer's recommended capacity-to-torque ratio of 10 ft^{-1} [33 m^{-1}].

Answer:

$$50{,}000/10 = 5{,}000\text{ft} - \text{lbs}\ [6{,}800\text{N} - \text{m}] \tag{8c.3}$$

A number of factors can affect the capacity-to-torque ratio. It is important to keep in mind that the capacity-to-torque ratio typically represents the average over many possible pile configurations and soil conditions. By virtue of the definition of mean or average, the actual capacity-to-torque ratio will be higher than the average in half of the instances and lower than the average in the other half of the instances.

Hoyt and Clemence (1989) found that the value of K_t is not strongly influenced by the number and size of helical bearing plates. Contradictory to this, Narasimha Rao et al. (1989) showed that the capacity-to-torque ratio increases with the number of helical bearing plates. Helical piles with 1.7-inch- and 2.4-inch- [44- and 60- mm-] diameter shafts were used in the study. Ghaly and Hanna (1992) and Ghaly, Hanna, and Hanna (1991b) tested miniature helical anchors with different geometries in a sand-filled testing tank equipped with stress transducers. They determined that helix geometry had a significant effect on the installation torque of the helical anchors but had little effect on the pullout capacity. In still other studies, K_t was found to be weakly correlated with number and size of helical bearing plates (Hargrave and Thorsten,

1992; Slemons, 2008). These results indicate that, under some circumstances, a test program may be appropriate for fine-tuning helical pile sizes in specific applications. An example of a circumstance warranting this type of test program is when a certain minimum length must be achieved on sites with debris, cobble, or other factors that make penetration difficult.

Traditional limit state theory can be used to determine the number and size of helical bearing plates necessary to achieve the required capacity at a particular site, as discussed. Care should be taken not to compound factors of safety (e.g., a factor of safety applied to the soil strength parameters, allowable bearing capacity, and service loads); otherwise, one can end up with a helical pile with too many bearing plates to penetrate the ground effectively (e.g., early refusal). Once preliminary helical pile sizing is completed, fine-tuning of helix sizes can be accomplished on sites with large numbers of piles by performing a series of field tests with different numbers and sizes of helical bearing plates. On many sites, the subsurface conditions are very complex, and sizing of helical piles can be difficult. Some examples include sites with highly random soil consistency, sites with limited geotechnical information, and sites with layered profiles. Performing a field test program to finalize pile selections is crucial in these circumstances.

In most cases, the installation torque increases with the addition of greater helical bearing plate surface area. An example of this is shown in Figure 8.11. At this site, several different helical pile configurations were attempted to obtain the best configuration for the project. The geometries of the test piles used in the study are shown in the table to the right of the graph. Actual torque measurements with depth are represented by the different symbols. Linear regression analysis was performed on each of the data sets to obtain a relationship between torque and depth for the various test piles.

A plot of helical bearing plate surface area and resulting torque for this site is shown in Figure 8.12. The data have been normalized by dividing by the installation

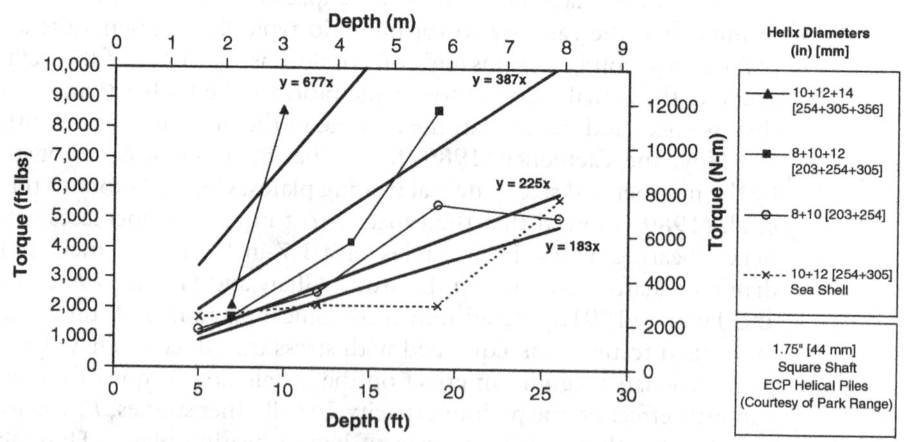

Figure 8.11 Installation torque for various helix configurations

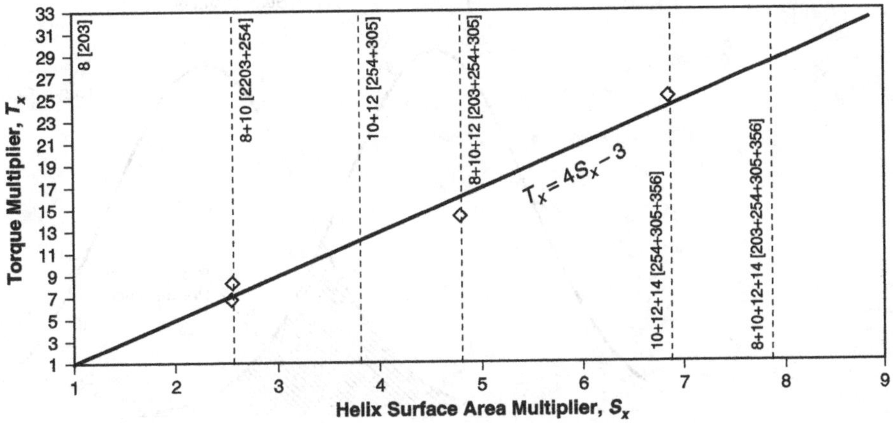

Figure 8.12 Increase in installation torque with helix surface area

torque and surface area by the corresponding values for a helical pile with single 8-inch-[203-mm-] diameter helical bearing plate. This type of analysis can aid in optimizing the helical pile configuration. When combined with load tests, a preliminary helical pile test program can improve the success of projects and provide opportunities for value engineering.

8.6 RELIABILITY

Hoyt and Clemence (1989) computed the Spearman coefficient of rank correlation for each method of analysis used to determine the capacity of a helical pile including cylindrical shear, individual bearing, and torque correlations. This test showed a very low degree of correlation between traditional limit state methods and the torque correlation method. This finding indicates that limit state methods and torque correlation methods are independent of each other. A high degree of reliability can be obtained using both methods in unison. That is to say, a helical pile sized for the subsurface conditions at a site using limit state theory whose capacity has been verified in the field through torque correlations is assured a high degree of success.

Reliability is the probability of long-term satisfactory helical pile performance. Reliability can be determined by overlapping the distributions between two independent variables at an assumed factor of safety, as in Figure 8.13. The probability of poor performance is the area under the intersecting curves. Best-fit normal distributions for the limit state method and the torque correlation method are shown in the figure. However, both variables exhibit a nonnormal distribution that is skewed toward conservatism.

A better approximation of the reliability of satisfactory helical pile performance can be made by calculating the probability of poor performance for each independent variable by hand. Of the 112 load tests described in Chapter 5, there were only 11 cases

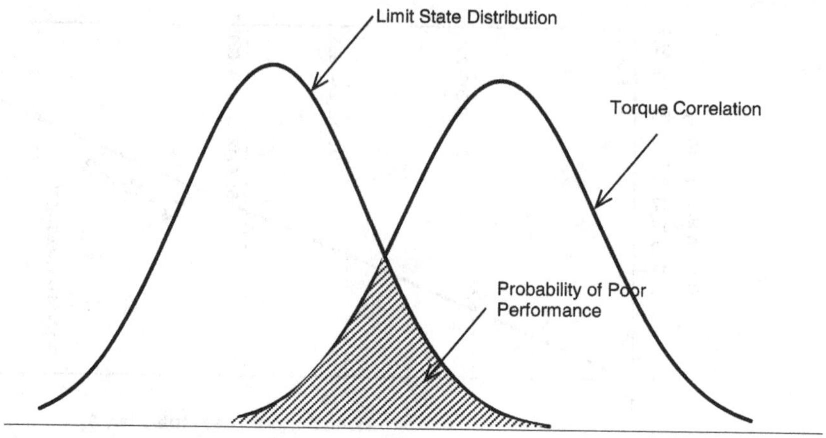

Figure 8.13 Normal distributions and probability

where the predicted capacity fell short of the measured capacity divided by a factor of safety of 2.0. This suggests a 9.82 percent probability of poor performance if limit state methods alone are used to calculate capacity with a factor of safety of 2.0. Of the 239 load tests described in Chapter 6 and shown in Figure 6.4, there were only 6 cases where the predicted capacity fell short of the measured capacity divided by a factor of safety of 2.0. This suggests a 3.05 percent probability of poor performance if torque correlation methods are used alone to calculate capacity with a factor of safety of 2.0.

If two independent methods of capacity determination are used together, then the probability of poor performance, p_f, is the product of the individual probabilities given by

$$p_f = p_1 p_2 \tag{8.6}$$

Where

p_1 and p_2 are the independent probabilities of poor performance.

Substitution of the probabilities for poor performance for the limit state method and torque correlation method into Equation 8.6 yields a combined probability of 0.3 percent. The reliability of using both methods is $1-p_f$, or 99.7 percent. According to standard methods of probability and reliability in geotechnical engineering, this represents an average target reliability and should generally result in satisfactory performance (Duncan, 2000).

In simple terms, the load test data presented in previous chapters suggest that roughly 10 out of 100 helical piles will exhibit unsatisfactory performance if limited state methods alone are utilized to size the helical pile. The load test data suggest that 3 out of 100 helical piles may have performance issues if torque correlations alone are used to verify helical pile performance. When the two methods are combined, the risk of poor pile performance drops to only 3 piles out of 1,000. All of this assumes that a factor of safety of 2.0 is used in the calculations. If higher target reliability is required for a project, load tests can be performed to obtain yet another variable to evaluate pile performance.

In conclusion, the pile designer should use traditional limit state methods to size helical piles and then use installation torque to verify capacity in the field. This combination of using two independent methods to establish the capacity of a helical pile should result in a high probability of success. Load tests can be used to add further confidence in the design of helical piles. The results of a few load tests or installation torque alone should not be used to "value engineer" helical piles for a particular site.

Chapter 9

Expansive Soil Resistance

Helical piles are an effective foundation system in expansive soil and bedrock. Helical pile foundations are used for both new construction and repair of existing foundations in areas with expansive soils. The bearing and pullout capacity of helical piles in expansive soils can be determined following the same procedures as given in Chapters 4 and 5. The minimum length of helical piles must be such that the helical bearing plates penetrate a stable bearing stratum so that soil movement along the shaft can be resisted. This chapter includes a summary review of expansive soil behavior and special precautions for design and application of helical piles in these conditions.

9.1 EXPANSIVE SOILS

The term "expansive soils" is used for those soil and bedrock types that exhibit volume change with variations in moisture content. Expansive soils have a strong affinity to water and can exert significant pressures on foundation elements when moisture becomes available. Expansive soils also can exhibit significant shrinkage if moisture is removed through drying. Expansive soil behavior is almost always associated with clay and mudstone with high clay content (hereinafter claystone bedrock). Swell pressure and volume change are correlated with plasticity. Coarse-grain soils as well as igneous and metamorphic bedrock are nonexpansive. Most highly cemented shale, clean sandstone, and limestone are nonexpansive. Fine-grain soils with high plasticity, claystone, very clayey sandstone, and some shale tend to be highly expansive.

Expansive soils can be found almost anywhere in the world. They are problematic in semiarid climates and climates that experience cycles of draught followed by wet seasons. Along the Front Range of the Rocky Mountains in Colorado, there exist many areas with highly plastic clays and claystone bedrock that have previously been buried by thousands of feet [meters] of overburden that has since weathered away through various geologic processes. These soils and bedrock are highly overconsolidated. The semiarid climate maintains high suction values and in many cases significant swell potential. Similar conditions exist in parts of Texas, California, Spain, Australia, and southern Africa. When land is developed for residential and commercial use, the ground surface is inevitably covered by impermeable materials such as slabs, structures, and pavement systems that block evapo-transpiration. This combined with irrigation of landscaping results in increased moisture availability and subsequent heave.

There are many other geographic areas where expansive soils exist at varying degrees of moisture content. Many parts of central Europe and central United States (e.g. Ohio, Indiana, Missouri, and Kentucky) have expansive soils that swell during wet seasons and shrink during dry seasons. Wetting and drying of expansive soils can be controlled to some extent with proper surface drainage precautions, but eventual changes in moisture content due to changes in land use are inevitable and have to be accounted for in design.

Further information regarding the geology and mineralogy of expansive soils along with a complete description of expansive soil behavior is contained in text books by Nelson and Miller (1992) and Chen (1988), as well as many articles and technical papers. A basic primer for homeowners, contractors, and professionals less familiar with expansive soils is provided in the excerpt adapted from Noe (2007). Proper identification and mitigation of expansive soils is critical in the design and performance of foundations, floor slabs, pavements, and other structures on expansive soils. The remainder of this text focuses on proper design of helical piles only.

Basic Questions and Answers about Expansive Soils for Homeowners (adapted from Noe, 2007)

What Are Expansive Soils?

Highly plastic clay soils and claystone bedrock are generally classified as 'èxpansive.'' This means they will tend to expand (increase in volume) as they absorb water and will shrink (decrease in volume) as water is drawn away. Expansive clays are naturally occurring materials that can be found almost anywhere the world.

What Damage Can Occur to Foundations from Expansive Soils?

When expansive soils supporting a foundation become moist, cracks may appear, windows and doors may stick, and floors may slope as the foundation becomes progressively more out of level.

What Can I Do to Minimize the Effects of Expansive Soils?

If homeowners discover their home is constructed over expansive soils, they can minimize cracking and possibly prevent major damage by controlling moisture levels around their foundation. Here are some suggestions that might help.

1. *Roof drainage.* Install rain gutters with downspouts well away from the foundation backfall via nonerodible surfaces.
2. *Planter and yard drainage.* All areas should drain properly. Even puddles are potential problems.
3. *Concrete and asphalt areas.* Where permitted, these areas should drain to the street. Otherwise, concrete and asphalt should flow to a yard or planter area.
4. *Subsurface drainage.* Install drains if necessary to eliminate ponding. Maintain all lines clean and free-flowing. Drain lines should discharge in accordance with local drainage plans.
5. *Repair plumbing leaks.* Monitor consumption. An unexplained increase in your water bill could indicate a leak. Repair immediately.
6. *Landscaping.* Plan carefully. Trees, small ones, can draw huge amounts of water from nearby soils. They should not be planted close to structures.

Homeowners are strongly encouraged to contact a foundation engineer for additional recommendations, preventative measures, and designs for repair.

If the Foundation Cracks, What Can a Homeowner Expect to Pay to Repair it?

"Selected annual U.S. losses from expansive soils were $798 million in 1970 and are expected to rise to $997 million by year 2000" (Nelson and Miller, 1992).

When problems with the foundation are discovered, they can be as small as a chipped plaster to as large as foundation wall cracks. While these problems may not cause the house to collapse, they may cost the homeowner anywhere from $5,000.00 to $300,000.00 dollars or more to repair.

What Can be Done to Repair Expansive Soil Damage?

Homeowners should take preventive measures immediately upon discovery of expansive soil damage to their foundation. A professional foundation engineer should make a site visit to diagnose the foundation's failure. After analyzing your particular situation, the engineer can design an engineered repair plan for your property. One solution is underpinning with helical piles.

9.2 FOUNDATIONS ON EXPANSIVE SOILS

The type of foundation used in areas with expansive soils depends on the estimated potential heave. In areas with low to moderate estimated heave potential, projected movements may be small enough that they can be tolerated by shallow foundation systems such as conventional footings or stiffened slabs with or without some reworking of the foundation subgrade. In areas with moderate to very high estimated heave potential, projected movements may be such that they cannot be tolerated by a shallow foundation and some type of deep foundation is usually necessary. Drilled shafts, micropiles, and helical piles have been used for deep foundations in expansive soils. Each has advantages and disadvantages.

Drilled shafts can be the most economical alternative provided groundwater and caving soils are not encountered. When properly designed and installed, they can resist expansive soils when bottomed well below the anticipated depth of wetting. Of the various deep foundation types, drilled shafts have the longest history of supporting structures on expansive soils. They also are advantageous because they can be designed to support heavy foundation loads as well as large lateral loads. Probably the biggest disadvantage of drilled shafts is the number of quality control issues that can affect their performance. Some construction issues include mushrooming of the top of the pile, arching of the concrete in the drill hole, caving or necking of drill holes, groundwater and contamination in drill holes, and disturbance ("mud caking") on the sides of drill holes. Other issues, which relate to reinforcement of drilled shafts, include incorrect reinforcing steel placement, damage to reinforcing steel, improper wet placement ("stabbing") of anchoring steel, and insufficient splices. Potential issues associated with the concrete include adding too much water in the field, excessive time elapsed between batching and placement, inadequate consolidation of concrete around reinforcing steel, improper curing, and contamination of concrete.

Micropiles have advantages in that they can be installed using smaller equipment with low headroom requirements. The portion of the micropile in the active zone can be fitted with a permanent steel casing to reduce side shear due to expansive soil movement. Micropile drilling equipment can penetrate practically any formation, including very dense soils, cobble, boulders, competent bedrock, and trash fill. Micropiles are perhaps the more expensive alternative discussed herein and can be subject to some of the same quality assurance issues described for drilled shafts. The smaller size of micropiles as compared to drilled shafts limits vertical and lateral capacity.

Helical piles are advantageous in expansive soils because adhesion along the central steel shaft is much less than that between cast-in-place concrete or grout and the surrounding soil or rock. Like micropiles, the diameter of the shaft is small so it has less surface area to be influenced by expansive soils. Helical bearing plates provide resistance to uplift so they can be installed to shallower depths than micropiles. One of the main disadvantages of helical piles that prevents their use on many projects is the inability to penetrate bedrock with a standard penetration test blow count greater than about 150 blows per foot. However, as is described in Section 9.5, early refusal may not necessarily be a problem in many cases. There also are several methods that can be incorporated to increase helical pile penetration as discussed in Chapter 2 and Chapter 16. Another disadvantage of helical piles is that like micropiles their smaller size limits vertical and lateral capacity compared to drilled shafts.

Deep foundation elements typically are connected directly to foundation walls and grade beams on expansive soils. A typical detail of a foundation on expansive soils is shown in Figure 9.1. It is important to isolate the structure from expansive soils insofar as practical thus minimizing the impact the soil can have on the structure. Isolation can be accomplished by the use of void form, minimizing the number of piles, and maximizing the span between piles. A collapsible cardboard void form (e.g., SureVoid™, by SureVoid Products, Inc.) is placed between piles to reduce expansive soil pressures on the underside of walls and grade beams. The void form is intended to carry the weight of the concrete during curing but degrade rapidly with time to prevent pressure on the underside of walls and grade beams due to expansive soils. The thickness of the required void depends on the heave potential at the site. Basement walls and grade beams are reinforced to span between piles. The lateral movement of basement walls typically cannot be resisted solely by connection to a wood floor system and has to be resisted by counterforts, dead men, anchors, or spanning horizontally between corners.

There are many aspects to the design of foundations on expansive soils besides shallow and deep foundation elements. A few of these are described in this paragraph. The reader is encouraged to consult the references cited earlier in this Chapter for more detailed information on design of foundations on expansive soils. The details contained in Figure 9.1 show a slab-on-grade basement floor. Basement slabs-on-grade are used only on sites with low estimated heave potential where movement can be tolerated and the owner understands the risks. An example would be commercial structures where the basement area is being used for parking or storage. If used, slabs-on-grade should be isolated from foundation walls, grade beams, and pile caps so that forces due to vertical movement of the slab are not transmitted to the foundation. Partition walls should be hung from the underside of floor joists and constructed with a void to allow for slab movement. Staircases should be hinged, and utilities should be flexible. A foundation drain should be placed along the base of the wall to prevent accumulation of water around the basement. Perhaps most importantly, the ground surface should slope away from the foundation on all sides to promote runoff, avoid infiltration, and control wetting of expansive soils. Often problems associated with expansive soil movement can be linked to improper surface drainage.

In areas with high to very high estimated heave potential or where floor movement cannot be tolerated, it is often necessary to use a structurally supported floor system spanning between deep foundation elements with collapsible void or air space beneath the floor. Figure 9.2 presents a photograph showing the construction of a home with a structural basement floor. The basement walls of this home are supported on helical piles with void between the piles similar to that shown in Figure 9.1. Structural steel floor support beams span between exterior walls and helical piles installed within the interior of the basement. The photograph in Figure 9.2 was taken immediately prior to placement of corrugated steel decking and the reinforced concrete floor slab. The

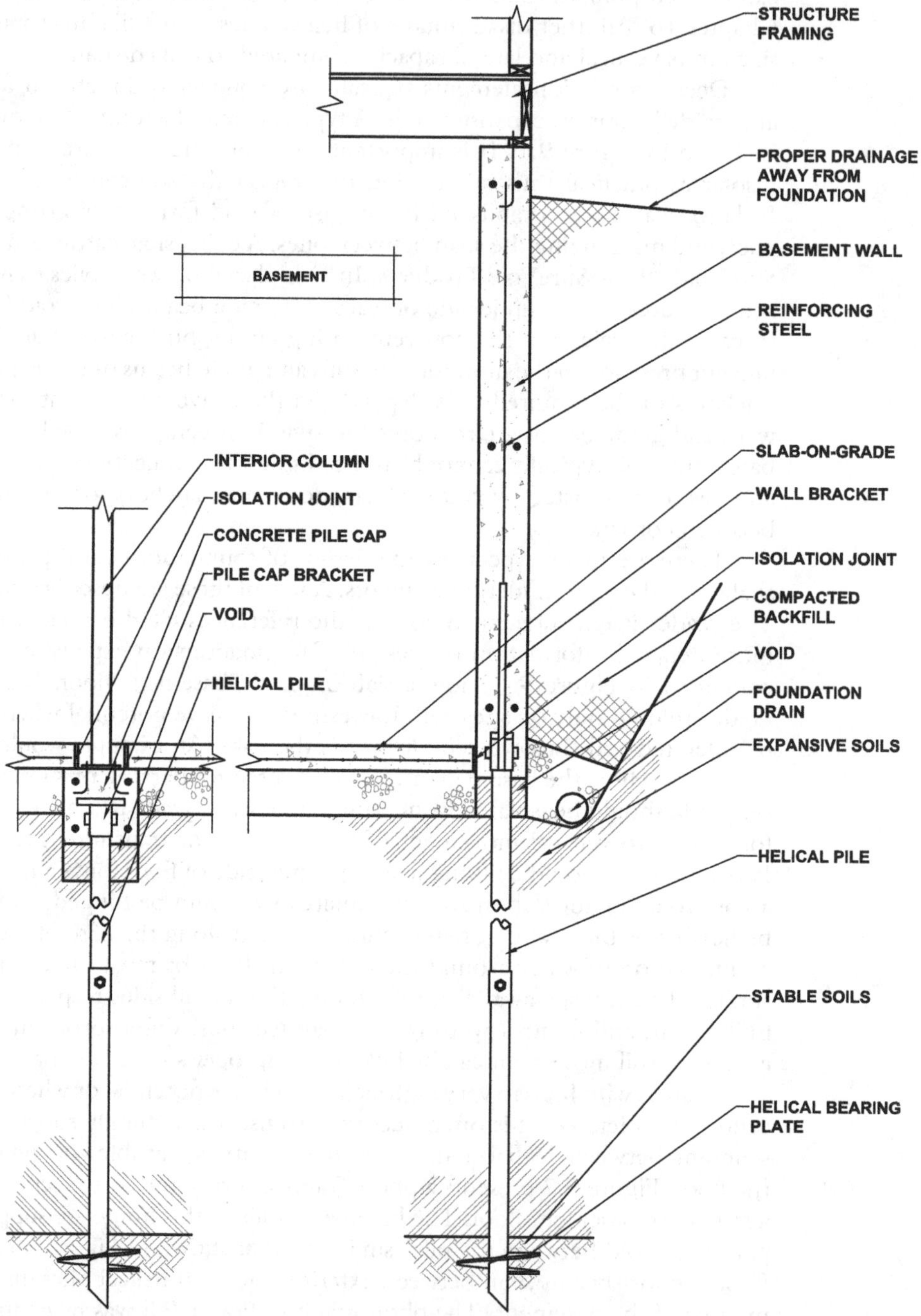

Figure 9.1 Typical basement wall construction on expansive soils

Figure 9.2 Structural floor construction on expansive soils

septic drain pipes under the slab are being laid out and will be hung from the underside of the steel beams to avoid movement in the event of heave. Other utilities that pass through the slab are being constructed with flexibility.

An example foundation plan for a residential structure on expansive soils is shown in Figure 9.3. This home has a walk-out-basement and attached two-car garage. The soil and bedrock at this site was judged to have high heave potential. To resist heave, the foundation is supported on helical piles. Each helical pile is represented by a circular symbol, and a number within a hexagon that indicates the pile type. The quantity, design capacity, minimum installation torque, manufactured cap, and minimum length of the different helical pile types are shown in the schedule along with some general notes regarding installation and inspection. Some practitioners suggest that better performance can be obtained by bottoming deep foundation elements at approximately the same depth for the entire structure. As a general rule, helical piles that support grade beams for garage walls, crawl spaces, or entryways should be designed with the same target installation depth as helical piles supporting basements and other below-grade habitable areas. Actual depth of helical piles in the field may vary as termination is defined by torque. Helical piles should be spaced apart to maximize economy and the dead load on the piles, which also helps to resist heave.

Due to the risk of slab-on-grade movement, the basement floor in Figure 9.3 is structurally supported using the construction methods similar to that shown in Figure 9.2. For ease of construction, a concrete pile cap is cast over the helical piles around the perimeter of the basement. The pile cap serves as a ledger to facilitate installation of floor beams. The ledger is nominally reinforced and is intended to span between the piles under the dead load of the basement floor and walls only. Live loads and loads

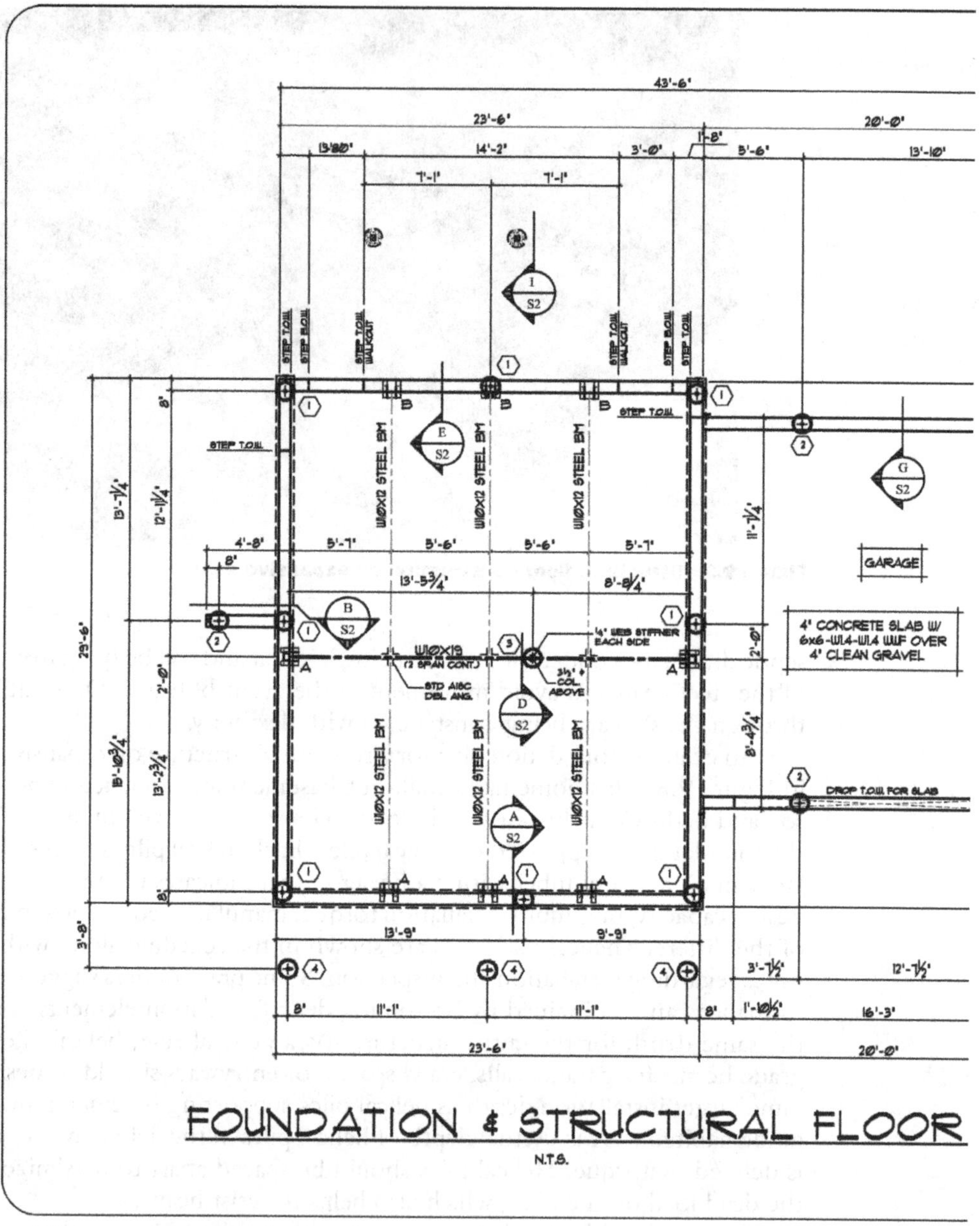

Figure 9.3 Example foundation plan on expansive soils

HELICAL PILE SCHEDULE						
SYMBOL	QUANITY	DESIGN CAPACITY	MIN. TORQUE	MIN. LENGTH*	PIER CAP CONNECTION	NOTES
1	8	40 KIPS	10 KIP-FT	11 FT	PLATE CAP 100	HELICAL PILES SHALL HAVE 3" DIAM. x 0.25 WALL SHAFT W/ SINGLE 12" HELICAL BEARING PLATE TYPICAL. MINIMUM OF 25% OF PIERS TO BE OBSERVED BY ENGINEER DURING INSTALLATION.
2	6	10 KIPS	2.5 KIP-FT	17' (FT) @ GAR. 14' (FT) @ CF.	1 5/8 GALVANIZED 'U' BAR (SEE DETAIL SHEET S2)	
3	1	40 KIPS	10 KIP-FT	14' (FT)	BEAM CAP 200	
4	3	5 KIPS	13 KIP-FT	20' (FT)	1 5/8 GALVANIZED 'U' BAR (SEE DETAIL SHEET S2)	

*MIN. DISTANCE BETWEEN TOP OF PIER AND UPPERMOST HELICAL BEARING PLATE

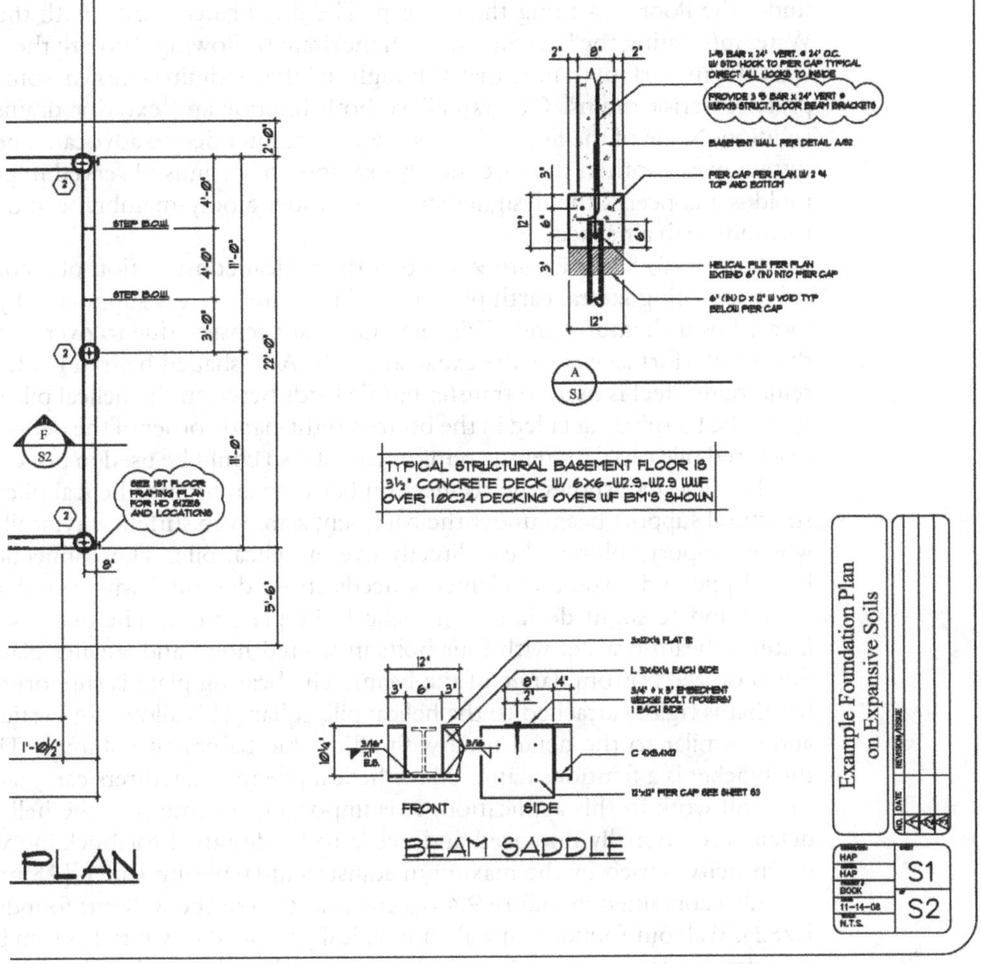

from the remaining structure are transferred to the piles by virtue of the completed basement walls.

Foundation sections and details associated with the example plan are shown in Figure 9.4. Section A/S2 shows a typical basement wall. The concrete pile cap and structural floor system are apparent design differences compared to the section shown in Figure 9.1. A cardboard void form is shown under the pile cap. Helical piles extend through the void to support the foundation.

An interior perimeter foundation drain appears in Figure 9.4. The drain is located under the floor and along the pile cap. The drain runs parallel with the grade beam. Water infiltrating the backfill can reach the drain by flowing through the void under the wall. It is important to note that although an interior drain is shown, some practitioners prefer exterior drains. Others believe both interior and exterior drains are the best solution. None of the figures in this chapter are intended to advocate one type of drain system over another, they are merely examples of designs observed in practice. When mold is a concern, the designer should consider a poly-membrane and ventilation of the underslab air space.

Section B/S2 in Figure 9.4 shows the typical construction of a counterfort that aids in resisting lateral earth pressures. The counterfort is supported by a helical pile located near the outer end. This pile must resist tension due to overturning forces on the counterfort as well as the expansive soils. A U-shaped hairpin made of galvanized reinforcing steel is used to transfer tensile loads between the helical pile and the structure. The hairpin is detailed in the bottom right-hand corner of the detail sheet. A manufactured pile cap with tensile reinforcement also could be used in place of the hairpin.

Detail D/S2 shows the connection between an interior helical pile and the main structural support beam under the basement slab. Web stiffeners typically are required where support columns bear directly over a helical pile. The connection between a helical pile and structural elements needs to be designed with some adjustment to accommodate slight deviations in helical pile placement. The detail shows a manufactured beam bracket with four bolts in slotted holes and smaller plates that clamp down on the bottom flange of the beam. The bearing plate is supported on a thread bar that is rigidly attached to the helical pile collar. This allows for vertical adjustment and is similar to the detail used with adjustable basement columns. The callout for the bracket is a fictitious name. Many helical pile manufacturers carry similar brackets that will work in this application. It is important to note that the helical pile in this detail is rotationally unbraced and needs to be designed for buckling with potential eccentricity caused by the maximum adjustment (typically 1 inch [25 mm]).

Also contained in Figure 9.4 is a cross-section of the walkout foundation (section E/S2). Walkout foundation walls are typically stepped down to bottom below frost. A step-down reinforcement detail is shown just below this section. In some jurisdictions, grade beams and pile caps are not required to bottom below frost provided the foundation spans over collapsible void and is supported on helical piles; in this way, any soil movement associated with frost is resisted. Still, it is considered advisable under certain conditions to extend the grade beam below frost depth or use an insulating barrier to reduce the potential for freezing in the under-slab air space.

Detail F/S2 shows a cross-section of the garage foundation. Support of garage slabs on structural floors is not a standard practice due to cost, lower risk of damage due to slab movement, relatively low cost of slab replacement, and the increased tolerance for movement of most owners. To allow for movement, the garage slab is isolated from foundation walls and interior columns, if any. Similarly, detail I/S2 shows shallow pier foundations for an exterior wood deck. These shallow foundations would be subject to potential heave in expansive soils. However, they often are used for better economy compared to a deep helical pile and because wood decks are fairly flexible and can be adjusted easily in the event of movement. Where less risk of movement is desired, wood decks can be supported on helical piles.

There are many methods of constructing helical foundations on expansive soils and basement structural floors. The foregoing examples provide a glimpse into the practice of foundation design on expansive soils for those less familiar with this type of construction. Those considering construction on expansive soils should engage the services of a design professional familiar with the standard of care exercised in the area of the project at that time.

9.3 ACTIVE ZONE

Deep foundations in expansive soils are generally extended to more stable soil and bedrock deep within the ground and are designed to resist expansive soil movement. The zone of soil and rock that is contributing to expansive soil movement at a particular time is termed the "active zone." The extent of the active zone below ground surface to be used in heave predictions and pile design, also known as design depth of wetting, is the subject of much debate.

Many researchers (McKeen and Johnson, 1990, McOmber and Thompson, 2000, Diewald, 2003, and Walsh, et al., 2009) believe that the active zone reaches a maximum depth where soil suction and moisture content are essentially in equilibrium with local climatic conditions and the availability of water at the surface. This depth depends on detailed geologic conditions such as the angle of dip of major bedding planes and the degree of fracturing. The design wetting depth is determined based on historic information from the same general geographic region where the project is located. Information used to arrive at the design depth of wetting may include performance of nearby structures, accumulation of moisture content and soil suction measurements, and the experience of local professionals.

Nelson, Overton, and Durkee (2001) believe that a more conservative approach should be taken in the design of structures on expansive soils. They advocate that foundations should be designed for the maximum depth of potential heave defined as the depth where the overburden vertical stress equals or exceeds the constant volume swelling pressure of the soil. This depth represents the greatest possible extent of the active zone. Constant volume swell pressure can be measured directly in laboratory, or it can be estimated from free swell odometer tests (Thompson, Rethamel, and Perko, 2006, Nelson, Chao, and Overton, 2007, Nelson and Miller, 1992). In general, the approach used by Nelson, Overton, and Durkee (2001) results in design depths of

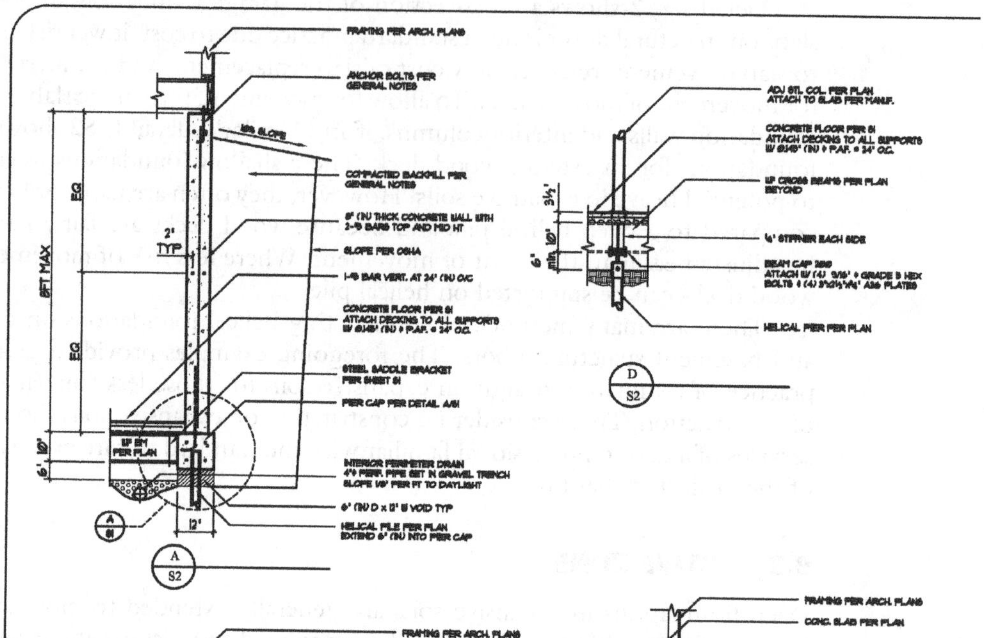

Figure 9.4 Example foundation details on expansive soils

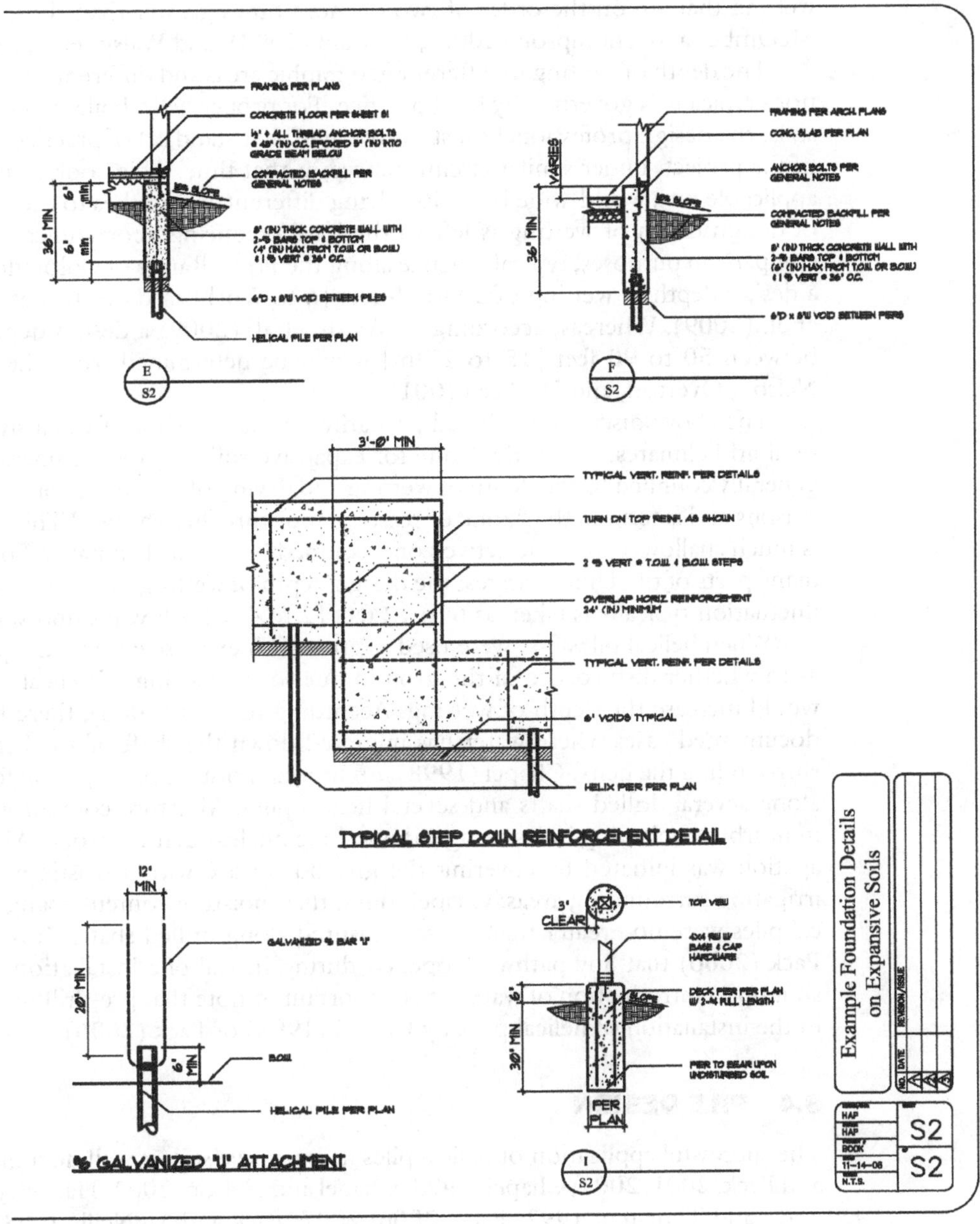
FRAMING PER PLANS
CONCRETE FLOOR PER SHEET S1
½" ♦ ALL THREAD ANCHOR BOLTS @ 48" (IN) O.C. EPOXIED 5" (IN) INTO GRADE BEAM BELOW
COMPACTED BACKFILL PER GENERAL NOTES
SLOPE
36" MIN
6" min
8" (IN) THICK CONCRETE WALL WITH 2-#5 BARS TOP & BOTTOM (4" (IN) MAX FROM T.O.W. OR B.O.W.) & 1 #5 VERT @ 36" O.C.
6"D x 8"W VOID BETWEEN PILES
HELICAL PILE PER PLAN
E
S2
FRAMING PER ARCH. PLANS
CONC. SLAB PER PLAN
VARIES
ANCHOR BOLTS PER GENERAL NOTES
COMPACTED BACKFILL PER GENERAL NOTES
8" (IN) THICK CONCRETE WALL WITH 2-#5 BARS TOP & BOTTOM (6" (IN) MAX FROM T.O.W. OR B.O.W.) & 1 #5 VERT @ 36" O.C.
6"D x 8"W VOID BETWEEN PIERS
F
S2
3'-0" MIN
TYPICAL VERT. REINF. PER DETAILS
TURN DN TOP REINF. AS SHOWN
2 #5 VERT @ T.O.W. & B.O.W. STEPS
OVERLAP HORIZ. REINFORCEMENT 24" (IN) MINIMUM
TYPICAL VERT. REINF. PER DETAILS
6" VOIDS TYPICAL
HELIX PIER PER PLAN
TYPICAL STEP DOWN REINFORCEMENT DETAIL
12" MIN
20" MIN
GALVANIZED #5 BAR "U"
6" MIN
B.O.W.
HELICAL PILE PER PLAN
#5 GALVANIZED "U" ATTACHMENT
3" CLEAR
TOP VIEW
4X4 RW W/ BASE & CAP HARDWARE
DECK PIER PER PLAN W/ 2-#4 FULL LENGTH
SLOPE
30" MIN
PIER TO BEAR UPON UNDISTURBED SOIL
PER PLAN
I
S2
Example Foundation Details on Expansive Soils
NO.
DATE
REVISION/ISSUE
HAP
HAP
BOOK
11-14-08
N.T.S.
S2
S2

wetting that are on the order of two or more times greater than those measured by McOmber and Thompson (2000), Diewald (2003), and Walsh, et al. (2009).

The depth of wetting in different geographic areas and different subsurface conditions typically is governed by local practice. For proper helical pile design in expansive soils, the design professional must be aware of that standard of practice in the vicinity of the project, under similar circumstances, at that time. This book is intended to be applicable on a world-wide basis. Resolving different approaches to the determination of design depth of wetting is left to the local practicing geotechnical engineer. For comparison purposes, typical practice along the Front Range of Colorado is to assume a design depth of wetting of 20 to 29 feet [6 to 9 m] based on the data from Walsh, et al. (2009). Whereas, according to Walsh, et al. (2009) a design depth of wetting between 50 to 90 feet [15 to 27 m] would be determined from the approach by Nelson, Overton, and Durkee (2001).

The above discussion is based primarily on the behavior of expansive soil sites in semi-arid climates. The active zone for expansive soils in more temperate climates is generally confined to the depth of wetting and drying of expansive soils with changing seasons and is termed the "zone of seasonal moisture fluctuation." This zone typically is much shallower than the active zone assumed in semiarid climates. For example, in many parts of the United States, the design depth of wetting due to seasonal moisture fluctuation typically is taken as 6 to 12 feet [2 to 4 m] below ground surface.

When helical piles first were used in areas with expansive soils, the question arose as to whether disturbance of the ground due to the passing of helical bearing plates would increase the depth of wetting. According to Pack (2006), there have been no documented cases where water has migrated down the shaft of a helical pile to soil surrounding the helix. Chapel (1998) monitored moisture propagation for three years along several drilled shafts and several helical piles. Moisture content was measured in nearby monitoring holes using a down-hole nuclear density probe. Moisture propagation was initiated by covering the ground surface with a plastic membrane and irrigating surrounding areas. Chapel found that moisture content changes along helical piles were no greater than those measured along drilled shafts. It is suspected by Pack (2006) that any pathways opened during helical pile installation quickly swell shut upon introduction of water. It is important to note that pre-drilling was not used in the installation of helical piles by Chapel (1998) or Pack (2006).

9.4 PILE DESIGN

The successful application of helical piles in expansive soils is well documented (Black and Pack, 2001, 2001; Chapel, 1998; Chapel and Nelson, 2002; Hardesty, 2007; Hargrave and Thorsten, 1992; Pack, 2006, 2007; Pack and McNeill, 2003). According to Pack (2006), it is estimated that 130,000 helical piles were installed for both remedial repair and new construction in expansive soils in the Front Range of Colorado over the 20-year period between 1986 and 2006. A diagram of a typical helical pile in expansive soils is shown in Figure 9.5. The depth of the active zone in expansive soils and bedrock is represented by the distance d_W. Uplift stresses within the active zone

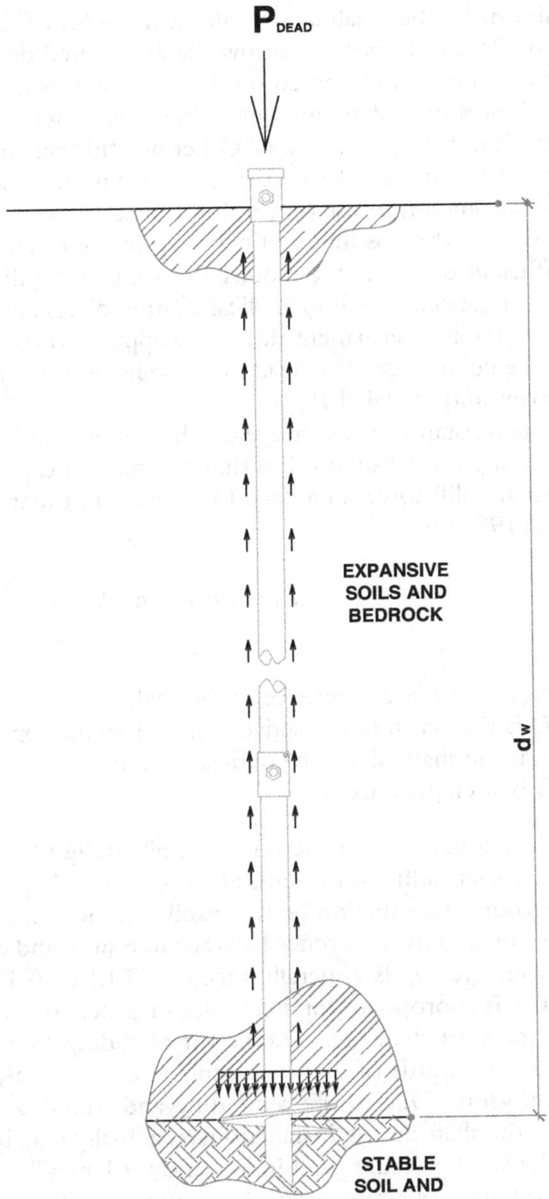

Figure 9.5 Helical pile in expansive soil and bedrock

are shown by the small arrows along the sides of the shaft. The bearing plates of a helical pile should bottom below the anticipated depth of the active zone plus some additional penetration based on the practitioners judgment and local practice. Some practitioners do not require any additional penetration beyond the design depth of wetting and have good results. Other practitioners use a distance of three to five feet [1 to 2 m] below the design depth of wetting to provide some margin for error.

It is common to use only a single helical bearing plate in expansive soils, because it maximizes the possibility of penetrating the often very stiff to hard expansive soils to sufficient depth. If the required bearing and pullout capacity cannot be achieved with a single helix, multiple helical bearing plates are permitted. It is also common to account for any permanent dead load applied to the top of the pile by subtracting it from the uplift force. The dead load applied to the pile in Figure 9.5 is represented by the arrow and the label P_{dead}.

The resistance to soil heave of a helical pile can be checked by ensuring the uplift force acting on the shaft is less than the pullout capacity of the helical bearing plates. The total uplift force on a deep foundation in expansive soils is given by Nelson and Miller (1992) as

$$P_u = \pi d_{eff} d_w \alpha_o P_s - P_{dead} \tag{9.1}$$

Where

πd_{eff} is the circumference of the shaft
d_w is the depth of the active zone below the top of the helical pile
α_o is the shaft adhesion coefficient, and
P_s is swell pressure.

The shaft adhesion coefficient used in pile design for expansive soils is different from ordinary shaft adhesion discussed in previous chapters. It is a parameter that takes into account the adhesion between soil and the shaft, the at-rest lateral earth pressure coefficient, and the difference between free swell and constant volume swell pressures. The parameter α_o is generally taken as 0.12 to 0.15 for drilled shafts. A value of $\alpha_o = 0.1$ is appropriate for steel piles in order to account for the reduced adhesion along the steel shaft. The coefficient of sliding friction of soil on steel is generally taken as two-thirds of that between soil and cast-in-place concrete or that between soil and grout. Due to the remolding and straining of soil that occurs immediately around the shaft during installation, α_o actually may be less than 0.1 for helical piles.

The swell pressure used in Equation 9.1 may be obtained from oedometer-type free swell/consolidation tests. An example swell/consolidation test on a sample of soil is shown in Figure 9.6. The procedure for conducting a swell/consolidation test involves placing a relatively undisturbed sample in an oedometer machine and applying a confining pressure. The confining pressure for the test shown in the figure is 1 ksf [48 kPa]. After a hold period, water is introduced around the sample and allowed to permeate the sample through porous stones. As water is absorbed into the sample, volume change typically occurs. Expansive soils will exhibit swell. Collapsible soils may

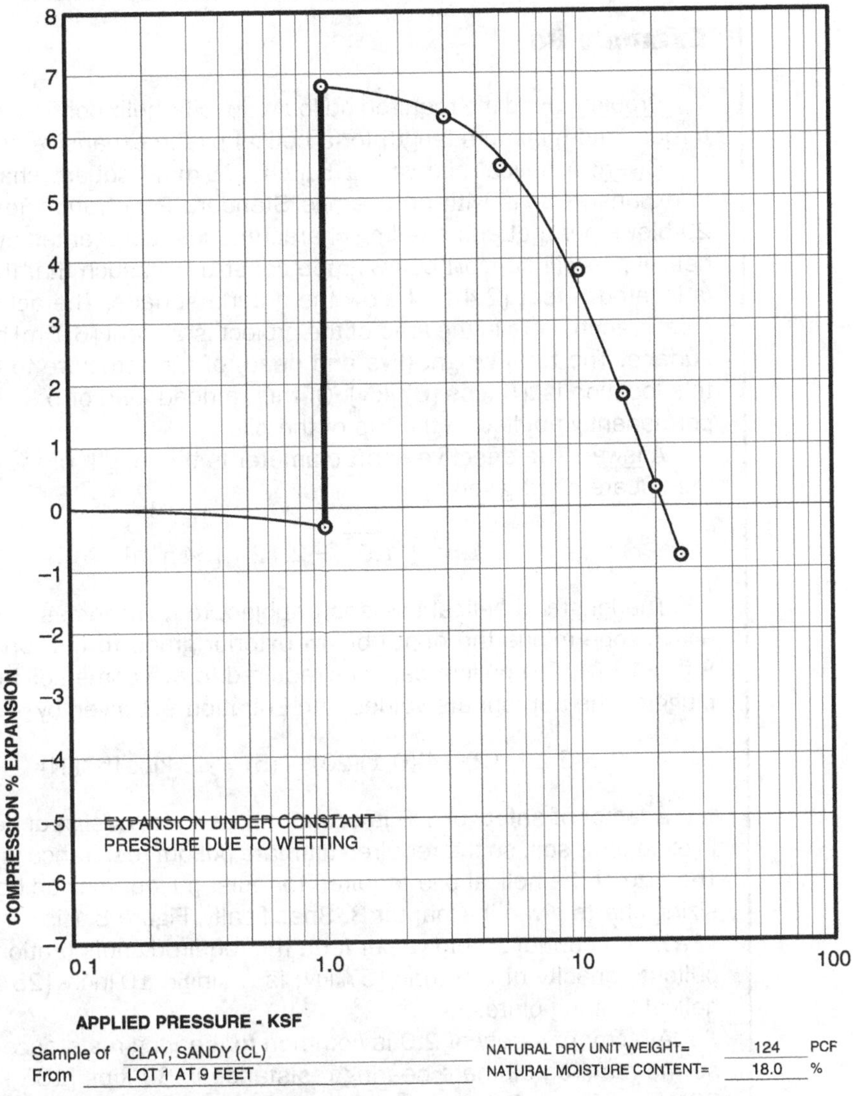

Figure 9.6 Example swell-consolidation test

exhibit collapse. The sample shown in the figure expanded approximately 7 percent due to wetting at constant pressure. Volume change measurements are taken over time. Once the sample appears to have stopped swelling, the confining pressure gradually is increased until the sample is forced back to its original volume. The pressure at this point is termed the "swell pressure." The swell pressure is approximately 23 ksf [1,100 kPa] for the sample shown in Figure 8.4. Several studies have shown that the swell pressure in the free swell oedometer test is approximately the same regardless of the initial confining pressure.

Example 9a

Problem: Find the required pullout capacity, helix configuration, installation torque, and minimum length for a helical pile in expansive soils.

Given: A helical pile with 1.5-inch- [38-mm-] square shaft is to be used in expansive soils with an average Standard Penetration Test blow count of 20 blows per foot and swelling characteristics represented by Figure 9.6. The helical pile will support below-grade construction such that the top of the pile is located 8 feet [2.4 m] below the ground surface. The active zone used by local practitioners in the area of the project is 22 feet [6.7 m] below the ground surface. The total weight (live and dead) of the structure to be supported at this location is 15 kips [67 kN]. Of this, a dead load of 5 kips [22 kN] will be permanently applied to the top of the pile.

Answer: The effective shaft diameter is the length of the diagonal across the square shaft given by

$$\sqrt{1.5^2 + 1.5^2} = 2.12 \text{ in} \, [54 \text{ mm}] \tag{9a.1}$$

The length of helical pile shaft subject to uplift forces is the depth of the active zone minus the depth below exterior grade to the top of pile (22 ft − 8 ft = 14 ft). The pullout capacity required to overcome soil heave is found by plugging the appropriate values into Equation 9.1 given by

$$\pi(2.12/12)(14)(0.1)(23) - (5) = 13 \text{ kips} \, [57\text{kN}] \tag{9a.2}$$

A factor of safety of 1.0 is commonly used for design of deep foundations in expansive soil, so the required ultimate pullout resistance is 13 kips [57kN]. The size of the helical pile required to resist pullout may be obtained from the sizing charts given in Chapter 8. Specifically, Figure 8.1 is for fine-grain soils. At a blow count of 20 blows per foot, the required configuration for an ultimate pullout capacity of 6.5 tons [57 kN] is a single 10-inch- [254-mm-] diameter helical bearing plate.

A factor of safety of 2.0 is common for bearing resistance of helical piles, so the required ultimate bearing resistance is 30 kips [133 kN] (2.0 × 15 = 30). From Figure 8.1, the size helical pile required to resist 15 tons at a blow count of 20 blows per foot has 10-inch and 12-inch diameter helical bearing plates. In this case, bearing capacity governs the helix configuration.

The required installation torque for pullout is determined by taking the ultimate pullout capacity and dividing by the capacity-to-torque ratio provided by the manufacturer or as described in Chapter 6. According to Figure 6.4, the empirical capacity-to-torque ratio for a helical pile with 2.12-inch [54-mm] effective diameter shaft is 11.2 ft-1 [37 m-1]. The required installation torque for pullout resistance is

$$13{,}000\ \text{lbs}/11.2\ \text{ft-1} = 1{,}200\ \text{ft-lbs}\ [1{,}600\ \text{N-m}] \tag{9a.3}$$

Similarly, the required installation torque for bearing resistance is

$$30{,}000\ \text{lbs}/11.2\ \text{ft-1} = 2{,}700\ \text{ft-lbs}\ [3{,}600\ \text{N-m}] \tag{9a.4}$$

Here again, the bearing capacity governs the design and sets the minimum installation torque.

The recommended minimum length of the helical pile is the depth of the active zone along the shaft given by

$$22 - 8 = 14\ \text{ft}\ [4.3\ \text{m}] \tag{9a.5}$$

The pile designer may desire to add a certain penetration beyond this length depending on local practice and his/her experience.

The first three parameters in Equation 9.1 give the outer area of the shaft, which is taken as circumference times length. The outer area of the shaft is multiplied by the uplift stress on the pile. Some researchers (Pack, 2006) suggest using the perimeter of the shaft rather than the effective diameter, which for square shafts is equal to 4 times the side dimension (4d). This correction reduces the total uplift on common square shaft sizes by approximately 10 percent. It is conservative to simply take the circumference of a circle circumscribed around the square shaft.

Proper design of helical pile foundations in expansive soils should consider minimum length, minimum installation torque, and helix size. Design depth of wetting and minimum penetration into a certain stratum may be used to establish recommended minimum pile lengths.

9.5 EARLY REFUSAL CONDITION

Refusal occurs when the soil or bedrock is sufficiently hard such that the helical pile cannot penetrate further. Occasionally, refusal can occur prior to reaching the design depth of wetting in expansive soils. This is termed early refusal. Under certain circumstances, helical piles that reach early refusal may be acceptable. Pack (2006) has published his professional opinion based on years of experience in areas with expansive soils that helical piles with a single small diameter helix installed to a certain torque have performed well even if installation is halted at shallow depths. Pack's experience can be explained by the fact that installation torque of a helical pile in weathered rock is related to hardness. Increased hardness can indicate changes from a more plastic claystone to a more brittle sandstone or shale, and hence may indicate a less expansive material in certain circumstances.

Helical pile refusal can be related to soil blow count. The chart shown in Figure 8.3 from Chapter 8 shows the predicted ultimate capacity of different helical bearing plate configurations in weathered bedrock. This chart is based on conventional limit state theory. As can be seen, a single 10-inch [254 mm] diameter helix encounters possible refusal at a blow count of approximately 100. The predicted ultimate capacity of the helix at this blow count is approximately 50 tons [440 kN]. Using the capacity-to-torque ratios from Chapter 6, this indicates a torque of 10,000 foot-pounds [13,600 N-m] for a 1.5-inch [38-mm] square shaft or 12,500 foot-pounds [16,900 N-m] for a 3-inch- [76-mm-] diameter round shaft. This is close to or above the maximum limit for some of the higher-strength shafts available on the market today. Hence, refusal would occur. If an 8-inch- [203-mm-] diameter helical bearing plate is used instead, practical refusal would be projected at a blow count of approximately 150.

The results of 634 swell/consolidation tests are plotted in Figure 9.7 with respect to blow count. Much of these data are from the private files of CTL|Thompson, Inc. Other data are from soil reports the author and his colleagues have collected over time. Most soil samples are from the Front Range of Colorado and include clayey sands, clays, clayey sandstone, and claystone bedrock. Scatter in the data indicate a lack of direct correlation between blow count and swell. However, most of the expansive soil data are located between a blow count of about 10 and 120 with the highest swells occurring between 20 and 100 blows. On many sites, there is a point at which the blow count becomes so high that swell drops off significantly. Above a blow count of 150, only 10 tests were above 1 percent swell. Most of the higher swell tests at a blow count of 200 were from a hard shale formation with interlayers of bentonite. Above a blow count of 200, there are only 2 swell tests above approximately 1 percent. In general,

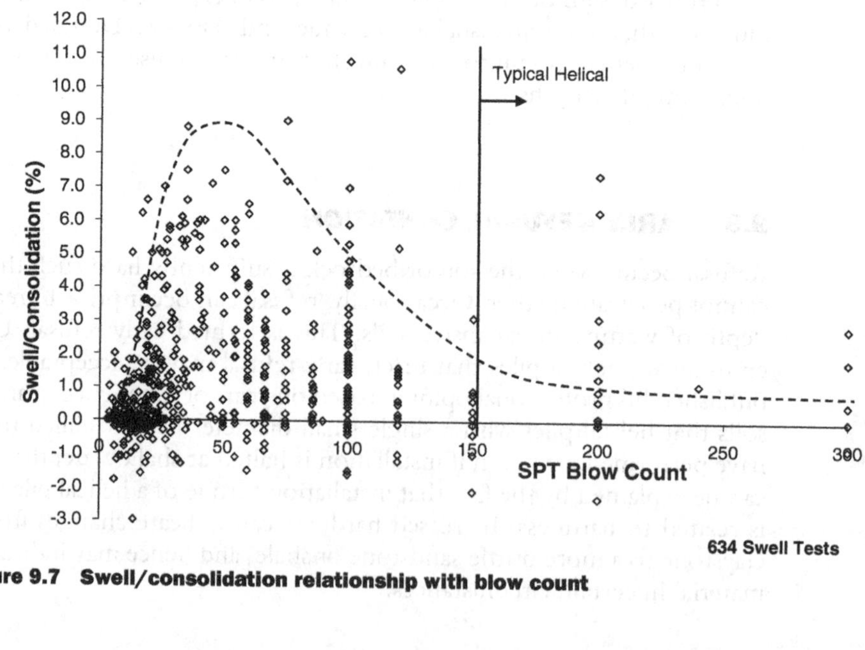

Figure 9.7 Swell/consolidation relationship with blow count

a high degree of hardness may indicate a transition in bedrock from an expansive claystone to a less expansive or nonexpansive very hard shale or competent rock.

For many commonly available shafts, a helical pile with single 8-inch- [203-mm-] diameter helix typically reaches refusal in bedrock with a blow count of 150. The results in Figure 9.7 suggest the helix of such a pile may in many cases be resting on low to nonexpansive material. Low swell tests in very hard material support to some extent the assertions of Pack (2006) that a helical pile can be stopped upon refusal in expansive soils even at shallow depth. However, as can be seen in the figure, there are some exceptions to this rule. The designer is cautioned that the characteristics of expansive soil and rock depend heavily on clay mineralogy and therefore geology and geography. The data shown in Figure 9.7 may be representative of certain formations in Colorado only. As can be seen, a few samples of very hard material exhibited higher swell. The hardness point at which certain formations become less expansive may differ. Local practitioners are encouraged to consult their records to establish if there is a point at which the materials in their area become so hard that they are less prone to swell. It is also important to be sure that the helical shaft being used is capable of achieving the required torque to embed the helical bearing plate into this very hard material.

In the preparation of this book, it was suggested that Standard Penetration Test blow count should be related to expansive potential (EP). Expansive potential (EP) was shown to be a good parameter for describing the level of risk of foundation movement (Nelson, Chao, and Overton, 2007). There was insufficient time to examine this correlation in detail for this text. It is clear that more research needs to be done to better understand the characteristics of expansive soils as they relate to material hardness.

Where project specific swell data indicate that hardness cannot be used as an acceptance criterion for helical piles, methods have to be incorporated to reach the design depth of wetting. Pre-drilling pilot holes to aid helical pile penetration should be avoided in expansive soils. To date, there have been no studies done to determine the effect of pre-drilling on depth of wetting in expansive soils. Some experts believe drilling a hole with a diameter larger than the helical pile shaft may create a path for moisture and alter the depth of wetting. If a pilot hole is necessary, it should be approximately the same size as the helical pile shaft and should only be used for an occasional obstruction. If subsurface exploration suggests a pilot hole is likely to be needed for most pile locations, a different deep foundation type with rock drilling capability such as drilled shafts or micropiles would be more sensible.

Chapter 10

Lateral Load Resistance

Structures are subject to lateral loads primarily from unbalanced earth pressures, wind, and earthquakes. Often in construction, foundations must be relied upon for lateral as well as axial support. The lateral load carrying capacity of helical piles has been documented in several studies (Hoyt, 2007; Perko, 2000, Perko, 2004b; Puri, Stephenson, Dziedzic, and Goen, 1984). Helical piles with tubular shafts and rigid couplings are best suited to support lateral loads.

Lateral load analysis can be subdivided in two categories: rigid and flexible. Rigid pile analysis may be performed on relatively short piles. The underlying assumption in this analysis is that the pile behaves as a rigid body and rotates about a fixed point in the soil. Relatively short helical piles with larger-diameter shafts have been used for many years to support a number of different light structures, such as flagpoles, lightpoles, signs, and sound walls, wherein the governing loads act in a lateral direction.

Flexible pile analysis generally deals with long and slender piles such as those used for buildings and other structures that require both significant axial capacity and some lateral capacity. In this analysis, the structural stiffness of the pile shaft must be taken into account along with its interaction with soil. The most common analysis of this type is termed the "p-y method."

This chapter contains a summary of some of the methods frequently used for lateral analysis of helical piles. A number of lateral load tests performed in various soil conditions are compared with engineering calculations. Lateral capacity predictions are shown to compare well with full-scale field test results. Effects of shaft coupling rigidity and soil disturbance during installation are discussed.

10.1 RIGID PILE ANALYSIS

Deep foundations with a length less than about 10 times the shaft diameter are generally considered rigid in lateral load analysis. Rigid foundations are assumed to rotate about a point. Rotation as a result of flexure of the foundation itself is assumed negligible. Examples of helical piles meeting this criterion are those specifically manufactured for lightpole bases, signpost bases, and sound wall foundations. These helical piles typically have 8-inch- to 24-inch- [203-mm- to 610-mm-] diameter shafts and fixed lengths from 3 feet to 20 feet [0.9 m to 6.1 m].

A typical "rigid" helical pile is shown in Figures 10.1 and 10.2. The helical pile has a large-diameter casing with single helix affixed to the leading end along with a smaller-diameter pilot point. Soil enters the casing through the opening in the helix.

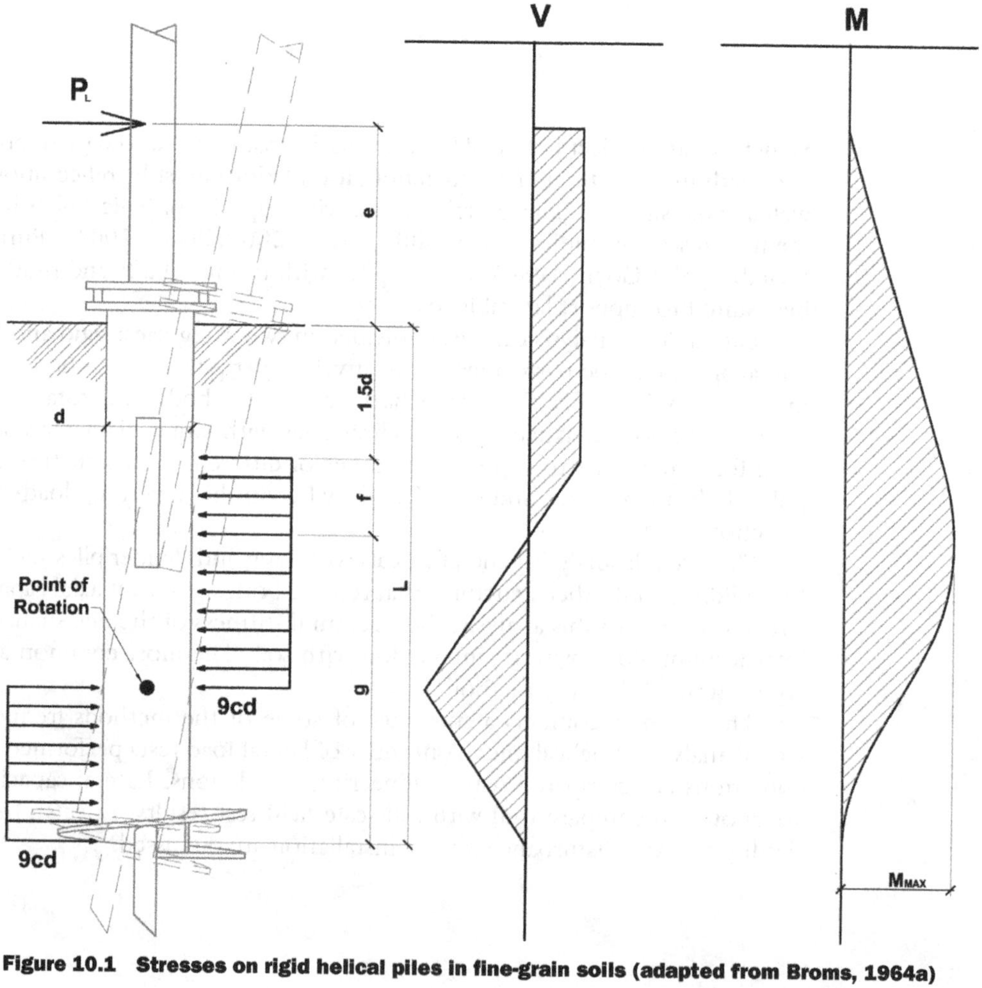

Figure 10.1 Stresses on rigid helical piles in fine-grain soils (adapted from Broms, 1964a)

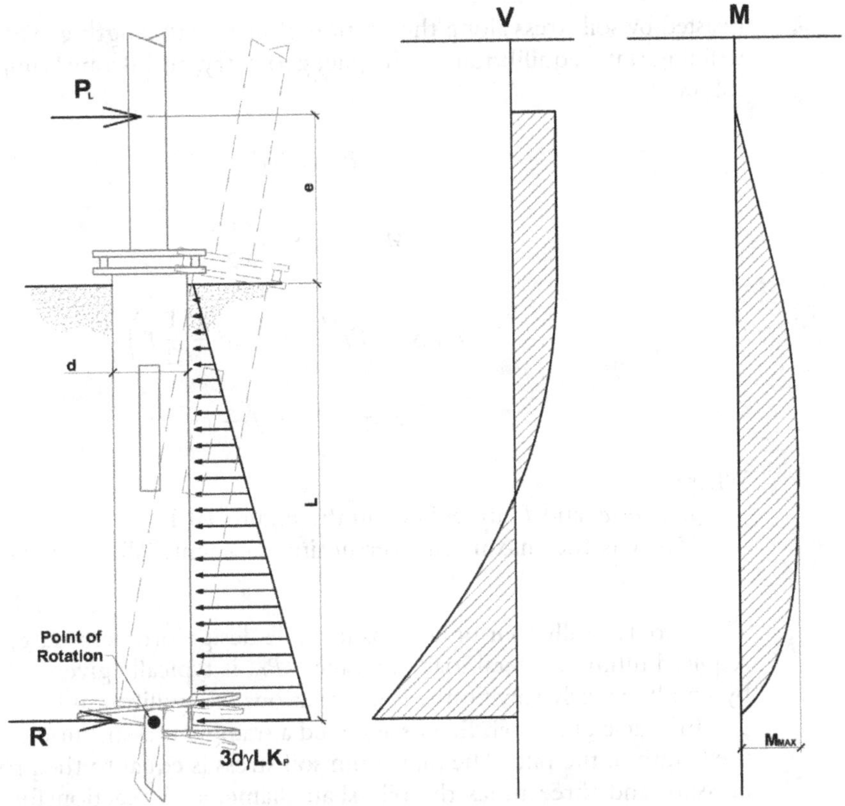

Figure 10.2 Stresses on rigid helical piles in coarse-grain soils (adapted from Broms, 1964a)

An optional slot is shown in the side of the casing. The slot can serve as a feedthrough for electrical conduit or the like. A cylindrical post, wide-flange beam, or other structure can be affixed to the top of the foundation via an adjustable bolted moment connection.

Several methods have been developed for design of shallow, rigid pile foundations. One method developed by Broms (1964a, 1964b) involves a simplified static analysis. Broms derived a separate and distinct treatment in fine-grain and coarse-grain soils. Diagrams showing the assumed soil stresses, shear, and moment diagrams by Broms are contained in Figures 10.1 and 10.2 for fine-grain and coarse-grain soils, respectively.

In fine-grain soil, Broms assumed lateral soil stresses have a rectangular distribution with a maximum pressure equal to the ultimate bearing capacity, given by $9cd$, where c is undrained cohesion and d is shaft diameter. Broms ignored lateral contributions of the top zone of soil to a depth of 1.5 times the shaft diameter. To simplify the static analysis, the lateral load on the pile, P_L, is assumed to be resisted by soil stress along the portion of the shaft with length f, and the overturning moment is assumed to be resisted by soil stress along the portion of shaft with length g. The resulting equations defining static equilibrium of the pile, geometry, and Broms' simplifying assumptions follow.

$$P_L = 9cdf \tag{10.1}$$

$$M_{\mathrm{MAX}} = 9cd\left(\frac{g}{2}\right)^2 \tag{10.2}$$

$$M_{\mathrm{MAX}} = P_L\left(e + 1.5d + \frac{1}{2}f\right) \tag{10.3}$$

$$L = 1.5d + f + g \tag{10.4}$$

Where

g, f, d, e, and L are defined in the Figure 10.1.

M_{MAX} is the maximum overturning moment. All other parameters have been defined.

There are typically four unknowns in a pile design problem, L, g, f, and M_{MAX}. The required ultimate lateral load resistance, P_L, is typically given. The solution is found by simultaneously solving the four equations of equilibrium.

In coarse grain soil, Broms assumed a triangular distribution of soil stresses along the length of the pile. The maximum soil stress is equal to the product of passive soil pressure and three times the pile shaft diameter. A reaction force, R, is applied to the bottom of the foundation to balance the applied lateral force and soil stresses. By summing the moments about the base of the pile, the resulting resistance to lateral loading is the closed form solution given by

$$P_L = \frac{\gamma d L^3 K_p}{2(e + L)} \tag{10.5}$$

Where

γ is the unit weight of soil, and

K_p is the passive earth pressure coefficient. All other parameters have been defined previously.

Another method for determining the lateral capacity of shallow, rigid pile foundations is that by Brinch-Hansen (1961). The Brinch-Hansen method determines the forces on a rigid pile foundation by considering the difference between three-dimensional passive and active earth pressures. Rankine-type lateral earth pressure theories consider only two-dimensional conditions and are appropriate for long, planar structures such as retaining walls. Larger earth pressures are possible in the case of driven pile, drilled shaft, and short helical pile foundations due to three-dimensional effects. The total lateral resistance acting on a cylindrical shaft in the soil involves a zone of soil much wider than the shaft diameter. The ratio of three-dimensional and two-dimensional soil resistance varies with soil shear strength and foundation depth but is usually on the order of 2 to 3.2 (Helmers, Duncan, and Filz, 1997). Brinch-Hansen's model is applicable to soils that have both cohesion and friction and was shown to most closely model field tests performed on sound wall foundations compared to other design methods.

The difficulty with the Brinch-Hansen method is that it requires an iterative approach. Virginia Polytechnic Institute adapted the Brinch-Hansen method to work within an Excel spreadsheet program called LCAP101, which is available on the Virginia Transportation Research Council Web site (Helmers et al., 1997). The spreadsheet uses an iterative approach to determine recommended foundation lengths based on input diameter, soil properties, wind load, and wall height. The user is allowed to input a sloping ground surface and to ignore a section of the pile shaft near the ground surface. The spreadsheet may be modified to include an additional resistance exerted by the helix and pilot point or lead section extending from the large-diameter rigid steel casing of a helical pile foundation (Perko, 2004b).

The distribution of soil stress about a pile assumed by Broms and Brinch-Hansen are idealistic. As a result, the maximum bending moment in the foundation determined using these methods is conservative (Hoyt, 2007; Perko, 2004b). Both the Broms and Brinch-Hansen methods are based on the limit state, the point at which driving forces and moments are exactly balanced by resisting forces and moments. As such, neither addresses serviceability or pile head deflection. Nonetheless, both methods are well established and usually provide for reasonable performance when used with a factor of safety of 1.5 or greater. Some building codes, such as AASHTO (2004 allow for a lower factor of safety. The types of structures generally supported on short, rigid piles typically are governed by strength and not serviceability (e.g., lightpoles, sound walls, and signs). When pile head deflection is a consideration, the flexible pile method should be used to predict performance. Good correlation has been found between these methods in most circumstances.

10.2 FLEXIBLE PILE ANALYSIS

The analysis of a deeply embedded pile differs from that given in the preceding section in that the structural properties and stiffness of the pile shaft are taken into account. An example diagram showing the soil stress distribution on a deeply embedded, flexible helical pile is shown in Figure 10.3. As can be seen in the figure, the pile shaft exhibits flexure rather than rigid body rotation.

One of the simple methods to perform analysis of deeply embedded, flexible piles is a software program called L-Pile™ from Ensoft, Inc. L-Pile was originally developed by a prominent researcher named Lymon C. Reese of the University of Texas at Austin. The software uses discrete elements to solve the conventional p-y method of analysis and has several predefined p-y curves for different soil and rock types. The user must

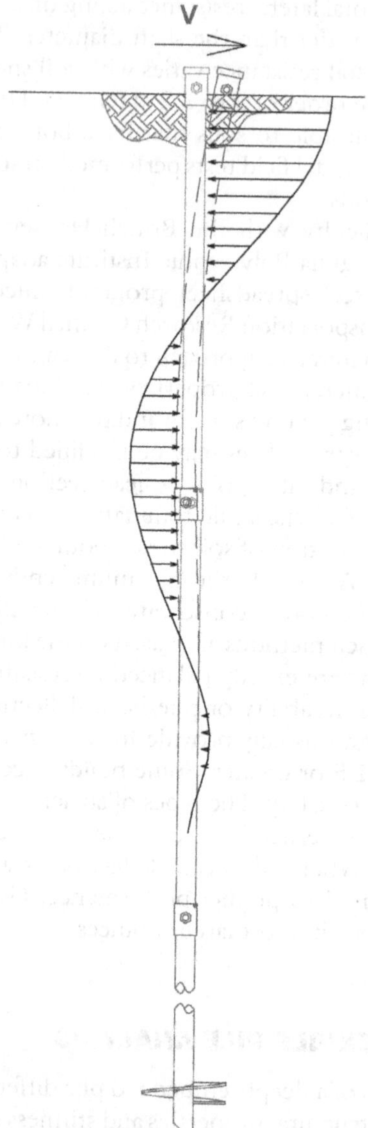

Figure 10.3 Example soil stress distribution on a flexible helical pile

input soil unit weight, shear strength parameters, strain at 50 percent stress, and a horizontal modulus. Some reference soil and rock properties for L-Pile^{TM} analysis are given in the tables in Chapter 3. The software is very powerful in that the effect of soil layering, sloping ground, and extension of the pile above ground can be accounted for.

The L-Pile™ user also must enter the structural properties of the helical pile shaft, including modulus of elasticity, area moment of inertia, shaft diameter, and gross area. The software displays the lateral deflection of the pile head and the pile shaft. The software also outputs bending moment, shear force, and soil resistance as a function of depth. If the lateral capacity exceeds the ultimate resistance of the pile, an error message is returned. Since the software does not solve for the lateral capacity directly, an iterative approach must be used to find the lateral capacity of a helical pile.

One approach to finding the lateral capacity using this software is to input several load cases at once. Example results from a run involving seven different load cases are shown in Figure 10.4. A plot of lateral deflection is shown on the left side of the figure, and a plot of bending moment is shown on the right side of the figure. Each solid curve in the plots represents a different load case. A legend can be manually created on the bottom of the plots to differentiate between the different load cases.

The International Building Code (IBC, 2006) states that the allowable lateral capacity of a pile is half of the load causing 1 inch [25 mm] of deflection. To find this value, the L-Pile™ user would incrementally increase the applied lateral loads in the software until the pile head reaches approximately 1 inch [25 mm] of deflection. The allowable capacity would be reported as half of this load. Another common practice is to compute the lateral load at which 1/2 inch [13 mm] of deflection occurs and take this as the allowable lateral load. AC358 (ICC-ES, 2007) states that the allowable lateral capacity of a helical pile shall be half the load causing 3/4 inch [19 mm] of deflection at the pile head.

The maximum deflection at the allowable lateral load differs with type of structure, sensitivity to movement, and load combination. Sometimes the design of structures does not require a specific deflection at the allowable load. Instead, they require only a certain factor of safety against the ultimate lateral capacity of the pile. In these cases, L-Pile™ software can be applied using successively higher load cases until an error message is obtained that pile deflections have exceeded software settings. The load at which this error message first occurs is taken as the ultimate capacity.

The load cases required by ASCE7 (American Society of Civil Engineers, 2006) for buildings can be input in the software to check deflection, bending moment, and shear. The software even allows an axial load to be applied to the top of the pile. Rather than apply full axial and lateral loads simultaneously, the pile designer is cautioned to use ASCE7 load cases as described further in Chapter 12. Full dead plus live axial load with full lateral loads may not be a required load case.

The pile designer should check the structural capacity of the pile by hand during various iterations and after final lateral pile analysis. The maximum bending moment from diagrams generated by L-Pile™ software can be compared with conventional ACI318(2005) or AISC (2001) calculations of the maximum flexural resistance of the shaft. The allowable lateral capacity determined by the software considers only soil-shaft interaction. Sometimes internal strength limitations govern pile design rather than ultimate soil resistance. This is especially true for slender shafts in firm, hard, or dense soils. One caution in checking combined flexure and axial load using conventional codes is that the moment values produced by L-Pile™ represent stresses for an

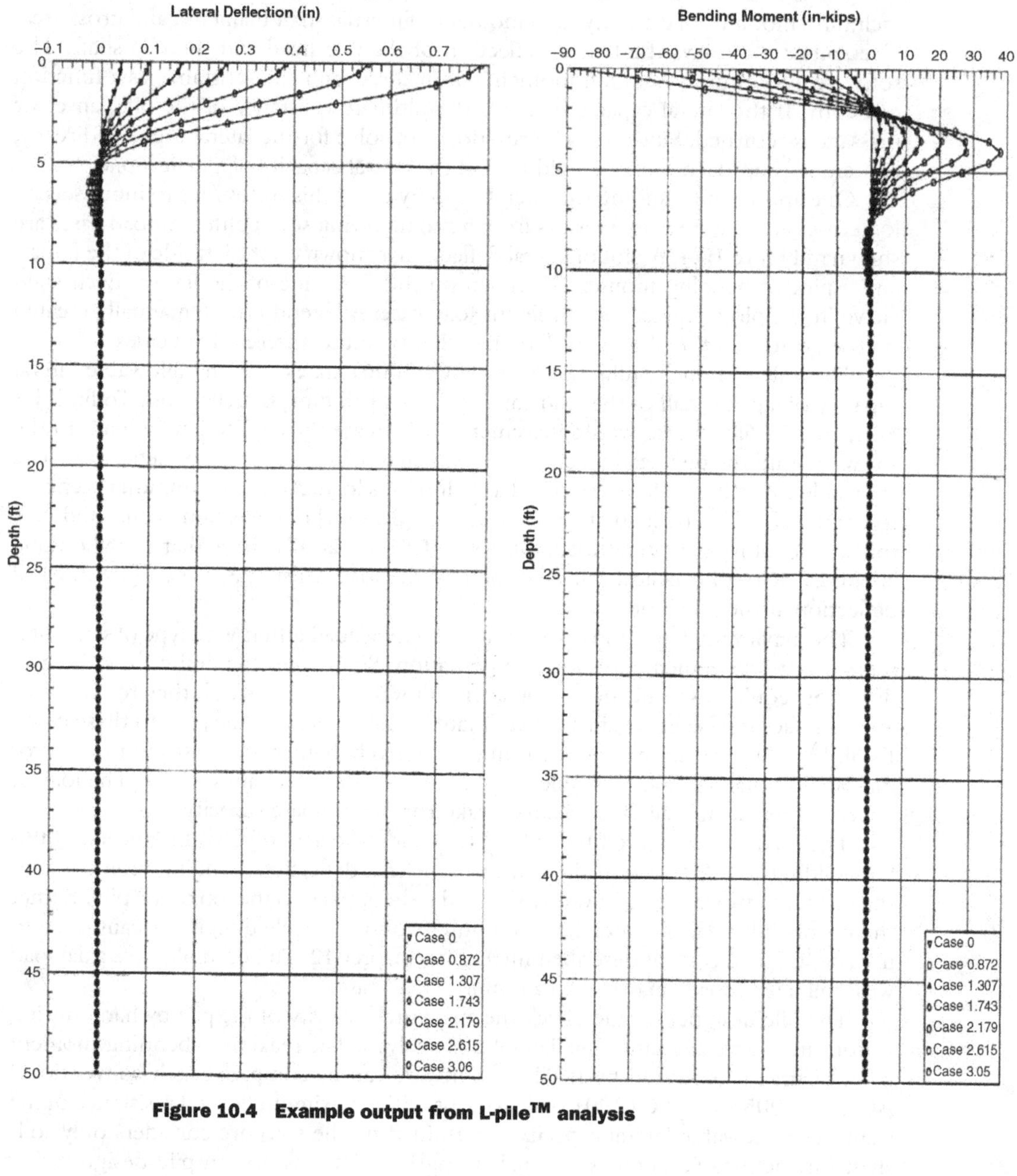

Figure 10.4 Example output from L-pile™ analysis

already deflected and partially "buckled" shaft. Further reduction in axial pile capacity using column buckling equations are not required for analysis of combined bending and axial loading using results from L-Pile™. The correct equation to use to compute nominal axial capacity in the combined stress equations is simply gross area of the shaft times yield strength.

Another feature of L-Pile™ software is that it allows for different conditions of pile head fixity. A conventional pile embedded only a few inches in the bottom of a concrete pile cap may best be represented by pinned end conditions. This is simulated in the software by zero moment with increments of shear. If the pile is deeply embedded in the concrete pile cap so as to form a moment connection, the pile head may be treated as fixed rotation-free translation. This is mimicked in the software by input of zero rotation with increments of shear. Which end conditions to use is generally a joint decision of the pile designer and the structural engineer.

At some depth, movement of the shaft under lateral loads is imperceptible. This occurs at a depth of approximately 10 feet [3 m] in the results shown in Figure 10.4. The point at which deflection becomes negligible is the minimum required length for a flexible pile under lateral loads. Design axial loads may require additional pile length.

Puri et al. (1984) performed calculations based on elastic theory and nonlinear p-y type analysis with L-Pile™ on flexible helical pile shafts deeply embedded in soil and compared the results with tests performed on small-scale laboratory samples and previously published full-scale lateral load tests. A main conclusion of their study is that even slender helical piles have some lateral capacity and that capacity can be validly estimated using nonlinear p-y type analysis and L-Pile™. A parameter was introduced to account for disturbance of the ground due to the installation process of helical piles. This parameter, termed C_u, was determined to be approximately 3.0 through correlations with lateral load test data. The effect of incorporating this parameter into p-y type analysis is to increase deflection under lateral loads directly through multiplication by C_u.

Perko (2000) used L-Pile™ to estimate the lateral load-deflection curves of helical piles. Helical piles with 3.0-inch O.D., 0.25-inch-thick wall, high-strength structural tube shafts were considered. Fixed-slope, free-translation pile head boundary conditions were incorporated in this analysis. These conditions are indicative of a pile that is rigidly fixed to a structure so as to resist maximum bending moments. However, the entire foundation could translate laterally. The soil conditions used in the analysis are shown in Chapter 3, Table 3.7. Shaft mechanical properties were obtained from the AISC (2001) code. The results of the analysis are given in Figures 10.5 and 10.6. The results indicate that between 3,000 and 6,000 pounds [13 and 27 kN] of lateral load generally can be applied in good soil conditions for $\frac{1}{2}$ inch [13 mm] of pile head deflection. These loads are not large, however even 3,000 pounds [13 kN] is sufficient to support lateral wind pressure of 28 psf [1.3 kPa] on an 8-foot ×8-foot [2.4 m ×2.4 m] section of wall or the lateral earth pressure behind a 4-foot [1.2-m] tall ×9-foot- [2.7-m-] long crawl space wall (with a factor of safety of 1.7). A parameter to account for disturbance of ground due to installation of helical bearing plates per Puri et al. (1984) was not taken into account by Perko (2000). However, the lateral loads were

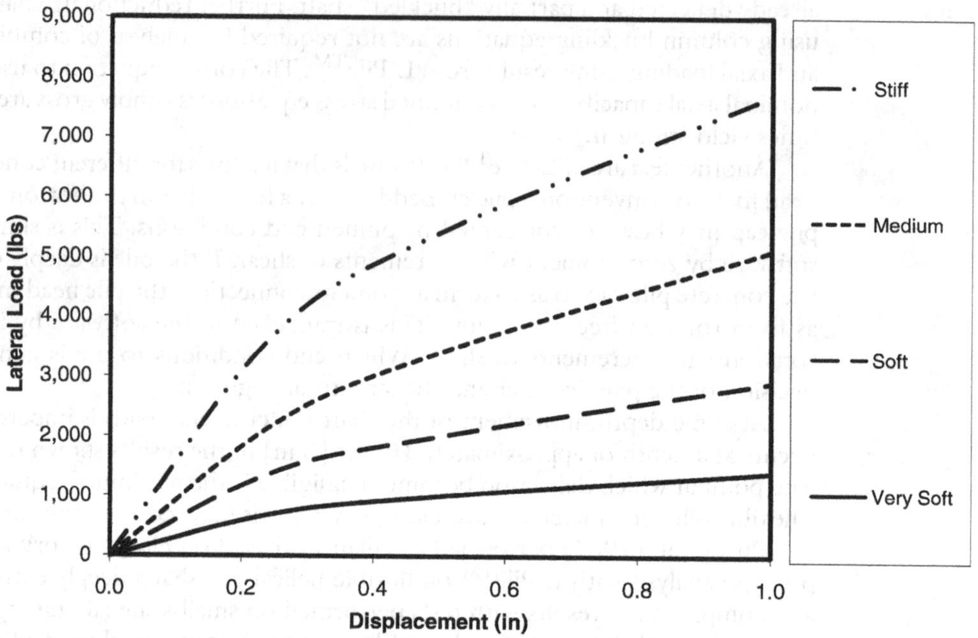

Figure 10.5 Lateral resistance of 3-inch- [76-mm-] diameter shaft helical pile in fine-grain soils (Perko, 2000)

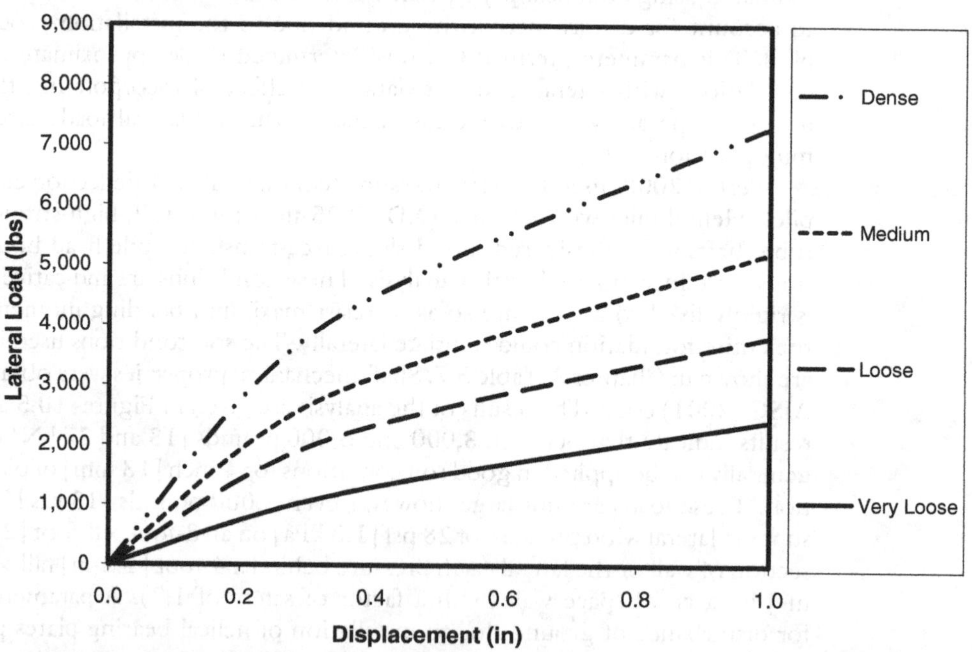

Figure 10.6 Lateral resistance of 3 inch- [76-mm-] diameter shaft helical pile in coarse-grain soils (Perko, 2000)

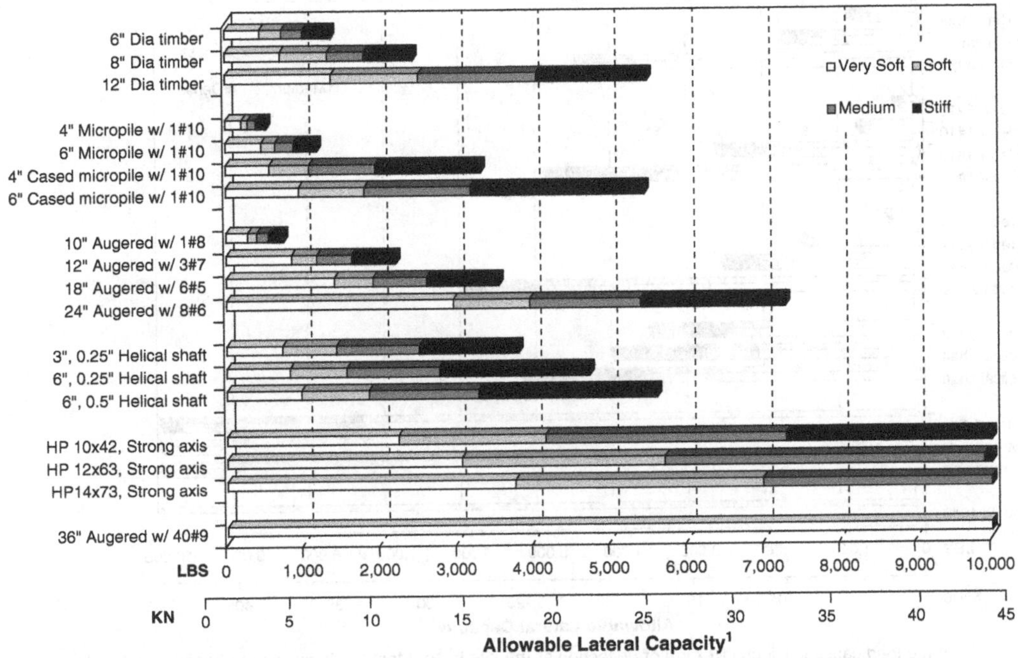

[1]Half the load causing 1 inch [13 mm] of deflection at the pile head. Most loads shown are based on allowable deflection. Some are limited by maximum allowable bending moment in shaft for given reinforcement.

Figure 10.7 Lateral load capacity comparison in fine-grain soils

confirmed in five separate load tests performed in stiff clays. Perko (2000) recommended more study is needed for various pier shaft configurations and soil conditions to determine positively if the soil disturbance parameter recommended by Puri et al. is justified.

The lateral capacity of several different pile types based on L-Pile^{TM} analysis is shown in Figures 10.7 and 10.8 for fine-grain and coarse-grain soils, respectively. In the construct of these figures, the allowable lateral load was defined as half the load causing 1 inch [25 mm] of deflection. Different soil consistencies and densities are represented by the different shades of gray in the three-dimensional bars. Various timber piles are shown at the top of the figure. These are followed by 4-inch- and 6-inch- [102-mm- and 152-mm-] diameter micropiles reinforced with a No. 10 reinforcing bar in the center. The lower micropiles have a permanent steel casing with 0.25 inch [6 mm] wall and an internal diameter matching the micropiles. Next, a series of drilled-shaft concrete piles are shown with various amounts of reinforcement. Three different helical pile shafts were analyzed with 3-inch- [76-mm-] and 6-inch- [152-mm-] diameter shafts having 0.25 inch [6 mm] and 0.5 inch [13 mm] wall thickness. Finally, three different H-piles and a larger-diameter, heavily reinforced, drilled-shaft concrete pile were analyzed. It should be noted that the lateral resistance of these piles was truncated to keep the scale of the chart to a level such that the capacities of the smaller piles can

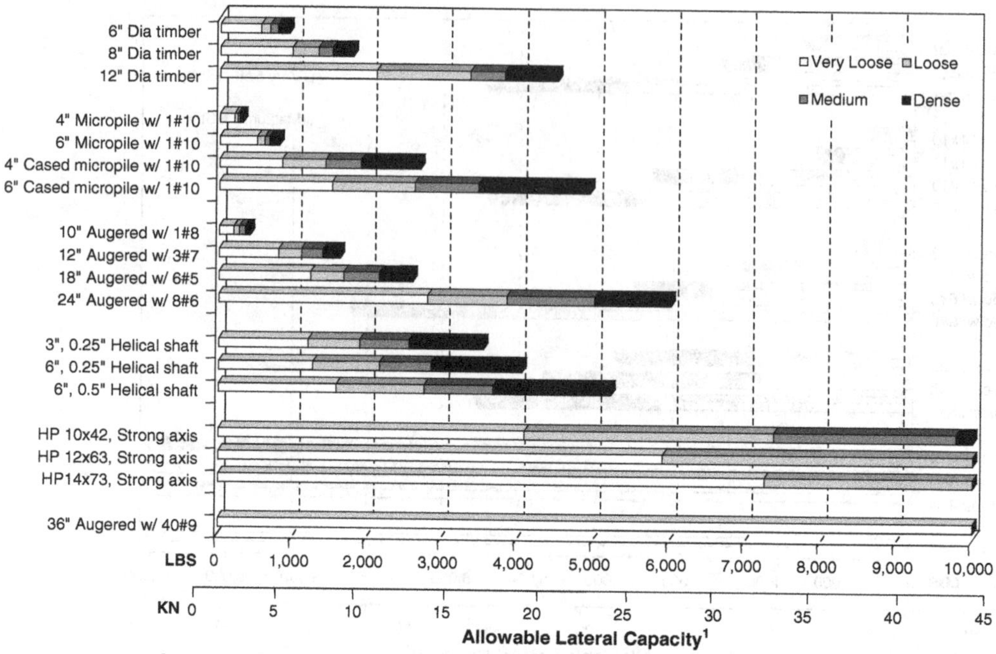

[1]Half the load causing 1 inch [13 mm] of deflection at the pile head. Most loads shown are based on allowable deflection. Some are limited by maximum allowable bending moment in shaft for given reinforcement.

Figure 10.8 Lateral load capacity comparison in coarse-grain soils

be read easily. The capacity of these piles in stiff or dense soils is much greater than shown by the scale in the charts.

The charts given in Figures 10.7 and 10.8 are intended as a guide to the pile designer in selecting the appropriate pile for the anticipated lateral loads. As can be seen, the anticipated lateral capacity of the helical pile is on the same order of magnitude as timber piles, micropiles, and smaller-diameter drilled-shaft piles in similar soil conditions. If larger lateral loads are required to be resisted, H-pile sections and large heavily reinforced augered concrete piles are more appropriate. It is for this reason that H-piles are used for soldier pile walls and large-diameter concrete piles are used in secant pile walls. Lateral loads experienced by many buildings typically can be accommodated by the lighter pile sections.

10.3 PILE GROUPS

Due to three-dimensional effects, lateral soil pressure on a single pile in ground under lateral loads is on the order of two to three times greater than the Rankine passive earth pressure. In fact, Broms's method for calculating the lateral capacity of rigid piles in sand uses three times the passive soil resistance. Another way to think about this is that the lateral resistance provided to one pile is equivalent to using passive soil resistance over three shaft diameters. If helical piles are closer together than three shaft diameters, then it is reasonable to expect their lateral stress distributions to overlap. The resisting pressure should approach Rankine-like behavior at a center-to-center spacing of one diameter, which is essentially a wall of piles (e.g., secant piles or tangent piles).

According to Fleming et al. (1985), the lateral soil pressure on piles also may be limited in a direction parallel with a line of piles. In this case, it is necessary to check for block failure between adjacent pile shafts by ensuring the shear strength of the rectangular block of soil between the shafts is sufficient to provide the required lateral soil pressures.

Helical piles are generally spaced several helix diameters on-center to avoid group efficiency effects, as discussed in Chapter 4. Since helical pile shafts are generally several times smaller than the helix diameter, their shafts are very far apart relative to their shaft diameter, and the lateral capacity of a group of helical piles will rarely be critical in design.

In situations where there is a large component of lateral load, it is important to ensure that the maximum bending moments induced in the piles do not exceed the allowable flexural resistance. The maximum bending moments of piles in a group can be estimated using the flexible pile analysis method for an individual pile (Fleming et al., 1985).

10.4 EFFECT OF HELICAL BEARING PLATES

The lateral capacity of long, flexible helical piles was studied by Prasad and Rao (1996). Laboratory tests were performed on small-scale helical piles embedded in clays. The ratio of length to helical bearing plate diameter in these tests varied from 12 to 18. It was found that the presence of helical bearing plates resulted in an increase in lateral capacity that was 1.2 to 1.5 times that of slender piles without helical bearing plates.

One of the conclusions of the study by Puri et al. (1984) on both small-scale laboratory tests and full-scale helical piles was that the lateral capacity of helical pile shafts is independent of the presence of helical bearing plates and is controlled almost exclusively by the mechanical properties of the shaft for depths of helical pile embedment greater than three to five times the critical stiffness factor (6 to 10 feet [1.8 m to 3.0 m] for commonly manufactured helix foundations).

Puri et al.'s work regarded both small-scale and full-scale helical piles in sand and clay. Prasad and Rao's work involved small-scale laboratory models in clay only. There may be a scaling phenomenon associated with the effect of helical bearing plates or differences caused by soil conditions. Until the discrepancy between the previous work of Prasad and Rao and that of Puri et al. is resolved, the effect of helical bearing plates on the lateral resistance of helical piles may be ignored to be conservative.

10.5 EFFECT OF COUPLINGS

The free movement of the couplings between helical pile sections was not taken into account in any of the foregoing computations as to the lateral capacity of helical piles. Couplings are uncommon in large-diameter, short helical piles; if used, these couplings are typically rigid, straight, and at least as strong as the shaft. As such, they do not need to be taken into account. The couplings used with more slender helical piles vary in strength and stiffness from manufacturer to manufacturer. Most are fairly rigid and at least as strong in shear and flexure as the shaft. Some of the more loose-fitting couplings can exhibit a slight free departure from straightness under lateral loads. This fact may in part explain some of the increased deflection measured in helical piles as discussed herein. Depending on the manufacturer, the forged upset couplings typically used with square-shaft helical piles can exhibit significant free departure from straightness under applied lateral loads. This combined with the fact that a square shaft creates a round annulus in the ground are reasons that square-shaft helical piles are considered to have negligible lateral load resistance.

Additional studies should be performed to determine the effect of couplings on lateral capacity. With the publication of AC358 (ICC-ES, 2007) and additional attention on the lateral capacity of helical piles, there is a trend in industry toward eliminating any free departure from straightness under lateral loads and enhancing the rigidity of helical pile couplings. The pile designer may find it necessary to specify minimum coupling rigidity in terms of free departure from straightness and strength when using helical piles to resist lateral loads.

10.6 LATERAL LOAD TESTS

Helical pile lateral load tests are typically conducted in accordance with ASTM D3966. In this method, a load frame can be constructed to apply lateral loads to a pile, or a simpler approach is to either push two piles apart or pull them together. In this way, two lateral load tests can be conduced simultaneously. Examples of a lateral load tests on slender-shaft helical piles are shown in Figures 10.9 and 10.10. The helical piles in Figure 10.9 are being pulled together with the use of a come-along and chain. Loads are being measured with a crane scale. The helical piles in Figure 10.10 are being pushed apart using a hydraulic jack mounted between two tubular struts. The struts slide freely in a lubricated housing attached to a steel channel to prevent buckling. Loads are being measured by pressure readings on the calibrated jack. Hydraulic pressure may be used to measure applied loads provided the ram, pressure gauge, and hydraulic pump are calibrated as a unit. According to the ASTM, the lateral load is to be applied as close to the pile cutoff elevation as practicable as shown in Figure 10.10. If piles are required to resist both lateral load and overturning moment, it may be necessary to apply the lateral load at some distance above the ground surface as shown in Figure 10.9.

ASTM D3966-07 requires that a primary and a secondary method of measurement be incorporated to monitor axial movement of the test pile. The primary method

Figure 10.9 Lateral load testing on helical piles by pulling together (Courtesy of Magnum Piering, Inc.)

Figure 10.10 Lateral load testing on helical piles by pushing apart (Courtesy of CTL|Thompson, Inc.)

Figure 10.11 Dial gage used to measure lateral displacements

typically consists of dial-gage extensometers as shown in Figure 10.10. A closeup view of one of the dial gages is shown in Figure 10.11. The photograph creates an optical illusion that suggests there is insufficient distance between the pile loading device and the reference beam. However, the reference beam is located at a distance above the loading apparatus such that there is a minimum clear distance greater than 2 inches [50 mm]. The dial gage in this photograph is located near the top of the pile shaft. The dial gage should contact the pile shaft closer to the pile cutoff elevation at the point of loading in a proper setup. ASTM D3966 requires that the supports for reference beams be at least 8 feet [2.5 m] away from the test piles.

The secondary method of pile head displacement measurement often consists of a ruler or engineer's scale affixed to the pile shaft as shown in Figure 10.9. Movement of the pile is detected through the use of an optical level or survey transit trained on the scale (not shown). Although these methods are shown separately in the figures, they should be combined to be in strict accordance with the ASTM. A wire, mirror, and scale system also may be considered as a secondary method of measuring pile head displacements.

ASTM D3966-07 contains several different loading procedures for lateral load testing of piles. The loading procedures have similar names as those for axial load tests but they differ in the load increments, hold time, and required displacement readings. The various procedures include standard, cyclic, excess load, surge loading, reverse loading, reciprocal loading, loading to a specified lateral movement, and combined loading. Unlike the axial load test, a quick test method is not provided in the ASTM for lateral load testing. The standard load procedure is used most often with helical

Table 10.1 Standard Lateral Loading Procedure (Adapted From ASTM D3966-07)

	Load[1]	Hold Time	Readings[2]
Loading	15%[3]	NA	zero
	25% 50% 75%	10 min 10 min 15 min	Immediately before and after each increment and at 5-min intervals
	100% 125% 150% 170% 180% 190%	20 min	
Test Load	200%	60 min	Immediately before and after each application of test load and at 15-min intervals
Unloading	150% 100% 50%	10 min	Immediately before and after each decrement and at 5-min intervals
	0%	30 min	15 min and 30 min after unloading

[1] Percent of design load
[2] Time, load, and movement
[3] Not specifically referenced in ASTM

piles. General rules for load increments, hold times, and readings for the standard load procedure are shown in Table 10.1. This method is described in more detail in the next paragraph. The reader is referred to the ASTM for other loading methods.

The standard load test procedure (Table 10.1) involves loading the pile in 25 percent increments to 200 percent of the design load and holding the test load for a minimum of 1 hour. After the hold period, the pile is unloaded in 50 percent decrements and a final reading is taken 15 minutes and 30 minutes after removing all loads. Time, deflection, and load readings are taken before and after applying each load increment/decrement and at regular intervals during hold times.

In some cases, it is permissible to test helical piles embedded in a concrete pile cap. The lateral load test shown in Figure 10.12 is an example of this. In this case, large diameter holes were augered to a depth of approximately 6 feet [2 m], the helical piles were installed in the center of the holes, and the holes were filled with concrete. A square pile cap was formed at the top of each pile. The load test was performed to evaluate the lateral resistance offered by the enhanced concrete pile caps in combination with the helical piles.

All of the preceding discussion regarded the free-head lateral load test method. ASTM D3966 also allows for a fixed-head version of the test. For the fixed-head test, a steel truss is to be connected to the pile shaft, which should extend several feet

Figure 10.12 Lateral load test on concrete pile caps formed over helical piles (Courtesy of Magnum Piering, Inc.)

[meters] above the ground surface. The truss is prevented from rotation using a steel roller and plate support system. A diagram of a fixed-head test is shown in Figure 10.13. According to the ASTM, a minimum clear distance of 10 feet [3 m] must be maintained between the test pile and the support for the end of the truss. For practical purposes, it can be assumed that the vertical reaction at the roller end of the truss is equal to the lateral load being applied to the test pile at the ground surface. The rollers must bear on a pile or footing of sufficient size to accommodate the applied loads with minimal deflection.

The fixed-head lateral test is difficult and costly to construct. In lieu of the fixed-head test, most practitioners allow a free-head test to be conducted. The results are compared with a free-head lateral pile analysis using L-Pile^{TM} or similar method. The calculations are adjusted to match the field test results. Then the results are extrapolated using the software to simulate the fixed-head condition.

10.7 EMPIRICAL RESULTS

The results of 32 lateral load tests performed on helical piles with 2.875-inch- to 10.75-inch- [73- to 273-mm-] diameter round shafts from 11 sites with fine-grain and coarse-grain soils were evaluated. The lateral capacity of these piles was estimated using L-Pile^{TM} software and the index soil properties contained in Chapter 3,

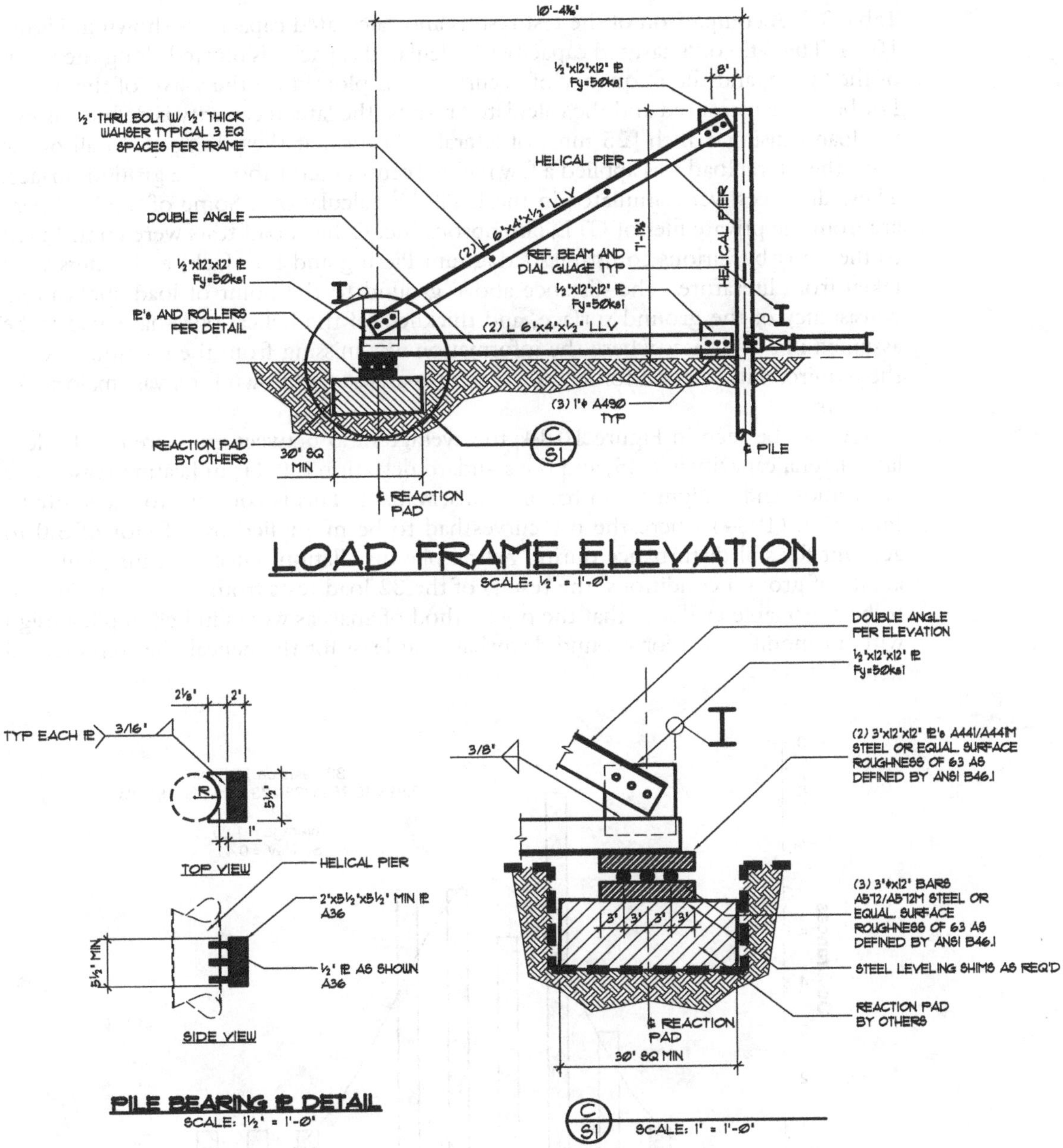

Figure 10.13 Example fixed-head helical pile lateral load test setup

Table 3.7. A comparison of the test results and calculated capacity is shown in Figure 10.14. The ratio of measured capacity to calculated capacity is plotted along the x-axis of the figure, and the frequency of occurrences is plotted on the y-axis of the figure. For both the measured and the calculated results, the lateral capacity is defined as half the load causing 1 inch [25 mm] of lateral deflection at the pile head. In all of the tests, the lateral load was applied a few inches [centimeters] above the ground surface. These distances were simulated in the L-Pile^{TM} calculations. Some of the load tests are from the private files of CTL|Thompson, Inc. Other Load tests were contributed to the book by various companies (Magnum Piering and RamJack) and others were taken from literature. The distance above ground to the point of load application, consistency of the ground surface, and thickness of the helical pile shafts had to be assumed in a few cases where the information was missing from the literature. All of the required information for lateral capacity analysis was known for a vast majority of the tests.

As can be seen in Figure 10.14, the average ratio between measured and calculated lateral capacity is 1.25, and the standard deviation is 0.41, indicating reasonable correlation and a slight trend toward conservatism. This is contrary to the study by Puri et al. (1984) where the p-y curves had to be multiplied by a factor of 3.0 to account for soil disturbance during helical pile installation. Given the inherent variability of ground conditions, the results of the 32 load tests from 11 sites are judged to be reasonable evidence that the p-y method of analysis works in helical pile design without modification for ground disturbance at least for the helical piles represented here.

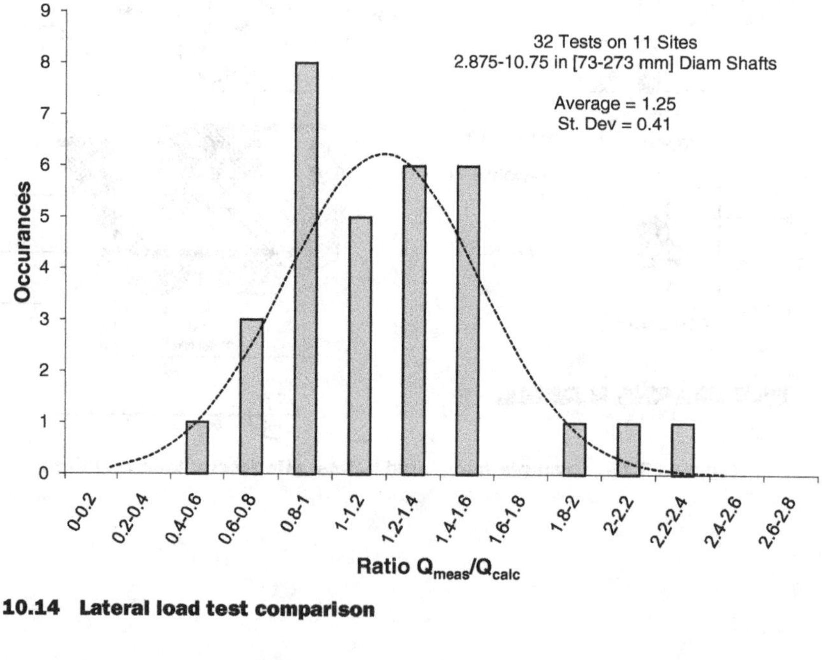

Figure 10.14 Lateral load test comparison

10.8 LATERAL RESTRAINING SYSTEMS

There are several methods for increasing the lateral resistance of helical pile–supported foundations. One method is to augment the helical pile with an augered concrete cap. A drawing showing the sequence of construction for an augered concrete cap is contained in Figure 10.15. First, a large diameter hole is drilled at each helical pile location. The hole should be of sufficient depth and diameter to resist design lateral loads. Next, a helical pile is installed in the center of each predrilled hole. The helical pile should be installed to sufficient depth and torque below the bottom of the predrilled hole to resist the anticipated axial bearing and pullout forces. A square pile cap can be formed at the ground surface to facilitate attachment to a structure. Reinforcing steel can be placed in the drill hole around the helical pile shaft if necessary to resist the anticipated shear and bending moments. Then the hole is filled with concrete. If the

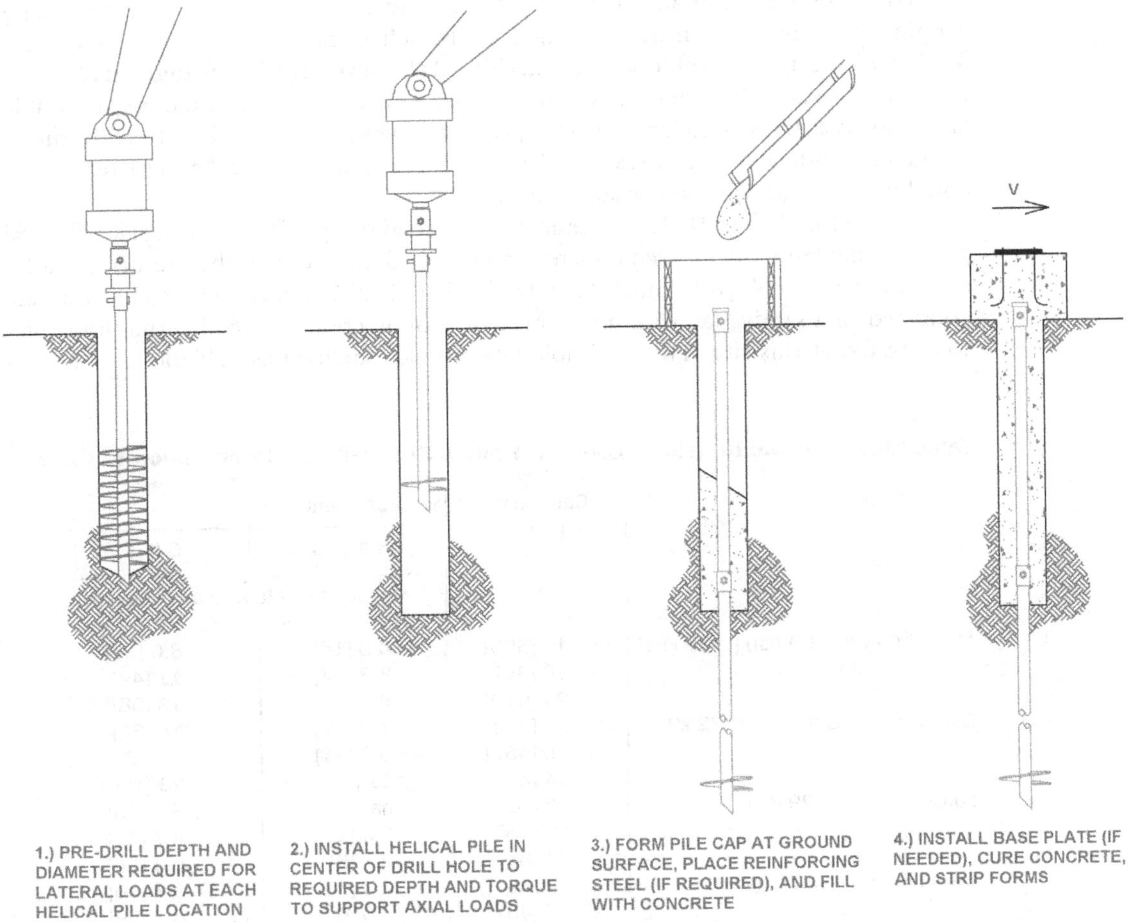

Figure 10.15 Helical pile with augered concrete cap

hole is narrow and fairly deep, a tremie pipe should be used for concrete placement to avoid segregation. Finally, a base plate or anchor bolts can be set in the concrete if needed. It is not always necessary to construct a formed cap at the ground surface. Attachment to the structure could be accomplished by extending the helical pile or reinforcing steel out of the top of the augered cap.

Construction of an augered cap simply requires a shallow post-hole auger. These augers are readily available at most equipment rental shops and can be easily adapted to work with the same hydraulic torque motor used for helical pile installation. In order for this method to work properly, the near-surface soils must have sufficient stand-up time to prevent caving while the holes are open and during helical pile installation. If the soils are clean coarse-grain materials and prone to caving, a temporary or permanent steel casing could be used to retain the soil during helical pile installation and concrete placement.

The required depth and diameter of the augered cap can be determined using a rigid pile analysis, such as the Broms or Brinch-Hansen method. For reference, Table 10.2 lists the allowable lateral load (F.S.=2) for several different augered concrete cap sizes based on the Brinch-Hansen method. As can be seen in the table, allowable lateral loads as high as 23 kips [100 kN] can be achieved using a 24-inch- [609-mm-] diameter by 6-foot 0-inch- [1.8-m-] deep augered cap in stiff clay. Larger lateral loads could be achieved with bigger augered caps.

The picture in Fig. 10.12 shows an example load test performed on a pair of helical piles augmented with augered concrete caps. The dimensions of the pre-drilled holes in this test were 24"[609 mm] diameter by 3'-0"[0.9 m] deep. The soil at this site consisted of a sandy clay with an average SPT blow count of 16. In total four piles were tested at this site. The allowable lateral loads, defined as half the lateral loads

Table 10.2 Allowable Lateral Load for Helical Piles with Augered Concrete Caps

	Cap Diameter in [mm]	Cap Length 3 ft [0.9 m]	Cap Length 6 ft [1.8 m]
		Allowable Lateral Load (kips) [kN]	
M. Stiff Clay (c=800 psf) [c=38 kPa]	12 [305]	4.0 [18]	8.0 [36]
	18 [457]	5.3 [24]	11 [49]
	24 [610]	6.7 [30]	13 [58]
Stiff Clay (c=1,500 psf) [c=72 kPa]	12 [305]	7.3 [32]	14 [62]
	18 [457]	9.3 [41]	19 [85]
	24 [610]	11 [49]	23 [100]
Loose Sand (phi=29 deg)	12 [305]	na	3.3 [15]
	18 [457]	0.5 [2]	4.7 [21]
	24 [610]	0.7 [3]	5.3 [24]
Dense Sand phi=39 deg)	12 [305]	0.7 [3]	6.7 [30]
	18 [457]	1.3 [6]	8.7 [39]
	24 [610]	2.7 [12]	11 [49]

causing 1"[25 mm] pile head deflection, ranged from 8.8 kips to 14.0 kips [39 to 62 kN] with an average of 11.2 kips [50 kN]. These results are a very good match to the predicted capacity of 11 kips shown on Table 10.2.

Another technique for increasing the lateral load carrying capacity of helical piles is to install them at a batter angle, as shown in Figure 10.16. Each helical pile contributes both a vertical component and a lateral component of force to the pile cap. In using this approach, the required allowable load of each pile has to be increased beyond that required to support the compressive load in order to take advantage of the lateral component of force. Referring to the vectors shown in the figure, when V is equal to zero, the horizontal components of both piles balance each other if their batter angles are the same. That is, P_1 equals P_2. The total vertical load carrying capacity is two times the axial load in each pile times the cosine of the batter angle (a in the figure). When a lateral load, V, is applied to the pile cap in the direction shown, the axial load in the first pile decreases and the vertical and lateral load increases on the second pile until the limit where P equals P_2 times cosine a. The maximum lateral load, V, is P_2 times sine a. The same is true in the opposite direction. One cannot take advantage of the lateral component of force in this configuration unless the load on one of the piles decreases and the load on the other increases. In order to keep the axial load on the pile battered in the direction of the applied lateral load below its maximum allowable load, it must be designed to carry much less load than its full capacity when in equilibrium with the other pile. The algebra and trigonometry involved in designing a pair of battered helical piles is further examined in the next example.

Example 10a

Problem: Determine the required axial capacity of a set of battered helical piles and the appropriate batter angle to carry the design compressive and lateral load combination.

Given: A compression load of 100 kips [445 kN] and a lateral load of 15 kips [67 kN] is to be carried by a pile cap with two battered helical piles.

Answer: There is an infinite number of possibilities. Under the applied vertical and horizontal loads, one answer is that all of the load is transferred to the helical pile battered in the direction of the applied lateral load such that

$$100\text{ kips} = P_2 \cos(a) \tag{10a.1}$$

$$15\text{ kips} = P_2 \sin(a) \tag{10a.2}$$

The simplest way to solve these equations simultaneously is using vectors by drawing a right triangle with one leg having length 15 and the other leg with

(Continued)

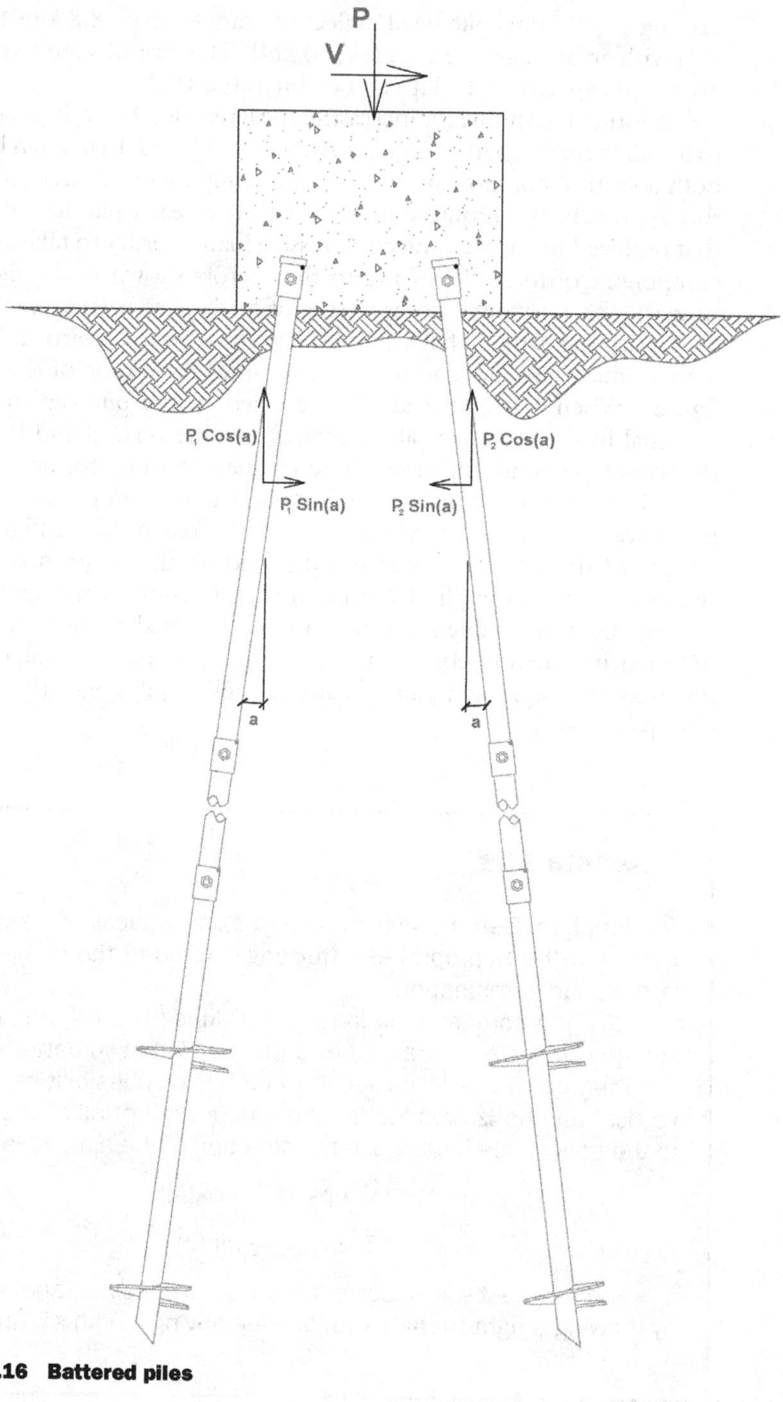

Figure 10.16 Battered piles

length 100. The short leg is opposite the angle a. The magnitude of P_2 is simply found by the Pythagorean theorem:

$$\sqrt{15^2 + 100^2} = 101 \text{ kips} \tag{10a.3}$$

The batter angle is obtained from trigonometry.

$$\arctan\left(\frac{15}{100}\right) = 8.5^\circ \tag{10a.4}$$

Each pile is required to carry a design load of 101 kips [449 kN] and should be installed at a batter angle of 8.5 degrees. What if the batter angle is increased? Can the required axial load on the piles be reduced? In order to answer these questions, we can look at simple statics. The equations governing equilibrium are given by

$$100 \text{ kips} = (P_1 + P_2)\cos(15^\circ) \tag{10a.5}$$

$$15 \text{ kips} = (P_2 - P_1)\sin(15^\circ) \tag{10a.6}$$

Solving for P_2 in terms of P_1 from Equation 10a.6 and substituting the result into Equation 10a.5 yields

$$P_1 = 22.8 \text{ kips} \tag{10a.7}$$

Substituting this answer into either of the equations yields

$$P_2 = 80.7 \text{ kips} \tag{10a.8}$$

Hence, the answer is yes; the required axial load on the piles can be reduced by increasing the batter angle to 15 degrees in this example. Since lateral loads typically act in opposite directions equally, both piles should be designed for an axial load of 80.7 kips [359 kN] at a batter angle of 15 degrees.

If increasing the angle reduces the load, at what batter angle is the required axial capacity a minimum? The answer is more difficult. The objective is to find batter angle a where P_2 is a minimum. One can combine Equations 10a.5 and 10a.6 and solve in terms of P_2:

$$P_2 = \frac{1}{2}\left(\frac{100}{\cos a} + \frac{15}{\sin a}\right) \tag{10a.9}$$

Those with excellent math skills or access to MathCAD™ or other software could take the derivative of Equation 10a.9 with respect to the batter angle and set the result equal to zero and search for minima. Otherwise, one could set up a spreadsheet with Equation 10a.9 and plot different combinations to

(Continued)

find the minimum value of P_2. Such an exercise yields this set of results:

a (deg)	P_2(kips)
8	104.4
10	94.0
12	87.2
14	82.5
16	79.2
18	76.8
20	75.1
22	73.9
24	73.1
26	72.7
28	72.6
30	72.7
32	73.1
34	73.7
36	74.6
38	75.6

The approximate minimum value is 72.6 kips [323 kN] at a batter angle of 28 degrees. The angle at which the axial load is a minimum depends on the combination of compressive and lateral loads. Note that the value is still roughly 50 percent higher than the allowable load required to carry the compressive load alone. Also note that beyond the minimum angle, the axial load actually starts to increase. Once the spreadsheet is set up, the helical pile designer can use it repeatedly to find optimum batter angles for different load combinations.

Still another method of increasing the lateral capacity of a foundation supported on helical piles is to use a "jack leg" or tie-back anchor installed at a shallow angle and attached to the foundation as shown in Figure 10.17. The vertical pile in this detail is designed to resist axial compression forces, while the tie-back anchor is used to resist lateral loads. One caution that the designer should be aware of is that under lateral loads, the vertical component of the tie-back will exert additional compression loads on the vertical helical pile. These loads need to be taken into account in sizing the vertical member. This technique often is used for thrust blocks on heavy buildings to resist seismic loads. As many as 25 helical anchors were used in each of the corner thrust blocks for a multistory reinforced concrete university building in Salt Lake City, Utah. Another example where this detail is common is for structures with walk-out basements on steep slopes where additional lateral support is required to resist sliding from unbalanced backfill forces around the foundation.

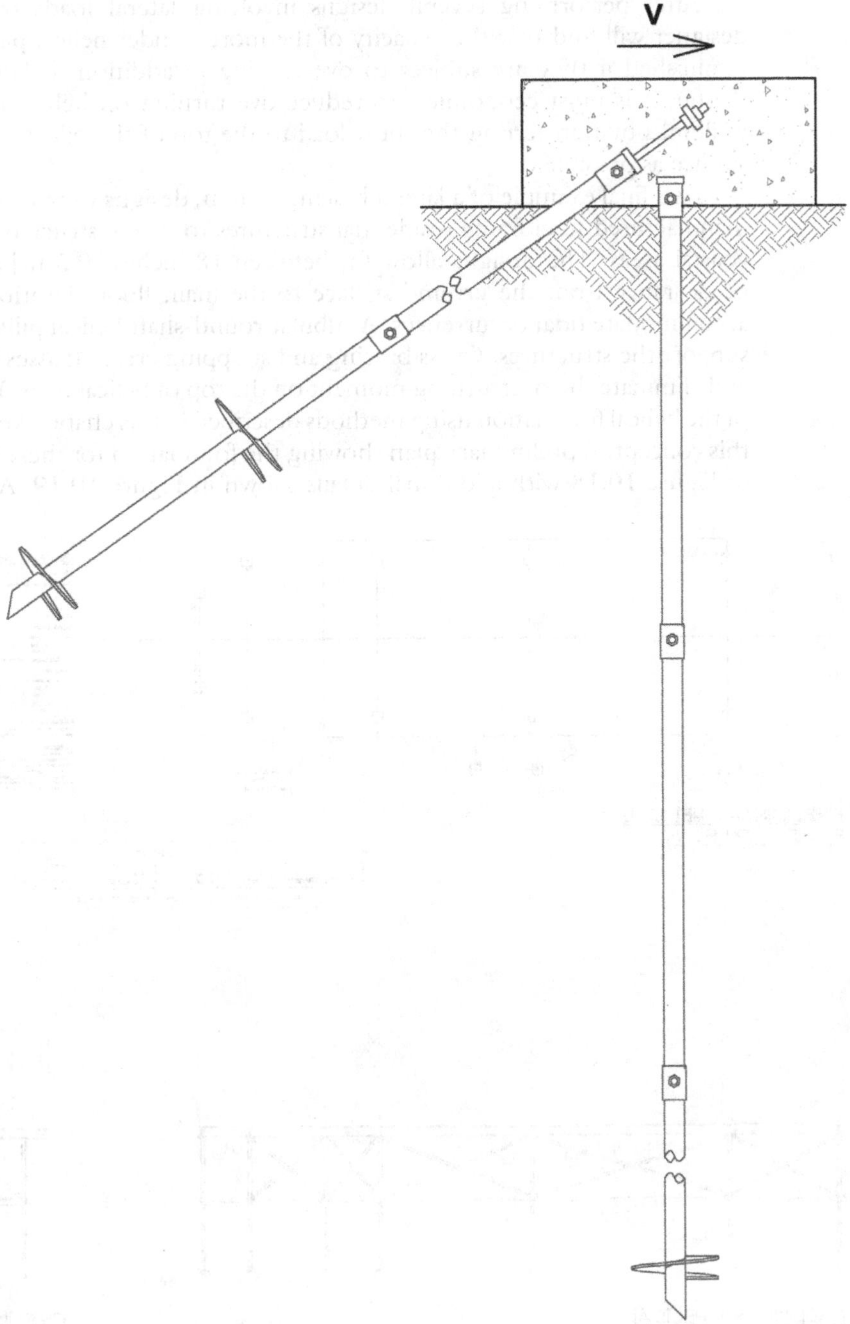

Figure 10.17 Tie-back application

After performing several designs involving lateral loads on helical piles, the designer will find that the capacity of the more slender helical pile systems is greatly diminished if they are subject to overturning in addition to lateral loads. For this reason, it is most economical to reduce overturning on helical piles for support of buildings by transferring the shear load to the top of the pile as close to the ground surface as possible.

As a final example of a lateral bracing system, designs were recently completed for manufactured steel frame residential structures to be constructed in hurricane-prone coastal areas. The designs allow for between 18 inches [0.5 m] and 7 feet [2.1 m] of clearance from the ground surface to the main floor elevation of the homes to accommodate tidal occurrences. A tubular round-shaft helical piling system is used to support the structures. Cross bracing and strapping create trusses beneath the homes and eliminate the overturning moment on the top of helical piles. Lateral load analysis of the helical foundation using methods described in this chapter verified the validity of this concept. A preliminary plan showing the foundation for these homes is contained in Figure 10.18 with additional details shown in Figure 10.19. A three-dimensional

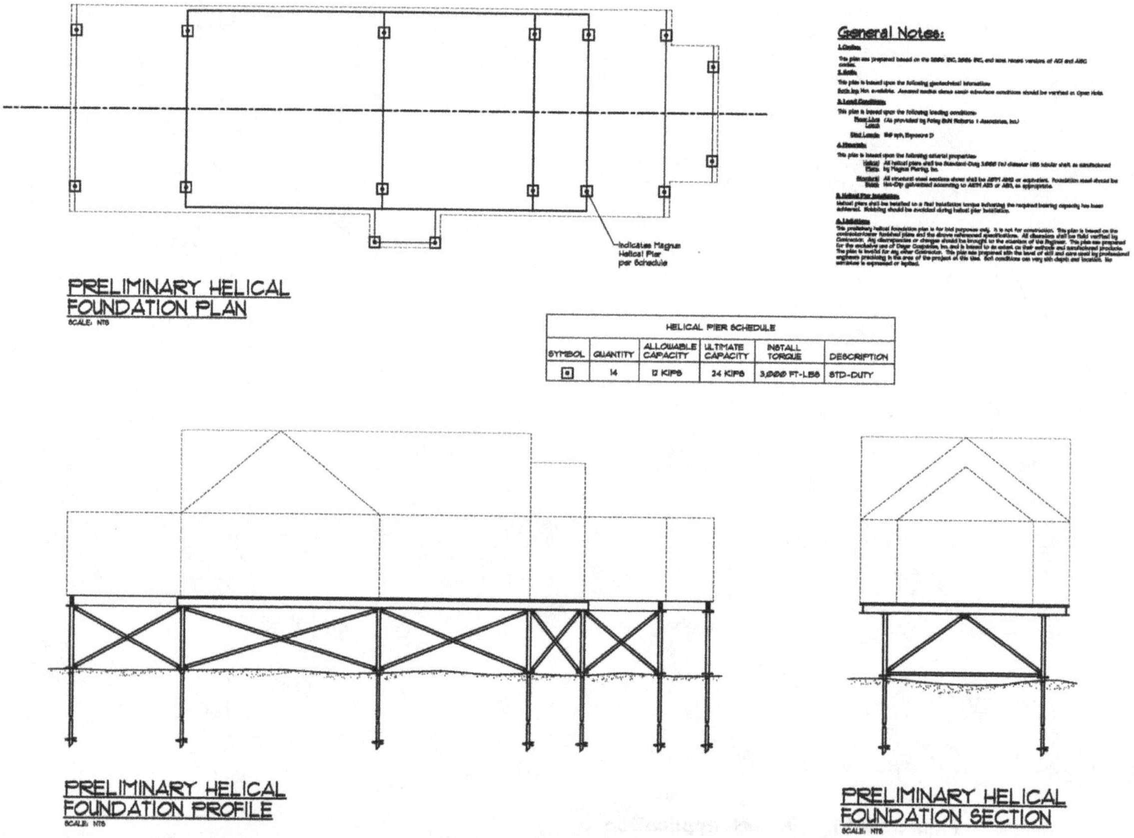

Figure 10.18 Elevated and braced foundation system on helical piles in hurricane-prone area (Courtesy of Magnum Piering, Inc.)

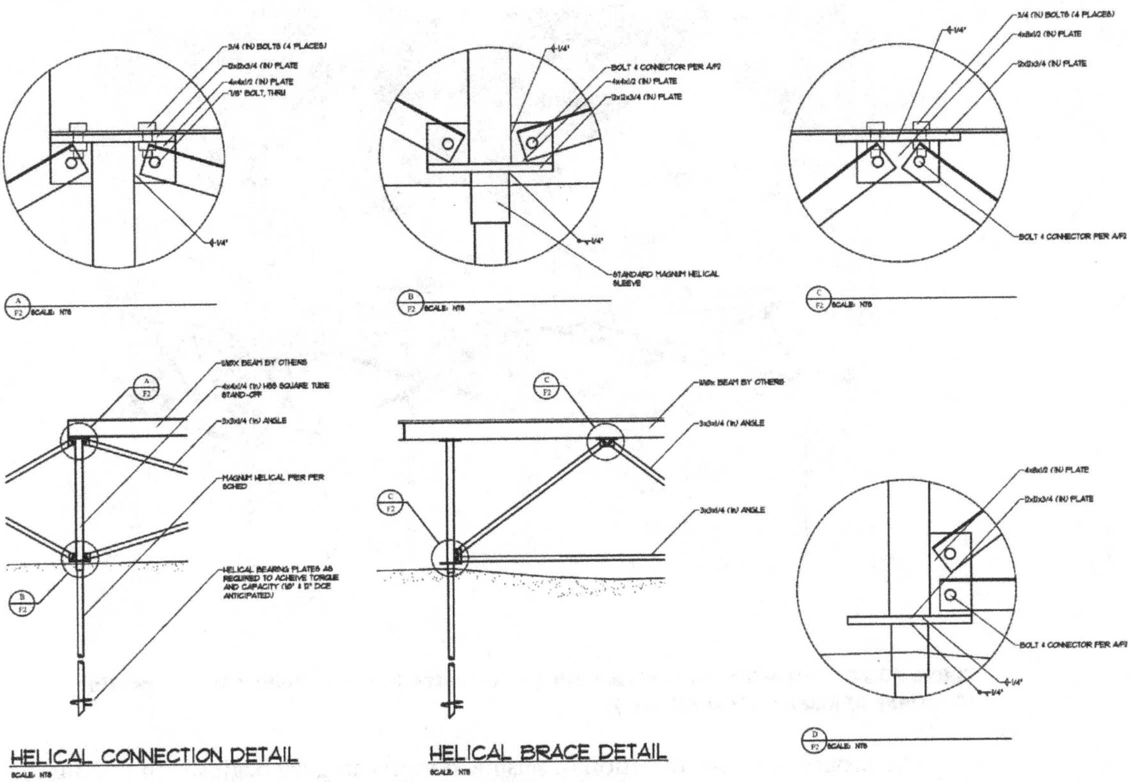

Figure 10.19 Details for elevated and braced foundation system on helical piles (Courtesy of Magnum Piering, Inc.)

rendering of an example home is contained in Figure 10.20. The helical piles and the structure's steel moment frame are being manufactured as a kit that can be easily transported to remote island areas where labor and construction materials are in demand. At the end of the project feasibility study, helical foundation systems were found to be more economical, reliable, and faster than the concrete piles or concrete footings. Helical foundations also provide unmatched wind uplift resistance (Perko, 2007a).

10.9 SEISMIC RESISTANCE

Helical pile foundations currently support many structures in seismically active regions of the United States. All foundations are required to resist earthquake loads except those for one- and two-family dwellings is seismic design categories A, B, or C where mapped short-period response is less than 0.4 times the acceleration due to gravity (g) (IBC, 2006). The *International Building Code* (IBC, 2006) requirements for design of pile foundations depend on the seismic design category. It is important for helical pile designers to determine this category (Perko, 2007b).

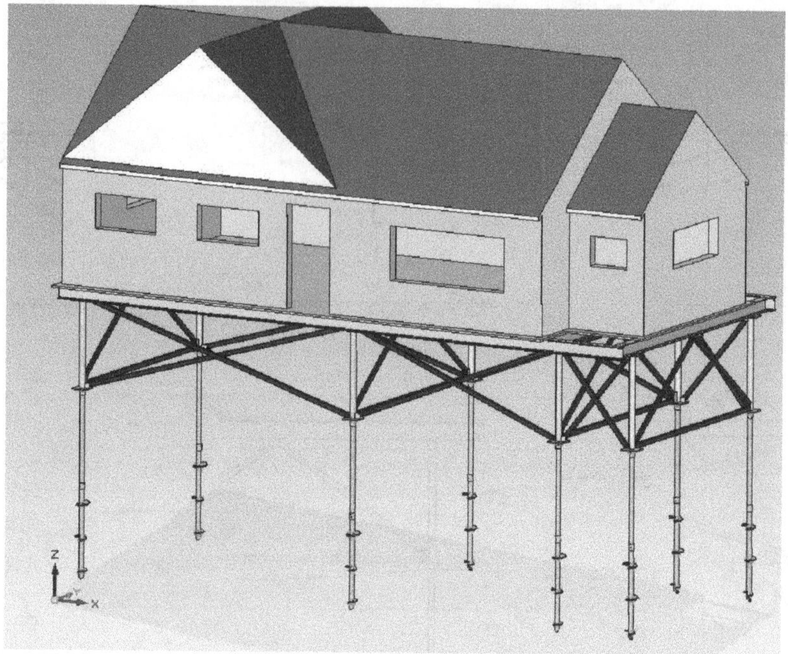

Figure 10.20 Three-dimensional rendering of elevated home on helical pile foundation (Courtesy of Magnum Piering, Inc.)

The process for determination of seismic design category begins with finding the maximum considered short period seismic acceleration at the site, S_S. This value may be obtained from maps contained in the IBC. An example for the contiguous United States is shown in Figure 10.21. The short-period seismic acceleration also may be provided by the local building official, geological survey, or project geotechnical engineer. The next step is to calculate the design acceleration, S_{DS}, given by the IBC (2006):

$$S_{DS} = \frac{2}{3} F_a S_S \tag{10.6}$$

Where

F_a is the site coefficient.

The IBC defines six site classes based on subsurface conditions and shear wave velocity in the upper 100 feet [30m] as shown in Table 10.3 (IBC Table 1613.5.2). Site coefficient, F_a, is found from Table 10.4 (IBC Table 1613.5.3(1)) by matching the site class with the mapped short-period seismic acceleration, S_S.

Seismic design category is determined from Table 10.5 (IBC Table 1613.5.6(1)) based on the design acceleration and the occupancy category. Category I and II are agricultural buildings, storage facilities, and other structures. Category III are high-occupancy structures, and Category IV are essential facilities. The IBC defines

Table 10.3 Site Class Definitions (IBC, 2006)

IBC Table 1613.5.2
Site Class Definitions

Site Class	Soil Profile Name	Average Properties in Top 100 Feet, See Section 1613.5.5		
		Soil Shear Wave Velocity, v_s (ft/s)	Standard Penetration Resistance, N	Strength, s_u (psf)
A	Hard rock	$v_s > 5{,}000$	N/A	N/A
B	Rock	$2{,}500 < v_s \leq 5{,}000$	N/A	N/A
C	Very dense soil and soft rock	$1{,}200 < v_s \leq 2{,}500$	$N > 50$	$s_u \geq 2{,}000$
D	Stiff soil profile	$600 < v_s \leq 1{,}200$	$15 \leq N \leq 50$	$1{,}000 \leq s_u \leq 2{,}000$
E	Soft soil profile	$v_s < 600$	$N < 15$	$s_u < 1{,}000$
E	—	Any profile with more than 10 feet of soil having the following characteristics: 1. Plasticity index $PI > 20$, 2. Moisture content $w \geq 40\%$, and 3. Undrained shear strength $s_u < 500$ psf		
F	—	Any profile containing soils having one or more of the following characteristics: 1. Soils vulnerable to potential failure or collapse under seismic loading such as liquefiable soils, quick and highly sensitive clays, collapsible weakly cemented soils. 2. Peats and/or highly organic clays ($H > 10$ feet of peat and/or highly organic clay where H = thickness of soil) 3. Very high plasticity clays ($H > 25$ feet with plasticity index $PI > 75$) 4. Very thick soft/medium stiff clays ($H > 120$ feet)		

For SI: 1 foot = 304.8 mm, 1 square foot = 0.0929 m^2, 1 pound per square foot = 0.0479 kPa, N/A = Not applicable.

Table 10.4 Site Coefficient, F_a (IBC, 2006)

IBC Table 1613.5.3(1)
Values of Site Coefficient, F_a^a

Site Class	Mapped Spectral Response Acceleration at Short Period				
	$S_s \leq 0.25$	$S_s = 0.50$	$S_s = 0.75$	$S_s = 1.00$	$S_s \geq 1.25$
A	0.8	0.8	0.8	0.8	0.8
B	1.0	1.0	1.0	1.0	1.0
C	1.2	1.2	1.1	1.0	1.0
D	1.6	1.4	1.2	1.1	1.0
E	2.5	1.7	1.2	0.9	0.9
F	Note [b]	Note [b]	Note [b]	Note [b]	Note [b]

[a] Use straight-line interpolation for intermediate values of mapped spectral response acceleration at short period, S_s.
[b] Values shall be determined in accordance with Section 11.4.7 of ASCE7.

two special seismic design categories, E and F, not shown in Table 10.5, for design earthquake accelerations over 0.75g.

There are no special requirements for pile foundations in seismic categories A and B except that the piles need to be designed for seismic load combinations on the structure. Helical piles can be designed using the same procedures for other deep foundations. In seismic category C, piles or pile caps are required to be interconnected with ties capable of tension or compression force, F_s, given by (IBC, 2006)

$$F_s = 0.1 P_c S_{DS} \tag{10.7}$$

Where

P_c is the higher column load.

The connections between piles and the pile cap need to be provided with tension and transverse steel as required for the column. Helical pile couplings need

Table 10.5 Seismic Design Category

IBC Table 1613.5.6(1)
Seismic Design Category Based on
Short-Period Response Accelerations

Value of S_{DS}	Occupancy Category		
	I or II	III	IV
$S_{DS} < 0.167g$	A	A	A
$0.167g \leq S_{DS} < 0.33g$	B	B	C
$0.33g \leq S_{DS} < 0.50g$	C	C	D
$0.50g \leq S_{DS}$	D	D	D

to develop the full tensile strength of the pile or be designed to resist seismic load combinations.

In seismic category D, E, and F, piles and pile caps must meet the requirements of category C. In addition, piles need to be designed to withstand maximum imposed ground motions. Helical piles often are required to resist lateral loads from seismic forces on the structure. Helical piles may be required to act as a column in liquefiable strata. The tensile strength of the connection between the piles and pile cap must resist the lesser of 1.3 times the ultimate tensile capacity of the pile and seismic load combinations. The moment connection between the helical pile and pile cap needs to be designed for the lesser of the flexural strength of the pile and seismic load combinations. The next example demonstrates determination of IBC seismic design category and the special requirements for helical pile design.

Example 10b

Problem: Determine the seismic design category and seismic design considerations for a helical pile foundation

Given: Low occupancy commercial building in Fort Collins, Colorado, on a site with very stiff clay having an average SPT blow count of 35

Answer: Fort Collins is located along the Rocky Mountains in northern Colorado. Based on Figure 10.21 it is very near the seismic contour corresponding to 0.2g. From Table 10.3, the blow count corresponds to site class D. The site coefficient, F_a, equals 1.6 (Table 10.4). The design short period acceleration is determined from Equation 10.6.

$$\frac{2}{3}(1.6)(0.2g) = 0.21g \tag{10b.1}$$

Finally, from Table 10.5, the seismic design category is B for occupancy category II. There are no special requirements for helical piles in seismic category B. According to the IBC simplified method, there are vertical, E_V, and horizontal, E_H, components of force on the structure due to earthquake loading given by

$$E_V = 0.25 S_{DS} P_{dead} \tag{10b.2}$$

$$E_H = \rho C_S W \tag{10b.3}$$

Where

P_{dead} is dead load

ρ is redundancy factor

$C_S W$ is the base shear (seismic response coefficient times weight of the structure).

(Continued)

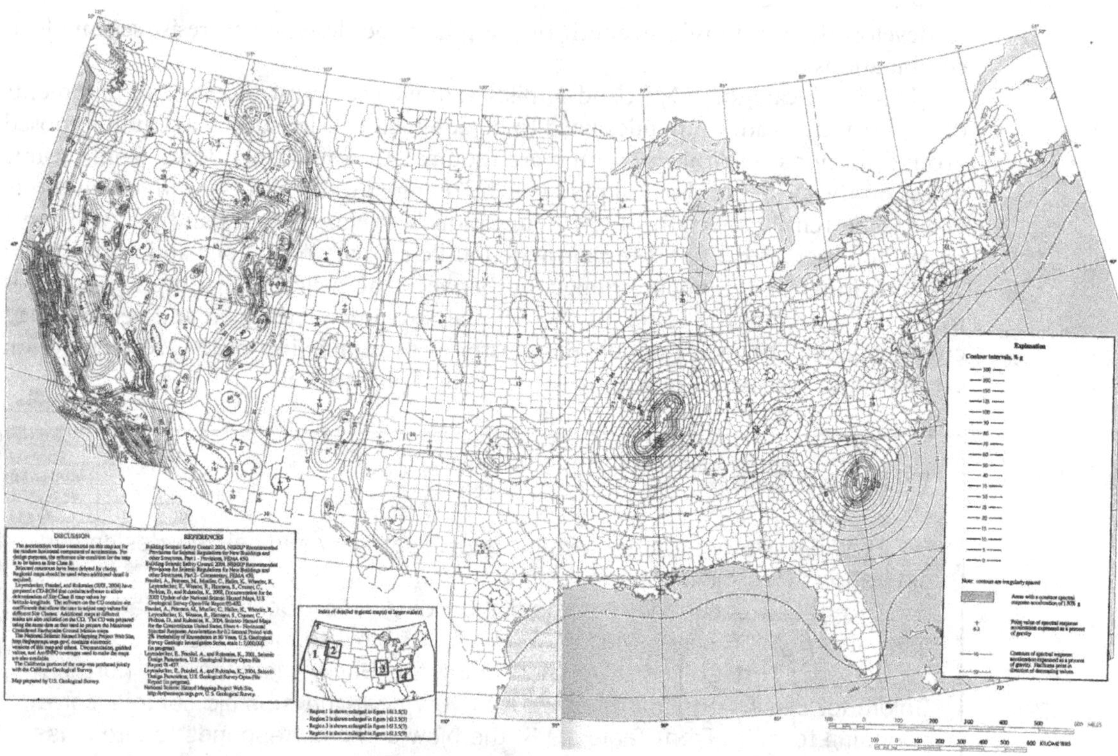

Figure 10.21 Mapped Short Period Seismic Accelerations (IBC, 2006)

> Further definition of these variables and better instructions on how they are applied are contained in the IBC (2006).
>
> The results of Equation 10b.2 and 10b.3 are used in ASCE7 load combinations to find the loads on helical piles. This simple example is given to familiarize the helical pile designer with seismic loads. On larger buildings, more sophisticated structural analysis is commonly done to derive the components of seismic force at each pile location.

Anecdotal evidence from engineers suggests that helical pile foundations perform well under seismic loads and actually protect structures from damage during seismic events. A California-based practicing engineer (Rupiper, 2000) visited several helical pile–supported structures after the Northridge earthquake of 1994 to assess their performance. Rupiper noted that the helical pile–supported structures performed better than those on other foundation types. The three-bedroom, wood-frame residence shown in Figure 10.22 was originally constructed with a slab-on-grade foundation. Prior to the earthquake, the portion of the home located to the right of the door was distressed due to the subsidence of fill. That area only was repaired with helical piles prior to the Northridge earthquake. After the earthquake, the portion of the home to

Figure 10.22 Residence, Northridge, CA (Rupiper, 2000)

the left of the front door, which was supported by the original slab-on-grade foundation, was badly damaged. The foundation was raised 3 inches [7.6 cm] in some areas. There were cracks in the concrete slab that opened 1 to 2 inches [2.5 to 5 cm]. The masonry fireplace chimney was badly damaged. The front door was severely racked and had to have the threshold removed in order to allow it to open. The area that had been underpinned by helical piles exhibited one minor radiating crack from a wood-framed pocket door. Otherwise, there was no damage in the area supported by helical piles.

The one-story shopping center shown in Figure 10.23 also was constructed with a slab-on-grade foundation. The end portions of the building were built on fill and the central portion was built in a cut area. The ends of the building were undergoing settlement and had been repaired using helical piles prior to the Northridge earthquake. After the earthquake, the end portions of the building exhibited very little damage, while the middle section was severely damaged (Rupiper, 2000).

Structures have been known to fail by many different mechanisms as a result of seismic shaking. Structural failures have been attributed to shear failure and buckling of columns, beam failure in bending (Housner, 1959; Read and Sritharan, 1993), excessive total and differential settlement, overturning, and excessive wall cracking (Read and Sritharan, 1993). Deep foundation failures have been attributed to

Figure 10.23 Shopping mall, Northridge, CA (Rupiper, 2000)

Table 10.6 Categories of pile failures (Mizuno, 1987)

Category	Cause of Pile Damage
1	Lateral spreading
2	Bridge approach embankment movement
3	Loss of bearing capacity by liquefaction
4	Shear and bending due to soil vibration
5	Building internal forces

liquefaction and subsequent loss of bearing capacity, broken pile shafts, and separation of pile from the pile cap (Kishida, 1966). Mizuno (1987) classified pile failures in Japan into five categories, which are summarized in Table 10.6.

Novak (1991) has classified the two most common pile foundation failures as cap uplift and tip uplift, as shown in Figure 10.24. Helical bearing plates near the bottom of helical piles provide tip anchorage even in homogeneous soils, thereby resisting tip uplift. Helical piles can be rigidly secured to grade beams to resist cap uplift. From a design standpoint, helical piles closely resemble fixed-head, small-diameter socketed piles. These types of piles best resist the two most common deep foundation failure modes.

A pile is considered "long" or "flexible" if the active length of the pile is less than the total pile length. The active length is defined as the length over which the pile deflects under lateral loading. Long or flexible piles cause the soil in their proximity (boundary zone) to exhibit nonlinear behavior due to the large strains that develop. Under large strains, the soil produces increased damping of seismic energy compared to a linear elastic soil model, and pile head accelerations are often reduced significantly. Tabesh and Poulos (1999) showed that pile head accelerations can be reduced as much as 60% due to soil yielding.

The slenderness ratio of a pile is defined as the ratio of the pile length to its diameter. Compared to driven piles and drilled shafts, typically helical piles have a larger slenderness ratio due to their small shaft diameter. A large slenderness ratio indicates a pile that is quite flexible in the direction perpendicular to the axis of the shaft. Parametric studies have been performed using the finite element method (FEM), boundary element method (BEM), analytical solutions, combinations of these approaches, and various approximate methods. It was shown in all cases that during dynamic lateral

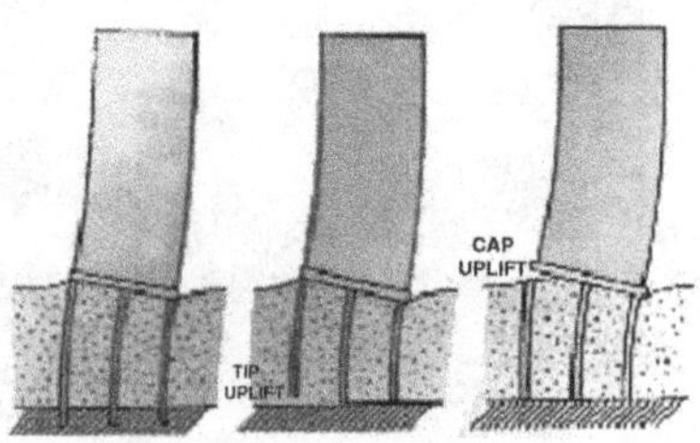

Figure 10.24 Common pile failures (Novak, 1991)

excitation, the more flexible the foundation (i.e., the larger the slenderness ratio), the greater the damping. The increased damping obtained results in smaller pile head displacements.

Due to the flexible behavior of helical piles, it is likely that they will provide considerable damping during seismic events relative to other foundation types, and thus a reduced amount of pile head displacement may be realized when helical piles are incorporated into the foundation of a structure. The reduction in the pile head displacement may contribute to a decreased amount of damage to a structure during seismic events and may prevent structural failure.

Past research related to dynamic pile analysis has been directed at identifying the pile's complex dynamic stiffness or impedance functions (Novak, 1991) due to their strong influence on structural response to seismic loading. The impedance functions are defined as amplitudes of harmonic forces or moments applied at the pile head that cause a harmonic motion of unit amplitude. It has been established (Gazetas and Dobry, 1984a; Novak and Aboul-Ella, 1978) that the dynamic stiffness of a laterally loaded pile (i.e., the real component of the impedance function) is generally insensitive to frequency over a wide range. Thus, the dynamic lateral pile stiffness has been assumed equal to the static lateral stiffness in some cases (Gazetas and Dobry, 1984a; Mylonakis and Gazetas, 1999; Nogami et al., 1992). Analyses incorporating this assumption have produced results similar to methods without this assumption, thus establishing confidence in its validity.

In contrast, the damping coefficient (i.e., the imaginary component of the impedance function) has been determined to be highly frequency dependent (Dobry et al., 1982; El Marsafawi et al., 1992; Gazetas and Dobry, 1984; Michaelides et al., 1997; Mylonakis and Gazetas, 1999; Novak, 1991; Novak and El Sharnouby, 1983; Sun and Pires, 1993;Velez et al., 1983). The deep foundation damping coefficient is comprised of material damping and radiation damping. Material damping is commonly assumed to be of the hysteretic type and is thus frequency independent. Radiation damping is due to the spreading of seismic waves (energy) away from the foundation element. Radiation damping cannot occur at frequencies below the fundamental natural frequency of the soil due to its inability to generate progressive waves (Novak, 1991). This phenomenon has been observed by various researchers (Gazetas and Dobry, 1984a; Michaelides et al., 1997; Novak and El Sharnouby, 1983; Sun and Pires, 1993; Velez et al., 1983). Michaelides et al. (1997) also showed that increasing the amplitude of the applied load increases the magnitude of material damping and reduces the magnitude of radiation damping.

The short-period spectral acceleration maps in the IBC were based on 5 percent dampening. It is possible one day that helical piles will be shown to Exhibit 25 percent dampening or more. This would not only reduce the seismic requirements for design of helical pile foundations but also would reduce the overall seismic forces transmitted to the structure (Perko, 2007b). The seismic performance of helical piles is expected to be frequency dependent. In order to gain an increased understanding of the seismic performance of helical piles, theoretical and experimental research must be performed that quantifies the seismic stiffness and damping coefficients over a wide range in excitation frequencies.

Chapter 11

Corrosion and Life Expectancy

Due to their slender steel shafts and thin helical bearing plates, helical piles often receive much scrutiny regarding corrosion and life expectancy. In order for the designer to verify that helical piles and anchors will provide long-term support for the design life of a structure, it is important that corrosion be contemplated. Helical piles are available in bare steel, zinc galvanized, or epoxy powder-coated. Batch hot-dip zinc galvanizing is the most often incorporated method of corrosion protection used by helical pile manufacturing companies today. Yet it is important to understand other zinc coating processes and their relevance to helical piling in case they are encountered. The use of powder coatings has been increasing in recent years; the process and properties of powder coating is of interest. In cases involving severe corrosion where protective coatings by themselves are insufficient, a sacrificial anode may be used to extend the life of the foundation.

In this chapter, a fundamental review of corrosion, epoxy coating, galvanizing, and sacrificial anodes is presented. The review is intended to aid foundation engineers in judging the expected life span of a helical pile and in understanding the benefits and drawbacks of different protective coatings and systems. Formulae are provided for determination of the design sacrificial thickness for corrosion and for sizing sacrificial anodes. Corrosion rates in air, water, concrete, organic soil, and contaminated soil as well as special precautions to avoid accelerated corrosion are discussed.

11.1 CORROSION BASICS

Corrosion is the exothermic chemical transformation of a metal or metal alloy to a nonreactive covalent compound such as an oxide or silicate that is often similar or even identical to the mineral from which the metals were extracted. Thus, corrosion has been called "extractive metallurgy in reverse" (Prayer et al., 1980).

"Rust" is a general term often used for the covalent compounds formed during the corrosion of iron and steel. The composition of rust depends on the abundance and species of other chemicals available during corrosion. Metallic corrosion occurs most prevalently in aqueous solutions, which conduct electric charge through ions. The electrochemical composition of the aqueous solution almost always governs the rate of corrosion and the composition of rust.

Of many possible reactions, one example of the net chemical reaction for the corrosion of iron and steel in the presence acid and water with ample dissolved oxygen is shown in Figure 11.1. The chemical reaction in this figure essentially indicates iron plus oxygen equals hydrated hematite. Water and acid (H^+) are essentially conserved on both sides of the reaction. Hydrated hematite is the reddish-brown mineral most often associated with the term "rust." It is also one of the minerals mined to produce iron. Although all hydrogen ions and some water are conserved on both sides of the chemical reaction, these molecules are important facilitators of the corrosion reaction.

With regard to helical piles, the rate of corrosion is generally governed by the flow of electrons in soil and is a function of moisture content, dissolved salts, acidity, soil mineralogy, and hydraulic conductivity. A measure of the rate of flow of electrical current is conductivity. Ground water by itself does not have very high conductivity. However, all ground water contains some soluble salts and other ions. Higher ion concentrations increase conductivity.

The acidity of ground water is a measure of the concentration of hydrogen ions. Recall that the concentration of hydrogen ions typically is represented by the pH (negative of the logarithm base 10 of the concentration of H+ ions). High acidity (low pH) indicates more hydrogen ions. Hydrogen ions remove electrons from iron,

$$4Fe + 3O_2 + 4H_2O + 8H^+ \rightarrow 2(Fe_2O_3{\cdot}H_2O) + 2H_2O + 8H^+$$

Figure 11.1 Corrosion chemistry

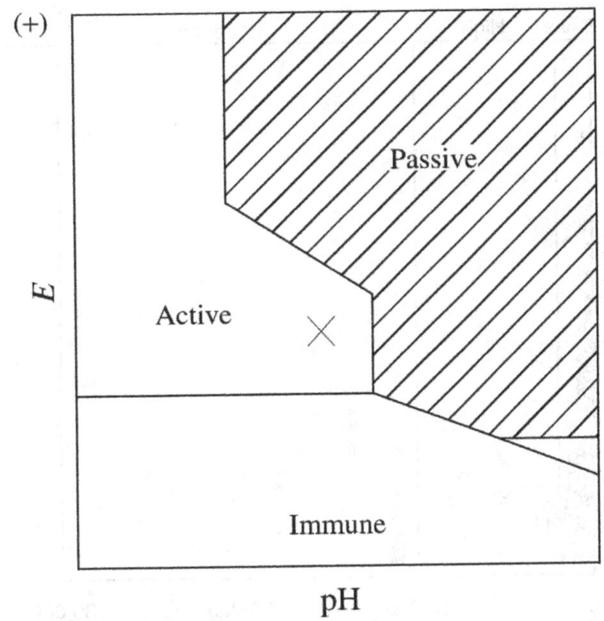

Figure 11.2 Pourbaix diagram (Mudali, Khatak, and Raj, 2007)

making it more chemically reactive. The rate of corrosion is also a function of the quantity of dissolved and free oxygen in the soil and the diffusion rate.

The electrochemistry of corrosion can be understood through the use of a generalized electrode potential-pH diagram (a.k.a. Pourbaix diagram) such as the one shown in Figure 11.2. Under normal conditions at the "X" marked in the figure, the metal and the electrolyte (the helical pile and the soil/water in this case) have an electric potential and pH combination in the active region, corrosion occurs, and the corrosion products are soluble so they can be removed. In the passive region, the pH is such that the corrosion products are insoluble and form a protective barrier resulting in passive corrosion resistance. If the electrode potential is reduced as in cathodic protection, the metal-electrolyte system is shifted to the immune region. Passivity and cathodic protection are discussed further in later sections. The boundaries of the three regions in the Pourbaix diagram are unique to the metal/electrolyte combination and hence depend on the properties of the helical pile and the surrounding subsurface conditions.

The opposite of electrical conductivity is resistivity. Either term is often used to describe the corrosivity of soil. Long-term soil testing programs conducted by the National Bureau of Standards between 1910 and 1955 comprise much of the data on soil corrosivity (Romanoff, 1989). These data are shown graphically in Figure 11.3 and represent the rate of corrosion as a function of soil resistivity for over 300 buried iron and steel samples from 54 locations across the United States. It can be seen in the figure that low resistivity (high electrical conductivity) is associated with high

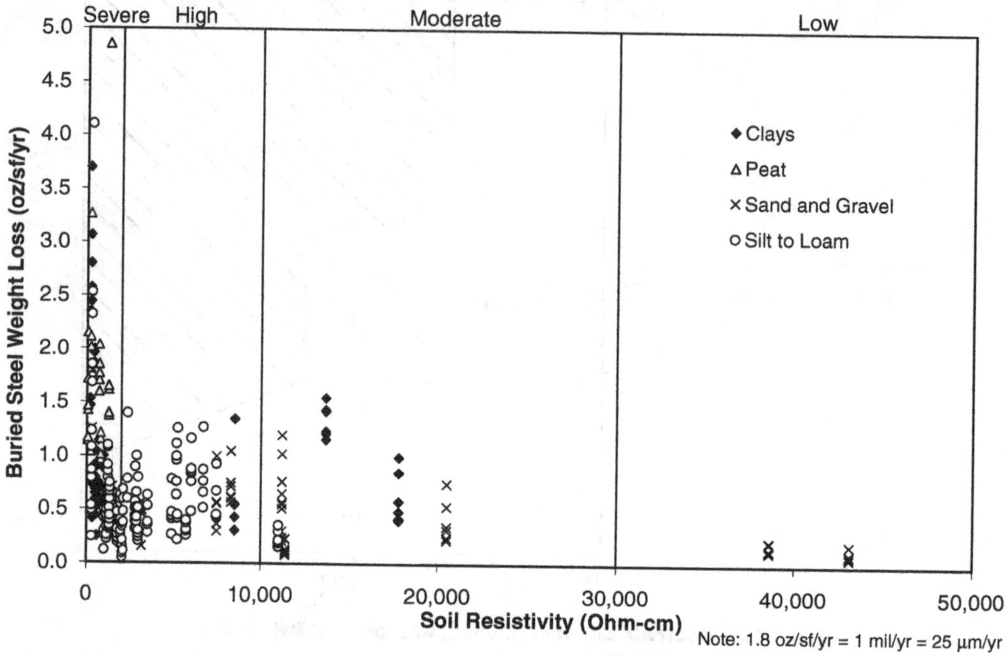

Figure 11.3 Corrosion rates for buried steel pipe (Perko, 2004a)

corrosion rates. General categories of soil corrosivity are shown across the top of the figure. Different symbols represent four general soil categories: clays, peat, sand and gravel, and silt to loam. The specimens buried in peat exhibited the highest rate of corrosion. Specimens buried in silt and loam materials are generally grouped in the high to severe corrosivity categories. The data for sand and gravel soils trend toward moderate to low corrosivity. The data for clay soils is more interspersed and lack general trend. In general, soils with high moisture content, higher permeability, ample supply of dissolved oxygen, considerable salt content, and high acidity have the least resistivity and are most corrosive.

On small projects, the general magnitude of soil resistivity may be estimated by the experienced geotechnical engineer based on soil type, ground water, and familiarity with local conditions. Municipal water districts often maintain a database of soil resistivity measurements for their area. These records are used as a guide for determining the degree of corrosion protection required for buried water distribution pipes. Resistivity data often can be obtained by contacting local municipal water districts. On larger projects or where corrosion is a particular concern, soil resistivity may be measured in common field or laboratory tests. The most common method of field measurement is the Wenner four-pin method. This method uses four electrodes driven into the soil and spaced at equal distances from each other. An electric current is sent through the outer electrodes. By measuring the voltage across the inner electrodes, soil resistance is calculated using Ohm's law. By varying the distance between pins, the resistivity measurement reaches different depths.

Soil resistivity does not provide a measure of the permeability, diffusivity, and therefore the residence time of water on buried surfaces. There is no single easily measured soil parameter that can be used to determine soil corrosivity. Rather, low soil resistivity values indicate areas of potentially high corrosivity that may warrant further study (Jones, 1986).

There are various forms of corrosion, including uniform, galvanic, crevice, pitting, intergranular, cracking, erosion, dealloying, and hydrogen damage (Jones, 1986). Uniform corrosion accounts for the greatest amount of metal transformation and will receive the most attention in this chapter. Galvanic corrosion poses both benefits and problems for helical piles so it also should be considered. Crevice and pitting forms of corrosion are insidious forms of corrosion. Protective coatings, such as hot-dip zinc galvanizing, can be used to protect helical piles from these types of corrosion. A possible mode of helical pile failure may be the corrosion-induced fracture of the shaft near the ground surface where there is increased oxidation. This effect may be diminished by proper site drainage and encapsulation of the helical pile shaft in concrete at the ground surface.

11.2 GALVANIC CORROSION

Galvanic corrosion occurs when two dissimilar metals are placed in contact within an electrolyte (soil and groundwater in this case). The metal or metal alloy with more negative electrode potential (anodic) will give up electrons and corrode before the metal with less negative electrode potential (noble). The galvanic series of metals is shown in Figure 11.4. Magnesium, zinc, and aluminum are the most anodic and will sacrifice themselves to protect other metals to which they are electrically coupled. Nickel, bronze, copper, and passive stainless steels are more noble and will cause increased corrosion of less noble metals, such as iron and steel, if placed in contact.

When the principles of galvanic corrosion are properly understood, many beneficial processes can be incorporated in the design of helical piles, including zinc coatings and sacrificial anodes. However, when galvanic corrosion is either misunderstood or ignored, problems with helical piles can arise. There have been some cases where helical anchors and other types of steel anchor systems have been decommissioned by galvanic corrosion. When dissimilar metals are placed in contact without consideration of galvanic corrosion, rates of corrosion can be inadvertently impacting.

Galvanic action can occur between a helical pile and anything that is electrically coupled to. According to AC358 (ICC-ES, 2007), zinc-coated steel and bare steel are not to be combined in the same helical piling system. It is important to avoid the helical pile becoming the sacrificial anode to a bare steel bracket or other structure. The electrical contact of a helical pile to a larger steel structure can result in unpredictable results. The helical piling system also should be electrically isolated from concrete reinforcing steel.

Corroded End
Anodic or less noble
(ELECTRONEGATIVE)
Magnesium
Zinc
Aluminum
Cadmium
Iron or Steel
Stainless Steels (active)
Soft Solders
Lead
Tin
Nickel
Brass
Bronzes
Copper
Nickel-Copper Alloys
Stainless Steels (passive)
Silver Solder
Silver
Gold
Platinum
Protected End
Cathodic or most noble
(ELECTROPOSITIVE)

Figure 11.4 Galvanic Series of Metals (AGA, 2000a)

There has been some discussion about combining galvanized steel with powder-coated steel; it is thought that the powder coating will act as an insulator and prevent galvanic effects, but this has to be examined carefully. The best practice is to construct all components out of the same material with the same protective coating and use one corrosion rate for the entire pile system.

The American Galvanizers Association (AGA, 2000a) cautions against placing copper or brass in contact with zinc galvanized steel. Apparently, even runoff water from copper or brass surfaces can contain enough dissolved copper to cause rapid corrosion. When the contact between a helical pile and another structure or metal of any type is required, precautions should be taken to prevent electrical contact. Joint faces should be insulated with nonconductive gaskets, paint, rubberized undercoating spray, or joining compounds. Connections should be made with insulating washers or grommets.

11.3 ZINC COATINGS

The zinc galvanizing process is the fortunate result of understanding the science behind galvanic corrosion of metals. The zinc helps protect iron and steel by galvanic corrosion

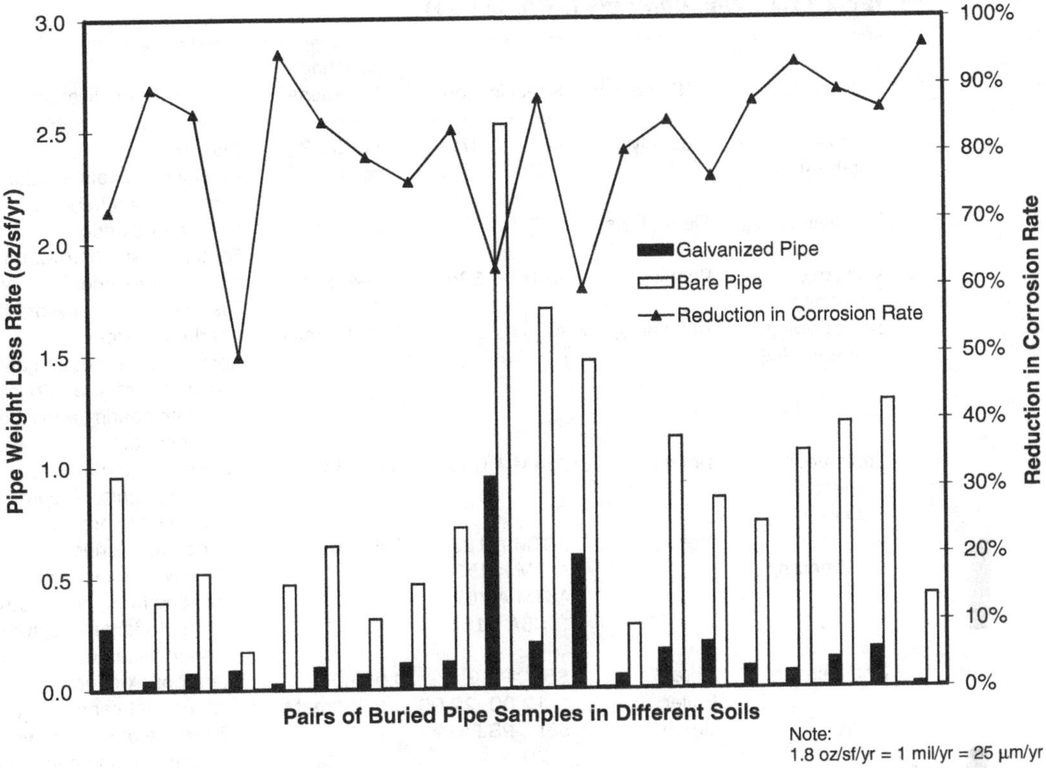

Figure 11.5 Effect of galvanizing on underground corrosion rates (Perko, 2004a)

in addition to other protective characteristics. Different types of iron, steel, and zinc have been found to corrode at essentially the same rate in most soil types (Uhlig and Revie, 1985). This finding caused some engineers, including the author at one time, to incorrectly believe that zinc coating of underground metallic structures was unimportant. Better understanding of the function of zinc coating and corrosion-related failure mechanisms leads to the conclusion that the zinc coating of helical piles is of significant benefit.

Figure 11.5 (constructed from the data in Uhlig and Revie, 1985) presents a graph showing the corrosion rates for pairs of bare and galvanized iron and steel pipe samples from 20 locations across the United States. Bare and galvanized steel samples are represented in the graph by the white and black bars, respectively. The reduction in corrosion rate due to galvanizing is represented by the line graph at the top of the figure and corresponds to the secondary y-axis. The data indicate that zinc galvanizing can reduce uniform corrosion rates by a factor of 50 to 98 percent.

There are several methods of coating metal with zinc, including painting, mechanical plating (ASTM B695), hot spraying or metallizing, electroplating (ASTM B633), the batch hot-dip process (ASTM A123 and A153), and continuous sheet

Table 11.1 Zinc Coatings (AGA, 2000b)

Method	Process	Specification	Coating Thickness	Application
Electro-galvanizing	Electrolysis	ASTM A 879	Up to 0.28 mils[1]	Interior. Appliance panels, studs, acoustical ceiling members.
Zinc plating	Electrolysis	ASTM B 633	0.2–1.0 mils[2]	Interior or exterior. Fasteners and hardware items.
Mechanical plating	Peening	ASTM B 695	0.2–4.3 mils[2]	Interior or exterior. Fasteners and hardware items.
Zinc spraying (metallizing)	Hot zinc spray	AWS C2.2	3.3–8.3 mils	Interior or exterior. Items that cannot be galvanized because of size or because on-site coating application is needed.
Continuous sheet galvanizing	Hot-dip	ASTM A 653	Up to 4.0 mils[1]	Interior or exterior. Roofing, gutters, culverts, automobile bodies.
Batch hot-dip galvanizing	Hot-dip	ASTM A 123 ASTM A 153 ASTM A 767 CSA G164	1.4–3.9 mils[3]	Interior or exterior. Nearly all shapes and sizes ranging from nails, nuts and bolts, to large structural assemblies, including rebar.
Zinc painting	Spray Roller Brush	SSPC-PS Guide 12.00, 22.00 SSPC-PS Paint 20 SSPC-PS 12.0	0.6–5.0 mils/coat	Interior or exterior. Items that cannot be galvanized because of size or because on-site coating application is needed. Large structural assemblies. Aesthetic requirements.

[1] Total for both sides of sheet.
[2] Range based on ASTM minimum thicknesses for all grades, classes, etc., encompassed by the specifications.
[3] Range based on ASTM and CSA minimum thicknesses for all grades, classes, etc., encompassed by the specifications.

galvanizing (ASTM A653). ICC-ES (2007) document AC358 recognizes only three of these methods in the acceptance of helical piles: mechanical plating, electroplating, and batch hot-dip galvanizing. Nonetheless, it is important for the helical pile designer to understand each of these methods because all may be encountered. Zinc coating methods are listed in Table 11.1 along with the recognized standard, coating thickness, and typical applications. These methods are described in detail with regard to helical pile applications in the remainder of this section. Micrographs of five zinc coatings are shown in Figure 11.6. The micrographs show the thickness of zinc deposits and their crystalline structure, which is useful in visualizing the various coatings as each is discussed. Additional information regarding zinc coatings can be found in AGA (2000b).

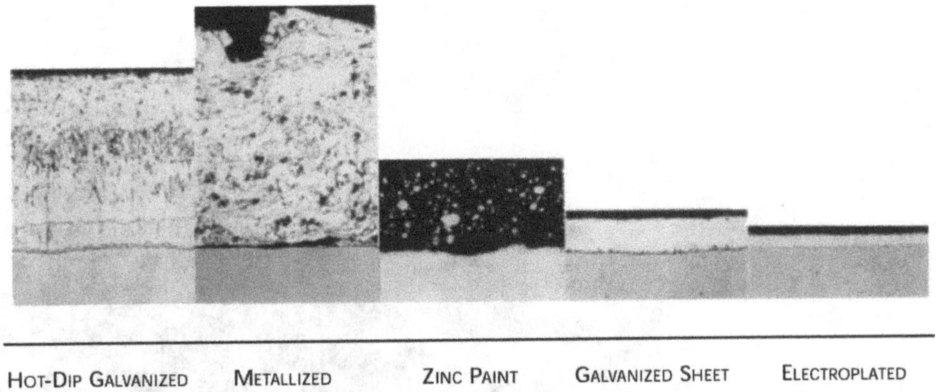

Figure 11.6 Zinc coating microstructures (AGA, 2000b)

Zinc paint is applied by brushing, rolling, or spraying and leaves a metallic zinc film on the steel after it dries. Zinc paint provides a protective barrier, but its function as cathodic protection is questionable. Zinc paint is used for touching up galvanized helical piles after cutting, welding, or damage. Mechanical plating is a process whereby small iron and steel parts are coated by tumbling in a mixture of promoter chemicals, zinc powder, and glass beads. Bolts, hex nuts, and other commercial fasteners used with helical piles are sometimes zinc coated by mechanical plating. Metallizing is the process of feeding zinc wire or powder through a heated applicator where it is melted and sprayed using air pressure onto steel. The resulting zinc coating is normally sealed with a polyurethane, epoxy, or resin. The use of metallizing historically has received little attention in the manufacture of helical piles. Its adhesion and abrasion resistance is lower; it does not coat the inside of hollow sections, and less is known about its performance. Electroplating consists of depositing a thin layer of zinc by electrodeposition. Other names for this are electrogalvanizing and zinc plating. The process is ordinarily used with sheet steel, light fasteners, or decorative parts and is not typically associated with the manufacture of helical piles or associated components although it is recognized by AC358 (ICC-ES, 2007).

Batch hot-dip galvanizing has been the most commonly used method of zinc coating in the production of helical piles. The batch hot-dip galvanizing process is shown in Figure 11.7. Figure 11.8 presents a photograph showing a set of helical pile

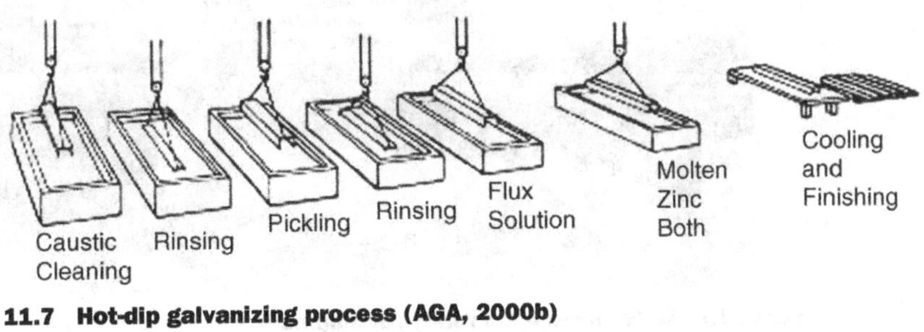

Figure 11.7 Hot-dip galvanizing process (AGA, 2000b)

Figure 11.8 Helical piles ready for galvanizing

extensions ready for galvanizing. The base steel is first cleaned with a hot alkali (caustic) to remove dirt, paint, and oils. Next the steel is subjected to an acid bath (pickling) and possibly abrasive cleaning to remove scale and rust. Fluxing is done after pickling to halt oxidation of the metal and prepare the surface for galvanizing. Finally, the materials are completely immersed in a bath of pure molten zinc at 840° F [450° C]. At this temperature, the zinc chemically bonds with the steel. A truckload of batch hot-dip galvanized helical piles ready for shipment to the project site is shown in Figure 11.9.

Batch hot-dip zinc galvanizing has several beneficial features with regard to the construction of helical piles. Batch hot-dip galvanizing coatings typically are 3.5 to 4 mils [80 to 100 microns] thick, moderately flexible, and result in a zinc iron alloy that

Figure 11.9 Galvanized piles ready for shipment

is harder than steel. Other types of zinc coatings are typically 0.5 to 1.0 mils [12 to 25 microns] thick, very flexible, and softer than steel (Industrial Galvanizers America, 1999). Hot-dip galvanizing protects iron and steel pipe surfaces inside and out, and the resulting coating is harder than any paint. Hot-dip zinc normally never flakes or peels.

Continuous sheet galvanizing is a hot-dip process that is used with long lengths of steel sheet, strip, or wire. It differs from the batch hot-dip process in that the molten zinc bath contains a small amount of aluminum, which suppresses the formation of zinc-iron alloys. An in-line heat treatment can be used to produce a fully alloyed zinc-iron compound. Continuous sheet galvanizing is not recognized by AC358 (ICC-ES, 2007) mainly because the committee that authored it was unfamiliar with the process. It is unknown whether this process can be adapted to helical piles.

11.4 PASSIVITY

When zinc corrodes, it forms zinc oxide in dry air and zinc hydroxide in moist air. These alkali compounds raise the pH along the surface of the metal, forcing passive conditions. This can be seen in the Pourbaix diagram shown in Figure 11.2; the passive state is entered as pH increases beyond the active state. The alkali compounds also react with carbon dioxide in the air to form zinc carbonate. The zinc carbonate film is the milky white coating often seen on aged zinc. The zinc carbonate adheres tightly to the underlying zinc and is practically insoluble. The products of zinc corrosion are non-conductive and exhibit passivity; they hinder the flow of electrical current and reduce the rate of corrosion. Carbon dioxide, which causes the layer to be durable, is present in atmosphere and almost always found in soil pore air and dissolved in ground water.

The process of passivity is often referred to as barrier protection. The effectiveness of a zinc coating in providing barrier protection depends on adhesion to the base metal and abrasion resistance. As helical piles are installed into the ground, pits and scratches may occur through the zinc coating. The process of galvanic corrosion protects bare iron and steel exposed in pits and scratches. In fact, zinc galvanizing will prevent corrosion of exposed areas of iron and steel up to $^1/_8$ inch [3 mm] wide (Industrial Galvanizers America, 1999).

After a zinc-coated helical pile is installed and then removed from the ground, it appears shinny and smooth, and it gives the false impression that the zinc has been rubbed off by the soil. However, for batch hot-dip galvanizing, the abrasion resistance increases with successively iron rich layers of zinc-iron alloy. A micrograph showing the different layers formed in the hot-dip galvanizing process is shown in Figure 11.10. The inner zinc-iron alloys (zeta, delta, and gamma) have higher abrasion resistance than the base metal. In fact, the delta and gamma layers have a higher hardness than quartz and virtually prevent abrasion by most soils. Zinc-iron alloys exhibit passive corrosion resistance and can create barrier protection. Other types of zinc coating processes do not produce these alloys. Because of its abrasion resistance, engineers often favor batch hot-dip galvanizing over powder coating and other zinc coatings.

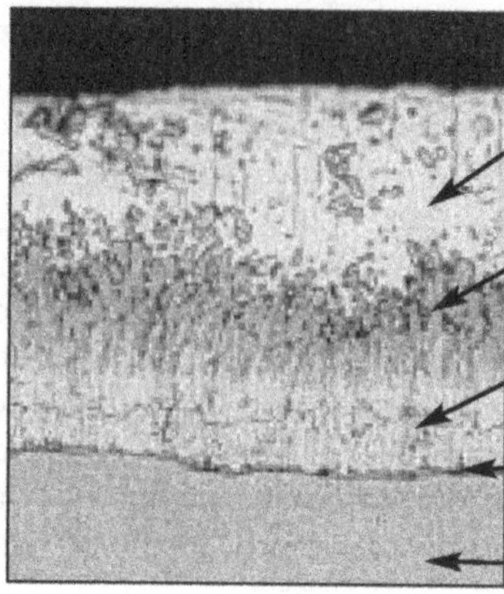

Figure 11.10 Batch hot-dip galvanized coating micrograph (AGA, 2000a)

11.5 POWDER COATING

Powder coating is the technique of applying thermoplastic or thermoset polymer powder to a part by lowering into a fluidized bed or by spraying electrostatically charged powder onto the part. After powder application the part is heated, and the powder melts to form a continuous film. Powder coating produces a uniform coating, which is tougher and more abrasion resistant than conventional paint and is not as prone to cracking, chipping, or pealing. Common polymers used in this process are polyester, polyester-epoxy, straight epoxy, and acrylics.

Powder coatings are beginning to appear more frequently in the production of helical piles. Acceptance criteria for powder coatings used with steel foundation systems are described in ICC-Evaluation Services (2003) document AC228. These criteria recognize polyethylene copolymer-based thermoplastic (ethylene acrylic acid/EAA) powder coatings only. ICC-Evaluation Services (2007) document AC358 for helical piles requires that powder coating meet the criteria in AC228 and that the coating thickness be at least 0.018 inch [450 μm].

Powder coatings are generally less expensive than zinc coatings and are environmentally friendly. Powder coatings emit zero or near-zero volatile organic compounds. Production lines produce less hazardous waste than conventional paints, and the overspray can be recycled. Powder coatings protect steel from corrosion by forming a nonconductive protective barrier. Because powder coatings are nonconductive, they protect against galvanic corrosion and may be placed in contact with other metals.

According to AC358 (ICC-ES, 2007), powder coatings provide for longer life expectancy for helical piles compared to bare steel, but not as long as zinc coatings. The minimum adhesion of powder coating to the base metal is 500 psi [3,400 kPa] compared to several thousand psi (several MPa) for batch hot-dip zinc galvanizing. Powder coatings mainly protect shallow portions of helical piles where the soil is more oxygen rich and corrosion is generally more insidious. Deeper helical pile sections and the helical bearing plates are less protected, as powder coatings are severely abraded by soil and rock during helical pile installation. Powder coating can be done after most types of zinc coating, thereby providing double protection if needed.

11.6 DESIGN LIFE

Corrosion can be accounted for in design by reducing the thickness of helical pile shafts, helical bearing plates, and other components by the sacrificial thickness over the design life in structural calculations. Sacrificial thickness may be computed using one of four methods described herein. These methods can be loosely termed: AC358, AASHTO, 98th percentile, and King. Descriptions and references for each of these methods are described in this section. Each has applications for helical piles depending on the type of structure to be supported, degree of corrosivity, and the applicable building code.

ICC-Evaluation Services (2007) document AC358 for helical piles contains corrosion provisions for bare steel, powder-coated steel, or zinc-coated steel. According to AC358, sacrificial thickness, T_s, may be computed by

Zinc-coated steel:

$$T_s = 25t_d^{0.65} \tag{11.1}$$

Bare steel:

$$T_s = 40t_d^{0.80} \tag{11.2}$$

Powder-coated steel:

$$T_s = 40(t_d - 16)^{0.80} \tag{11.3}$$

Where

t_d is the design life span in years.

Note that T_s from these formulae is in units of μm. For conversion, 2.54 μm equals 1 mil. A mil is 0.001 inch. These rates were derived from FHWA-NHI-00-044 (Elias, 2000) for soils with moderate corrosivity. The rate of corrosion is the least for zinc-coated steel due to passivity and cathodic protection. The corrosion rate for powder-coated steel is the same as for bare steel except the powder coating extends the life by 16 years. AC358 specifies a minimum zinc coating thickness of 3.4 mil [86 μm].

According to the AC358 method, the thickness of each component of a helical pile should be reduced by 1/2 T_s on each side, for a net reduction in thickness of T_s.

$$T_d = T_n - T_s \tag{11.4}$$

The parameter T_n is the nominal thickness, and T_d is design thickness. For bare and powder-coated steel, T_n is simply the base steel thickness. For zinc-coated steel, T_n may be taken as the sum of the base steel thickness plus the zinc coating thickness.

According to AASHTO (2004) Section 11.10.6.4.2a, the corrosion loss of galvanized steel in moderately aggressive soils shall be computed from

$$\begin{aligned}\text{Loss of galvanizing} &= 0.58\ \text{mil/yr}\ [15\ \mu\text{m/yr}]\ \text{for first 2 years}\\ &= 0.16\ \text{mil/yr}\ [4\ \mu\text{m/yr}]\ \text{for subsequent years}\\ \text{Loss of carbon steel} &= 0.47\ \text{mil/yr}\ [12\ \mu\text{m/yr}]\ \text{after zinc depletion}\end{aligned}$$

This AASHTO specification applies to batch hot-dip galvanized steel straps used for slope reinforcement. However, practitioners have commonly used AASHTO rates of corrosion in the design of other underground structures. These rates represent the corrosion loss on each surface, so sacrificial thickness is determined by applying these rates to both sides of a structural member (e.g. multiplying by 2).

Both AC358 and AASHTO specifications explicitly exclude potential pile corrosion situations defined as exposure conditions with resistivity less than 1,000 Ω-cm, pH less than 5.5, high organic content, sulfate concentrations greater than 1,000 parts per million, landfills, or mine waste. A corrosion engineer should be consulted when potential pile corrosion situations are encountered. It may be appropriate to provide additional corrosion protection to extend the life of the helical pile system. A possible solution is incorporation of a sacrificial anode as discussed in a forthcoming section.

Perko (2004a) analyzed the National Bureau of Standards (NBS) data shown in Figure 11.3 and Figure 11.5 in order to determine the life expectancy of bare steel and galvanized helical piles with tubular shafts having 0.25-inch [6-mm] wall thickness. The algebraic mean and standard deviation of the data in each of the four different categories of corrosivity based on electrical resistivity were determined. Life expectancy was calculated based on the time required to corrode one-third the thickness of the base steel. The basis for this determination is that the structural capacity of helical piles has a minimum factor of safety of 1.5 and one-third the thickness could be removed before impending instability. Corrosion loss rates measured by the NBS were applied to both sides of a structural member to find the loss in thickness.

Perko's (2004a) results are shown in Table 11.2. The corrosivity of soils is separated into four major categories based on electrical resistivity. Example soil types comprising each of the major soil categories are given in the table. Minimum life expectancies were calculated based on the highest measured corrosion rate in each category and are shown by the top number in parentheses in the right-hand columns. Average life expectancies are based simply on the mean corrosion rate in each category and are depicted by the bottom numbers in these columns. Life expectancies with 98 percent probability were computed assuming normal distributions and taking 2 standard deviations from the mean corrosion rate in each category. The bold numbers in

Table 11.2 Life Expectancy of 0.25-inch- [6-mm-] Thick-wall Tubular Helical Piles (Perko, 2004)

Soil Resistivity (Ohm-cm)	Corrosivity Category	Example Soils	(Minimum) **98% Probability** *(Average)* Helix Foundation Life Expectancy	
			Bare Metal	Galvanized
0–2,000	**Severe**	Soil in marine environments; organic soils and peat; soft, wet silts and clays; wet shales	(15) **30** *(80)*	(40) **75** *(200)*
2,000–10,000	**High**	Stiff, moist clays; medium-dense silts and loams; wet clayey to silty sand; wet sandstone	(55) **70** *(135)*	(140) **170** *(340)*
10,000–30,000	**Moderate**	Dry to slightly moist clays; dry silts and loams; sand and gravel; limestone	(50) **55** *(140)*	(125) **140** *(350)*
>30,000	**Low**	Dry shales; dry sandstone; clean and dry sand and gravel; slate and granite	(345) **325** *(555)*	(865) **810** *(1385)*

the last two columns in the figure show the resulting 98th percentile life expectancies. All life expectancies in Table 11.2 are given in units of years.

Perko's results indicate that a galvanized helical pile of the configuration described above has a 98 percent probability of a lifespan between 75 and 810 years depending on soil type. The results show a longer life expectancy for high corrosivity compared to those for moderate corrosivity. The reason for this apparent discrepancy stems from the NBS data, which has a series of outliers in the moderate category that raise the mean corrosion rate and standard deviation, thereby lowering the life expectancy. In practice, these categories could be averaged together and reported at a single rate.

The rate of loss in thickness for each of the categories shown in Table 11.2 can be backed out by dividing the corroded thickness ($^1/_3 \times 0.25$ inch $= 0.083$ inch [2.1 mm] by the life expectancy for that scenario. These calculations were performed for the data representing 98 percent probability, and the results are given next. High- and moderate-corrosivity categories were combined to obtain these rates.

Bare Steel

Severe ($<2{,}000\ \Omega$-cm) $= 2.8$ mil/yr [71 μm/yr]

Moderate- High (2,000-30,000 Ω-cm) $= 1.3$ mil/yr [33 μm/yr]

Low ($>30{,}000\ \Omega$-cm) $= 0.3$ mil/yr [8 μm/yr]

Galvanized Steel

Severe (<2,000 Ω-cm) = 1.1 mil/yr [28 μm/yr]
Moderate-High (2,000-30,000 Ω-cm) = 0.5 mil/yr [13 μm/yr]
Low (>30,000 Ω-cm) = 0.1 mil/yr [3 μm/yr]

Yet another method of determining sacrificial thickness is that by King (1977) shown in Figure 11.11. In the King method, soil resistivity and pH are measured in soils at the site. The nomograph is entered at the soil resistivity value on the left axis. A trace is made along the curved lines representing different soil resistivities until a line corresponding to the measured pH is reached. Depending on whether the soil is acidic or alkaline, the corrosion rate corresponding to that combination of resistivity and pH is shown in either the axis above or below the chart. According to A.B. Chance (2003), the King nomograph can be used to estimate the rate of corrosion of galvanized steel helical pile shafts. Rates obtained from the nomograph are for each surface so they must be multiplied by a factor of 2 to obtain sacrificial thickness. This method predicts negligible rates of corrosion for alkaline soils with moderate to high resistivity and very high rates of corrosion in acidic soils with low resistivity.

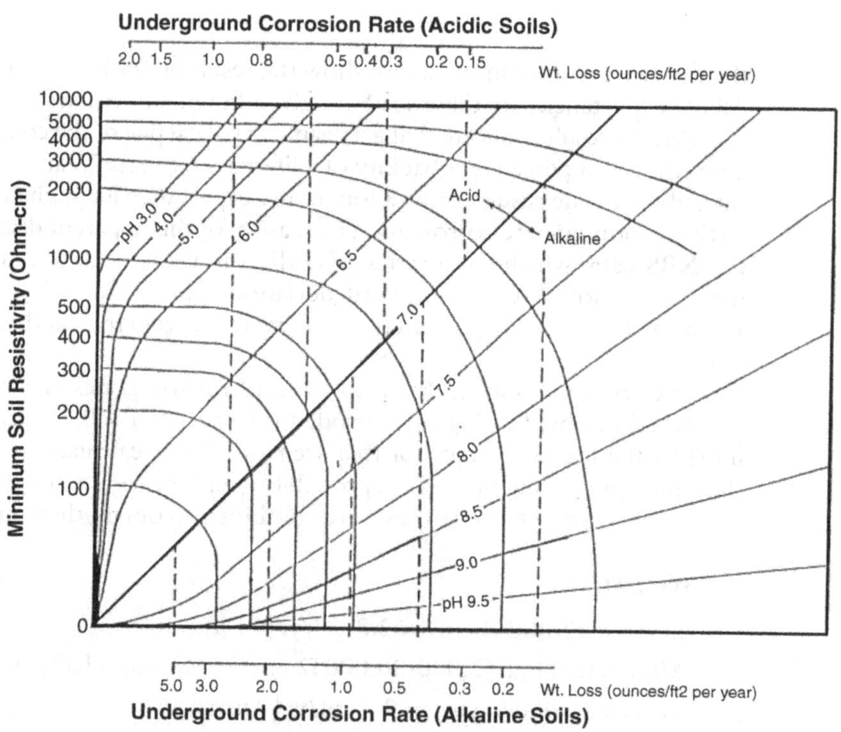

Figure 11.11 Nomograph for estimating underground corrosion rates (King, 1977)

The corrosion rates of AC358 (ICC-ES, 2007), AASHTO (2004), Perko (2004a, hereinafter the 98th percentile method), and King (1977) are compared in Figure 11.12. Rates from AASHTO and King were multiplied by 2 to obtain the corrosion loss in terms of material thickness. In general, corrosion losses in severe environments found by the 98th percentile method and the King method (500 ohm-cm) are the

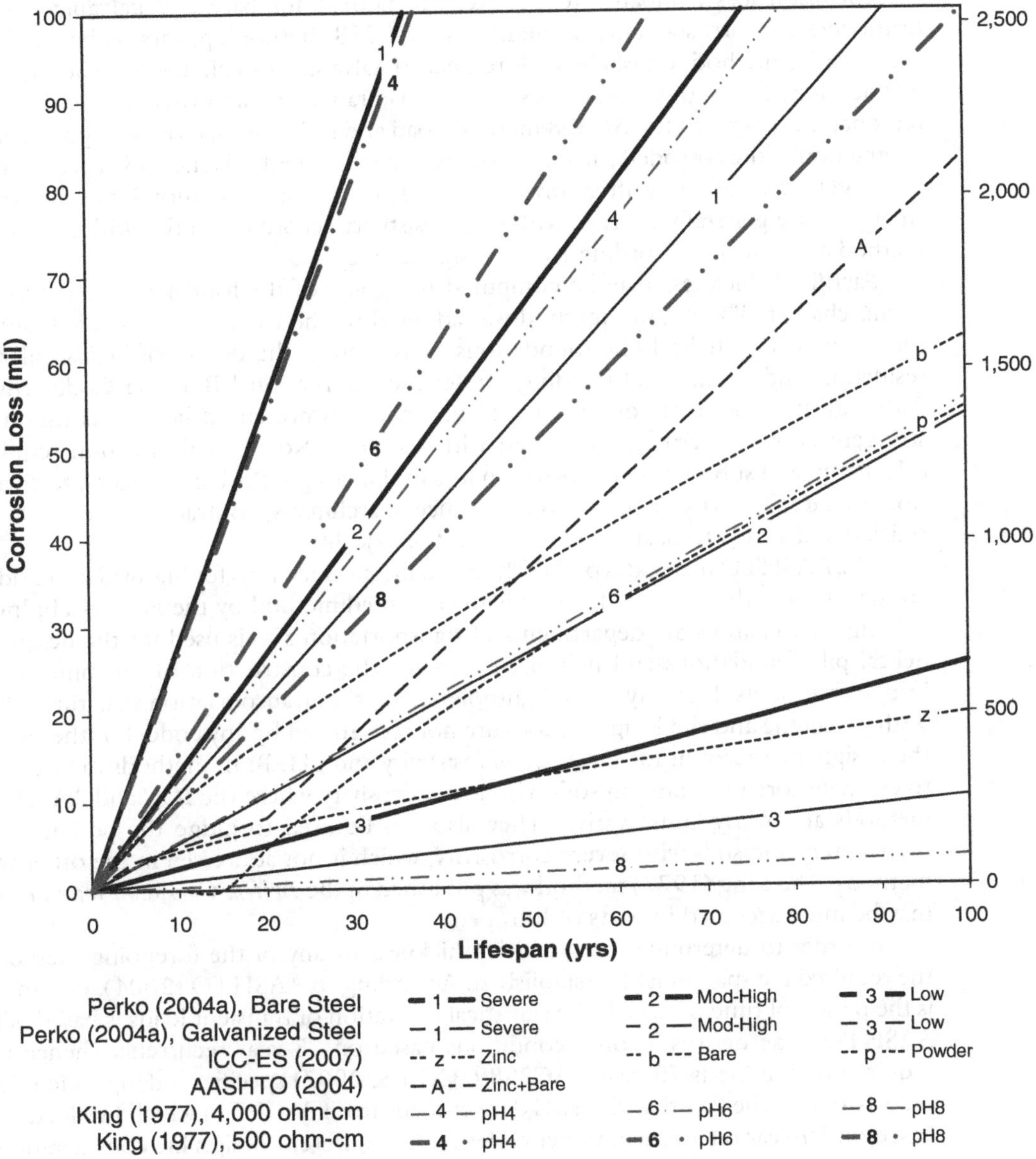

Figure 11.12 Sacrificial thickness computed with various methods

highest and range between 50 mil [1300 μm] to over 100 mil [2500 μm] in 50 years. The corrosion losses determined by AASHTO fall just below the severe conditions. The AASHTO method was developed for galvanized steel straps buried in moderately aggressive, oxygen-rich fill soils and may be conservative for undisturbed soils.

As shown in Figure 11.12, the 98th percentile method predicts roughly twice the corrosion loss compared to the AC358 method for bare and galvanized steel in moderately corrosive environments. The AC358 method appears to be the least conservative method, especially with respect to galvanized steel. The AC358 method was developed by regression analysis over a broad range of corrosivity and represents a reasonable average. In the AC358 method, load and resistance factors are applied after subtraction of the corroded thickness, whereas in other methods (i.e., 98th percentile), the target factor of safety after corrosion is 1.0. The lowest corrosion losses shown in the figure are generally for sites with high resistivity according to the 98th percentile method and low pH according to the King method.

Sacrificial thickness may be computed using any of the four methods described in this chapter. The AC358 method was adopted by the International Code Council for acceptance of helical pile foundations. It is used in the design of helical piles in residential and commercial buildings under the International Building Code (IBC). Although it is the least conservative of the methods presented here, it is the most stringent corrosion criteria associated with the IBC. No other foundation requires calculation and subtraction of corrosion loss in the design. Problems related to corrosion are rare. The IBC is made up of engineers, architects, contractors, and building officials and strives to balance economy and practicality.

The AASHTO method is part of the standard specifications for highway and bridge construction in the United States. It is the accepted method by the Federal Highway Administration and state departments of transportation and is used for the design of helical pile foundations and helical anchors for the construction of retaining walls, bridge abutments, highway signs, lightpoles, and other transportation structures. The 98th percentile and the King methods are not recognized by any code, but they allow the designer to account for varying soil resistivity and pH. Both methods can be used to estimate corrosion rates in soils with low corrosivity where the ICC and AASHTO methods are overly conservative. They also can be used to judge the potential for corrosion loss in soils with severe corrosivity, which is not addressed in the other two methods. The King (1977) method was published in the *British Corrosion Journal* and may be more accepted in parts of Europe.

In order to determine the sacrificial thickness by any of the foregoing methods, the required life span must be established. According to AASHTO (2004), design life is the period of time on which the statistical derivation of transient loads is based. The AASHTO code derives loading conditions based on 75-year recurrence; hence the required design life is 75 years. AC358 (ICC-ES, 2007) specifies a design life of 50 years based on the fact that the IBC derives loads from ASCE7 in which live loads are based on 50-year recurrence. Other codes require different design life considerations. Some local municipalities require a design life of 100 years.

Example 11a

<u>Problem:</u> Find the design dimensions of a helical pile shaft taking into account corrosion for a commercial building under the purview of International Building Code.

<u>Given:</u> Helical piles with 5.722-inch [145 mm] outside diameter, 0.25-inch-[6-mm-] thick tubular shafts are to be used in soils with an average resistivity of 8,000 ¨-cm. The piles are bare steel.

<u>Answer:</u> The soil resistivity does not suggest a severe pile corrosion situation. Since this project is under the IBC, corrosion rates referenced in AC358 apply. Based on the AC358 method and 50-year design life, the sacrificial thickness for the bare steel pile is given by

$$T_S = 40(50^{0.80}) = 915\,\mu\text{m} = 36\,\text{mil} \tag{11a.1}$$

The design thickness of the shaft is

$$T_d = 0.25 - 0.036 = 0.214\,\text{in}\,[5.4\,\text{mm}] \tag{11a.2}$$

The design diameter is

$$d = 5.722 - 0.036 = 5.686\,\text{in}\,[144\,\text{mm}] \tag{11a.3}$$

These dimensions should be used to compute gross area, section modulus, and other necessary properties. Simple compression, tension, and buckling (if required) should be computed following AISC (2001) or more recent code with the reduced sections and all applicable load and resistance factors.

Example 11b

<u>Problem:</u> Find the design dimensions of a helical pile shaft taking into account corrosion for a pedestrian bridge abutment under the purview of AASHTO specifications.

<u>Given:</u> Helical piles with 3.5-inch [89-mm] outside diameter, 0.22-inch-[5.6-mm-] thick tubular shafts are to be used in soils with an average resistivity of 2,000 ¨-cm. The piles are hot-dip galvanized according to ASTM A123 and A153 as applicable and have a zinc coating thickness of 3.9 mil [100 μm].

<u>Answer:</u> Despite the low resistivity, this is not a severe pile corrosion situation as defined by AASHTO. Based on the AASHTO method, the sacrificial thickness for the galvanized steel piles for 75-year design life is given by

$$T_S = 2\,[2\,\text{yrs}\,(15\,\mu\text{m=yr}) + (n-2)\,(4\,\mu\text{m=yr}) + (75\,\text{yrs} - n)\,(12\,\mu\text{m=yr})] \tag{11b.1}$$

Where

n is the number of years to zinc depletion which is found by recognizing that the first two terms in Equation 11.6a have to equal the thickness of the zinc coating, t_O.

$$t_O = 100\,\mu m = [2\,yrs\,(15\,\mu m{=}yr) + (n - 2)\,(4\,\mu m{=}yr)] \quad (11b.2)$$

$$n = 19.5\,yr \quad (11b.3)$$

Plugging this result in Equation 11.b1 yields the sacrificial thickness.

$$T_S = 2\,[30 + 17.5\,(4) + 65.5\,(12)] = 1{,}772\,\mu m = 70\,mil \quad (11b.4)$$

The design thickness of the shaft is

$$T_d = 0.22 + 2(0.0039) - 0.070 = 0.158\,in\,[4.0\,mm] \quad (11b.5)$$

The design diameter is

$$d = 3.50 + 2(0.0039) - 0.070 = 3.438\,in\,[87\,mm] \quad (11b.6)$$

These dimensions should be used to compute gross area, section modulus, and other necessary properties. Simple compression, tension, and buckling (if required) should be computed with AASHTO load cases.

Example 11c

Problem: Find the design dimensions of a helical pile shaft taking into account corrosion for support of a structure based on the judgment of the engineer.

Given: Helical piles with 1.75-inch [44-mm] square shafts are to be used in soils with an average resistivity of 35,000 ¨-cm. The piles are hot-dip galvanized according to ASTM A123 and have a zinc coating thickness of 3.9 mil [100 μ]. A design life of 100 years is assumed.

Answer: Due to the high resistivity, this site falls in the low-corrosivity category. Based on the 98th percentile method, the rate of corrosion is 0.1 mil/yr [3 μm/yr] and the sacrificial thickness is

$$T_S = 3\,(100) = 300\,\mu m = 11.8\,mil \quad (11c.1)$$

The design thickness and side dimension of the square shaft is

$$d = 1.75 + 2(0.039) - 0.0118 = 1.816\,in = 46\,mm \quad (11c.2)$$

The sacrificial thickness is less than the thickness of the galvanized coatings. This amount of corrosion is negligible so the gross area, section modulus, and other necessary properties of the original section can be used in design.

11.7 SACRIFICIAL ANODES

Cathodic protection reduces the rate of corrosion by providing an excess supply of electrons to a corroding metal surface. The excess electrons slow the rate of metallic ionization. The effect of cathodic protection can be visualized in the Pourbaix diagram shown in Figure 11.2; cathodic protection causes a downward shift in electrode potential. If the shift is sufficient, the helical pile steel and soil/water system moves from normal corrosive conditions, marked by the "X" to the immune region. In addition to reduction in uniform corrosion, cathodic protection protects against stress corrosion cracking, pitting corrosion, corrosion fatigue, and erosion.

There are two general types of cathodic protection: using an impressed current or attachment to a sacrificial anode. The most common method of cathodic protection of helical piles is a sacrificial anode. Sacrificial anodes consist of an anodic or less noble metal or metal solution that is buried separately from the helical pile and attached via a conducing wire. Use of an impressed current involves wiring the helical pile foundation to a current pump that provides a steady supply of electrons. The system must be supplied electrical power from an external source perpetually throughout the life of the structure. Impressed current cathodic protection is not often associated with helical pile foundations. This section focuses on the external sacrificial anode.

The sacrificial anode was invented by Sir Humphrey Davy in 1824. Davy attached small zinc buttons and iron nails to the copper sheathing on the hulls of wooden warships to arrest the rapid decay of the copper. Zinc and iron are less noble than copper and will corrode preferentially. Unfortunately, the suppression of toxic ions allowed the growth of marine organisms, which impaired the sailing speed. Since speed was desirable over longevity, cathodic protection was discontinued. Cathodic protection was first used extensively in the 1920s for buried steel oil pipelines along the Gulf of Mexico in the United States (Mudali, Khatak, and Raj, 2007).

Similar to zinc coatings, the sacrificial anode is a beneficial application of the principal of galvanic corrosion. Construction of a sacrificial anode is accomplished by electrically connecting an anode to a helical pile. In addition, there must be a path for ionic exchange. This typically occurs in an aqueous solution. In the case of helical piles, the soil and pore water acts as the medium for ion flow and the effectiveness of the anode depends on soil resistivity. The anode is buried near the helical pile so that electrons can flow through the soil/water to complete the galvanic cell.

The sacrificial anode must have more negative electrode potential than the structure to be protected. Magnesium, aluminum, and zinc are more negative (less noble) than steel, according to the galvanic series shown in Figure 11.4, and are frequently used as anodes to protect steel today. Magnesium is the most common for protection of helical piles. In salt water, zinc and aluminum anodes are more common because magnesium does not form a passive barrier and has a relatively short life. Zinc works well in fresh and marine water. Aluminum and aluminum-zinc alloys deplete at the slowest rate of the possible anodes. As such, they may not protect helical piles as reliably as magnesium. For underground use, anodes are prepackaged in a mixture of 75% gypsum, 20% bentonite, and 5% sodium sulfate in a permeable cloth bag. The purpose of the mixture is to absorb corrosion products, maintain conductivity by capturing water from the soil, improve reliability, and keep the anodes active. (Mudali et al., 2007)

Sacrificial anodes must be of sufficient size and have sufficient efficiency to provide the necessary usable life. The electrode potential difference must be large enough to provide the necessary current. Proper sizing and selection of anodes depends on the surface area of the helical pile(s), required life expectancy, soil type, resistivity, pH, and location of the water table. To provide full protection of the helical pile in soil, the electric potential with respect to a copper/copper sulfate reference electrode must be −850 mV. In the sacrificial anode system, the maximum driving voltage is controlled by the electric potential difference between the anode and the steel structure. With known anodes, the maximum driving voltage cannot be less than about −1,000 mV so they are unlikely to overprotect helical piles. Partial protection can be offered at lower electric potential. At a minimum, it is generally recognized that a potential shift of −300 mV must occur upon connection to the anode (A.B. Chance, 2003).

There are two ways of establishing the current requirements to achieve the necessary shift in electrode potential. The first applies to existing construction and consists of applying a current to a helical pile and measuring the change in pile-to-soil potential. This is discussed further in Mudali et al. (2007). For new construction, the required current, I_{req}, can be calculated from Equation 11.5:

$$I_{req} = i_o A \tag{11.5}$$

Where

i_o is the required current density, and
A is the surface area of the helical pile.

The parameter i_o depends on the properties of soil and ground water surrounding the helical pile and is based to a large extent on the designer's experience.

Table 11.3 lists common ranges of current density requirements for bare steel in different soil and aqueous conditions. This table can be used as a guide for initial design. Based on Ohm's law, the required current density is inversely proportionate to soil resistivity. Values in the range from 1 to 3 mA/ft^2 [11 to 32 mA/m^2] typically are used to protect galvanized helical piles where soil resistivity is 5,000 Ω-cm or less (A.B. Chance, 2003).

The estimated useful life of the sacrificial anode, L_t, can be calculated from Equation 11.6 (Mudali et al., 2007):

$$L_t = \frac{T_h W E_f U_f}{h_y I_{req}} \tag{11.6}$$

Where

T_h is the theoretical output
W is the weight of the anode

Table 11.3 Typical Current Density Requirements for Cathodic Protection of Bare Steel (Zaki, 2006)

Environment	Current Density (mA/ft^2)	Current Density [mA/m^2]
Soil	0.75–5.0	40–58
Fresh water	1–3	11–32
Seawater	4–5	43–64
Moving seawater	1–3	11–32
Sea mud	1–3	11–32

E_f is efficiency
U_f is utilization rate, and
h_y is the number of hours in a year (8766 hr/yr)

The utilization rate is usually taken as 85%, which indicates that once the anode is 85% consumed, its effectiveness will decay rapidly. The parameters, T_h and E_f, depend on the type of anode utilized.

Table 11.4 lists the characteristics of several anodes. As stated previously, magnesium anodes are incorporated most often in helical pile design in soil. There are two types of magnesium anode alloys, standard (H-1) and high potential. Generally, high-potential anodes are desirable if the soil resistivity exceeds 8,000 Ω-cm. Zinc anodes are generally limited to environments with resistivity below 1,500 Ω-cm because of their low driving voltage (Mudali et al., 2007). Theoretical output is affected by the surface area of the anode (A.B. Chance, 2006). It is best to contact a vendor of cathodic protection for assistance with selection and sizing anodes.

Table 11.4 Common Sacrificial Anode Materials (Mudali et al., 2007)

Material	Theoretical Output T_h [A-h kg^{-1}]	Actual Output [A-h kg^{-1}]	Efficiency E_f	Consumption Rate [Kg A-yr^{-1}]	Potential to CSE
Zinc type I	860	781	90%	11	1.06
Zinc type II	816	739	90%	12	1.10
Magnesium standard H-1 alloy	2205	551–1279	25–58%	6.8–16	1.40–1.60
Magnesium high potential	2205	992–1191	45–54%	7.3–8.6	1.70–1.80
Aluminum Al/Zn/Hg	2977	2822	95%	3.1	1.06
Aluminum Al/Zn/In	2977	2591	87%	3.3	1.11

The required weight of anode material can be estimated by rearrangement of Equation 11.6 and solving for *W*. According to A.B. Chance (2003), anodes are readily available in weights ranging from 5 pounds to 60 pounds [2 to 27 kg]. A sacrificial anode typically can extend the life of a bare steel or galvanized helical pile beyond the life of its coating by 10 to 20 years. At the end of its usable life, a sacrificial anode can be replaced, thereby extending the life further. An example calculation of sacrificial anode weight is shown in Example 11d.

Example 11d

<u>Problem:</u> Find the required weight of a sacrificial anode to add 20 years to the life span of a helical pile and the number of helical piles that can be attached to an anode weighing 60 pounds [27 kg].

<u>Given:</u> Helical piles with 1.5-inch- [38-mm-] square shafts are to be used in wet marine clays with an average resistivity of 5,000 ¨-cm. The average depth of the helical piles is 20 feet [6.1 m].

<u>Answer:</u> A midrange value for the required current density is assumed for the soil from Table 11.3: 2 mA/ft^2 [22 mA/m^2]. The surface area of one helical pile is estimated by multiplying the perimeter of the shaft times its length. Surface areas of the helical bearing plates are ignored for simplicity.

$$A = 4\,\text{sides}\,(0.038\,\text{m=side})\,(6.1\,\text{m}) = 0.93\,\text{m}^2 = 10.0\,\text{ft}^2 \qquad (11\text{d}.1)$$

The required electrical current is found from Equation 11.5.

$$I_{req} = 22\,(0.93) = 20.5\,\text{mA} \qquad (11\text{d}.2)$$

Because the soil resistivity is less than 8,000 ¨-cm and greater than 1,500 ¨-cm, a standard (H-1) magnesium anode with average efficiency (35%) is selected. The theoretical output of the anode is obtained from Table 11.4. Required weight of anode per helical pile is estimated by rearrangement of Equation 11.17 with L_t = 20 years, T_h = 2205 A-h/kg, E_f = 35%, U_f = 85%, and h_y = 8766 hr/yr.

$$W = \frac{20\,\text{yr}\,(8766\,\text{hr=yr})\,(0.0205A)}{2205A\,\text{hr=kg}\,(0.35)\,(0.85)} = 2.7\,\text{kg} = 6\,\text{lbs} \qquad (11\text{d}.3)$$

The number of helical piles that can be attached to an anode weighing 60 pounds is determined by simple division.

$$i = \frac{60\,\text{lbs}}{6\,\text{lbs=pile}} = 10\,\text{piles} \qquad (11\text{d}.4)$$

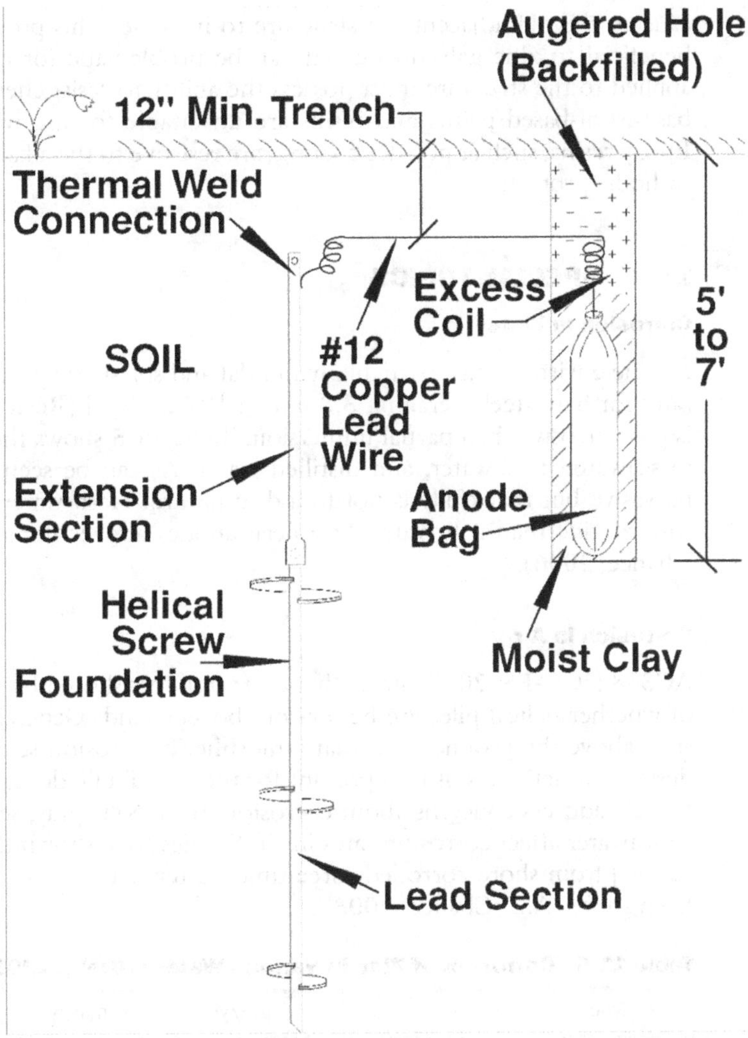

Figure 11.13 Application of a sacrificial anode (A.B. Chance, 2003)

An example of a sacrificial anode application for helical piles is shown in Figure 11.13. The anode bag is placed in an augered hole and backfilled with moist clay. The anode bag must be in intimate contact with the soil and can be moistened to begin the reaction. The anode is connected to the helical pile or piles by an insulated copper lead wire coiled at each end to prevent overtensioning and breakage in the event of soil movement (A.B. Chance, 2003).

When a metal is acting as a cathode, it causes hydrogen evolution or oxygen reduction, which consumes hydrogen ions (acid) or releases hydroxyl ions (base), causing the pH adjacent the structure to increase. This promotes passivity, which is beneficial to zinc galvanizing but can be problematic for other coatings. Coatings applied to the structure must possess the ability to resist chemical attack from strong bases. Oil-based paints and resins are unsuitable for use with cathodic protection. Epoxy resins, such as powder coating, are resistive to the alkali and could be used with cathodic protection.

11.8 SPECIAL TOPICS

Corrosion in Water

Marine environments, particularly in tidal and splash zones, can cause high corrosion rates for bare steel averaging 3.8 mil/yr [97 μm/yr] (Revie, 2000). Full immersion is less corrosive than partial immersion. Table 11.5 shows the corrosion rates of zinc in seawater, freshwater, and distilled water. As can be seen, distilled water is most pervasive but thankfully is not found in nature. Hard water is usually less corrosive toward zinc than soft water. Sacrificial anodes are recommended in seawater (A.B. Chance, 2006).

Corrosion in Air

AC358 (ICC-ES, 2007) states that corrosion loss should be accounted for regardless of whether helical piles are below- or aboveground. Generally zinc coatings protect steel above the ground such that atmospheric corrosion seldom governs helical pile design. Nonetheless, it is important for the helical pile designer to be able to quantitatively address concerns about corrosion in air. Salt spray, sea breeze, and proximity to seawater affect corrosion rates in air. Studies have shown that zinc exposed 80 feet [24 m] from shore corroded three times faster than zinc exposed 800 feet [244 m] from shore (A.B. Chance, 2003).

Table 11.5 Corrosion of Zinc in Various Waters (Craig, 1995)

Water Type	μm/yr	mils/yr	oz./ft^2
Seawater			
Global oceans, average	15–25	0.6–1.0	0.385–0.642
North Sea	12	0.5	0.308
Baltic Sea and Gulf of Bothnia	10	0.4	0.257
Freshwater			
Hard	2.5–5	0.1–0.2	0.064–0.128
Soft river water	20	0.8	0.513
Soft tap water	5–10	0.2–0.4	0.128–0.25
Distilled Water	50–200	2.0–8.0	1.284–5.13

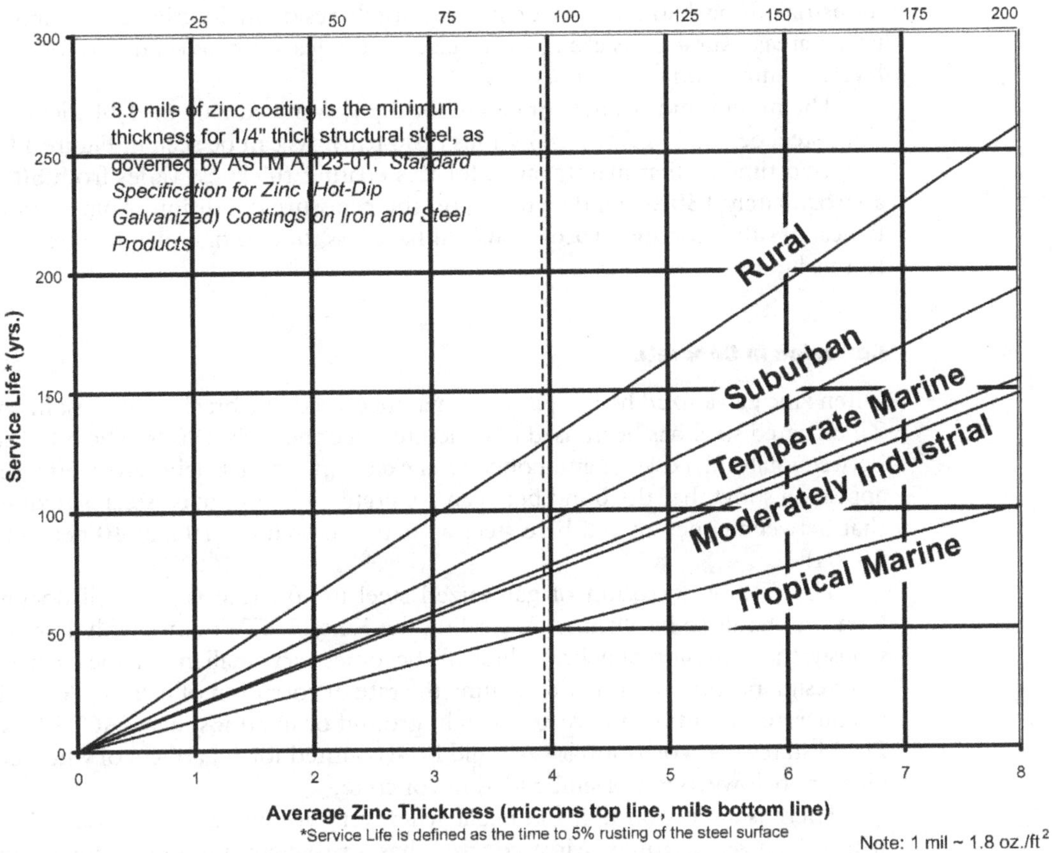

Figure 11.14 Time to first maintenance of hot-dip galvanized coatings in atmosphere (AGA, 2000a)

Since 1926, ASTM committees and others have been collecting records of zinc coating behavior under various categories of atmospheric conditions (AGA, 2000a). Relationships between time to first maintenance and hot-dip galvanized coating thickness in atmosphere under different exposure conditions are shown in Figure 11.14. The chart displays the time the coating will last until 5% rusting of the substrate steel, which is not the design life for the structure, but rather the time when maintenance should be considered. Tropical marine areas are located in close proximity to coastline in warm climates. Temperate marine areas are similar except they experience changing seasons. In marine air, zinc corrosion occurs by a different mechanism. Chlorides from sea spray react with protective zinc oxides and zinc carbonates to form zinc chlorides, which may be washed away. Moderately industrial areas have air emissions that may contain sulfides and phosphates that cause rapid zinc coating consumption. Most cities and urban areas have atmospheric conditions that are considered moderately industrial. Suburban areas are defined as largely residential-perimeter communities of urban areas. Rural areas are characterized by the least corrosion due to relatively low levels of sulfur and other emissions.

The minimum thickness of batch hot-dip galvanized zinc on 1/4-inch- [6-mm-] thick steel per ASTM A123 is 3.9 mils [100 μm]. As can be seen in Figure 11.14, the expected time to first maintenance for this coating thickness varies from 50 years to approximately 130 years, depending on the exposure condition. Thinner zinc layers typical of other coating processes would have less time to first maintenance according to this figure.

Corrosion in Concrete

Often zinc galvanized helical piles and manufactured pile brackets are cast in concrete. Zinc-coated steel has been used in concrete since the early 1900s. The bond strength between galvanized steel and concrete is excellent. In fact, laboratory and field tests appear to show that the bond between concrete and galvanized steel is stronger than that between concrete and bare steel by between roughly 10 and 40 percent (AGA, 2000a).

The rate of corrosion of galvanized steel in concrete is less well documented. Concrete tends to be alkaline. According the King (1977) nomograph described previously, the corrosion of helical pile shafts becomes very small in alkaline environments. For design purposes, it is fair to assume the rate of corrosion of helical piles embedded in concrete should be no greater than in ground or in atmosphere. AC358 (ICC-ES, 2007) states that corrosion loss should be accounted for regardless of whether helical piles are belowground or embedded in concrete.

There is a reaction that can occur when very alkaline concrete comes in contact with galvanized coatings. Most concrete has a sufficient amount of chromate which prevent this reaction. The reaction is only a concern while the concrete is wet. Hydrogen evolution from this reaction can reduce the bond stress. After initial curing the reaction stops so it has minimal effect on the life expectancy of the steel and zinc coating. If bond stress is a concern and the concrete is known to be chromate depleted, the production of hydrogen can be handled by chromate quenching the galvanized steel as specified in ASTM A780 or by adding 100 ppm chromate to the concrete mix. (AGA, 2001)

Corrosion in Organic Soil

Saturated organic soils and peat create an anaerobic condition in which bacteria can thrive. Corrosion can be very severe under such conditions even though oxygen, which is normally needed for corrosion, is not available. The reason is that a different type of corrosion occurs, one that involves sulfate-reducing bacteria. The resulting by-product is iron sulfide, visually identified as an odorous black slime.

Bare steel or galvanized helical piles can be used in organic soils by accounting for sacrificial thickness in severe corrosion conditions using either the 98th percentile method or the King method. If the life span needs to be increased, cathodic protection is possible, but the required electrode potential is higher. In organic soil with anaerobic bacteria, the electrode potential for full protection is −0.95 V (Mudali et al., 2007). From Ohm's law, the required current density and electrode potential are proportionate. Hence, current densities described previously may be increased by the ratio 0.95/0.85 to achieve similar protection in organic soil.

Corrosion in Contaminated Soil

According to the American Galvanizers Association (AGA, 2000a), zinc galvanizing generally performs well in solutions of pH above 4.0 and below 12.5. In fact, since many liquids fall within this range, galvanized containers are used for storage of many chemical solutions including hydrocarbons, alcohols, halides, cyanides, esters, phenols, amines, and amides. In the case of contaminated ground conditions on helical pile project sites, it is useful to understand some of the chemicals that are compatible with zinc. A more comprehensive list of compatible chemicals is contained in the AGA reference.

Chapter **12**

Foundation Systems

In this chapter, a number of special considerations for design of helical pile supported foundations are presented. A simple foundation design for a multifamily residence is described. The basic features of the plan are discussed including example sections, details, and general notes. Determination of foundation loads through service load combinations is discussed. A case study involving a commercial parking garage supported on helical piles and simple computer analysis for applying load combinations to more complex, statically indeterminate pile caps are presented. Design procedures for concrete pile caps including punching shear and plate bearing are summarized.

12.1 BASIC FOUNDATION PLAN

A basic helical pile foundation plan for a multifamily residence is shown in Figure 12.1. This plan contains many of the common features in helical pile foundation drawings. Open circles on the plan represent the locations of helical piles. Dimensions are given for the foundation walls, grade beams, and for each helical pile position. A title block at the lower right corner of the plans contains plan sheet number, project name, revision number, revision history, date, scale, and individuals responsible for the plan.

A schedule of helical pile properties is shown at the lower left corner of Figure 12.1. Quantity, design capacity, minimum depth, final torque, and manufactured cap are listed for each helical pile type. The minimum helical pile length shown on the schedule is to ensure that the piles bottom below a thick layer of soft material encountered in the soil borings. Other reasons for providing a minimum depth could include trash fill, frost, expansive soils and minimum embedment for pullout. The minimum shaft diameter and wall thickness are based on buckling calculations in the soft soils. Minimum final torque is based on correlations between capacity and torque given in

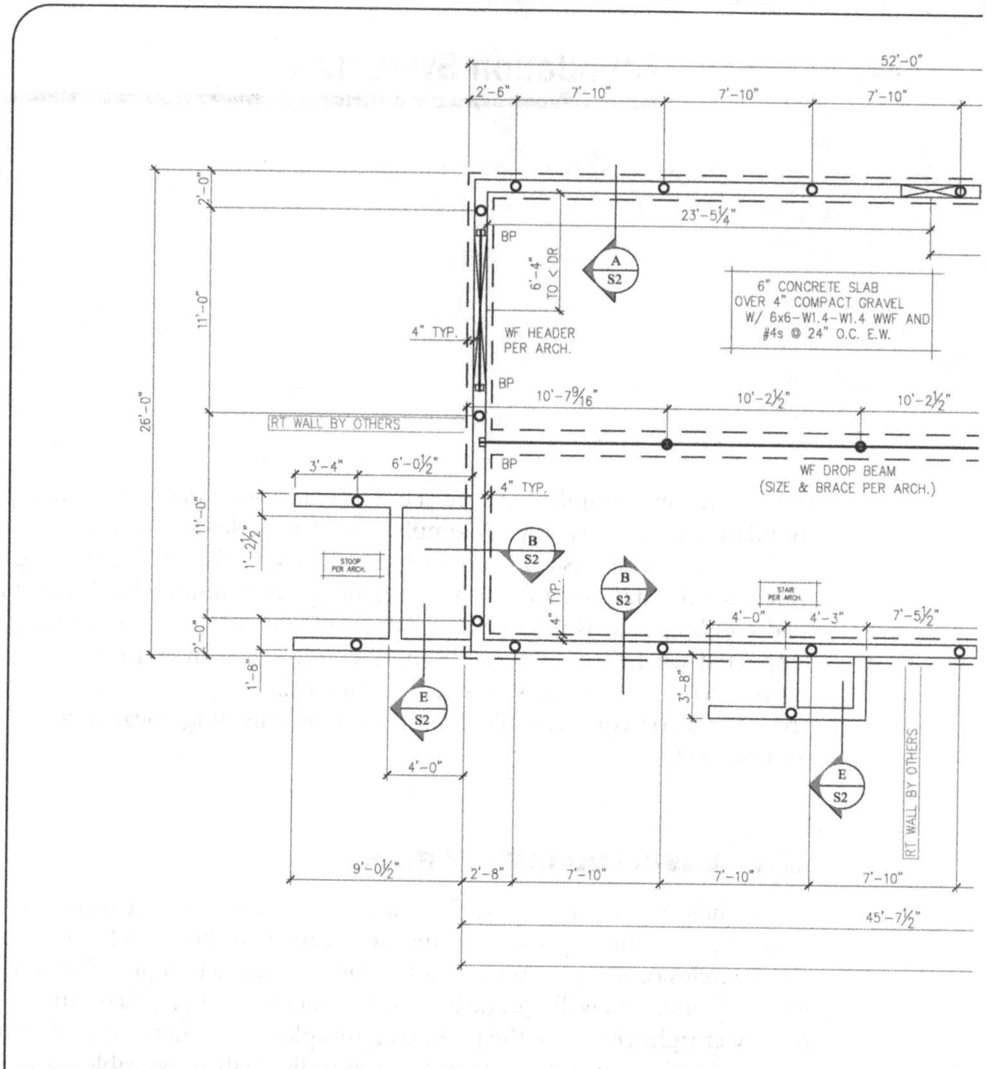

HELICAL PILE SCHEDULE						
SYMBOL	QUANITY	DESIGN CAPACITY	MIN. LENGTH *	MIN. FINAL TORQUE	MANUF. CAP	NOTES
O	30	28 KIPS	15' (FT)	7,000 FT-LBS	MIN. 5"X5"X½" PLATE CAP	MIN. 0.250 WALL 3.00" HSS SHAFT

* MIN. DISTANCE BETWEEN GROUND SURFACE AND UPPERMOST HELICAL BEARING PLATE

Figure 12.1 Example foundation plan for multifamily residence

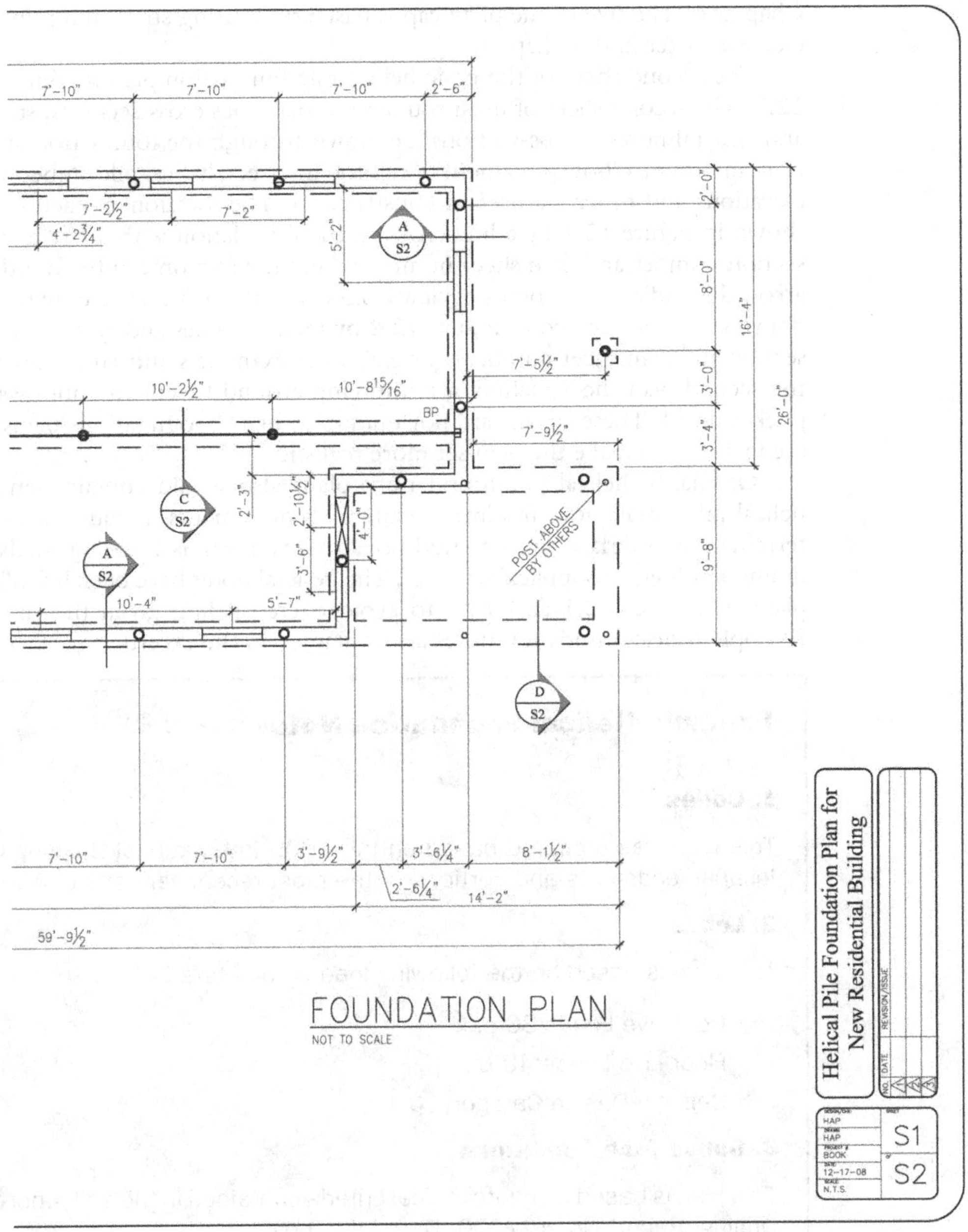
7'-10"
7'-10"
7'-10"
2'-6"
7'-2½"
7'-2"
4'-2¾"
5'-2"
A
S2
2'-0"
8'-0"
16'-4"
7'-5½"
3'-0"
26'-0"
10'-2½"
10'-8 15/16"
BP
7'-9½"
3'-4"
2'-3"
2'-10½"
C
S2
4'-0"
3'-6"
POST ABOVE BY OTHERS
9'-8"
A
S2
10'-4"
5'-7"
D
S2
7'-10"
7'-10"
3'-9½"
3'-6¼"
8'-1½"
2'-6¼"
14'-2"
59'-9½"
FOUNDATION PLAN
NOT TO SCALE
Helical Pile Foundation Plan for New Residential Building
NO.
DATE
REVISION/ISSUE
HAP
HAP
BOOK
12-17-08
N.T.S.
S1
S2

Chapter 6. The size of the plate cap is based on bearing stress and punching shear as discussed later in this chapter.

The second sheet of the basic helical pile foundation plan is contained in Figure 12.2. The second sheet of most foundation plans has cross-sections, standard details, and general notes. Cross-sections are drawn through the foundation at various locations in order to show placement of reinforcing steel, helical pile embedment, heights, elevations and other features of construction. The location of each cross-section is shown in Figure 12.1 by a line crossing the foundation with a circle containing the section number and plan sheet number where the section can be found as well as an arrow that indicates the point of view from which the section was drawn. Cross-section drawings are designated in Figure 12.2 by identical lines and symbols containing the section and plan sheet numbers. Several other example standard details are shown on the second plan sheet including reinforcing around basement windows and a beam pocket detail. These details are not unique to helical foundations plans; rather, they are included to make the plan set more realistic.

Ordinarily, helical pile foundations plans also would contain general notes on helical pile installation, minimum material requirements, torque measurement, and termination criteria as well as traditional items, such as building loads, foundation drains, grading, and applicable codes. The general notes have been left off the example plans in Figures 12.1 and 12.2 to avoid overcrowding given the size of the text. Example helical foundation notes are contained in the excerpt.

Example Helical Foundation Notes

1. Codes

This plan was prepared based on the 2009 International Building Code with local amendments and portions of the most recent versions of ACI318.

2. Loads

This plan is based on the following load parameters:

Roof Live Loads:30 psf

Floor Live Loads:40 psf

Seismic Design Category: C

3. Subsurface Conditions

This plan is based on geotechnical criteria contained in the soil report by Lithos Drilling, Report No. 127300, Dated 5/21/09.

Backfill Equivalent Fluid Density: 50 pcf

Bearing Stratum: Dense Sand (N = 40-50 bpf)

(Continued)

4. Materials

Concrete: Cast-in-place concrete shall consist of Type II cement, 6%±1% air entrainment, and a minimum 28-day compressive strength of 2,500 psi (UNO).

Reinforcing Steel: All concrete reinforcing steel shall be deformed Grade 60 and shall conform to ASTM A615.

Helical Piles: All helical piles and manufactured plate caps shall have a current ICC-ES evaluation report indicating the materials can support the design loads shown on the plans. Helical piles and plate caps shall be furnished by a manufacturer with ISO9001 accredited quality control program.

5. Foundation Construction

All concrete and reinforcing steel shall be placed and detailed in accordance with ACI detailing manual and ACI code, latest editions. All reinforcing steel shall be wired in place to maintain the required location during concrete placement. Minimum concrete cover over reinforcing steel shall be 2 inches (UNO). Overlaps shall be 40 bar diameters but not less than 24 inches.

6. Helical Pile Installation

Helical piles shall be installed to the location, inclination, orientation, elevation, minimum torque, load bearing capacity, and minimum depth shown on the plans. Standard tolerances are as follows: horizontal position ±1 inch, elevation ±0.5 inch, inclination and orientation ±5 deg. Installation torque shall be monitored using equipment calibrated to IAS standards within the past 12 months.

7. Controlled Inspections

Slab subgrade, concrete reinforcing steel placement, and a minimum of 20% of helical pile installations should be observed by the Foundation Engineer. Contractor shall notify Foundation Engineer 24hrs prior to placing concrete or starting helical pile operations to arrange for inspection.

8. Excavation and Backfill

All excavations shall be per OSHA. Non-standard sloped or braced excavations or excavations taller than 20 feet shall be designed by an engineer. Backfill materials, placement, and compaction shall conform to project soil report. The final grade around the structure shall be sloped to conduct water away from the foundation and backfill per local codes.

(Continued)

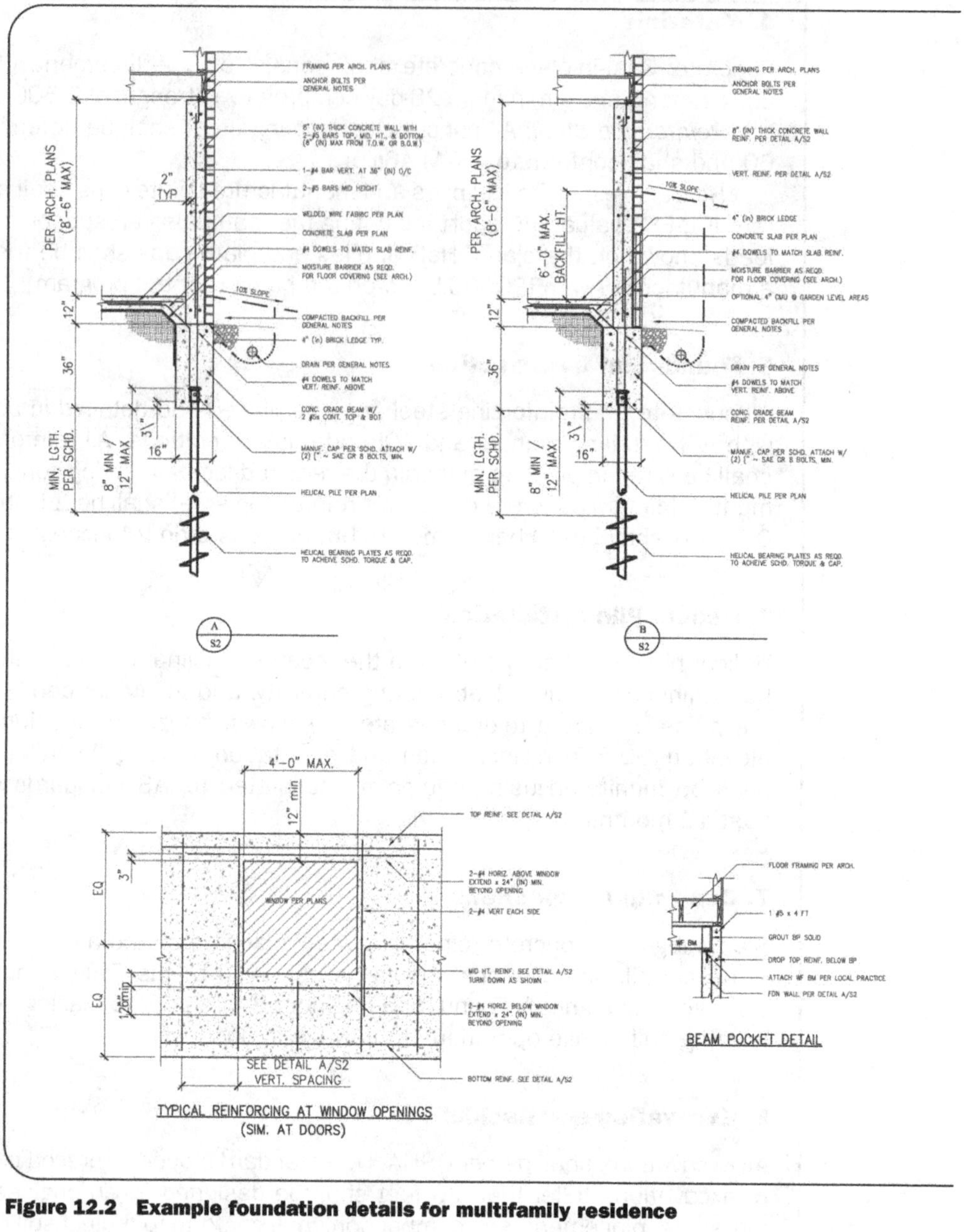

Figure 12.2 Example foundation details for multifamily residence

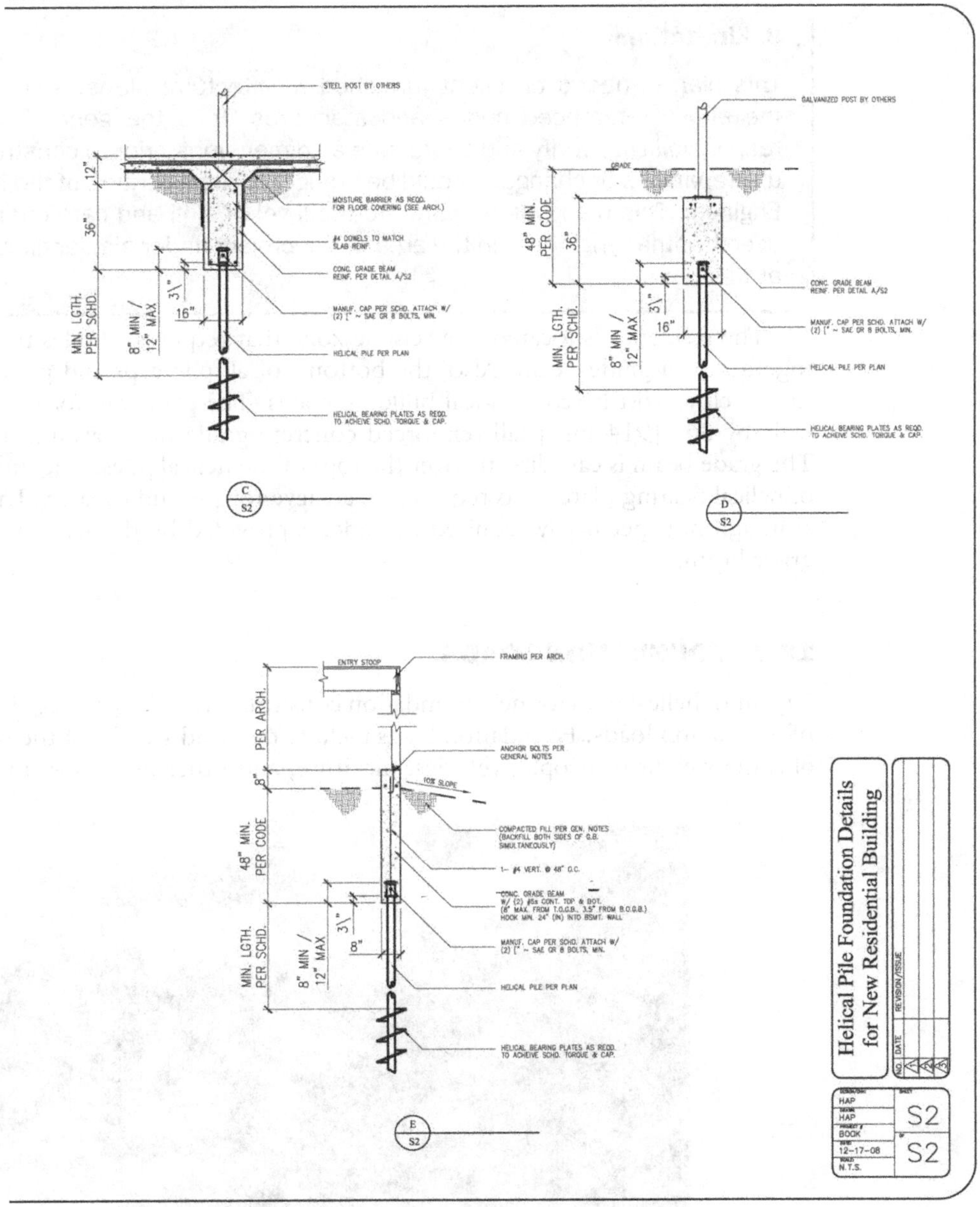
STEEL POST BY OTHERS
MOISTURE BARRIER AS REQD. FOR FLOOR COVERING (SEE ARCH.)
#4 DOWELS TO MATCH SLAB REINF.
CONC. GRADE BEAM REINF. PER DETAIL A/S2
MANUF. CAP PER SCHD. ATTACH W/ (2) ["~ SAE GR 8 BOLTS, MIN.
HELICAL PILE PER PLAN
HELICAL BEARING PLATES AS REQD. TO ACHEIVE SCHD. TORQUE & CAP.
12"
36"
16"
MIN. LGTH. PER SCHD.
8" MIN / 12" MAX
C
S2
GALVANIZED POST BY OTHERS
GRADE
48" MIN. PER CODE
D
S2
ENTRY STOOP
FRAMING PER ARCH.
ANCHOR BOLTS PER GENERAL NOTES
PER ARCH.
COMPACTED FILL PER GEN. NOTES (BACKFILL BOTH SIDES OF G.B. SIMULTANEOUSLY)
1- #4 VERT. @ 48" O.C.
CONC. GRADE BEAM W/ (2) #5s CONT. TOP & BOT. (8" MAX. FROM T.O.G.B., 3.5" FROM B.O.G.B.) HOOK MIN. 24" (IN) INTO BSMT. WALL
E
S2
Helical Pile Foundation Details for New Residential Building
NO. DATE REVISION/ISSUE
HAP
HAP
BOOK
12-17-08
N.T.S.
S2
S2

9. Limitations

This plan is based on client furnished architectural plans, soil report, and the above referenced codes and standards. It is the general contractors responsibility to verify and coordinate all dimensions prior to construction. Any discrepancies or changes should be brought to the attention of the Foundation Engineer. This plan was prepared to the level of skill and care ordinarily practiced by other engineers in the area of the project under similar circumstances at this time.

The residence is located in a seismic zone that requires all piles to be connected together by a grade beam. Also, the bottoms of all pile caps and grade beams have to be below frost based on local building codes. The plans call for a 16" [406 mm] wide by 36" [914 mm] tall reinforced concrete grade beam around the perimeter. The grade beam is cast directly over the tops of the helical piles. The number and size of helical bearing plates is as required to achieve torque and capacity. Lateral bracing, although not specifically required by code, is provided by the slab that ties into the grade beam.

12.2 FOUNDATION LOADS

Layout of helical piles for new foundation construction and repair requires knowledge of foundation loads. Foundation loads include the dead weight of the structure itself plus the weight of people, vehicles, furniture, and other live loads. In addition, the

Figure 12.3 Installation of residential foundation (Courtesy of ECP)

loads from flooding, earthquakes, wind, snow, ice, and other environmental factors need to be taken into consideration. Calculation of various foundation loads is covered in most structural engineering and architecture books. It is beyond the scope of this text to reiterate how to determine tributary areas and the weights of various structures. Foundation loads typically are given to the pile designer by the structural engineer or architect on most projects. The focus of this section is to provide instruction for the proper treatment of these loads and the resolution of axial and lateral loads on individual helical piles.

The design of foundations should be based on load combinations provided in applicable building codes. For bridge and highway structures, load combinations in AASHTO (2004 or more recent) are commonly used. The designer needs to check with local building officials to determine the applicable code for residential and commercial buildings.

The *International Building Code* (IBC, 2006) references load combinations in the American Society of Civil Engineers (2006) *Minimum Design Loads for Buildings and Other Structures* (ASCE7). This manual contains load combinations for load and resistance factor design and for allowable strength design. There has been some advocacy of load and resistance factor design load combinations in foundation engineering. Historically, pile design has been based on allowable strength design load combinations. A proper approach to load and resistance factor design has not been developed yet. As a result, this text focuses primarily on allowable strength load combinations.

ASCE7 allowable strength design load combinations are provided in Table 12.1. The helical pile designer should calculate structure loads based on all eight different load combinations given in the left column. The working loads that govern helical pile design should be from whatever load combinations produce the maximum compression, tension, and lateral loads on the piles and their connection to the structure.

Each of the load types used in Table 12.1 is defined in the right column. Dead load includes the weight of foundation walls, grade beams, and floor slabs. Soil load includes lateral earth pressure and the weight of ground water and buoyant effects. According

Table 12.1 ASCE7 ASD load combinations (ASCE, 2006)

ASD Load Combination	Load Types
1. D + F	D = Dead load
2. D + H + F + L + T	L = Live load
3. D + H + F + (Lr or S or R)	Lr = Roof live load
4. D + H + F + 0.75(L + T) + 0.75(Lr or S or R)	S = Snow and ice load
5. D + H + F + (W or 0.7E)	E = Earthquake load
6. D + H + F + 0.75(W or 0.7E) + 0.75L + 0.75(Lr or S or R)	W = Wind load
7. 0.6D + W + H	R = Rain load
8. 0.6D + 0.7E + H	H = Soil load
	F = Fluid load
	T = Thermal load

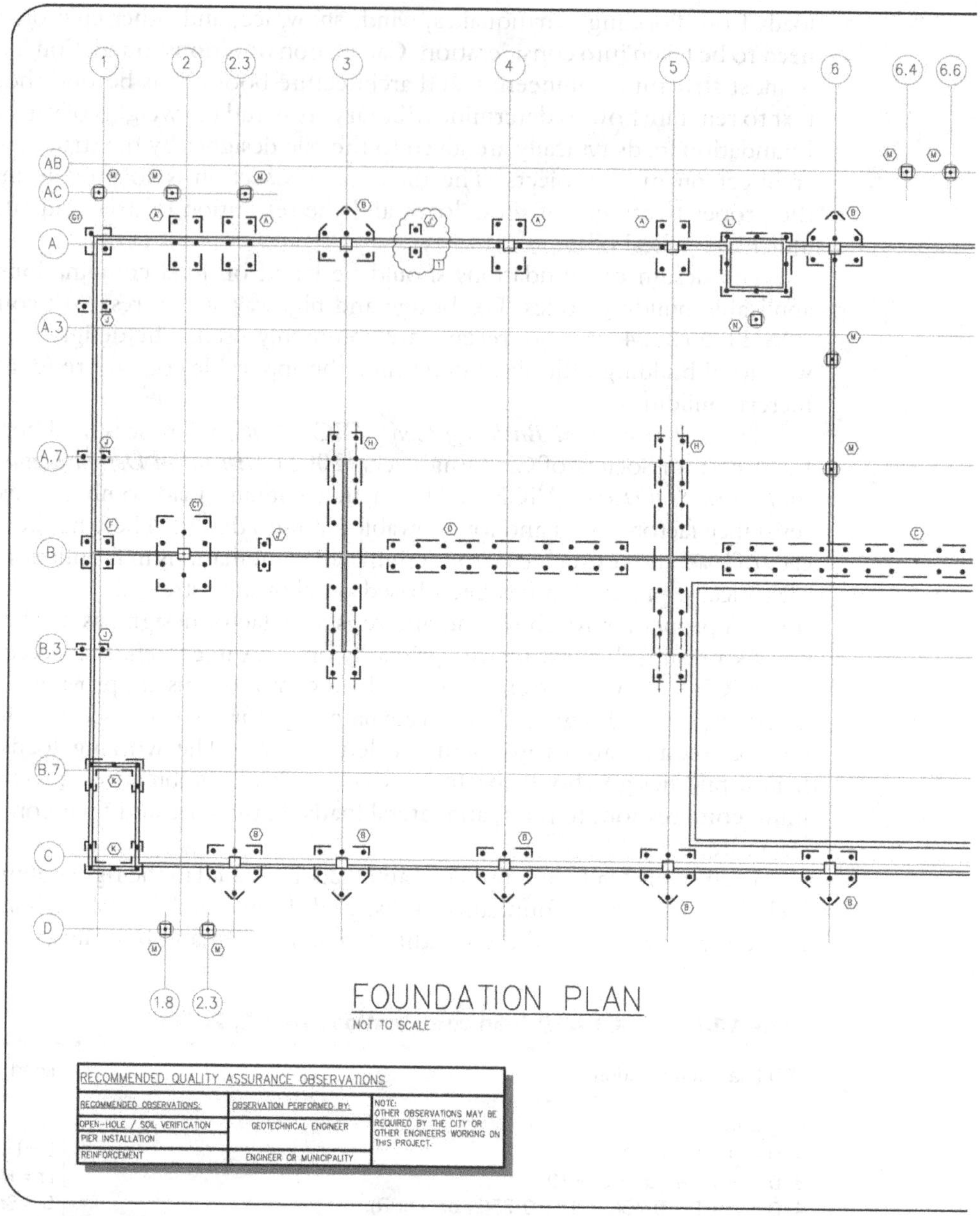

Figure 12.4 Foundation plan for parking garage

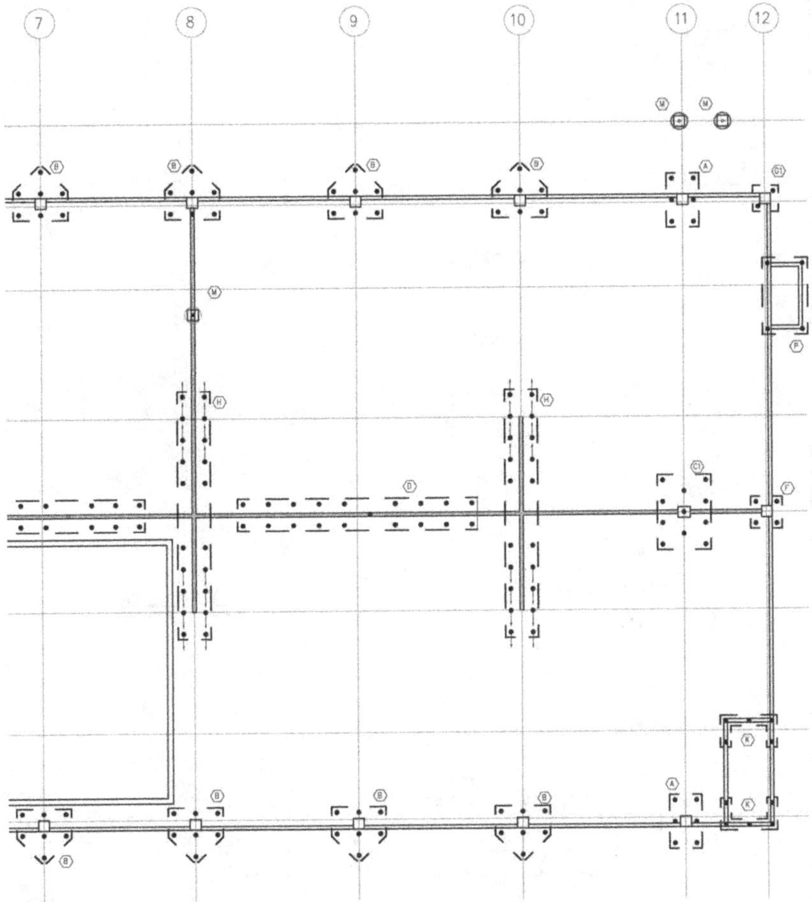

HELICAL PIER SCHEDULE							
SYMBOL	DESCRIPTION	REQ'D CAPACITY			REQ'D FINAL TORQUE	DIMENSIONS	COMMENTS
		COMP.	TENSION	LATERAL			
⊙	3" HELICAL PILE	30 KIPS	N.A.	N.A.	7,500 FT–LBS	3.00" O.D., 0.250 WALL HSS STEEL TUBE, Fy=64 ksi	GALVANIZED
•	6" HELICAL PILE	100 KIPS	50 KIPS	7.5 KIPS	50,000 FT–LBS	5.875" O.D., 0.280 WALL HSS STEEL TUBE, Fy 64 ksi	BARE STEEL

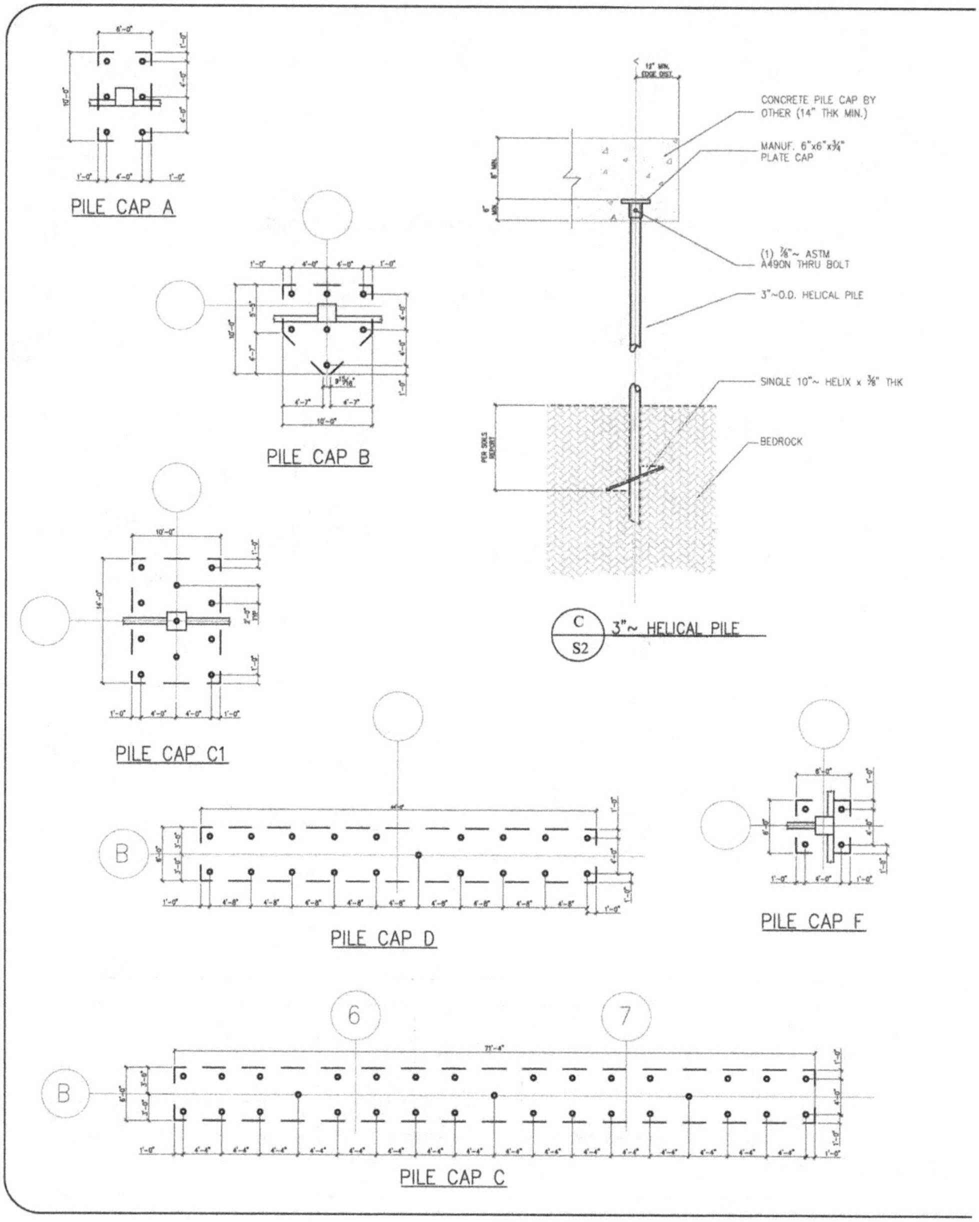

Figure 12.5 Foundation details for parking garage

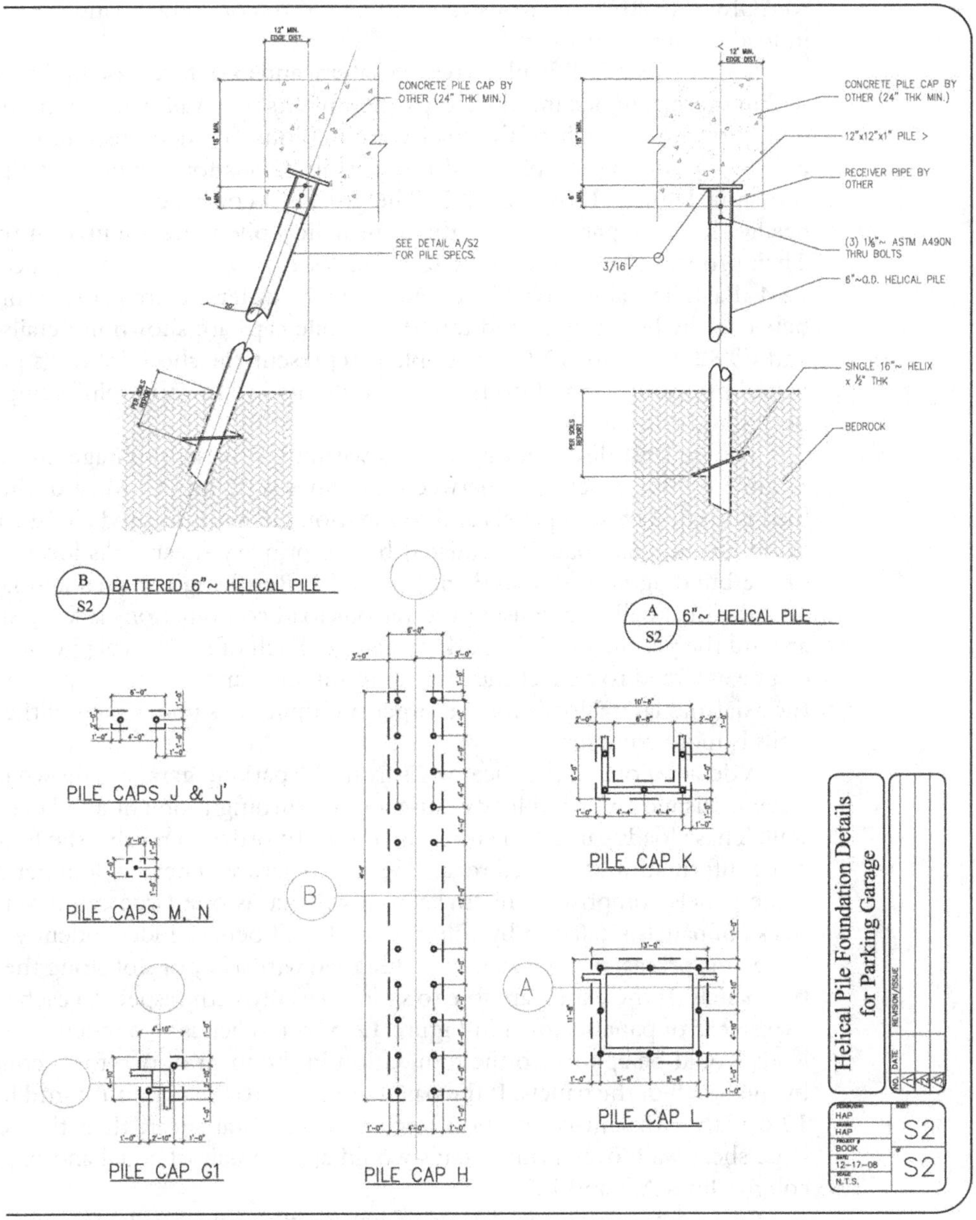
CONCRETE PILE CAP BY OTHER (24" THK MIN.)
SEE DETAIL A/S2 FOR PILE SPECS.
B S2 BATTERED 6"~ HELICAL PILE
12"x12"x1" PILE
RECEIVER PIPE BY OTHER
(3) 1⅛"~ ASTM A490N THRU BOLTS
6"~O.D. HELICAL PILE
SINGLE 16"~ HELIX x ½" THK
BEDROCK
A S2 6"~ HELICAL PILE
PILE CAPS J & J'
PILE CAPS M, N
PILE CAP G1
PILE CAP H
PILE CAP K
PILE CAP L
Helical Pile Foundation Details for Parking Garage
HAP
HAP
BOOK
12-17-08
N.T.S.
S2
S2

to ACI318 (2005), soil loads cannot be used to counteract wind or earthquake loads in load combinations 7 or 8.

Sometimes it is difficult to resolve lateral and axial forces on individual helical piles within more complex multipile caps and under shear walls without modeling load and deflection under each of the load combinations. To demonstrate this, consider the helical pile foundation plan and foundation details for a commercial parking garage shown in Figures 12.4 and 12.5. The garage is constructed of precast concrete and has five levels of parking. Two different helical pile types are used in the foundation. Their properties are given in the schedule located at the bottom right corner in Figure 12.4. Each helical pile type is represented by a different circular symbol. Connections between the helical piles and foundation pile caps are shown in details A/S2, B/S2, and C/S2 in Figure 12.5. These plans represent the shop drawings provided by the helical pile designer and do not include design of concrete reinforcing steel in grade beams and pile caps.

The general distribution of loads within the parking garage are discussed here. Double T floor panels span between grid lines A, B, and C. Most of the live and dead load is carried by wall panels and foundation elements located along these grid lines. Wind and seismic loads are resisted by the primary shear walls located in the center of the building at grid lines 3, 5, 8, and 10. Resolving the axial and lateral loads for an individual helical pile using the various load combinations is fairly straightforward around the perimeter of the parking garage. Each of the helical piles in a multipile cap can be assumed to deflect the same amount and carry roughly equal loads. Resolving the axial and lateral loads for the larger multipile caps under each of the primary shear walls is more complex.

A detail of one of the shear walls from the parking garage is shown in Figure 12.6. Lateral seismic and wind loads cause an overturning moment and lateral shear on the wall. These loads can act in either direction. In order to resolve the foundation loads, more information is needed regarding the structure. The pile designer needs to know if the panels comprising this shear wall will act as one large cantilever wall or if the stacks of panels separated by column line B will behave independently.

Some precast panel systems are designed with a key or slot along the joint between two panels. If the panels are free to slide vertically with respect to each other, then the two stacks of panels shown in Figure 12.6 would act as two independent shear walls; loads would be applied to the foundation in the form of two force couples separated by the width of the panels. If the panels are fixed to the column at grid line B in Figure 12.6 with sufficient connection to resist horizontal shear, then they can act as one large shear wall; overturning loads would apply a pair of equal and opposite forces at column lines A.7 and B.3.

One approach to the design of helical piles under tall shear walls is to resolve foundation loads at each of the pile caps shown at the bottom of Figure 12.6. Under seismic and wind load combinations, overturning may be resisted by reducing the axial force in one pile cap and increasing it in another. The pile caps can be assumed to deflect uniformly such that the axial loads are divided evenly among the piles. Lateral loads may be resisted by the interaction of vertical pile shafts with the soil. The resistance

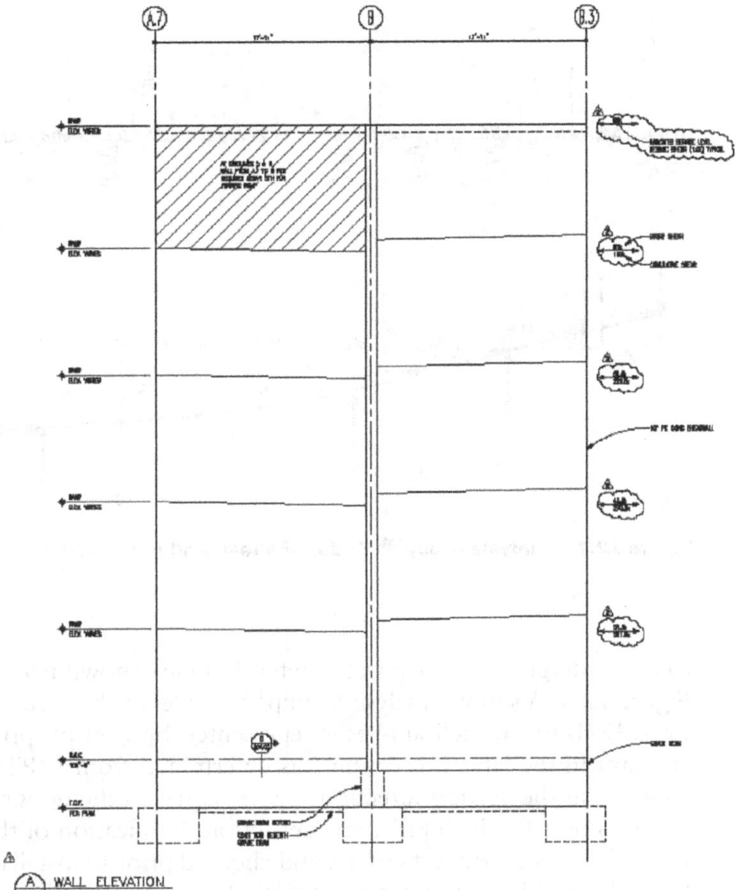

Figure 12.6 Profile view of primary shear walls for parking garage

to lateral and axial loads could be checked using L-Pile™ software as discussed in Chapter 10.

Difficulty arises if the lateral resistance of vertical helical piles is insufficient, if battered piles are added to the design, or if a continuous pile cap is used. One cannot have piles in compression and tension within the same pile cap without accounting for pile deflection. There has to be strain compatibility between adjacent piles and the pile cap itself. In more complex cases, it is necessary to use a computer model to account for pile deflection and to resolve the maximum axial and lateral loads on helical piles.

A detail of the helical pile configuration used to support the primary shear walls in the parking garage is shown in Pile Cap H of Figure 12.5. As can be seen, the pile cap consists of a narrow rectangle supported by 10 pairs of helical piles. The last four pairs of piles at the ends of the pile cap are battered. The pile caps had to be kept narrow due

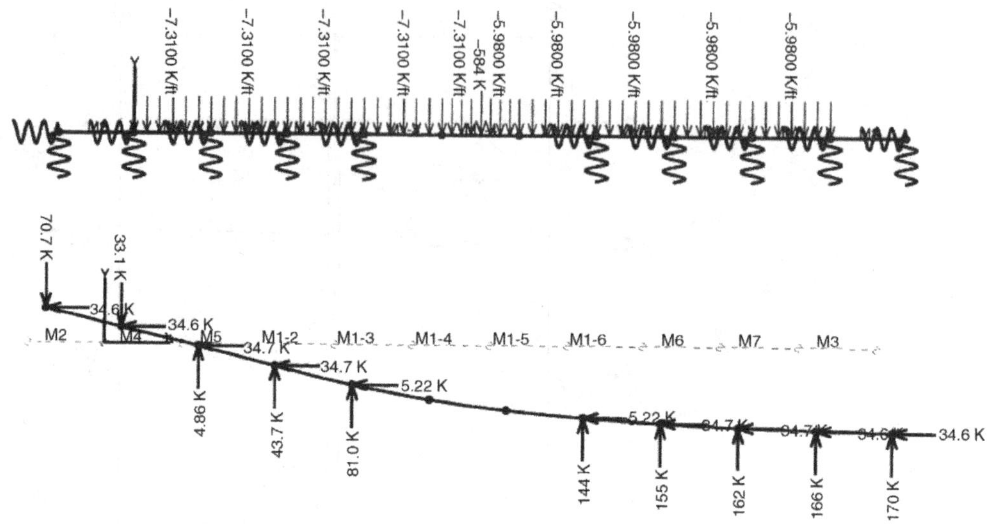

Figure 12.7 Analysis Group^{TM} Model of shear wall grade beam

to the underground storm water detention tank shown between grid lines 5 and 8 in Figure 12.4. A simple Analysis Group^{TM} model of the grade beam is shown in Figure 12.7. Each pair of helical piles is represented by a set of spring supports. The spring constant in the lateral direction was determined from L-Pile^{TM} analysis. The spring constant in the vertical direction was set equal to the design allowable axial load on the piles over 0.5 inch [13 mm] deflection. Verification of the spring constants were worked into the load test criteria and checked prior to installation of production piles. For taller buildings and more complex designs, it would be more rigorous to orient the spring supports equal to the batter angle of the piles.

Foundation loads on the pile cap were separated into distributed and point loads in order to model actual conditions as close as possible. Analysis Group^{TM} software was used to apply the eight load combinations given in Table 12.1 plus associated permutations as a result of changing load directions. Due to the mass of the precast structure, the load combinations involving seismic forces resulted in the highest pile loads. The resulting deflected shape of the pile cap and the forces on each pair of piles under the most severe seismic load combination are shown at the bottom of the figure. Under these loads, two pairs of piles are expected to go in tension at the end of the grade beam and the remaining piles are in compression. The loads on each pair of piles can be compared to the required working loads on individual helical piles as given in the schedule in Figure 12.4. By varying the number of piles and pile locations, pile cap designs can be optimized for the fewest number of piles.

The deflection results of Analysis Group^{TM} or other computer software can be used to verify foundation performance against applicable building code standards. Shear and bending moment results from the software also can be used design of the concrete pile cap. Bearing, simple shear, and punching shear are discussed in the next

Figure 12.8 Installation of helical piles for parking garage (Courtesy of Hayward Baker, Inc.)

section. Flexure can be designed in accordance with ACI318. More information about the design of the parking garage and the foundation analysis can be found in Perko (2008a).

A photograph of a helical pile being hoisted for installation at the parking garage is shown in Figure 12.8. Here again, it is important to understand the capability and configuration of helical pile installation equipment for better design of helical pile foundations. In the photograph, a fixed-mast drill rig is used to install the helical piles in a single "stroke". An extension was added to the lead section in anticipation of the required depth of installation. This machine was incapable of installing the battered piles so a large excavator with pendulum style torque motor had to be mobilized to finish the installation of battered piles.

12.3 PILE CAP DESIGN

The proper design of pile caps is an important subject that is somewhat neglected in structural engineering texts because it involves foundations and in foundation engineering texts because it involves concrete design. In this section, the basic provisions of ACI318 (2005) are reviewed with respect to the design of concrete pile caps supported on helical piles. The size and depth of concrete pile caps are based primarily on the shear strength of concrete. The reinforcing steel ratio required for flexure is usually near or controlled by the minimum ratios required by ACI318. The size of the steel bearing plate on a helical pile is governed by the bearing strength of the concrete. Shear strength of concrete is discussed next. Bearing strength of concrete is discussed later in this section. Both are important considerations for the helical pile designer.

Shearing failure of a concrete pile cap occurs when adhesion of the cement and friction between the aggregate is overcome; the cap splits and sideways movement occurs along the common fracture plane. Concrete pile caps should be checked for beam shear (one-way action) and punching shear (two-way action). There are two design approaches for analysis of concrete shear strength. The first is based on nominally reinforced concrete. According to ACI318, the design shear strength of nominally reinforced concrete for one-way action is given by

$$\phi V_n = \phi 2\sqrt{f'_c} b_o d_s \tag{12.1}$$

and for two-way action is given by

$$\phi V_n = \phi 4\sqrt{f'_c} b_o d_s \tag{12.2}$$

Where

ϕ is the resistance factor,
f'_c is concrete compressive strength
b_o is the perimeter of the critical shear section, and
d_s is the depth to the reinforcing steel.
For reinforced concrete shear, ϕ is equal to 0.75.

Equations 12.1 and 12.2 are unit sensitive and have to be calculated in terms of pounds and inches.

The parameter b_o is the length or perimeter of critical shear sections, which are defined in Figure 12.9 for a typical nominally reinforced concrete pile cap. The detail at the top of the figure shows a plan view of the pile cap. One-way shear action through the pile cap is represented by the dashed line acting at a distance d_s from the load bearing column. In this case, b_o is equal to the width of the pile cap. Two-way shear action is represented by the dashed square drawn at an offset distance of $d_s/2$ around the load bearing column. In this case, b_o is the perimeter of the dashed square. The detail at the bottom of the figure is a profile view of the pile cap, which shows one-way and two-way shear planes extending out from the load bearing column at 45 degree angles. These details are for helical piles in compression only.

The Concrete Reinforcing Steel Institute (CRSI) handbook (1996) suggests that the minimum distance from the edge of the cap to the center of a pile should be 15 inches [381 mm] for piles with design loads less than 60 tons [430 kN]. The CRSI handbook also states that hooks should be provided when necessary to ensure proper anchorage of the reinforcing steel and prevent premature tied-arch shear mode failure. Deep beam shear should be checked instead of one-way beam shear for piles closer

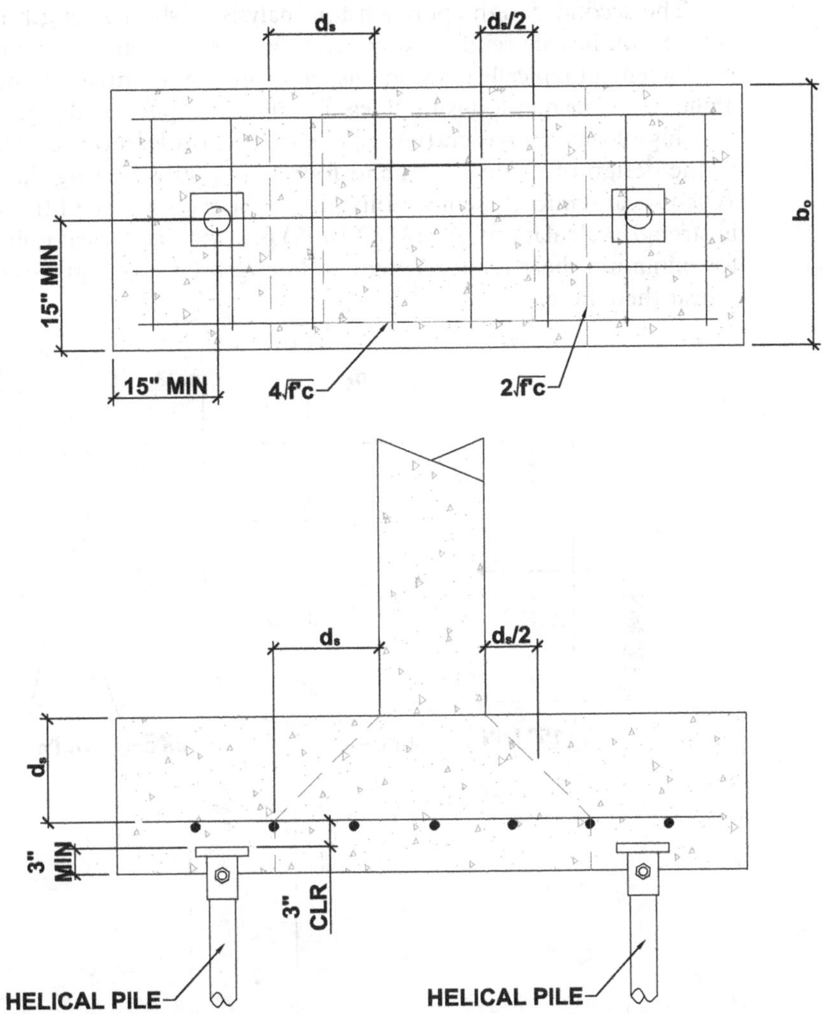

Figure 12.9 Plain concrete pile caps

than d_s from the load bearing column. The CRSI handbook sample calculations do not include evaluation of two-way shear around the piles. It is believed that this is due to the fact that two-way shear around the column and one-way shear through the pile cap likely govern the pile cap design.

In ACI318, the minimum area of shear reinforcement in flexural members is exempt for pile caps. The shear strength of concrete given in Equations 12.1 and 12.2 is not limited to half the computed values per section 11.5.5 of ACI318 (2005). The designer should use discretion for pile caps spanning multiple piles that need to be designed for significant flexure. For such caps, it is prudent to design for minimum shear reinforcement, which is discussed next.

The second design approach for analysis of shear strength in concrete pile caps is based on heavily reinforced concrete with shear reinforcement. Detailing of shear reinforcement typically is accomplished by including stirrups in both directions within a pile cap. This method is used less frequently in pile cap design except in the case of very high loads. However, the use of stirrups in at least one direction is very common in the design of grade beams and foundation walls bearing directly on helical piles. When shear reinforcement is utilized, the shear resistance of the steel can be included in strength calculations. ACI318 (2005) provides fairly clear guidelines and equations for minimum shear reinforcement and design shear strength so it is not necessary to repeat them here.

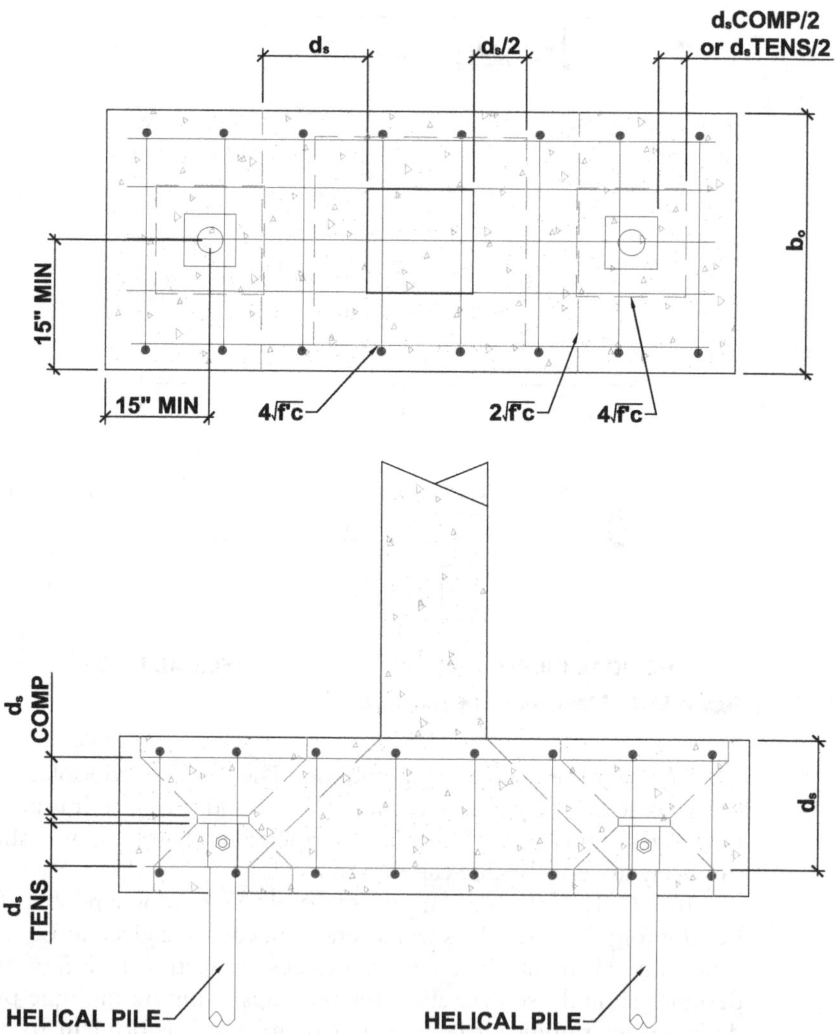

Figure 12.10 Reinforced concrete pile caps

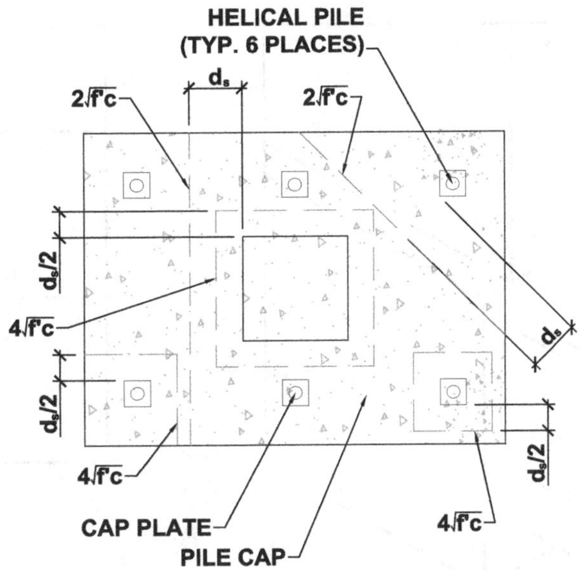

Figure 12.11 Critical shear sections for multiple pile cap

A typical heavily reinforced pile cap is shown in Figure 12.10. This pile cap differs from that shown in Figure 12.9 in that a mat of reinforcing steel is shown both at the top and the bottom of the pile cap. As in the previous figure, the detail shows one-way shear action through the pile cap and two-way shear action around the perimeter of the load bearing column. The helical piles are shown extending past the bottom reinforcement in order to provide a tension connection. In this case, two-way shear should be checked around the helical pile base plate. The values of b_O are similar to those shown in Figure 12.9. The details differentiate between the different depths of reinforcing steel, d_S, used for compression and tension analysis. Critical shear sections are shown for both compression and tension.

Up to this point, the discussion has been focused on design of pile caps spanning between two piles. Figure 12.11 shows a multiple pile cap with example critical shear sections. Potential modes of shear failure for multiple pile caps include break-out of a single pile or a small group of piles located near the edge of the pile cap, shearing down the center of the pile cap, and punching of individual piles and the load bearing column through the pile cap. One-way and two-way shear is differentiated by the shear strength formulae that label each critical shear section. The designer needs to select the path of least resistance for calculation of all possible beam and punching shear combinations for a specific design. The example shown here is intended to show some possibilities so that the designer is better equipped in the analysis of pile groups. The number of possible configurations of multiple pile caps is limitless. This example is neither all-inclusive nor comprehensive.

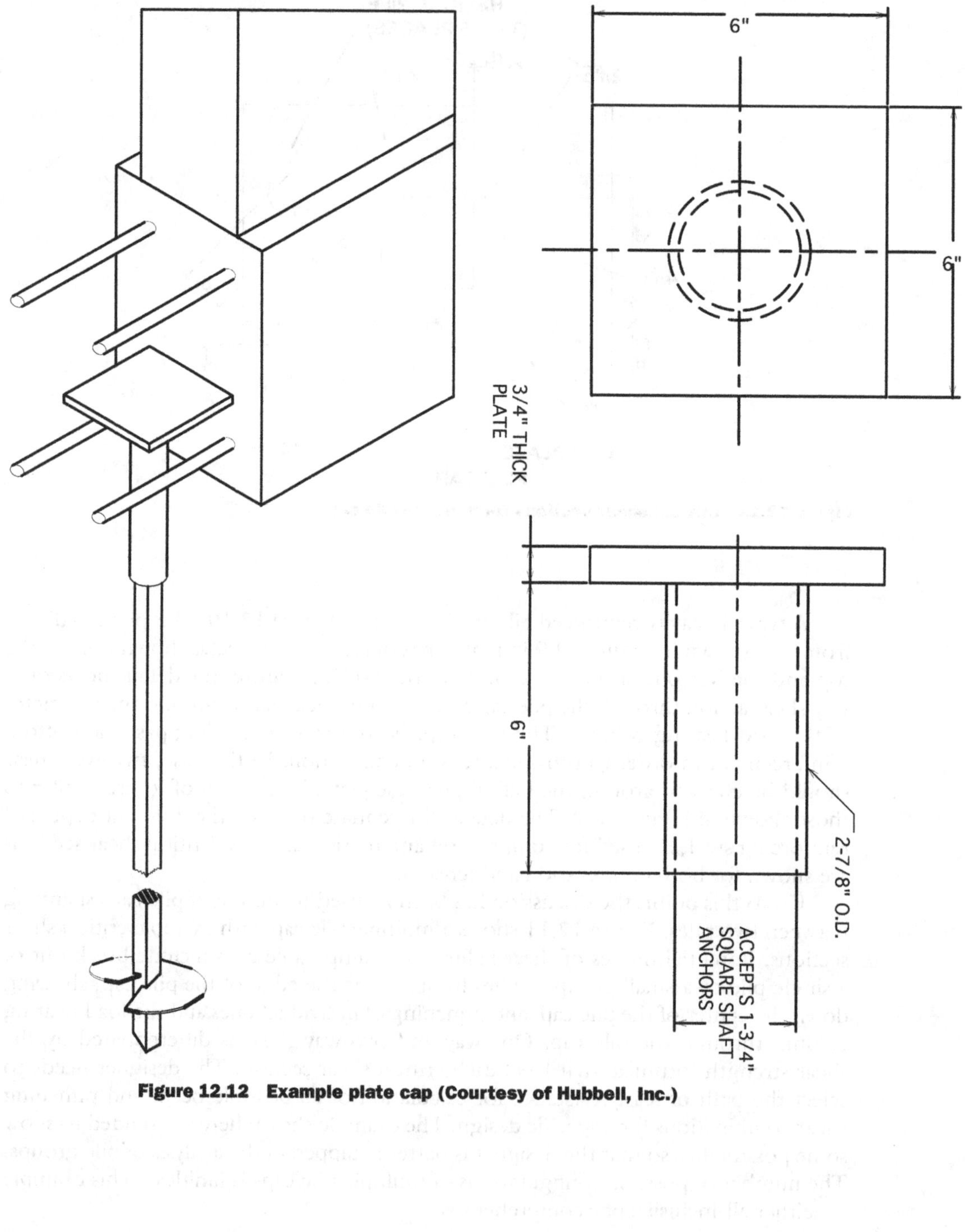

Figure 12.12 Example plate cap (Courtesy of Hubbell, Inc.)

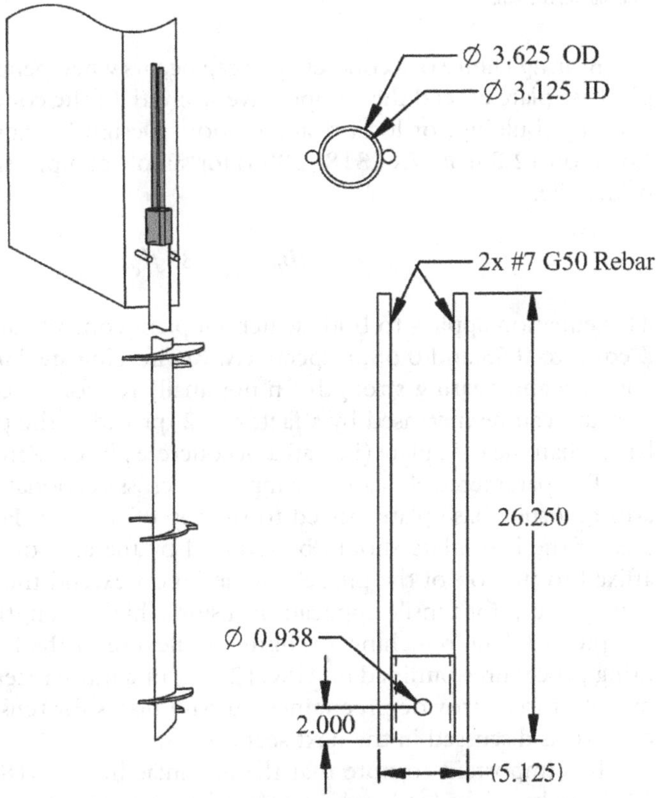

Figure 12.13 Example rebar cap (Courtesy of Magnum Piering, Inc.)

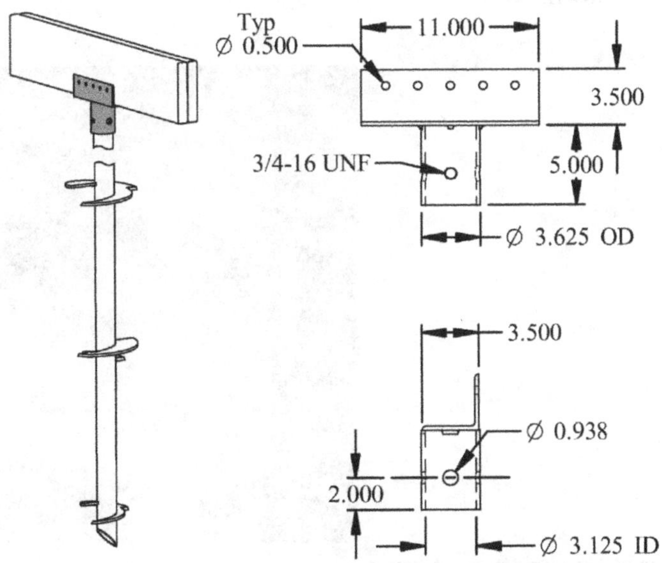

Figure 12.14 Example wood beam cap (Courtesy of Magnum Piering, Inc.)

Bearing failure of a concrete pile cap occurs when penetration stresses at the helical pile base plate exceed the compressive strength of the concrete, which can cause radial cracking, bulging, or local disintegration. Design bearing strength is checked using Equation 12.3 from ACI318 (2005) for simple compression on the bearing area, A_1, of the pile.

$$\phi B_n = \phi 0.85 f'_c A_1 \tag{12.3}$$

This equation applies to both structural plain concrete and reinforced concrete with ϕ equal to 0.55 and 0.65, respectively. Reinforcing steel increases confinement of the concrete and bearing strength. In nominally reinforced concrete, the design bearing pressure can be increased by a factor of 2, provided the pile cap is at least four times larger than the base plate (i.e., area of concrete pile cap/area of the pile base plate ≥ 4).

The parameter A_1 can be simply the cross-sectional area of the pile butt or the area of a steel base plate affixed to the top of the pile. For tensile loads, the bearing area of the base plate should be reduced by the area of the pile shaft if it is directly affixed to the top of the pile. Some designers extend the bearing plate into the concrete pile cap for tensile applications using a high-strength thread bar, as discussed in Chapter 13. The punching strength and flexure of the base plate should be checked using procedures outlined in AISC (2001) or a more recent edition. Many helical pile manufacturers provide preengineered base plates for tension and compression applications, as discussed in the next section.

It is important to note that the formulae in ACI318 (2005), namely Equations 12.3, are based on load and resistance factor design and should be divided by a factor of safety of 1.5 to obtain allowable capacity if the factored load combinations are unknown.

Figure 12.15 Example flitch beam cap

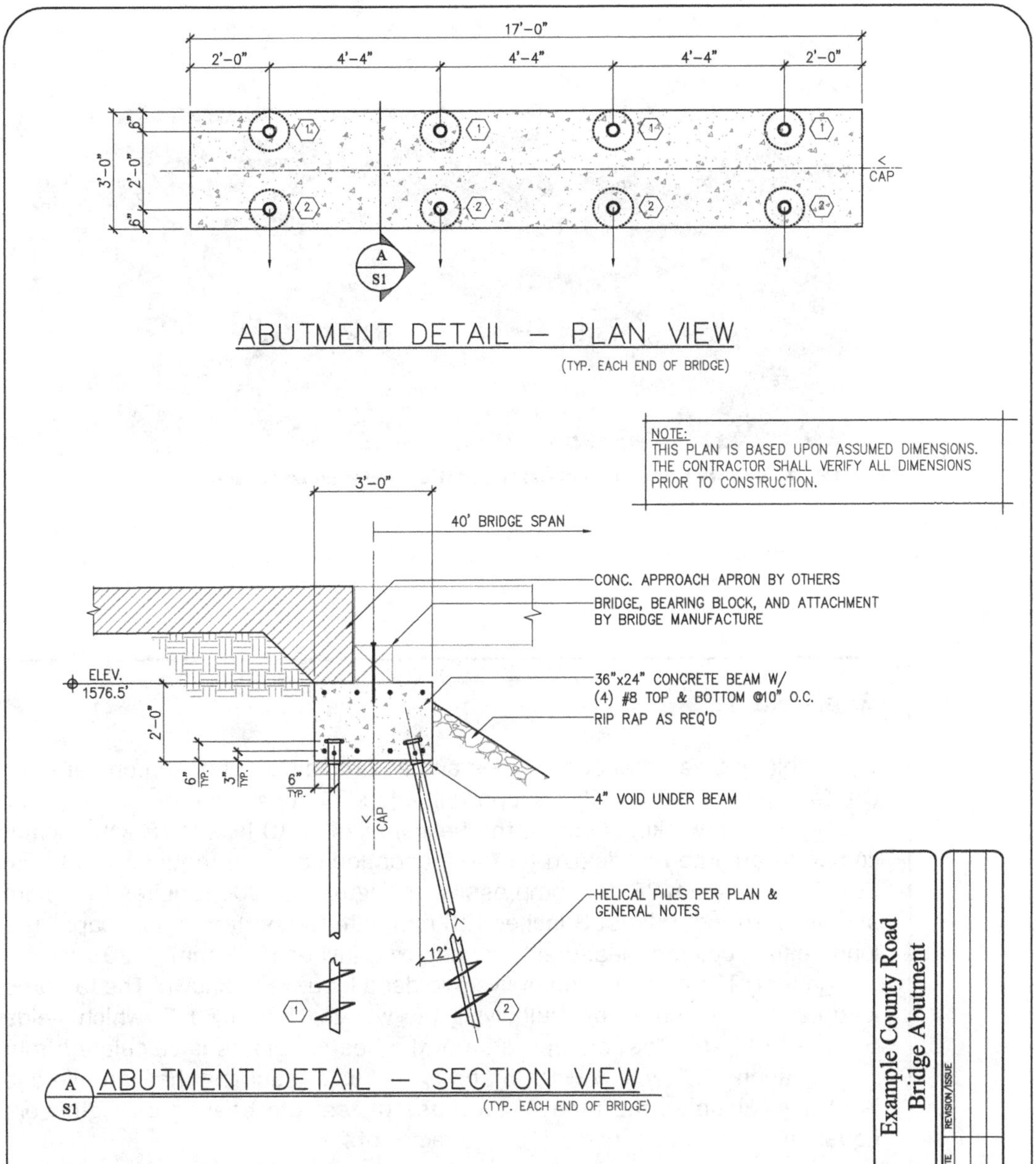

HELICAL PILE SCHEDULE

SYMBOL	QTY	ALLOW. CAP.	ULT. CAP.	REQ'D TORQUE	BATTER	CAP	MIN. LENGTH ✳	NOTES
1	4 PER ABUTMENT	24 KIPS (C)	48 KIPS (C)	6,000 FT-LBS	0^	5x5x1/2 PLATE CAP	6'-0"	MINIMUM OF 25% OF PIERS TO BE OBSERVED BY GEOTECHNICAL ENGINEER DURING INSTALLATION.
2	4 PER ABUTMENT	24 KIPS (C)	48 KIPS (C)	6,000 FT-LBS	12^			

✳ MIN. LENGTH = FINISHED GRADE TO UPPER CUTTING BLADE

Figure 12.16 Example county road bridge abutment

Figure 12.17 County bridge on helical foundation under construction

Example 12a

Problem: Size steel base plates and a nominally reinforced concrete pile cap for a pair of helical piles supporting a load bearing column.

Given: The working load on the helical piles is 40 kips [178 kN] each in compression. The mix design for the foundation concrete requires 4,000 psi [28 MPa] minimum 28-day compressive strength. IBC2006 requires that piers and piles extend at least 3 inches [76 mm] into the bottom of pile caps. The load bearing column measures 8 inches by 8 inches [203 mm × 203 mm].

Answer: The combination of live and dead loads is unknown. The factored load can be estimated by multiplying the working load by 1.5, which yields 60 kips [267 kN]. The required area of the bearing plates is calculated from

Equation 12.3 with $\phi = 0.65$ and $f'_c = 4$ ksi. Assuming that the pile cap is at least 4 times the area of the base plates, the bearing strength from Equation 12.3 can be increased by a factor of 2.

$$\begin{aligned} 60\,\text{kips} &= 2(0.65)(0.85)(4\,\text{ksi})A_1 \\ \therefore A_1 &= 14\,\text{in}^2[90\,\text{cm}^2] \end{aligned} \tag{12a.1}$$

A 4-inch × 4-inch [10-cm × 10-cm] plate has a bearing area of 16 square inches [103 cm^2].

The required pile cap thickness can be determined considering punching shear and Equation 12.2 The length of the critical shear section, b_o, is the perimeter of a square drawn offset a distance h/2 from the column base.

$$b_O = 4\left(8'' + d_S\right) = 32 + 4d_S \tag{12a.2}$$

Substituting Equation 12.2 into Equation 12.2 with $\phi = 0.75$ yields

$$\begin{aligned} 120{,}000\,\text{lbs} &= 0.75\,(4)\,\sqrt{4{,}000\,\text{psi}}\,(32 + 4d_S)\,d_S \\ \therefore 632 &= 32d_S + 4d_S^2 \end{aligned} \tag{12a.3}$$

which can be solved using the quadratic formula. The result is d_S= 9 inches [23 cm]. The thickness of the pile cap is this distance plus the 3 inch minimum pile embedment and the 3 inch clear cover above the bottom mat of reinforcement for a total of 15 inches [381 mm].

The required minimum width of the concrete pile cap can be determined considering beam shear, Equation 12.1, and solving for the length of a shear section, b_O = B, passing through the center of the pile cap offset a distance d_S from the load bearing column. The pile cap is required to resist the load on one of the piles in one-way shear. From Equation 12.1, the one-way concrete shear strength is given by

$$\begin{aligned} 60{,}000\,\text{lbs} &= 0.75\,(2)\,\sqrt{4{,}000\,\text{psi}}\,(B)\,9'' \\ \therefore B &= 70''[1.8\,\text{m}] \end{aligned} \tag{12a.4}$$

Where horizontal space is limited, the pile cap can be made thicker in order to reduce the required width. Realistically, it may be more reasonable to set the minimum width of the pile cap to 30 inches [762 mm] in order to satisfy CRSI edge distance recommendations and solve for the depth of the pile cap. Then, punch shear of the column can be checked. The required minimum thickness of the narrower concrete pile cap is given by

$$\begin{aligned} 60{,}000\,\text{lbs} &= 0.75\,(2)\,\sqrt{4{,}000\,\text{psi}}\,(30'')\,d_s \\ \therefore d_S &= 21''[0.5\,\text{m}] \end{aligned} \tag{12a.5}$$

Helical piles may support other materials besides concrete. In the case of wood structures, the designer needs to check the crushing strength of the wood member in order to properly size the steel plate cap. National Design Standard (NDS) codes for wood design can be consulted to determine the allowable crushing strength depending on the wood species and orientation of the grain.

The IBC (2006) has general provisions for the design of pile caps. According to this code, helical piles must extend a minimum of 3 inches [76 mm] into the bottom of pile caps. Pile caps must extend a minimum of 4 inches [102 mm] beyond the

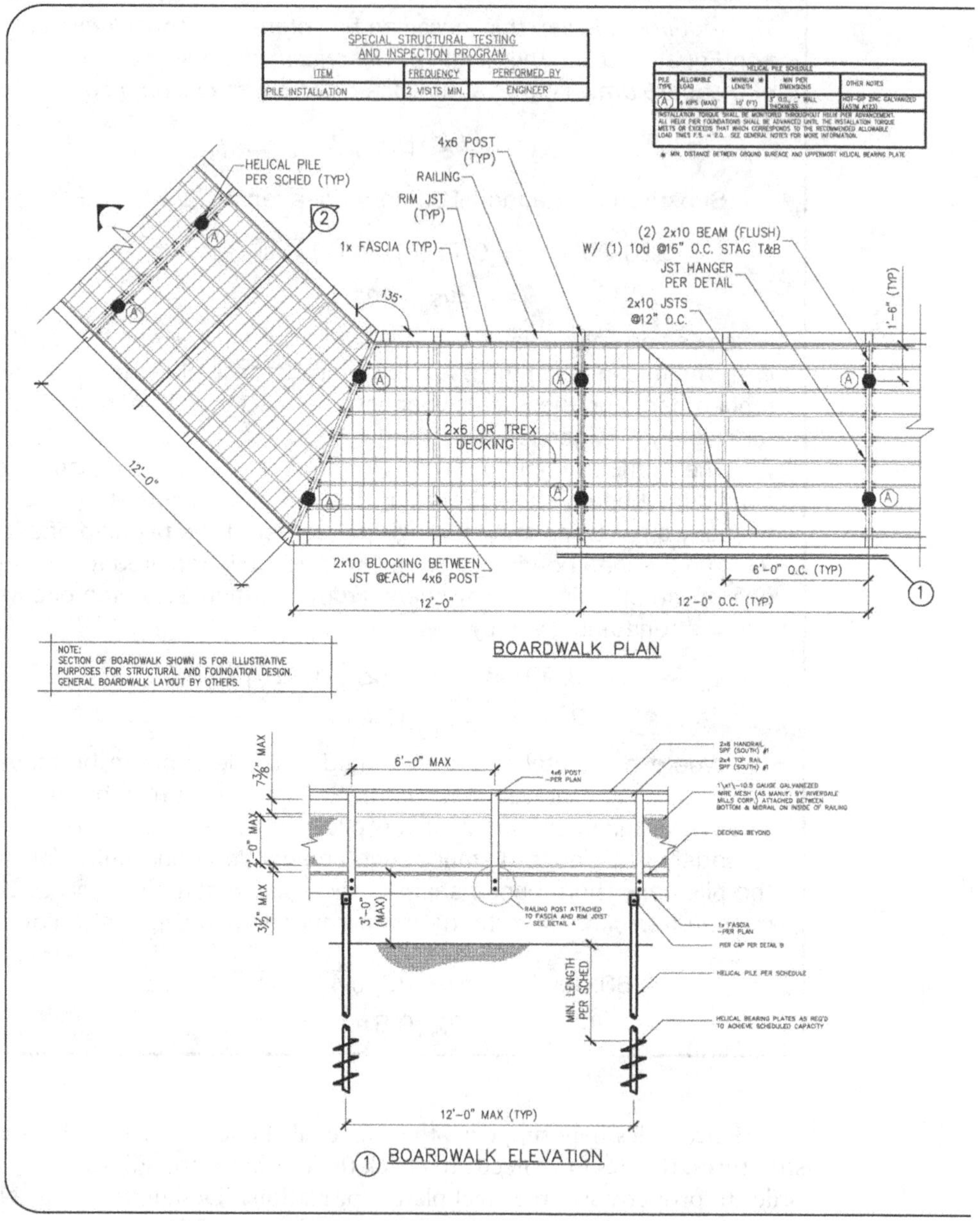

Figure 12.18 Example boardwalk foundation design

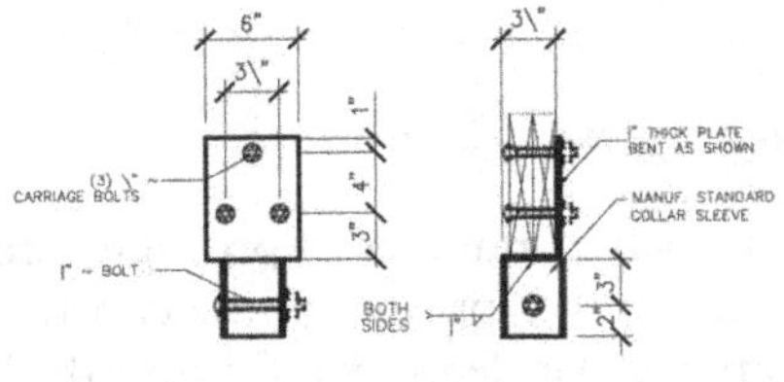

DETAIL B

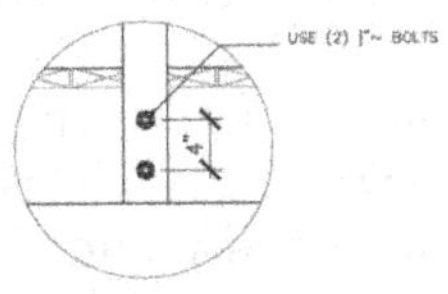

DETAIL A

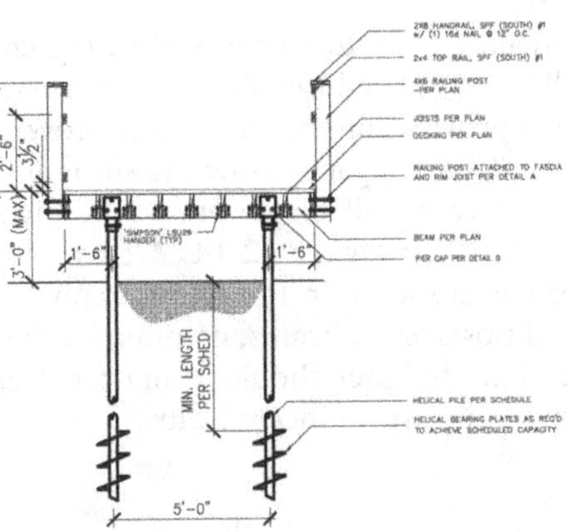

② BOARDWALK CROSS SECTION

General Notes:

1. Codes:

This plan was prepared based on 2003 IBC codes and portions of the most recent versions of AISC Steel Construction Manual 13th edition, and the NDS for wood construction.

2. Loads:

This plan is based upon the following load parameters (provided by Civil):
Boardwalk: Live Load = 60 psf
Handrail: Live Load = 50 plf, Live Point Load = 200 lb
Wind: Speed = 90 mph Exposure C
Seismic: Zone A

3. Materials:

This plan is based upon the following material properties:

Wood: All dimensional lumber shall be pressure treated Spruce-Pine-Fir (South) #2 or better unless noted on the plan.

Fasteners and connectors: All fasteners and connectors in contact with pressure treated lumber shall be G185 hot-dip galvanized, type 304 stainless steel or type 316 stainless steel or ACQ approved. All carriage bolts to be ASTM A307 or better.

4. Helical Piles:

All Helix foundation piers shall be 3"O.D. with ⅛" wall thickness. The number and size of blades shall be per the installation contractor so as to achieve appropriate installation torque and capacity. All pier components shall be hot dip zinc galvanized per ASTM 123 or 153. Helix pier installation should be observed by a geotechnical engineer to verify installation torque and minimum depths. Provide a minimum of 72 hours notice prior to installation work. The loads shown on the plan are minimum design loads. The manufacturer's recommendations should be followed regarding the torque and bearing capacity relationship for the particular helix pier selected.

5. Framing:

All framing shall be in accordance with the provisions of applicable building code. All connections or members not shown are per code or the general contractor/owner. Provide solid blocking to transmit loads to the foundation as necessary. Refer to the code for additional requirements.

6. Limitations:

This plan is only a foundation and framing design. It is the contractors/owners responsibility to verify and coordinate all dimensions prior to construction. This plan is based on the above referenced specifications. Any discrepancies or changes should be brought to the attention of the engineer.

Example Boardwalk Foundation

NO.	DATE	REVISION/ISSUE

HAP	S1
HAP	
BOOK	
12-18-08	S1
N.T.S.	

edges of piles. In seismically active areas, pile caps must be connected together by a foundation ties, as discussed in Chapter 10. Minimum pile spacing is discussed in Chapter 4.

12.4 MANUFACTURED PILE CAPS

Most helical pile manufacturers provide a number of preengineered helical pile caps. One of the more common manufactured caps used in the construction of new foundations is a simple plate attached to a collar sleeve that fits over the helical pile butt. An example is shown in Figure 12.12. This type of cap generates support by direct bearing against concrete or other material. Manufactured plate caps should be designed with sufficient area for concrete or other material bearing as well as sufficient thickness for punching and flexure of the steel plate itself. Pile caps such as the one in Figure 12.12 generally also work in tension. Pullout resistance is provided by concrete bearing against the bottom of the cap.

Another example of a common manufactured cap used in new foundation construction with helical piles is shown in Figure 12.13. This cap consists of a collar sleeve welded to one or more sections of deformed reinforcing steel bar. The cap resists compression and tension axial loads by direct bond to the concrete. Manufactured rebar caps should be designed following the procedures for reinforcing steel bar axial strength and required bond length in the ACI318 code. Standard hooks also could be used in tension applications.

Concrete shear and punching need to be checked by the engineer of record on a project-by-project basis following procedures similar to those given in the preceding section. Some manufacturers provide minimum concrete cover and edge distance to help ensure adequate factors of safety against concrete shear and punching.

Preengineered steel pile caps are available for practically any application. An example of a wood beam cap is shown in Figure 12.14. A smaller beam cap for support of steel and wood flitch beams is shown in the photograph in Figure 12.15. There are caps for support of wood posts, steel beams, communication towers, boardwalks, columns, and much more. The designer should contact helical pile manufacturing companies for catalogs of available preengineered caps.

12.5 BRIDGES AND BOARDWALKS

Helical piles have been used for new construction and repair of pedestrian bridges, boardwalks, county road bridges, and larger highway bridge piers. An example of a county road bridge abutment design is shown in Figure 12.16. A plan view of the abutment with helical pile layout is shown at the top of the figure. A section view through the abutment is shown in the middle of the figure, and a schedule of helical pile properties is contained at the bottom of the figure.

The abutment consists of a rectangular reinforced concrete grade beam supported on eight helical piles. The piles are spaced along two rows for lateral bracing. One row of helical piles is battered at a 12-degree angle to provide additional lateral resistance. A collapsible cardboard void form is used under the grade beam to resist frost heave. Rip-rap is placed along the slope adjacent the abutment beam for scour protection. A photograph of the completed abutments and bridge structure is shown in Figure 12.17. The photograph was taken prior to backfilling and approach ramp construction.

An example of a helical pile foundation for a boardwalk is shown in Figure 12.18. The plan view of the boardwalk shows flush beams spaced at regular intervals along the length of the structure. Each flush beam is supported on two helical piles. The boardwalk sections were manufactured in a factory setting and delivered to the site to reduce the amount of work in adverse weather conditions. The decking and railing were constructed on site after assembly of the boardwalk sections. The helical pile design professional developed the custom bracket shown in detail B for use with this project. General notes regarding applicable codes, load conditions, construction materials, and helical piles are provided on the right side of the figure. An L-Pile™ analysis was performed on the piles to verify lateral resistance. Boardwalks also can be supported by drop beams on helical piles, as shown in the photograph in Figure 12.19.

Figure 12.19 Boardwalk on helical piles

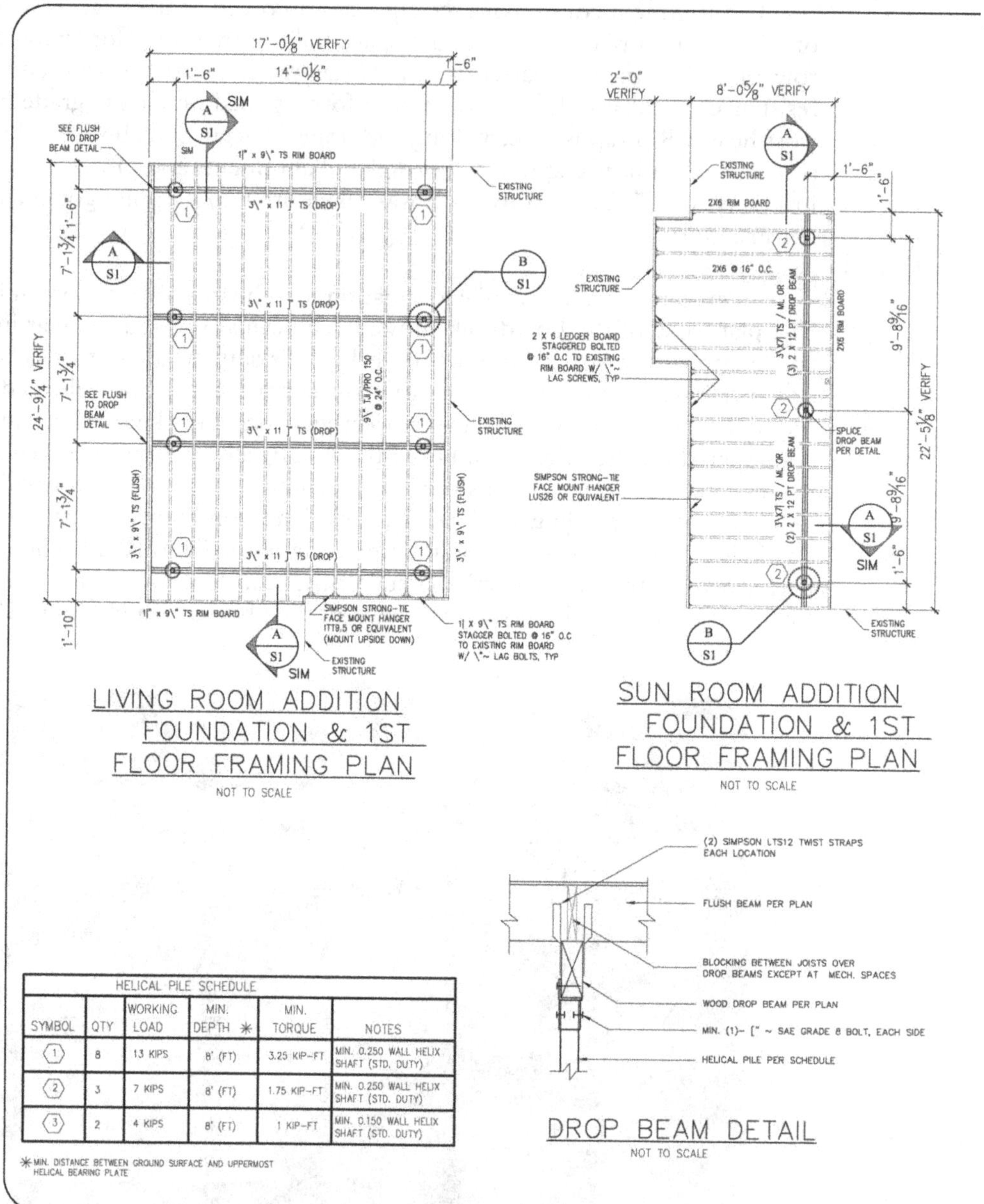

HELICAL PILE SCHEDULE

SYMBOL	QTY	WORKING LOAD	MIN. DEPTH ✱	MIN. TORQUE	NOTES
1	8	13 KIPS	8' (FT)	3.25 KIP-FT	MIN. 0.250 WALL HELIX SHAFT (STD. DUTY)
2	3	7 KIPS	8' (FT)	1.75 KIP-FT	MIN. 0.250 WALL HELIX SHAFT (STD. DUTY)
3	2	4 KIPS	8' (FT)	1 KIP-FT	MIN. 0.150 WALL HELIX SHAFT (STD. DUTY)

✱ MIN. DISTANCE BETWEEN GROUND SURFACE AND UPPERMOST HELICAL BEARING PLATE

Figure 12.20 Foundation plan for concreteless residential additions

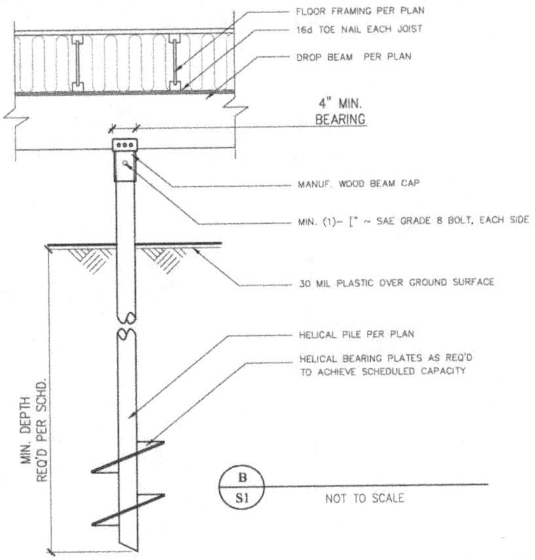

General Notes:

1. All helix foundations and pier caps shall have ICC-ES evaluation report documenting product meets required capacity.
2. Slope ground surface to drain away on all sides. In areas where sidewalks or paving do not immediately adjoin the structure, this slope should have a minimum grade of 10% for at least 10 feet.
3. Exterior slabs shall be isolated from the structure.
4. Helical pile foundation installation should be observed by Engineer to verify installation torque and minimum depth.
5. All framing lumber shall be Hem-Fir No. 2 or better, Fb=850 psi, Fv=75 psi, Fc=405 psi, E=1,300,000 psi.
6. TJI Joists shall be by Trus Joist MacMillion or equivalent.
7. TimberSrand and Microllam beams and headers shall be by Trus Joist MacMillion or equivalent.
8. Sheath all exterior walls with minimum 7/16" thick shear panels conforming to UBC grade structural 1. Nail panels with 8d nails at 4" o.c. at all edges and 12" o.c. at intermediate members.
9. All connections shall be Simpson Strong-Tie or equivalent.
10. All construction shall be in conformance with the latest edition of the Uniform Building Code.
11. Structural design is based on the following loads and general design criteria.
    Live Loads: Roof=30psf Floor=40psf
    Wind Shear: 90 mph wind, Exposure B
    Seismic Zone=1
12. Provide cross-ventilation under floor system. Total of all vent openings shall exceed 0.3 sf open area.
13. Provide minimum 3/4" thick plywood tongue and groove floor sheathing attached with 8d nails spaced at 6" o.c. along beams, joists, and intermediate framing members.
14. All interior and exterior walls to be framed with 2x4 studs at 16" o.c.. Exterior walls shall have double top plates. Use minimum (2) studs built-up and nailed with 16d nails at 24" o.c. at each exterior wall corner. Studs are to be Hem-Fir Larch Stud Grade and better,
    Fb=650psi, Fv=75psi, Fc=500psi, E=1,300,000 psi.
15. All TimberStrand beams shown can be replaced with equivalent size Microllams, 1.9E.
16. Use Wolmanized TimberStrand for all drop beams with in 12" from ground surface.
17. Contractor to verify all measurements prior to construction.
18. Obtain utility locations prior to installing piles.

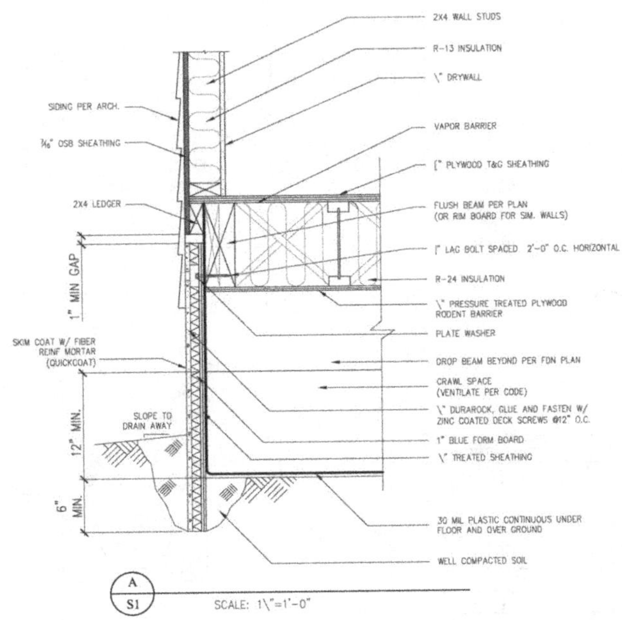

Concrete-less Foundation for Residential Additions

NO.	DATE	REVISION/ISSUE
1		
2		
3		

HAP	SHEET S1
HAP	
BOOK	OF S1
12-18-08	
N.T.S.	

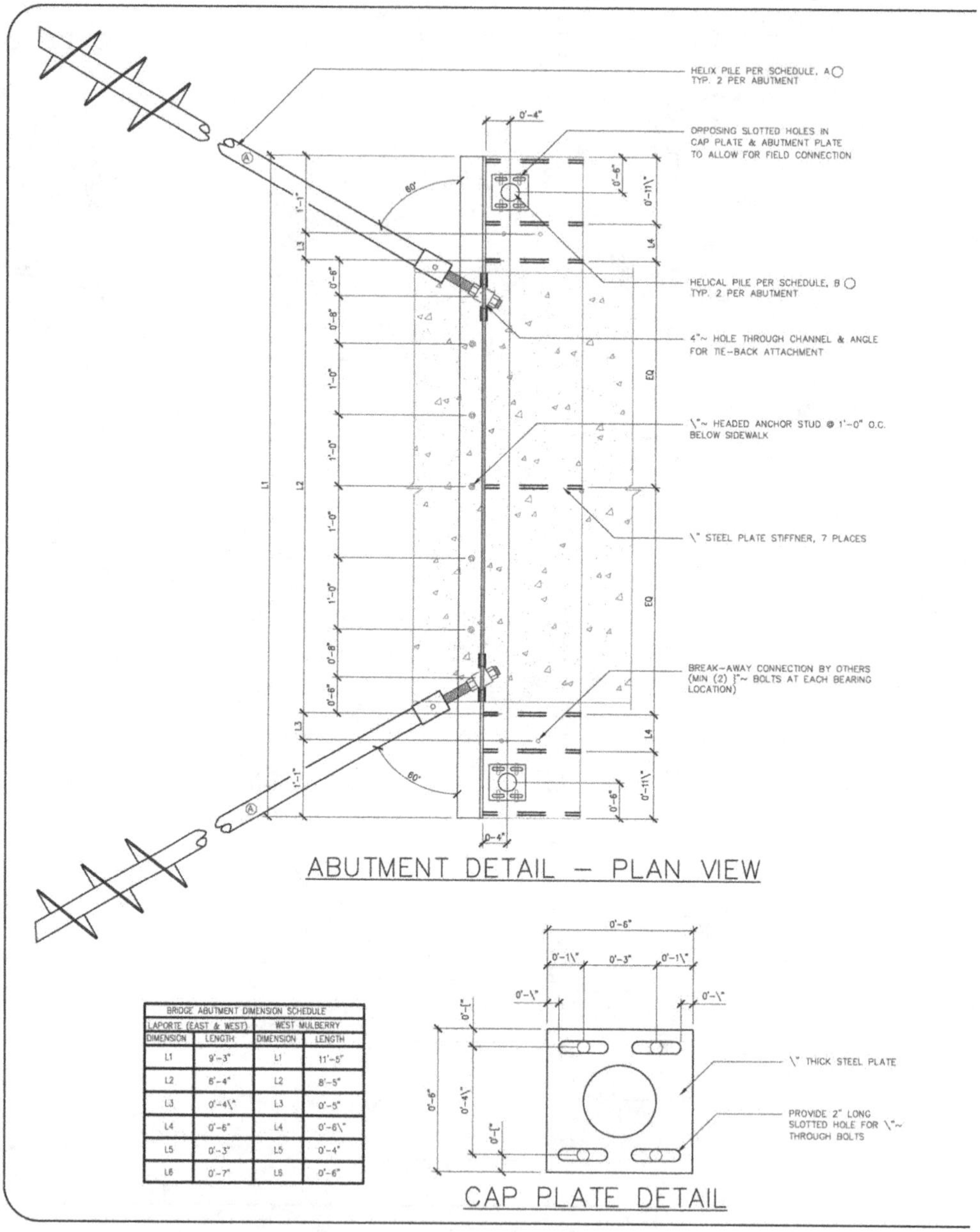

Figure 12.21 Example concreteless pedestrian bridge abutment

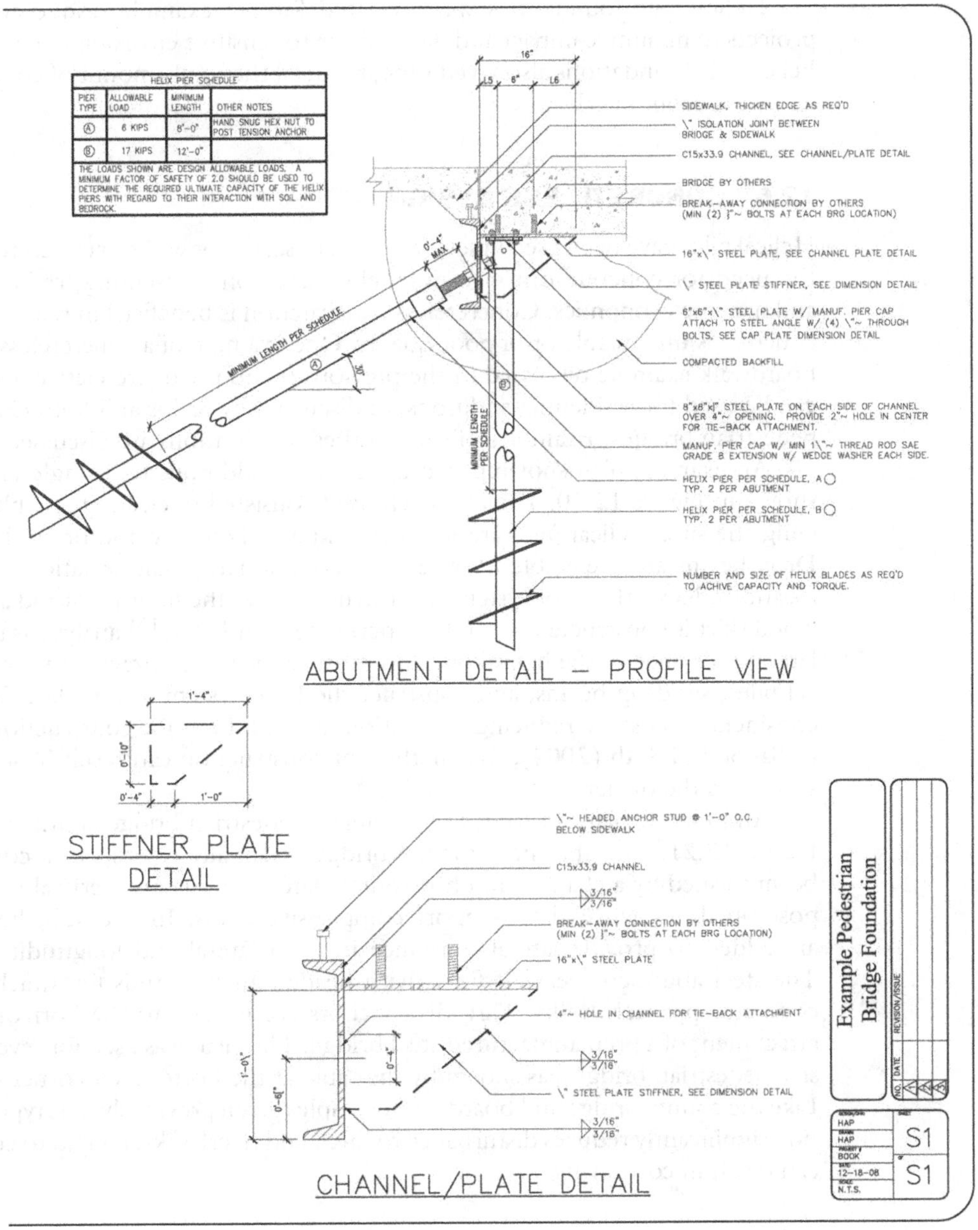
HELIX PIER SCHEDULE
PIER TYPE | ALLOWABLE LOAD | MINIMUM LENGTH | OTHER NOTES
A | 6 KIPS | 8'-0" | HAND SNUG HEX NUT TO POST TENSION ANCHOR
B | 17 KIPS | 12'-0" |
THE LOADS SHOWN ARE DESIGN ALLOWABLE LOADS. A MINIMUM FACTOR OF SAFETY OF 2.0 SHOULD BE USED TO DETERMINE THE REQUIRED ULTIMATE CAPACITY OF THE HELIX PIERS WITH REGARD TO THEIR INTERACTION WITH SOIL AND BEDROCK.
16"
6"
0'-4" MAX
MINIMUM LENGTH PER SCHEDULE
30°
MINIMUM LENGTH PER SCHEDULE
SIDEWALK, THICKEN EDGE AS REQ'D
\" ISOLATION JOINT BETWEEN BRIDGE & SIDEWALK
C15x33.9 CHANNEL, SEE CHANNEL/PLATE DETAIL
BRIDGE BY OTHERS
BREAK-AWAY CONNECTION BY OTHERS (MIN (2) }"~ BOLTS AT EACH BRG LOCATION)
16"x\" STEEL PLATE, SEE CHANNEL PLATE DETAIL
\" STEEL PLATE STIFFNER, SEE DIMENSION DETAIL
6"x6"x\" STEEL PLATE W/ MANUF. PIER CAP ATTACH TO STEEL ANGLE W/ (4) \"~ THROUGH BOLTS, SEE CAP PLATE DIMENSION DETAIL
COMPACTED BACKFILL
8"x8"x|" STEEL PLATE ON EACH SIDE OF CHANNEL OVER 4"~ OPENING. PROVIDE 2"~ HOLE IN CENTER FOR TIE-BACK ATTACHMENT.
MANUF. PIER CAP W/ MIN 1\"~ THREAD ROD SAE GRADE 8 EXTENSION W/ WEDGE WASHER EACH SIDE.
HELIX PIER PER SCHEDULE, A TYP. 2 PER ABUTMENT
HELIX PIER PER SCHEDULE, B TYP. 2 PER ABUTMENT
NUMBER AND SIZE OF HELIX BLADES AS REQ'D TO ACHIVE CAPACITY AND TORQUE.
ABUTMENT DETAIL – PROFILE VIEW
1'-4"
0'-10"
0'-4"
1'-0"
STIFFNER PLATE DETAIL
\"~ HEADED ANCHOR STUD @ 1'-0" O.C. BELOW SIDEWALK
C15x33.9 CHANNEL
3/16"
3/16"
BREAK-AWAY CONNECTION BY OTHERS (MIN (2) }"~ BOLTS AT EACH BRG LOCATION)
16"x\" STEEL PLATE
4"~ HOLE IN CHANNEL FOR TIE-BACK ATTACHMENT
3/16"
3/16"
\" STEEL PLATE STIFFNER, SEE DIMENSION DETAIL
3/16"
3/16"
1'-0½"
0'-4"
0'-6"
CHANNEL/PLATE DETAIL
Example Pedestrian Bridge Foundation
NO. DATE REVISION/ISSUE
HAP
HAP
BOOK
12-18-08
N.T.S.
S1
S1

Helical pile foundations were selected for the example bridge and boardwalk projects to minimize impact and disturbance to sensitive environmental wetlands. The helical pile foundations also saved time, reduced the total amount of concrete needed, and were economical.

12.6 CONCRETELESS DESIGN

Helical piles have been used in several different scenarios without concrete. Eliminating the need for concrete can save time, eliminate concrete curing, reduce excavation, and manage economics. Concreteless construction is beneficial in remote areas where concrete is unavailable or of poor quality. One example of a concreteless design is the boardwalk example discussed in the previous section. Concreteless construction also can be used for residential additions, as discussed by Sailor and Soth (2004), and for pedestrian bridges. Examples of each of these applications are discussed next.

An example plan showing two concreteless additions to a single-family home is shown in Figure 12.20. The approach used consisted of clearing, grubbing, and leveling the site. Helical piles are installed and attached to wood or steel drop beams. Drop beams are preferable in order to accommodate small deviations in helical pile locations. Next, the floor system is constructed over the floor joists and an all-weather wood skirt is constructed around the perimeter. An L-Pile™ analysis is used to check lateral resistance of the helical piles. In most cases, the contractor can install the helical piles, set drop beams, and construct the floor system in one day, thereby saving considerable cost by reducing excavation and the time for construction. According to Sailor and Soth (2004), this method of construction can result in significant cost savings to the owner.

An example plan showing a concreteless pedestrian bridge abutment is shown in Figure 12.21. The abutment for the bridge essentially consists of a composite steel beam formed by a channel and horizontal plate with gussets. Vertical helical piles are positioned at each end to support compressive loads. In addition, helical anchors are added to provide lateral resistance in both lateral and longitudinal directions. The steel abutment beam is fitted with headed anchor studs for attachment of the concrete approach walks. Threaded anchors are welded to the horizontal plate for attachment of a premanufactured steel bridge. This plan was used for several different-size pedestrian bridges, as shown by the table at the bottom left corner of the figure. Like the county bridge and boardwalk examples given previously, this type of construction significantly reduces disturbance to ditch and riverbanks, is rapid to construct, and can result in cost savings.

12.7 LATERAL BRACING

The IBC (2006) requires that helical piles be laterally braced in all directions. Three or more piles are required for bracing in an isolated pile cap. A minimum of two piles are considered braced in a pile cap, provided the cap is supported by a wall or grade beam in a direction perpendicular to the axis connecting the two piles. Helical piles should be installed in lines spaced at least 1 foot apart and located symmetrically under the center of continuous walls unless other measures and types of bracing can be used to provide for eccentricity and lateral forces. Single rows of piles can be used in one- and two-family structures and in lightweight construction less than 35 feet (10.7 m) in height as long as the piles are within the width of the wall.

Some practitioners rely on passive soil resistance against pile caps to brace the tops of piles for lightly loaded structures in areas with low seismic hazard. If this approach is taken, the pile cap needs to be designed for overturning due to load eccentricity. Procedures for rigid or flexible pile analysis per Chapter 10 can be used to check stability. Other methods of lateral bracing include using floor slabs or other diaphragm systems, tie beams, and tie back anchors.

Chapter 13

Earth Retention Systems

Helical piles and helical anchors have numerous applications in the design and installation of earth retention systems from retaining walls to excavation and slope stabilization. This chapter begins with a review of lateral earth pressures used in earth retention design. Eight different configurations of retaining walls and excavation shoring utilizing helical anchor technology are discussed, including tie-back walls, reticulated helical pile walls, counterfort walls, buttress walls, soldier pile and lagging, sheet pile walls, shotcrete walls, and braced excavations. Several actual case studies are provided along with general procedures for designing with helical piles. A new theory on the pressure behind timber lagging in soldier beam and lagging systems is presented. The importance of proper site grading and drainage in earth retention system design is summarized. Procedures for post-tensioning are presented including conventional use of a hollow-core ram as well as a method for predicting the clamping force from simply tightening the lock nut on helical anchors. The chapter concludes with a brief section on repair of existing retaining walls using helical tie-backs.

13.1 LATERAL EARTH PRESSURE

This section contains a brief overview of lateral earth pressure. Practically all foundation and soil mechanics texts contain a chapter dedicated to this subject (i.e., Bowles, 1988; Das, 1990; Peck, Hansen, and Thornburn, 1965). For the practicing geotechnical engineer, this section will be a basic review. The goal of this section is simply to provide a baseline for later discussion regarding types of earth retention systems used with helical piles.

Lateral earth pressure can be described by three general cases: at rest, active, and passive. At-rest earth pressure describes the state of horizontal stress in undisturbed ground. The active case arises when a retaining wall is free to rotate such that the internal strength of the backfill can be mobilized. Passive earth pressure is the horizontal stress that must be overcome in order to force a retaining against the earth. According to Peck, Hansen, and Thornburn (1965), laboratory tests and experience have demonstrated that the amount of movement necessary to reduce earth pressure from at rest to active is on the order of 0.1 percent of the wall height. Most retaining walls can and do deflect this distance without undesirable consequences so it is reasonable to base the design of ordinary retaining walls on the active case. If movement cannot be tolerated, the higher at-rest case should be assumed. Passive earth pressure can be used in earth retention design to account for the resistance provided by backfill along the front of a retaining wall, provided the backfill cannot be removed.

Basic lateral earth pressure diagrams and equations governing each case are shown in Figure 13.1. These diagrams were adapted from the references listed at the

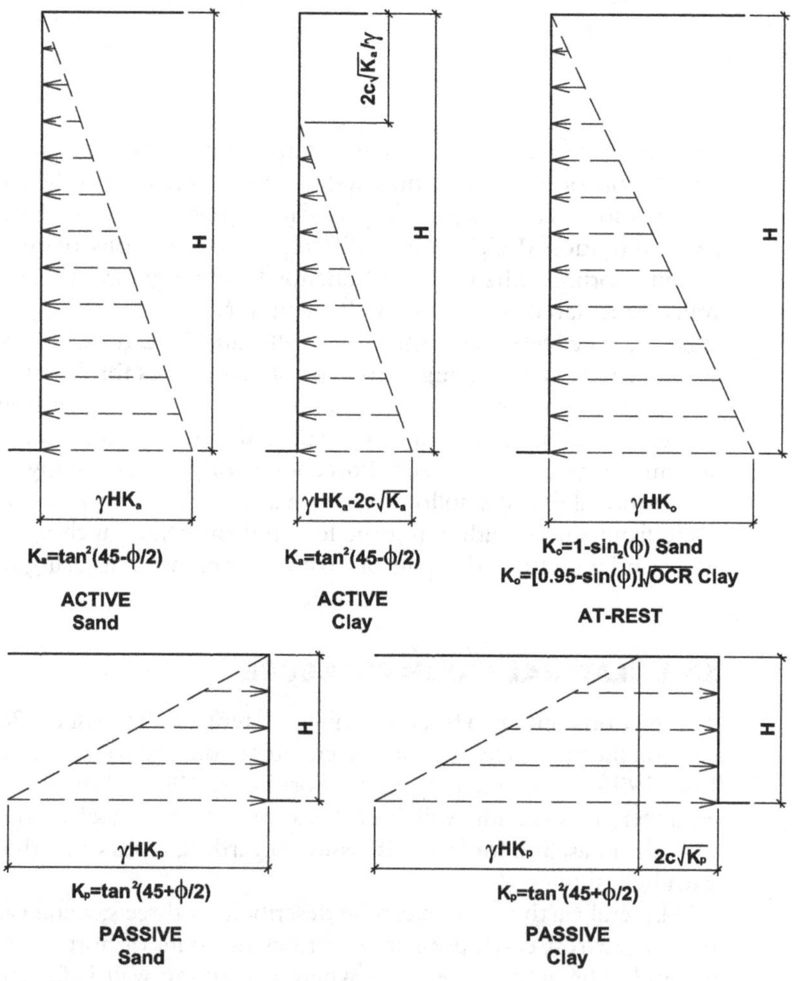

Figure 13.1 Lateral earth pressure diagrams

beginning of this section. As can be seen, the at-rest earth pressure case has significantly higher pressures than the active earth pressure case. The first active earth pressure diagram labeled sand in the figure is used most frequently in retaining wall design. For clayey soils, cohesion can be taken into account using the second diagram shown in the figure. A small amount of cohesion assumed in design can greatly reduce the total earth pressure on a retaining wall. Caution should be used when applying cohesion, as changes in moisture content and confining stress can negate apparent cohesion, as discussed in Chapter 3. It is conservative when dealing with clayey soils to assume the higher active earth pressure given in the first diagram for sandy soils.

Field measurements and experience have shown that the lateral earth pressure for braced or tie-back walls for support of excavations differ from the at-rest and active cases. Peck (1969) presented the three modified earth pressure diagrams shown in Figure 13.2 for excavation support. These pressure distributions often are referred to as Peck's apparent earth pressure diagrams and are used frequently in the design

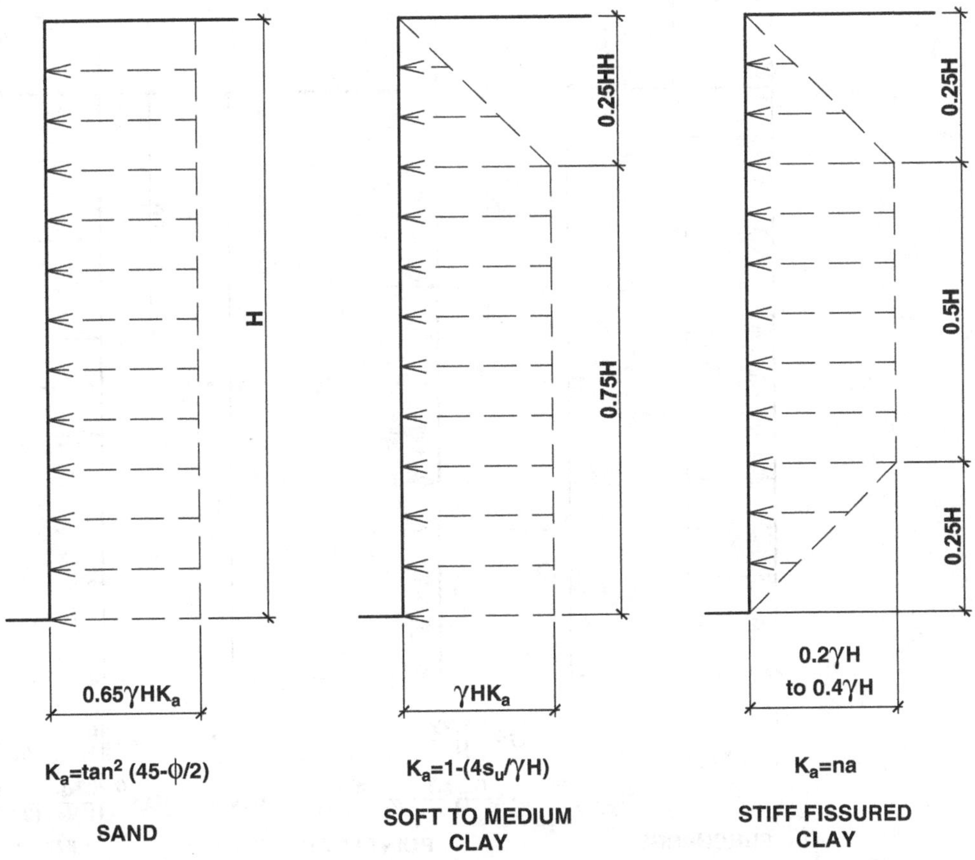

Figure 13.2 Lateral earth pressure diagrams for braced or tie-back excavations

of excavation support systems. The first diagram shown in the figure is for coarse grain soils. The second and third diagrams are for fine-grain materials of different consistency.

In addition to the lateral earth pressures just given, it is sometimes necessary to account for the effect of uniform surcharge pressures, point loads, and line loads at the ground surface behind the earth retention system. The earth pressure distributions from these additional loads are shown in Figure 13.3. Earth retention systems located along a roadway typically are designed for a uniform surcharge of 100 to 200 psf [4.8 to 9.6 kPa] depending on local codes. The effect of uniform surcharge is a rectangular lateral stress distribution with magnitude equal to the active earth pressure coefficient, K_a, times the surcharge pressure as shown in the first diagram in Figure 13.3. The point load stress distribution shown in the second diagram is useful for modeling the effect of point loads due to construction equipment operating in the vicinity of the wall. A continuous footing from an adjacent building can be simulated using the formulae

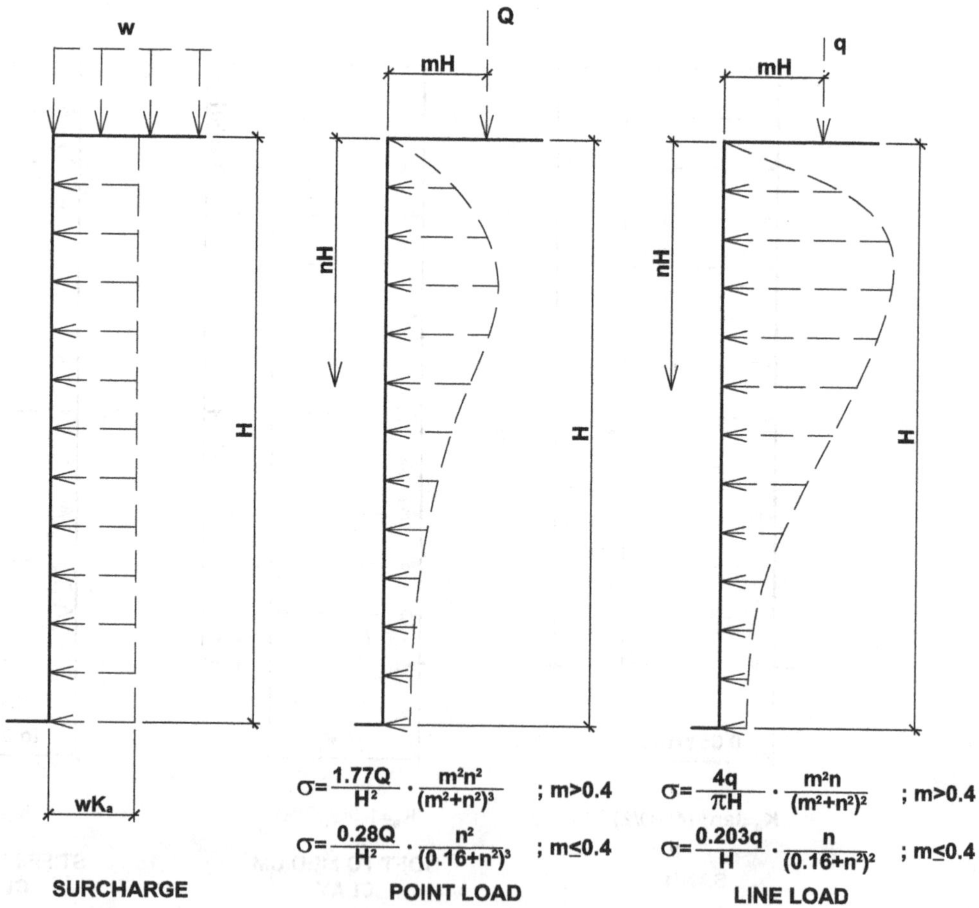

Figure 13.3 Lateral earth pressures from uniform, point, and line loads

for line loads shown by the third diagram in the figure. These formulae were adapted from Das (1990).

Another important factor to consider in determining lateral earth pressure is ground water. In many cases, ground water can be controlled by drainage systems installed in conjunction with the retaining wall. In systems without ground water control, hydrostatic pressure is a vital consideration. Examples include sheet pile and slurry wall systems that are used as cofferdams and buildings that are designed below ground water without permanent dewatering. Ground water is taken into account by adding a hydrostatic pressure diagram to the lateral earth pressure diagram. The unit weight of soil for that portion of the lateral earth pressure diagram located below water should be calculated using the buoyant unit weight of soil. Buoyant unit weight, $\gamma\prime$, can be estimated for most soils from

$$\gamma' = 0.623\gamma_d \tag{13.1}$$

Where

γ_d is dry unit weight.

Equation 13.1 assumes the soil has a specific gravity of 2.65, which is very common.

Other important factors that can affect the lateral earth pressures on a retaining wall include sloping backfill and friction along the back of the wall. These factors are discussed at length in other geotechnical texts. There are no special considerations for helical piles. Designers should follow the same procedures when accounting for these factors in earth retention systems utilizing helical piles as they would for other types of systems. Sloping backfill can result in significantly higher lateral earth pressures and always should be taken into account. Consideration of friction between retaining walls and backfill generally results in lower lateral earth pressures. It is conservative to ignore friction on the back of the wall.

13.2 RETAINING WALLS

Retaining walls commonly constructed using helical piles generally can be categorized as one of four types: tie-back, reticulated, counterfort, and buttress. Each of these wall types is shown in Figure 13.4. Helical tie-back walls comprise the most predominant use of helical anchors for construction of retaining walls. They deserve the most emphasis and are reserved for discussion until the end of this section. Reticulated helical pile walls are commonly used for bridge abutments. They also may be used on hillsides where access to the front of the wall with heavy equipment is problematic. The nature of reticulated helical pile walls is such that the maximum height of earth retained is on the order of 6 to 8 feet [1.8 to 2.4 m]. Reticulated helical pile walls generally consist of two or more rows of helical piles supporting a cast-in-place, reinforced, concrete mass. One or more rows of helical piles are installed at a batter angle to provide lateral resistance. The rows of helical piles are spaced apart to provide lateral and rotational bracing. There are several methods for design of reticulated helical pile walls. A finite element software program can be used to model the soil-concrete-pile

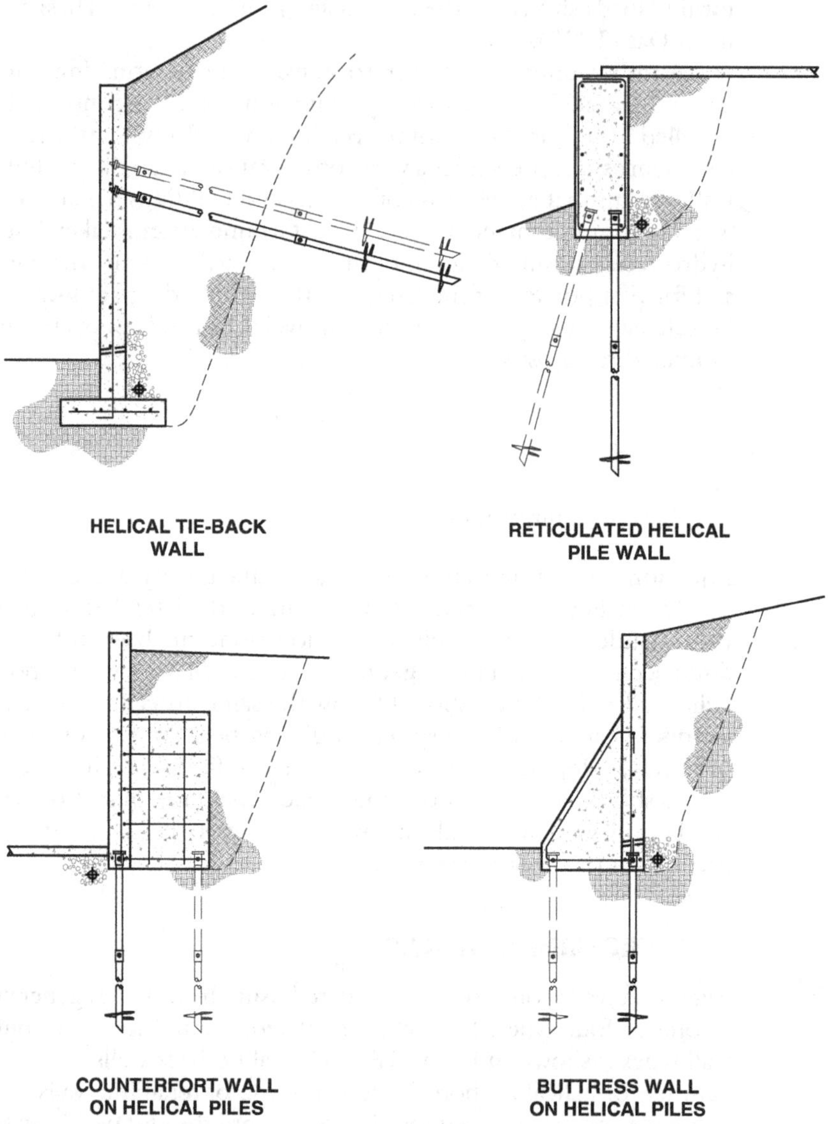

Figure 13.4 Retaining walls

interactions. After shear and bending stresses are obtained from the finite element analysis, conventional steel and concrete design procedures are used to size the pile shafts and reinforcement for the concrete mass.

Counterfort and buttress retaining walls are similar in that they generally consist of a long primary wall with short secondary wall sections oriented perpendicular to the primary wall and spaced at regular intervals along its length. The secondary wall sections are buried behind the wall in the counterfort system. In the buttress system, the secondary wall sections are exposed at the front of the wall. On good ground, these wall systems are commonly supported on a large-spread footing foundation. On poor ground or when excavation for an appropriate footing is limited by groundwater or other problematic soil conditions, counterfort and buttress walls can be founded very effectively on helical piles. A row of helical piles is placed under the primary wall system. Helical piles are added at the end of each counterfort or buttress.

Helical tie-back walls generally consists of cast-in-place, reinforced concrete walls with spread footings and one or more rows of helical tie-backs. When a single row of tie-backs is used, it is good practice to alternate the position of the tie-backs slightly for additional stability as shown in the top-left detail in Figure 13.4. Tie-backs differ from soil nailsin that they generate pullout resistance from well past the active zone of soil behind the wall, whereas the design intent of helical soil nails is to reinforce the entire active zone so that it becomes a stable soil mass. Tie-backs are typically longer and higher capacity than soil nails. Fewer tie-backs are generally required for stability than soil nails. Tie-backs also are often post-tensioned.

The active zone of soil behind a retaining wall is that portion of the soil that is mobilized by wall movement. It can be visualized as that portion of soil that would slide down the slope in the event that the retaining wall tipped over. For the active earth pressure case, the active zone is generally assumed to be a triangular wedge extending at an angle of $45+\Phi/2$ from horizontal, where Φ is the angle of internal friction of the soil. If the wall is to be designed for the at-rest earth pressure case, the active zone should be taken as a triangular wedge of soil at a 45-degree angle from horizontal.

Helical tie-backs are typically installed at a 10- to 15-degree angle from horizontal for two primary reasons. First, the distance required to reach past the active zone of soil is shorter when the tie-backs are installed at a downward angle. The second reason for installing helical tie-backs at an angle is that the resistance to pullout provided by many soils increases with depth; installation at a downward angle increases the depth of the helical bearing plates below the ground surface and therefore the pullout capacity. If job site conditions merit, there is no reason that helical tie-backs cannot be installed horizontally. The tie-backs would have to be slightly longer and more helical bearing plates may have to be used on the top row of tie-backs to generate sufficient pullout. The preferred use of inclined tie-backs stems from the grouted anchor industry. Installation at a 10- to 15-degree angle helps to ensure that the entire anchor hole is filled with grout.

Helical tie-backs can be installed at an angle much steeper than 15 degrees if site conditions merit. However, one of the important considerations in the design of tie-back walls is the downward component of force caused by the helical tie-backs. A force diagram for a typical helical tie-back wall is shown in Figure 13.5. Each tie-back results in a horizontal and a vertical component of force in the wall facing. The magnitude of each component with respect to the other depends on the angle of installation. The foundation of a tie-back wall has to be designed not only for the weight of the wall facing but also for the downward component of force induced by the tie-backs. Otherwise, a bearing capacity failure may occur. Bearing capacity failure may be one of the most predominant forms of failure of tie-back walls. It is often confused with a rotational failure. As the foundation settles, the wall will rotate outward because slack will be created in the inclined anchors.

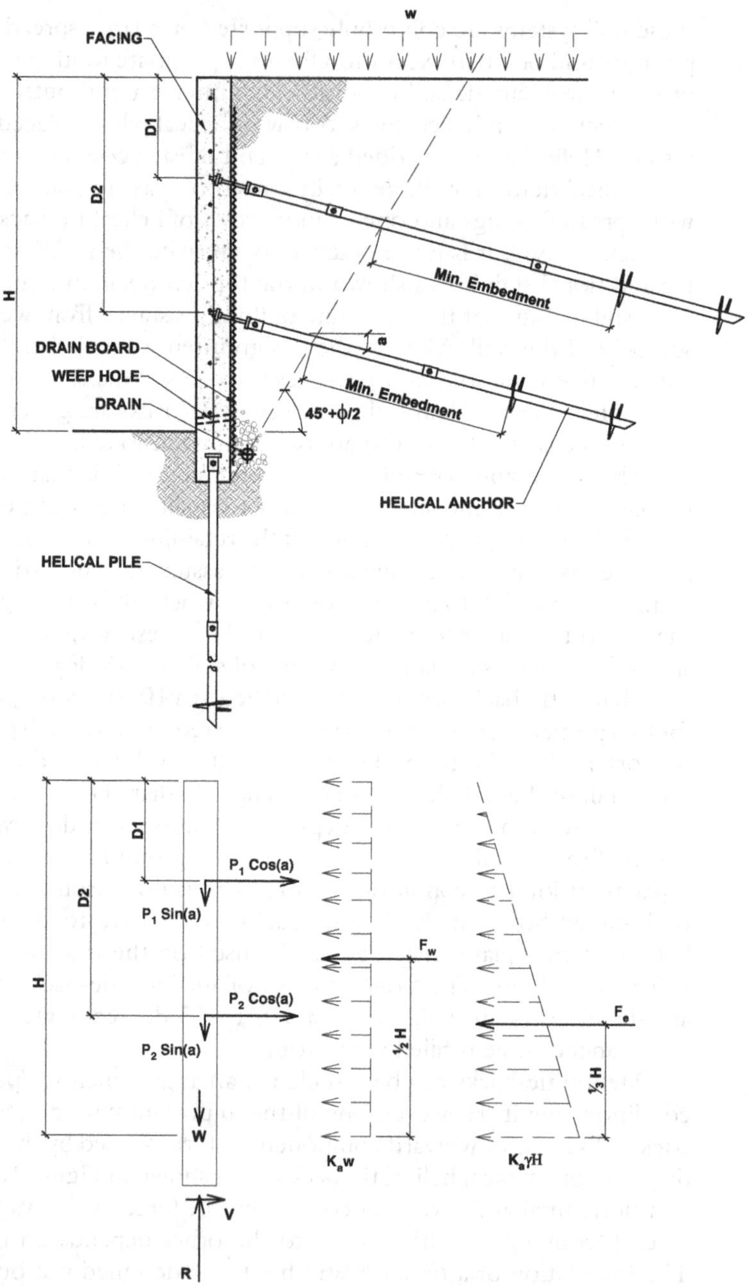

Figure 13.5 Force diagram for tie-back retaining wall with helical anchors

Minimum embedment of helical anchors in tension is discussed extensively in Chapter 5. As discussed in that chapter, it was shown in previous research that the minimum embedment of inclined anchors can be estimated using the same procedures as those for vertical anchors. However, in the case of earth retention, the minimum embedment is measured as the distance from the active soil plane to the shallowest helical bearing plate. If the minimum embedment were taken from the wall facing, there is a possibility that the tie-backs will not penetrate sufficiently past the active zone and a wall failure could occur as a combination of sliding of the active wedge and pullout of the anchors.

By their nature, helical tie-backs require a rigid wall facing to resist the full lateral earth pressure. Wall facing can be designed using a simple finite element plate analysis similar to the Analysis Group™ model shown in Figure 13.6. A drawing of the finite element mesh and support conditions for the example wall is contained in the box at the upper left corner of the figure. The wall in this example had two rows of helical tie-backs. Each helical tie-back was modeled as a spring support as shown by the numbered square symbols on the finite element mesh. The wall is supported on both ends by a 90-degree corner. Pinned conditions were used to model the wall corners, which assumes the concrete will crack. Shear is resisted at the corners by the reinforcing steel alone. Tie-back wall facing corners could be modeled as fixed end supports, but this usually results in an unreasonable amount of reinforcing steel. The spring constants used to model the helical anchors were based on experience and verified through pullout testing. If desired, each tie-back can be post-tensioned to a certain percentage of its capacity in order to reduce wall deflections. This can be modeled in a finite element program by placing a post-tensioning force at each spring support.

Results of the finite element analysis of the wall facing are shown at the bottom of Figure 13.6. Shear and moment in the wall facing are represented by the four contour plots. The speed of finite element modeling allows the designer to perform a number of iterations in order to optimize the spacing and layout of helical anchors so as to balance shear and bending stresses across the facing. As can be seen in the results, both shear and moment exhibit stress concentrations at the tie-back locations. Extra reinforcement in the form of a small matt of reinforcing steel can be added at the tie-back locations to improve performance and reduce the amount of steel in free field areas. In addition to the finite element analysis, a conventional punching shear analysis, as described in Chapter 12, is typically performed at tie-back locations to size the bearing plate and minimum concrete cover.

The connection between a helical tie-back and the wall facing can be accomplished in several different ways. One way is to place a bearing plate between two hex nuts at the end of a threaded anchor cap and embed the entire assembly within the reinforced concrete. In some cases, it is desired to use a thinner wall section such that punching resistance can be achieved only if the bearing plate is placed on the outside of the wall. For this circumstance, the bearing plate typically is oriented flush with the wall facing and at an angle off perpendicular from the inclined anchor. It is imperative that wedge washers be used to account for the angle between the lock nut and the bearing plate. If only one edge of the lock nut bears on the plate, it will induce combined tension

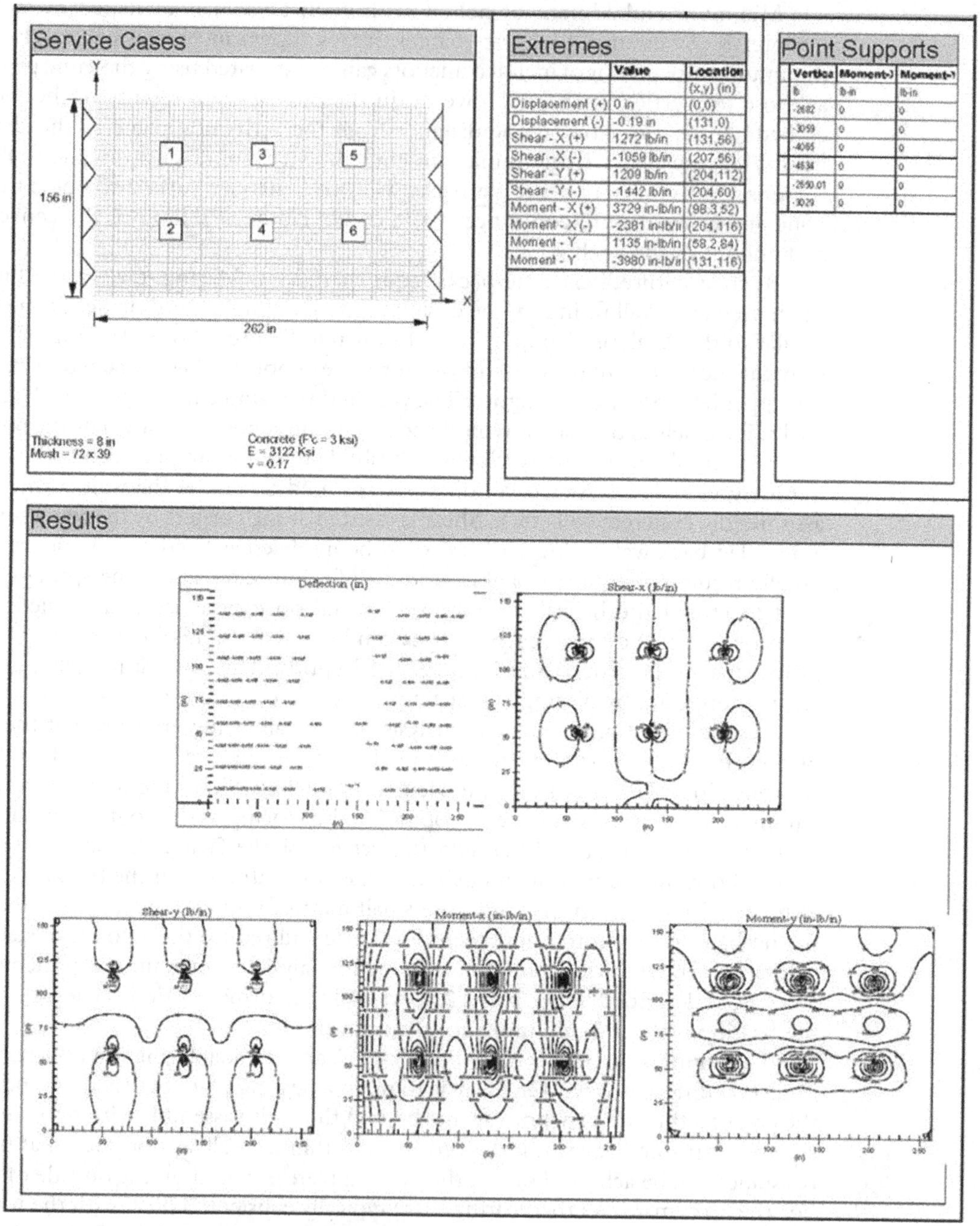

Figure 13.6 Structural analysis of wall facing

and bending in the threaded bar attached to the helical anchor. There have been cases where tie-back walls of all types have failed due to rupture of the anchor rod because of combined tension and flexure.

The design of retaining walls requires an analysis of anchor pullout, lateral earth pressures, overturning stability, sliding, and bearing capacity. Typically, the active earth pressure case is used for design of the basic wall types shown in Figure 13.4. Pullout, bearing, and lateral capacity of helical piles have been discussed extensively in preceding chapters of this book. The general procedures for designing retaining walls with helical piles are the same as other types of retaining wall systems. Numerous texts have been written on the subject of retaining wall design, so these procedures are not repeated here. On sloping sites or when multiple retaining walls are stacked above one another, the retaining wall designer's focus needs to include global stability. Global stability can be checked using one of the commercially available slope stability software programs. An example of a Slope/W® analysis of a series of helical tie-back walls and concrete gravity walls is shown in Figure 13.7. An analysis of anchor pullout, lateral earth pressures, overturning, sliding, and bearing capacity resolves the local and internal stability of the retaining wall only.

Besides the four main categories of retaining walls, countless variations and hybrid retaining wall configurations can be created utilizing helical pile technology. Helical tie-backs can be added to exceptionally tall reticulated helical pile walls, counterfort walls, and buttress walls for additional stability. A helical pile can be used to replace the footing in a conventional tie-back wall when bearing capacity is a concern. The other variations that can be accomplished using helical piles are left to the imagination of the designer.

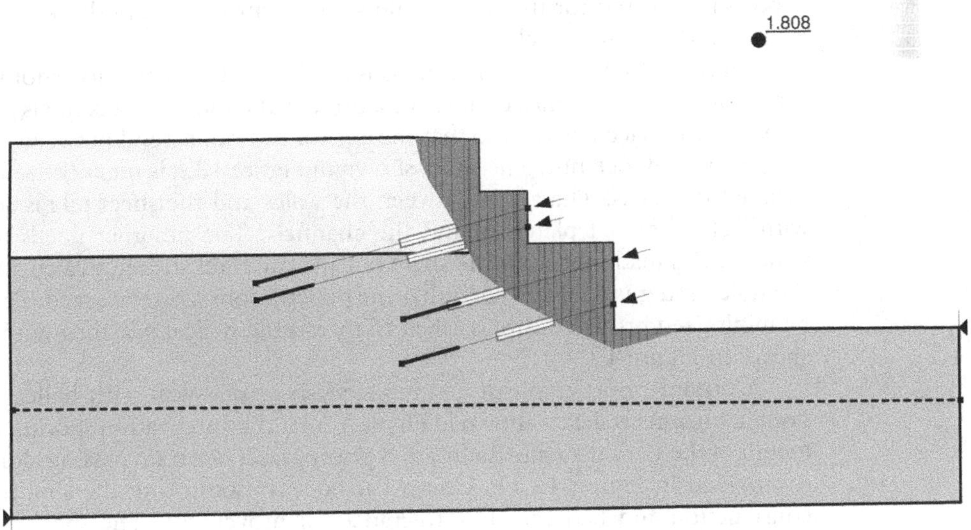

Figure 13.7 Global stability analysis

13.3 EXCAVATION SHORING

Helical anchors have been used extensively in excavation shoring. Four of the more common categories of excavation shoring are shown in Figure 13.8. These include shotcrete walls, sheet pile walls, soldier pile and lagging walls, and braced walls. Shotcrete walls are created by excavation, installation of helical tie-backs, and construction of shotcrete facing in several stages. This type of shoring system is best suited for cohesive soils that have sufficient stand-up time to allow for the work that needs to be accomplished at each stage without excessive raveling. The soil also needs to have sufficient strength to support the shotcrete without caving during the curing period. Closely spaced, augered, concrete mini-piles have been used to support coarse-grain soils so that a shotcrete wall can be constructed.

The example shotcrete wall shown in the figure has three rows of helical tie-backs. An optional cast-in-place concrete waler is shown at each anchor row location. The concrete walers are cast against the earth after installation of the helical anchors and prior to excavation for the next lift. Concrete walers can reduce the required thickness of shotcrete for the remaining facing. The walers also improve punching resistance at the helical tie-back locations. A case study involving a shotcrete excavation shoring system with helical tie-backs and helical pile underpinning is presented in Chapter 14.

Sheet pile shoring systems are constructed by driving or vibrating continuous corrugated sheet piling from the ground surface along the perimeter of the excavation. After installation of sheet piling, the excavation is made to the depth of the top row of helical anchors. Penetrations are made through the sheet piling using a torch in order to facilitate helical anchor installation. After the helical tie-backs have been installed, a steel waler is placed over the anchors and welded to the sheet pile. Following waler installation, the anchors are proof load tested and, if necessary, post-tensioned. The process is repeated for the second and subsequent rows of tie-backs. Post-tensioning is discussed later in this chapter.

Many different waler configurations can be used in excavation shoring. One of the most common assemblies consists of a double channel. The channels are placed back to back with space between so that the anchor rod can extend between the channels to a cap plate and lock nut. The waler shown in Figure 13.8 is set at the same angle as the inclined tie-backs. The angle between the waler and the sheet pile is accommodated with welded gusset plates behind the channels. The designer needs to account for penetrations made through the sheet pile in structural strength calculations or design the waler and gusset plates such that the penetrations are reinforced. Penetrations can be minimized provided care is taken to thread the helical pile through the opening as shown in Figure 13.9.

A distant photograph of a completed sheet pile wall with helical tie-backs and double-channel walers is shown in Figure 13.10. The excavation shoring runs along the length of the Union Pacific Railroad. A photograph of a train passing along the shoring is provided in Figure 13.11. Union Pacific was monitoring the shoring throughout construction and had very low tolerance for movement. The sheet pile and helical tie-back system performed very well. Sheet pile and helical tie-back systems can be

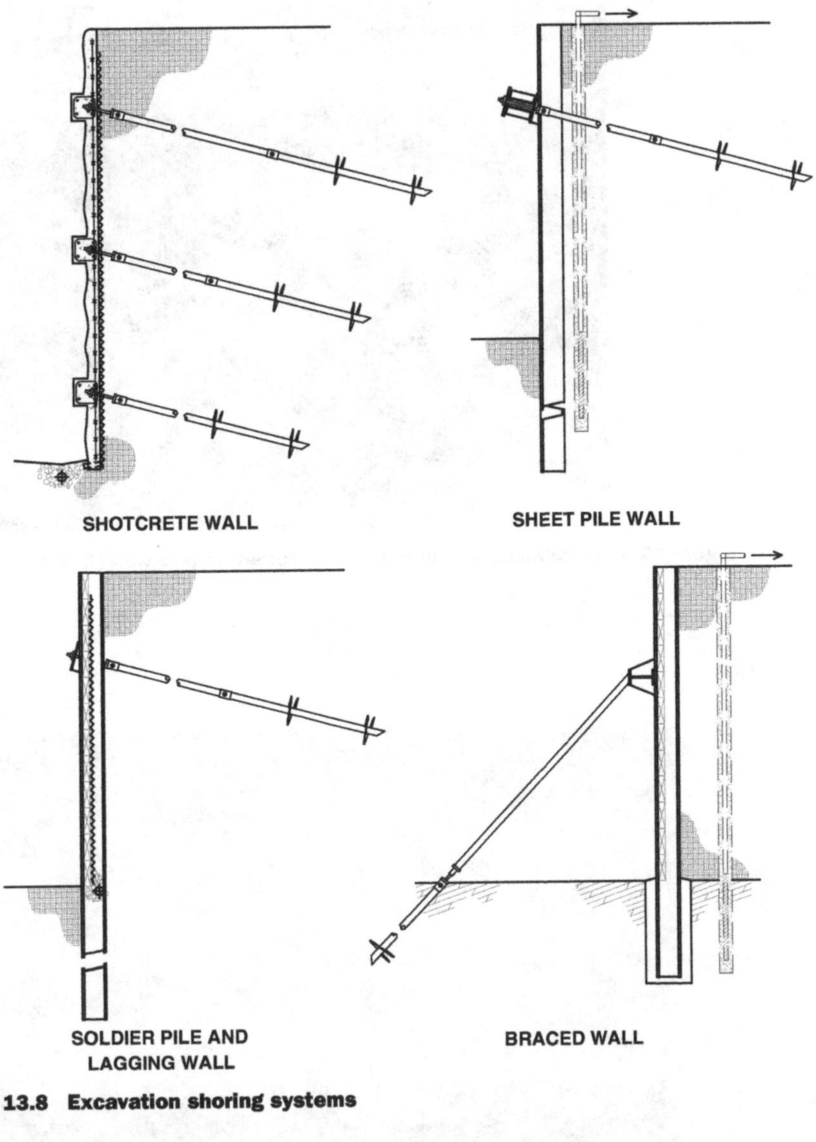

Figure 13.8 Excavation shoring systems

designed using Pilebuck™ or other standard sheet pile software. Designing sheet pile with helical tie-backs requires no additional consideration beyond other types of tie-backs other than that already discussed in this text.

Soldier pile and lagging systems consist of driving or vibrating steel H-pile from the ground surface at regular intervals around the perimeter of the excavation. After pile installation, the excavation is made and timber lagging is placed between the H-piles. The excavation is halted at each row of helical tie-backs to allow for installation

Figure 13.9 Installation of helical anchor for sheet pile wall (Courtesy of Magnum Piering)

Figure 13.10 Sheet pile and helical anchor wall (Courtesy of Magnum Piering)

Figure 13.11 Union Pacific train passing over helical anchor wall (Courtesy of Magnum Piering)

and construction of a waler. A photograph showing an example of soldier pile and lagging system construction is contained in Figure 13.12. The completed system is shown in Figure 13.13. In this case, double square tube steel was used for the walers. The walers are mounted flush with the face of the soldier piles. Base plates span the square channel at each tie-back location. Wedge washers are used to take up the angle between the tie-back anchor rod and the base plates. The whole waler system is welded together for stability. One of the welders is working in the lower left-hand corner of the photograph in Figure 13.13.

The design of soldier pile and lagging systems with helical tie-backs can be performed using P-y Wall™ software following procedures used for just about any other tie-back system. Peck's apparent earth pressure diagrams typically are used when multiple rows of helical tie-backs are incorporated in the design. Here again, there are no special considerations for helical anchors that have not been discussed previously in this text.

Braced wall systems can consist of practically any wall system, including shotcrete, sheet pile, or soldier piles and lagging. The main difference between this general category of excavation shoring and those described previously is that lateral stability is achieved through a series of internal braces. An example of a braced soldier pile wall system is shown at the bottom right corner of Figure 13.8. A braced wall system generally consists of several square or round tube rakers extending from the waler back into the excavation to a thrust block, or in this case a helical pile installed at an

Figure 13.12 Installation of helical anchor for soldier pile and lagging system (Courtesy of ECP)

angle similar to that the raker angle. Braced walls are disadvantageous because they get in the way of construction within the excavation. However, braced wall systems are fairly economical compared to tie-backs, and they are sometimes necessary when tie-backs cannot be installed due to property line restrictions, easement issues, utilities, or nearby structures.

Figure 13.13 Soldier pile and lagging wall with helical tie-backs (Courtesy of ECP)

Helical piles in this application simply need to be designed for the bearing force on the raker. Buckling of the helical pile and raker combination should be considered. Once again, there are no special provisions for using a helical pile to create a braced excavation. As in all braced wall designs, the raker angle has to be such that the uplift force transferred to the wall can be resisted by the pullout resistance of the piles under the wall and/or the weight of the wall. In other words, the designer should include an analysis whereby overturning about the raker is checked. More information about the design of braced wall systems can be found in MacNab (2002).

Soldier pile wall systems that are placed by drilling instead of pile driving can consist of a pair of C-channels rather than conventional H-pile. Similar flexural strength can be achieved with double channels as compared to H-pile with similar weight. Significant savings in the amount of steel required for the shoring system can be realized by installing the helical tie-back between the channels and eliminating the waler. However, this can result in the channels being unbraced for the entire length of the cantilever. Bracing needs to be provided by other means, or the channels have to be sized for lateral torsional buckling due to the unbraced length. The H-piles shown in the braced excavation detail in Figure 13.8 also need to be designed for the requisite unbraced length if the waler is eliminated and a raker is placed at each soldier pile. See the American Institute of Steel Construction (AISC) code (most recent version) for information on lateral torsional buckling.

13.4 TIMBER LAGGING

Timber lagging used in soldier pile and lagging systems is typically on the order of 3 to 4 inches [76 to 102 mm] thick. A quick analysis using conventional NDS wood design codes would show that the lagging is insufficient to support active earth pressure or Peck's apparent pressure. Yet these systems have worked very well throughout history. The design of timber lagging to span between steel H-piles typically has been based on the practitioners experience, limited Goldberg-Zoino tables from the Navy (U.S. Navy, 1988), a modified version of active earth pressure (MacNab, 2002), or a simplified calculation based on soil arching over a trapdoor (Spencer, White, and Prentis, 1986). These methods often come under criticism by structural engineers. They are not well founded, are difficult to apply to alternate lagging materials, and do not take into account surcharge pressure that could exist behind braced excavations.

The reason that timber lagging works in soldier pile and lagging systems is that the soil arches between the piles as shown in Figures 13.14 and 13.15. It is also known by practitioners that the pressure on lagging approaches a constant at some depth for most soil conditions. A model to predict the maximum earth pressure on timber lagging was recently derived based on a simple limit state analysis of the silo-shaped sliding wedge shown in Figure 13.16. The model shows that lateral pressure reaches a constant at a depth of roughly 1.4 times the spacing between H-piles. The maximum

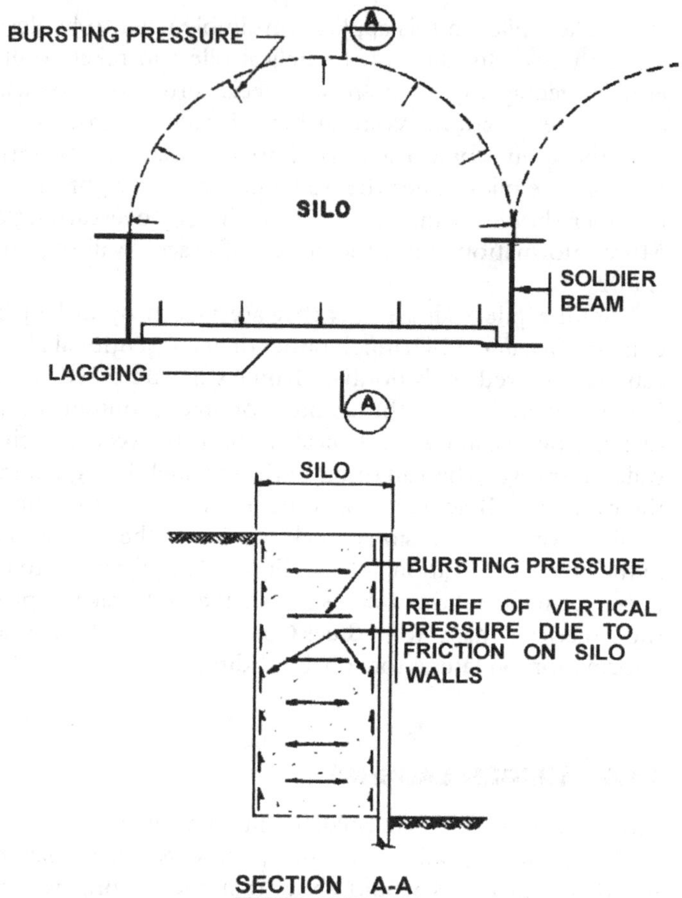

Figure 13.14 Silo effect for soldier pile and lagging (US Navy, 1988)

pressure on the lagging for coarse-grain soils at this depth is given by Perko and Boulden (2008).

$$P_{\max} = K_a\,(w + 1.2l\gamma) \tag{13.2}$$

Where

the parameter l is the spacing between the soldier pile flanges, and

w is uniform surcharge pressure at the ground surface. All other parameters have been defined previously.

A comparison between the model and the Goldberg-Zoino U.S. Navy tables is shown in Table 13.1. The model compares well with these tables. Timber lagging thickness

Figure 13.15 Soil silo formed by caving (Perko, 2008)

recommendations for walls with uniform surcharge load are given in the lower portion of Table 13.1 for two different wood strengths.

Perko and Boulden (2008) also considered fine-grain soils with cohesion in the derivation of the model. It was found that a small amount of cohesion significantly reduced the computed pressure on the lagging. In fact, the predicted pressure became negligible at a cohesion given by

$$c = \frac{l\gamma}{2} \tag{13.3}$$

This is confirmed by the fact that cohesive soils typically do not come into contact with lagging, even in deep excavations. However, the designer is cautioned to be conservative when it comes to cohesion in earth retention design for the reasons discussed previously in this chapter and Chapter 3.

13.5 HELICAL SOIL NAILS

A detail showing the general features of helical soil nail reinforced soil mass is shown in Figure 13.17. A completed project is pictured in Figure 13.18. Helical soil nails differ from helical tie-backs in that soil nailing involves the reinforcement of earth behind a retaining wall or on a slope so that the soil itself is stable without the need for significant structural facing. Soil reinforcement is facilitated by grouting helical soil nails continuously along the shaft or by using a continuous series of spaced-apart helical bearing plates along the entire length of the shaft as shown in Figure 13.17.

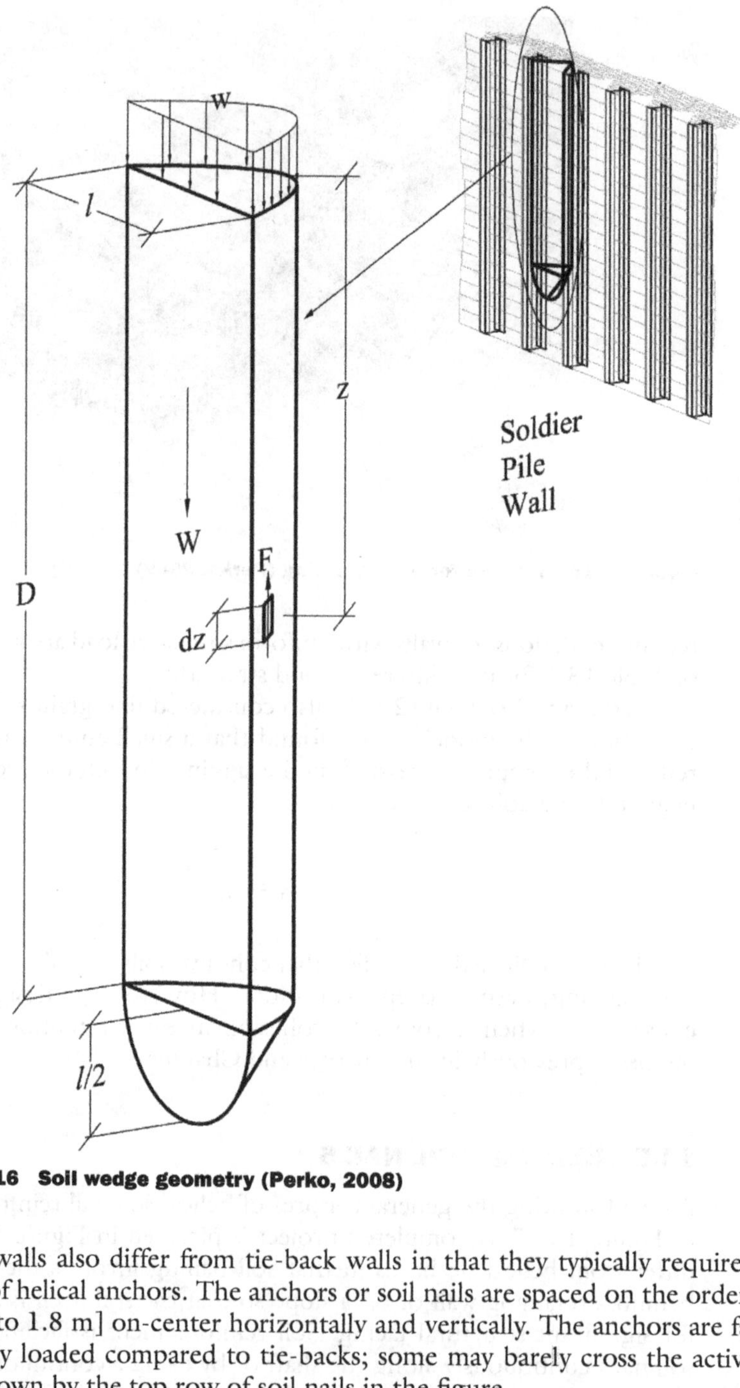

Figure 13.16 Soil wedge geometry (Perko, 2008)

Soil nail walls also differ from tie-back walls in that they typically require a greater number of helical anchors. The anchors or soil nails are spaced on the order of 4 to 6 feet [1.2 to 1.8 m] on-center horizontally and vertically. The anchors are fairly short and lightly loaded compared to tie-backs; some may barely cross the active zone of soil, as shown by the top row of soil nails in the figure.

Table 13.1 Recommended Lagging Thickness (Perko, 2008)

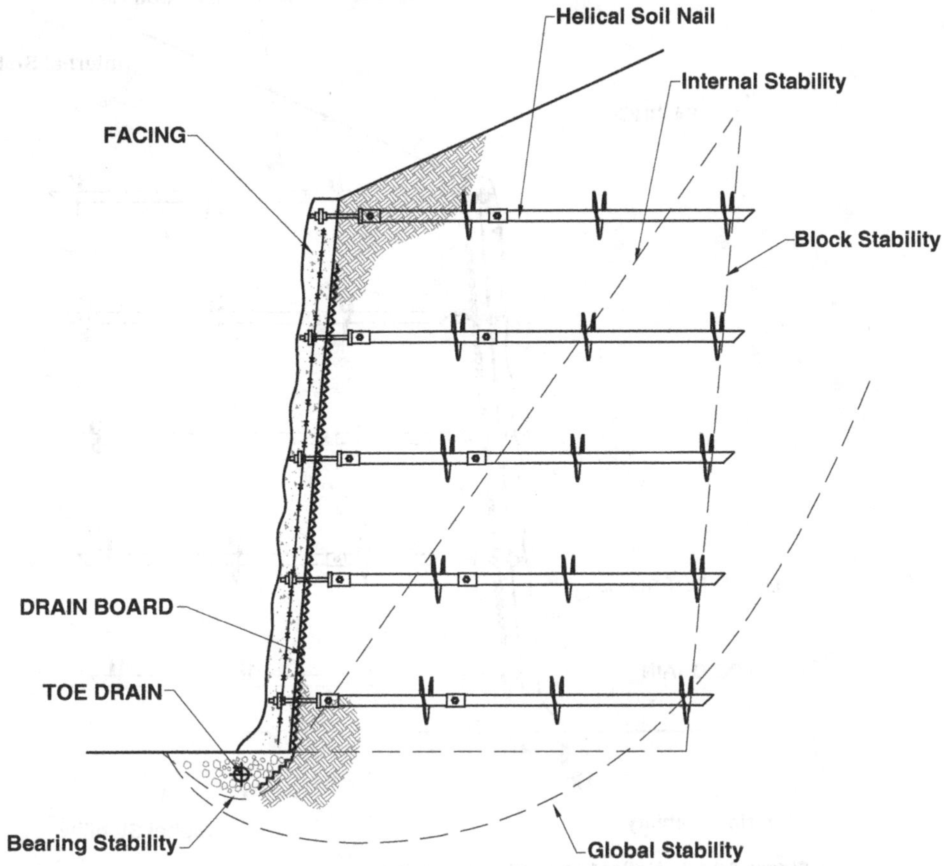

S-Nail™ or Gold-Nail™ software can be used to design helical soil nail retaining walls. An example output from S-Nail™ for a helical soil nail wall is shown in Figure 13.19. The diameter of the nail may be set equal to the diameter of the helical bearing plates provided the bearing plates are spaced 2 to 3 plate diameters on-center along the entire length of the helical soil nail shaft. The aforementioned software provides factors of safety for internal and external stability of the soil nail wall input by the user. Optimization of the design requires an iterative approach.

The simplest way to estimate the required configuration of a helical soil nail retaining wall or earth embankment and avoid the rigors of numerous software iterations is to perform a simple sliding block analysis on the reinforced soil mass. The reinforced soil mass is the block of soil contained within a line drawn around all of the soil nails, as outlined in Figure 13.17. Active earth pressures are appropriate for estimating driving force behind the block of soil but may result in a tension crack in the ground surface immediately beyond the extent of the soil nails. Using at-rest earth pressure is more

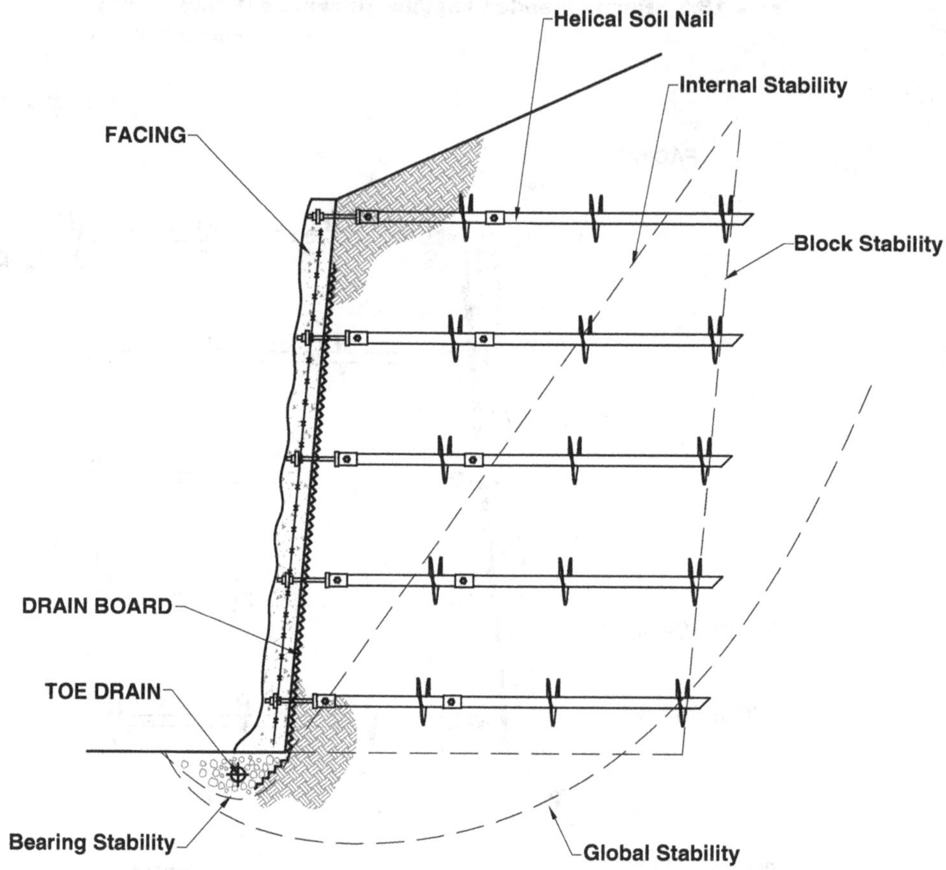

Figure 13.17 Helical soil nail application

conservative and will minimize deflections. Factor of safety for sliding block stability is given by the ratio of resisting force to the driving force as in

$$FS = \frac{\sum F_{resisting}}{\sum F_{driving}} \tag{13.4}$$

The resistance to sliding of the reinforced soil mass is the friction between the bottom of the block and the foundation soil, which is simply the weight of the block times the coefficient of friction. Weight is the volume of soil in the block times the unit weight of the soil. The coefficient of friction for soil-on-soil is the tangent of the internal angle of friction, tan(Φ). The driving force pushing against the back of the block is the result obtained from summing or integrating the pressure in the at-rest earth pressure diagram. Hence, the factor of safety against sliding is

Figure 13.18 Helical soil nail reinforced slope (Courtesy of ECP)

given by

$$F_S = \frac{LH\gamma \tan(\Phi)}{\frac{1}{2}\gamma K_o H^2} \tag{13.5}$$

which can be reduced to

$$FS = \frac{2L \tan(\Phi)}{K_o H} \tag{13.6}$$

Where

L is the minimum length of the soil nails.

All other parameters have been defined previously.

If the height of the slope and soil properties are known, Equation 13.6 can be used easily to solve for L, required to provide the desired factor of safety. Plugging some typical soil properties into Equation 13.6 results in a ratio of L/H on the order of 0.6. This equation provides a reasonable first approximation for soil nail reinforcement of sloping ground as well since the volume of a parallelogram is still base times height. The weight of soil in the sloping backfill behind the wall shown in Figure 13.17 is not taken into account in Equation 13.6 but could be incorporated easily by

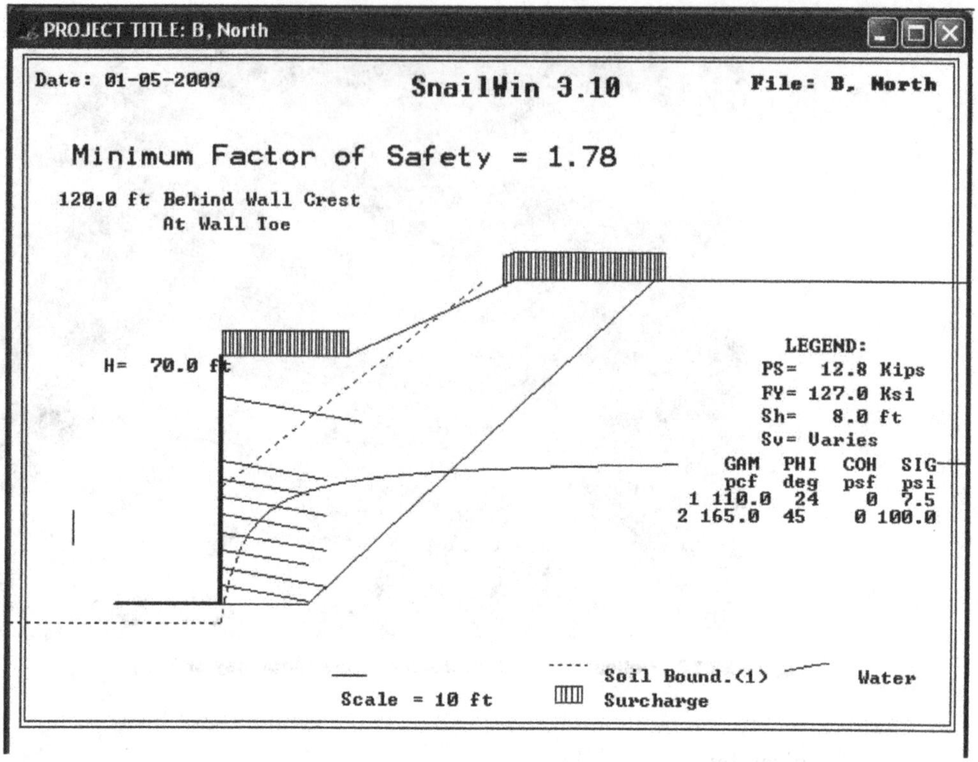

Figure 13.19 Sample S-Nail™ output for helical soil nail project

calculating the volume of the wedge of soil above the top row of anchors and adding that to the base friction term. Also, the at-rest earth pressure would be computed using the height of soil at the back of the soil nails. With practice, the designer will find that these simple back-of-the-envelope calculations save much time. When hand calculations are used to construct the geometry in S-Nail™ or Gold-Nail™, the first iteration with respect to the nail lengths will be very close to the correct factor of safety.

Soil nail walls typically have nominal facing thicknesses. The purpose of the facing is merely to prevent raveling of the slope face. Manufactured chain link-mesh for geoapplications has been used successfully in conjunction with helical soil nails for slope stabilization. The software described previously provides an estimate of the punching resistance required in the facing. Chain-link mesh manufacturers provide different-size shear connections for fastening the mesh to helical soil nails, depending on the required punching resistance. If shotcrete is used, the punching shear resistance can be calculated following procedures similar to those outlined in Chapter 12 with applicable provisions from the ACI specifications for shotcrete.

13.6 GRADING AND DRAINAGE

One of the most important factors contributing to the success or failure of earth retention projects is proper grading and drainage. The ground surface above and below an earth retention system should be graded to provide rapid runoff of storm water. Water should not be allowed to pond behind or at the base of retaining walls and even temporary construction shoring systems. Backfill should be properly compacted behind retaining walls to reduce the risk of settlement, which can ruin site grading with time. A 12-inch- [300-mm-] thick covering of fine-grain soils is recommended above backfill behind a retaining wall to reduce infiltration where possible.

All of the retaining wall and excavation shoring systems depicted in Figures 13.4 and 13.8 contain an example of a drain system located behind or at the base of the wall. A slotted drain pipe should be incorporated into the drainage system. The drain pipe should slope to daylight or to a collection area where water can be removed by pumping. The drain pipe should be encased in uniformly graded, free-draining sand or gravel.

For permanent retaining walls, considerations should be given to the compatibility of the slots in the drain pipe and the drainage material and also between the drainage material and the surrounding soils. Based on Terzaghi's filter criteria, the average diameter of the drainage material should be no more than 4 to 8 times greater than the average diameter of the backfill and subgrade material to avoid transport of fines and clogging. If this criterion cannot be satisfied in fine-grain soils, a geosynthetic filter fabric should be used between the drainage material and the surrounding soils. The maximum width of the slots in the drain pipe should be such that no more than 10 to 25 percent of the drainage material particle sizes can pass through the openings.

In addition to the drain system, weep holes are commonly recommended through the base of free-standing retaining walls. Weep holes generally consist of 2-inch- to 4-inch- [50- to 100-mm-] diameter pipes spaced roughly 8 feet [2.5 m] on-center along the length of the wall. The openings should be screened to prevent rodents and the accumulation of debris. Weep holes have been used since Roman times to relieve hydrostatic pressure behind retaining walls and should be incorporated in earth retention design whenever possible.

For excavation shoring, it may be impractical to install a drain system behind the wall. In these cases, dewatering wells should be used whenever there is a potential for accumulation of ground water above the base of the wall system. Two of the details contained in Figure 13.8 show examples of dewatering wells located behind the shoring systems.

13.7 POST-TENSIONING

Helical anchors used in earth retention systems can be post-tensioned to reduce total deflections. The simplest way to apply a post-tensioning load is to tighten the lock nut to a predetermined torque. The clamping force provided by the lock nut preloads the helical anchor. The amount of post-tensioning provided by the nut can be estimated from Jones et al. (2004):

$$P_c = \frac{T}{(0.159p) + (1.156\mu d_b)} \tag{13.7}$$

Where
- P_c is the post-tensioning force
- T is the torque applied to the lock nut by a wrench
- p is the pitch of the anchor threads
- d_b is anchor rod diameter, and
- μ is the coefficient of friction between the anchor rod and nut

Pitch may be determined by taking the inverse of the number of threads per unit length. The coefficient of friction of the steel threads and the lock nut can be taken as 0.35. If the threads are lubricated, the coefficient of friction can be much less than this.

The formula given in Equation 13.7 yields a post-tension force of 10 kips [44 kN] for a lock nut without lubrication on a $1\frac{1}{8}$-inch [29-mm] anchor rod with UNC threads tightened to 350 foot-pounds [475 N-m], the limit of some common impact wrenches. If common grease is used to reduce the friction coefficient to 0.1, then the post-tensioning force based on Equation 5.9 would increase to 27 kips [120 kN]. If a Teflon-based lubricant is used, the post-tensioning force can be even greater. These results show that the clamping force provided by a lock nut is substantial, but it depends heavily on the condition of the threads and the friction between threads. Using a lock nut and wrench to post-tension a helical anchor is somewhat inexact but can be beneficial in removing some of the initial deflection for less critical wall systems.

An example showing a more exact method for post-tensioning a helical anchor is shown in Figure 13.20. The example depicts a soldier pile earth retention system; however, one can imagine how the method could be adapted to many other tie-back systems. After anchor installation to the required depth and torque, a tie-back cap with threaded rod is attached to the anchor and the waler is installed (if used). The waler typically consists of double channels that are stitch-welded to the solder piling. Steel gusset plates are used to position the waler perpendicular to the tie-back anchors as discussed in a preceding section. A base plate and lock-off nut is used to attach the tie-back to the waler. A wedge washer can be used to correct any remaining discrepancies between the angle of inclination of the anchor and the waler.

When the system is ready for post-tensioning, a chair assembly is placed over the tie-back and fastened temporarily to the waler. The chair assembly serves as a base for the hydraulic ram and allows access to the lock-off nut. Hydraulic rams are available with a hollow core so that they may be placed directly over and centered on the tie-back anchor rod. A cap plate, washer, and nut are positioned over the ram. Post-tensioning may be applied at one time or in increments, depending on project specifications. Once the post-tensioning force is established, the lock-off nut is snug-tightened and the pressure is released from the ram. The final step is removal of the

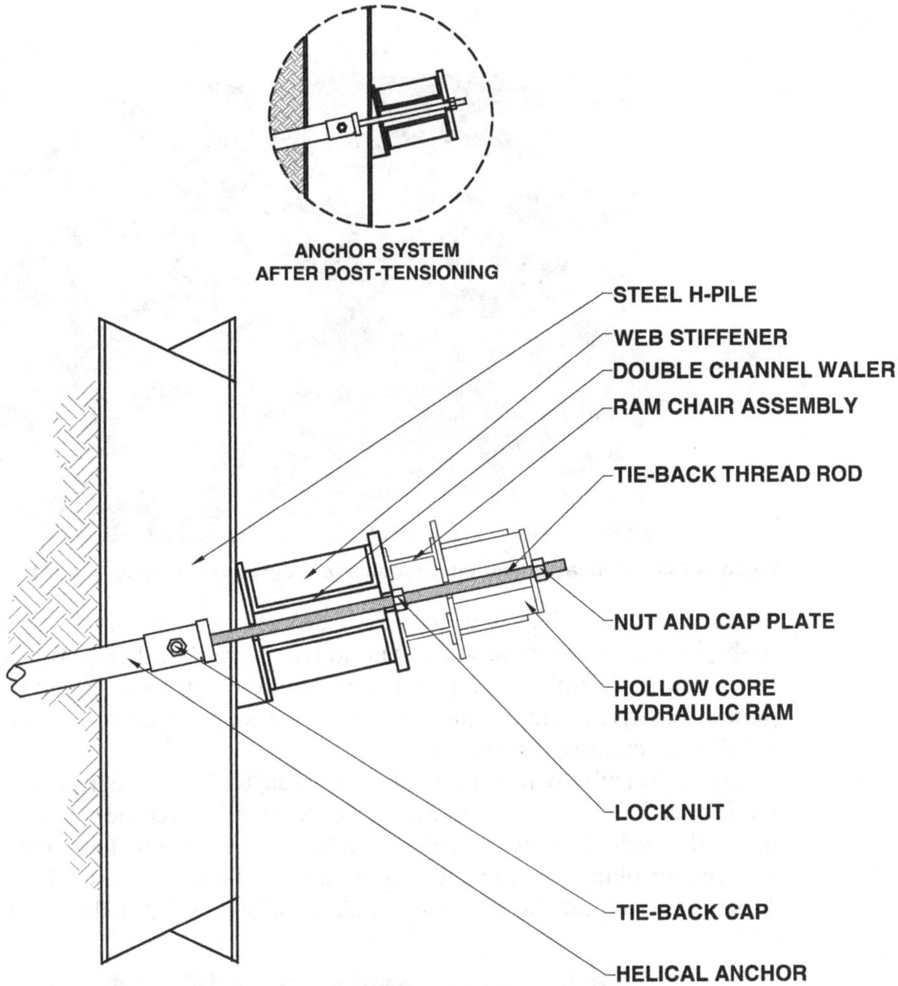

Figure 13.20 Example post-tension ram assembly

ram and chair assembly. If desired, the tie-back rod can be cut off just above the lock nut. Post-tensioning can be combined with proof load tests and performance tests. Proof load tests and performance tests are discussed in Chapter 7. A photograph of a post-tensioning exercise is shown in Figure 13.21.

13.8 WALL REPAIR

Helical tie-backs and helical piles have been used extensively for retaining wall repair. For concrete gravity and cantilever walls, a hole is cored at each helical tie-back location. The hole has to be slightly larger diameter than the helical bearing plates if the

Figure 13.21 Helical anchor post-tension in progress (Courtesy of ECP)

tie-backs are to be installed from the front of the wall. For this reason, often helical tie-backs used in retaining wall repair utilize the smallest helical bearing plate available. If necessary, multiple bearing plates of the same diameter can be used for additional pullout resistance and torque. Helical tie-backs installed in this way can be used to stabilize an existing retaining wall.

It is difficult to restore a retaining wall to its original position through post-tensioning because the tie-backs would have to overcome passive earth pressure to move the wall. However, there have been cases where basement walls have been restored to plumb by repeated tightening of lock nuts over long periods of time. The theory behind this exercise is that some soils exhibit shrinkage with changes in

Figure 13.22 Repair of counterfort wall with helical anchors

Figure 13.23 Repair of timber pile and lagging wall with helical anchors

season and moisture content. Apparently, from the anecdotal evidence that exists in the industry, it is possible to move a wall in this way.

If a wall has to be rotated back to its original position, a more direct approach is to excavate behind the wall as shown in Figure 13.22. The photograph in the figure shows soil excavated from behind a conventional counterfort wall system. In this case, the helical anchor leads can be started from behind the wall so that the size of the core hole required through the wall can be reduced to just large enough to accommodate the bolted couplings between successive sections of helical anchor. After post-tensioning and repositioning of the wall, the area behind the wall can be backfilled with light equipment to restore grades.

Using helical tie-backs for repair of existing retaining walls is typically more economical than replacing the wall entirely. Helical pile installation equipment is small and more maneuverable than conventional tie-back drills. This allows helical anchors to be installed even in very tight site conditions, as shown in Figure 13.23. In this image, the torque motor has been mounted sideways on a skid-steer loader. The overhang from an adjacent structure can be seen in the photograph, indicating the limited access. Helical anchors also can be installed using hand-operated equipment suspended by a makeshift block-and-tackle system.

Chapter **14**

Underpinning Systems

There has been tremendous growth in the use of helical piles for underpinning existing structures. Underpinning is done for repair of distressed foundations, to augment load carrying capacity, and to accommodate excavations alongside structures. This chapter focuses on design of underpinning systems and includes a discussion of foundation repairs, underpinning brackets, pile bracing, floor slab support, and braced excavations. Methods for concrete design are summarized for analysis of overturning due to underpinning eccentricity and for support of floor slabs on helical piles. Example plans are provided for foundation repair and excavation bracing. The importance of interim stability during construction is discussed.

14.1 FOUNDATION REPAIR

Foundation distress may be due to expansive soils, settlement, or slope movement. In some cases, the goal is simply to stabilize the foundation in its current state so that the risk of further distress is minimized. In other cases, a foundation may be lifted and restored to an elevation as close to its original elevation and condition as practical (Hoyt, 2007). In any case, the design of foundation repairs requires the same fundamental approach.

Typically, design of foundation repairs begins with a diagnosis of the current condition of the structure. It is usually necessary to perform subsurface exploration to characterize the likely cause of foundation distress. Once the affected areas of the foundation are identified, risks of foundation movement and potential repairs are discussed with the owners so that they can make the decision whether to proceed with partial or complete underpinning. The design professional estimates foundation loads and required helical pile spacing. Performance criteria are established for underpinning brackets, which may include capacity, maximum eccentricity, and connection to the structure. The design professional also should evaluate lateral and rotational bracing of the piles. In a design-bid-build process, the contractor's helical pile designer usually selects the helical pile and bracket and provides either calculations or load tests verifying capacity of the connection between the underpinning bracket and the structure.

The process involved in the design of foundation repairs is best described using an example. A sample repair plan for an aircraft hanger is shown in Figure 14.1. The contours on the plan indicate the elevation of the floor slab in inches as determined from a conventional survey. The survey showed that the slab was over 5.5 inches [140 mm] out of level. The high point in the slab was located in the northeast corner of the building. Structural as-built plans and an exploratory boring indicated the building was constructed on a footing foundation over expansive soils. It was concluded that differential heave was the likely cause of foundation distress.

Unless foundation problems can be linked to a specific zone of soil under a building or a certain loading condition (i.e., settlement of more heavily loaded section), it is prudent for the designer to recommend the lowest-risk approach consisting of complete underpinning of the foundation with helical piles. However, economics and the owner's tolerance for foundation movement often result in compromise. In the foregoing example, the owner's primary concern was proper operation of the hanger door located at the north end of the building. Apparently, the hanger door required almost weekly readjustment to function. The owner was less concerned about the rest of the building and cracking of the floor slab, which could be tolerated with periodic maintenance. Based on the owner's needs and understanding of the risks, a partial foundation repair plan was developed.

The repair plan, shown in Figure 14.1, consists of excavation along the entire north end of the building. Eleven helical piles were installed on the exterior of the foundation. The piles were attached to the foundation using lifting brackets as shown in detail A/S1. Building loads were determined by summing the weight of the footing, grade beam, building walls, and tributary area of the roof as well as a portion of floor slab. Helical piles were spaced so as not to exceed the maximum flexural stress in the existing grade beam and also to maintain approximately the same load on each pile. In order to balance loads, the piles were concentrated at the most heavily loaded areas of the building located at each end of the hanger door. Design capacity and final installation torque for the helical piles is shown in the schedule near the bottom of the repair plan.

After the piles were securely attached to the existing foundation, a 6-inch [152-mm] air space was excavated below the existing footing to allow for the additional potential heave and relieve pressure on the foundation. The air space was filled with a degradable, cardboard void form to prevent contamination during the backfill process, as shown in detail C/S1. After the void-forming process, lifting brackets were used to relevel the hanger door support columns. Then a reinforced concrete sister wall was cast along the existing grade beam where the foundation was most severely compromised. The excavation along the perimeter of the foundation was backfilled, and proper drainage was established through site grading and repaving.

The expansive soils at this site were underlain by a highly plastic clayey mudstone locally termed claystone. In order to increase the likelihood of penetrating the bedrock to sufficient depth to bottom below the expansive soil active zone, a single 12-inch- [305-mm-] diameter helix was specified in Figure 14.1. Sizing charts similar to those provided in Chapter 8 were used to estimate the bearing capacity and uplift resistance of this helix. A 3-inch- [76-mm-] diameter helical pile shaft was specified based on availability. A capacity-to-torque ratio of 8 ft^{-1} [26 m^{-1}] was used to determine the required final installation torque, which is consistent with the information contained in Chapter 6. Other shaft sizes and associated capacity to torque ratios could have been substituted for the combination used in this project.

14.2 UNDERPINNING BRACKETS

Most helical pile manufacturers supply preengineered foundation underpinning brackets for attachment of helical piles to existing foundations. Underpinning brackets can be divided into two general categories: plate brackets and angle brackets. Plate brackets consist of a vertical plate only and attach the face of an existing footing, grade beam, or foundation wall. Angle brackets have a horizontal plate that supports existing foundations from underneath. They also may or may not have a vertical plate, angle, or straps that attach to the face of the foundation.

An example of a plate bracket is shown in Figure 14.2. This bracket consists of a flat plate attached to a collar tube that fits over a round-shaft helical pile. The plate has a number of penetrations for attachment to this existing structure using either expansion bolts or epoxy anchors. The bracket supports the existing foundation through direct shear on the anchor bolts. An example angle bracket is shown in Figure 14.3. This bracket includes an angle reinforced by two vertical gusset plates. A cylindrical sleeve with square tube or “T-pipe” fits over a square-shaft helical pile. The T-pipe is held in the bracket by a cross bolt and two vertical adjustments. Typical installations of the plate and angle brackets are shown on the left side of both figures.

Design of the connection between an underpinning bracket and an existing foundation is difficult for two reasons. First, the strength of the concrete, reinforcing steel, and structural integrity of the existing foundation often is unknown. Second, the interaction of the bracket and the foundation is statically indeterminate. Forces within the system depend on the relative stiffness of the helical pile shaft and the existing foundation.

The capacity of the connection between underpinning brackets and existing foundations can be verified through field load testing on every pile, bracket, and its connection to the structure. Load testing is advantageous because capacity is verified with a high degree of certainty. However, the method of testing usually involves applying load to the pile while lifting up on the structure. The structure’s dead load provides the reaction for the load test. Load tests typically are conducted to 150 percent of the design load (live plus dead) on the pile and bracket. If the existing foundation can safely span a sufficient distance so that the required dead loads for

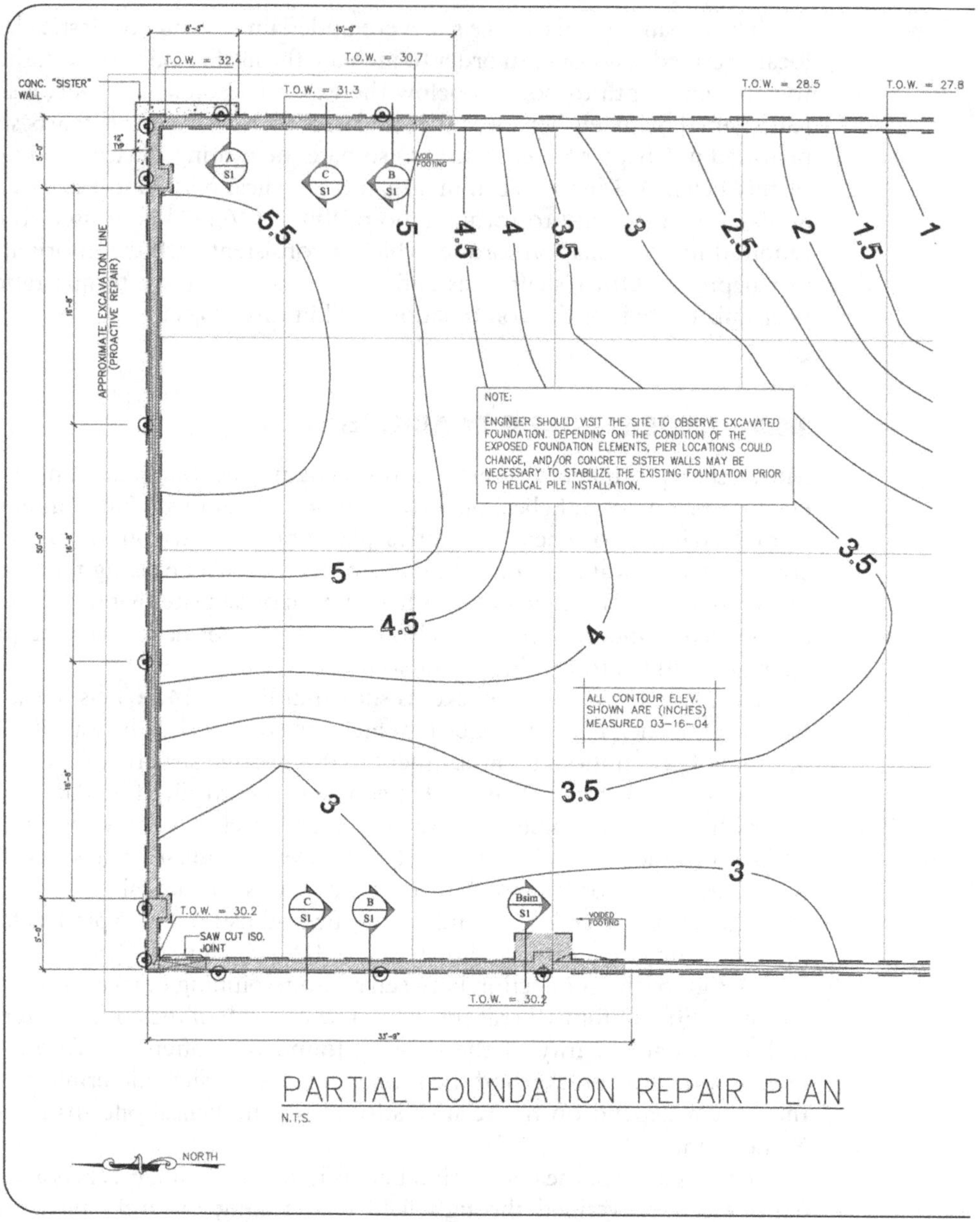

Figure 14.1 Example Partial Underpinning Plan for Aircraft Hanger

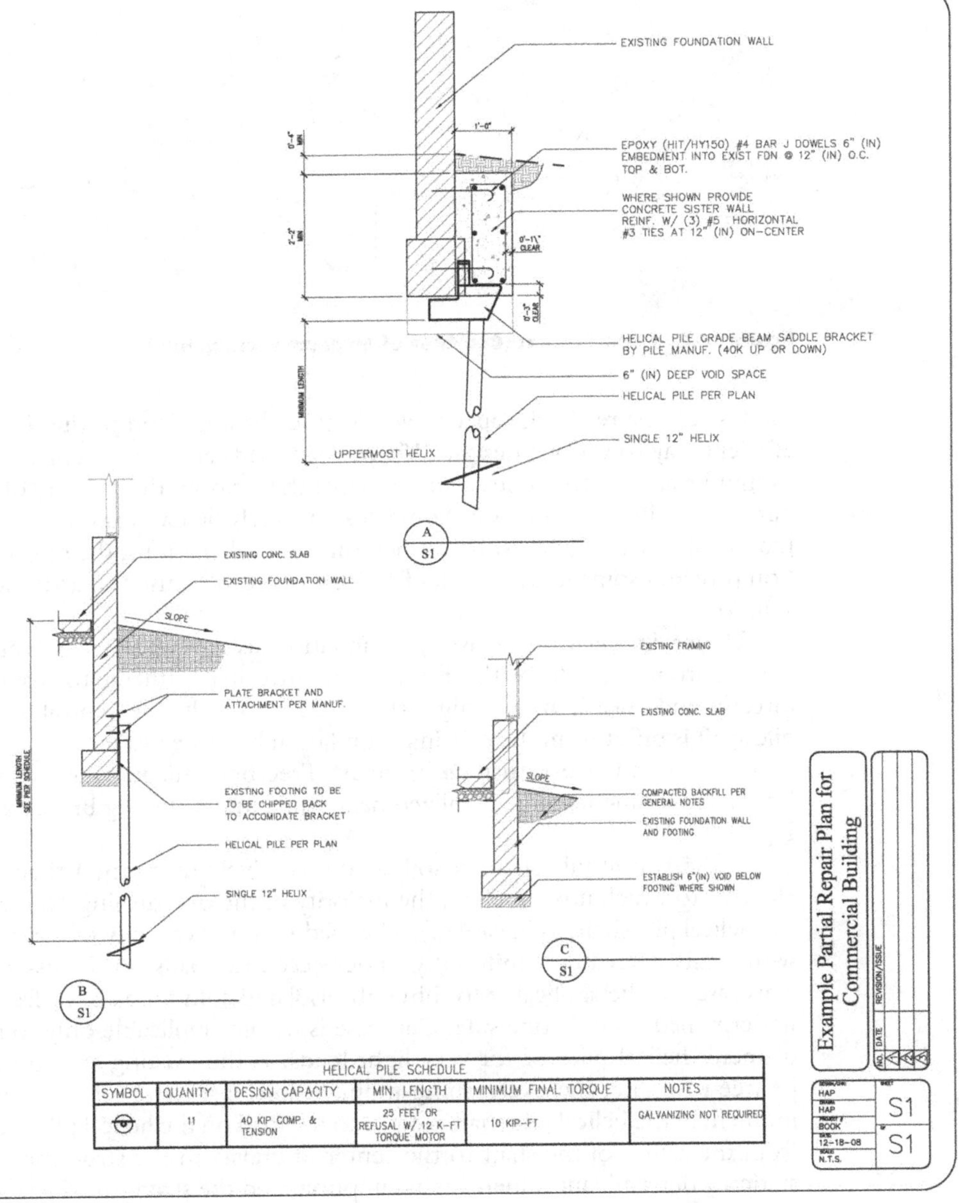

HELICAL PILE SCHEDULE					
SYMBOL	QUANITY	DESIGN CAPACITY	MIN. LENGTH	MINIMUM FINAL TORQUE	NOTES
⊖	11	40 KIP COMP. & TENSION	25 FEET OR REFUSAL W/ 12 K-FT TORQUE MOTOR	10 KIP-FT	GALVANIZING NOT REQUIRED

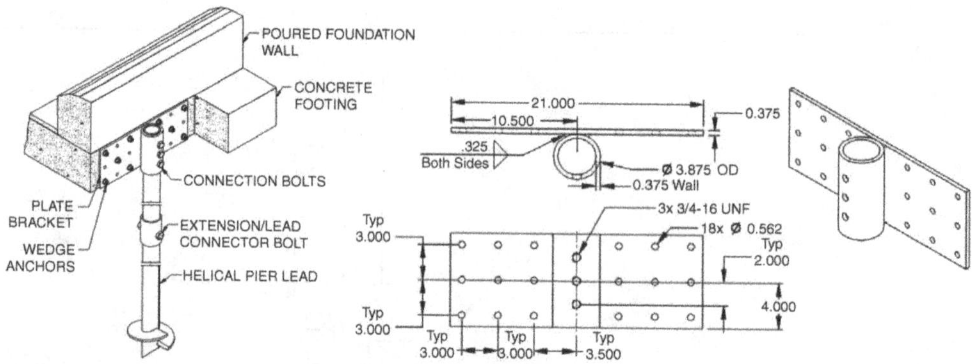

Figure 14.2 Plate Bracket (Courtesy of Magnum Piering, Inc.)

the test can be realized, capacity verification through load testing is an effective and efficient way to verify a design. Where dead loads are insufficient or when test loads cannot be applied to the structure for any other reason, the capacity of the connection has to be verified another way. Some designers rely on past experience and engineering judgment. If capacity must be verified through calculations, the remainder of this section provides some general simplifications that can be used to arrive at a closed form solution.

Underpinning of an existing foundation using helical piles typically is accomplished from the side of the foundation since it is difficult to install a helical pile directly underneath an existing structure. As a result, the central axis of the helical pile shaft is offset from the existing foundation by some distance. The resulting eccentricity causes a net overturning moment. Free-body diagrams showing the resulting forces and moments on generalized helical pile underpinning brackets are depicted in Figure 14.4.

If the helical pile shaft in soil is comparatively more rigid than the foundation element to which it is attached, the majority of the overturning moment is carried by the helical pile shaft, as in case (a). The load-carrying capacity of common helical piles with shafts decreases significantly under eccentric loads, as discussed in Chapter 7. This case may be applicable to thin slabs, isolated column pads, or free-standing walls underpinned on only one side. This case is usually applicable only to relatively large-diameter helical piles or for very light loads. If the existing structure is assumed to be free to rotate, the overturning moment in the structure is zero and the bending moment in the helical pile shaft is equal to the load on the helical pile times the distance from the center of the shaft to the center of gravity in the structure. The problem is statically determinant under this assumption, and the strength of various components can be solved using conventional design.

If the helical pile shaft in soil is comparatively more flexible than the foundation element to which it is attached, the majority of the overturning moment is carried by the structure, as in case (b). This case may be applicable to reinforced concrete grade beams, thickened slabs, walls, or column pedestals supported on both

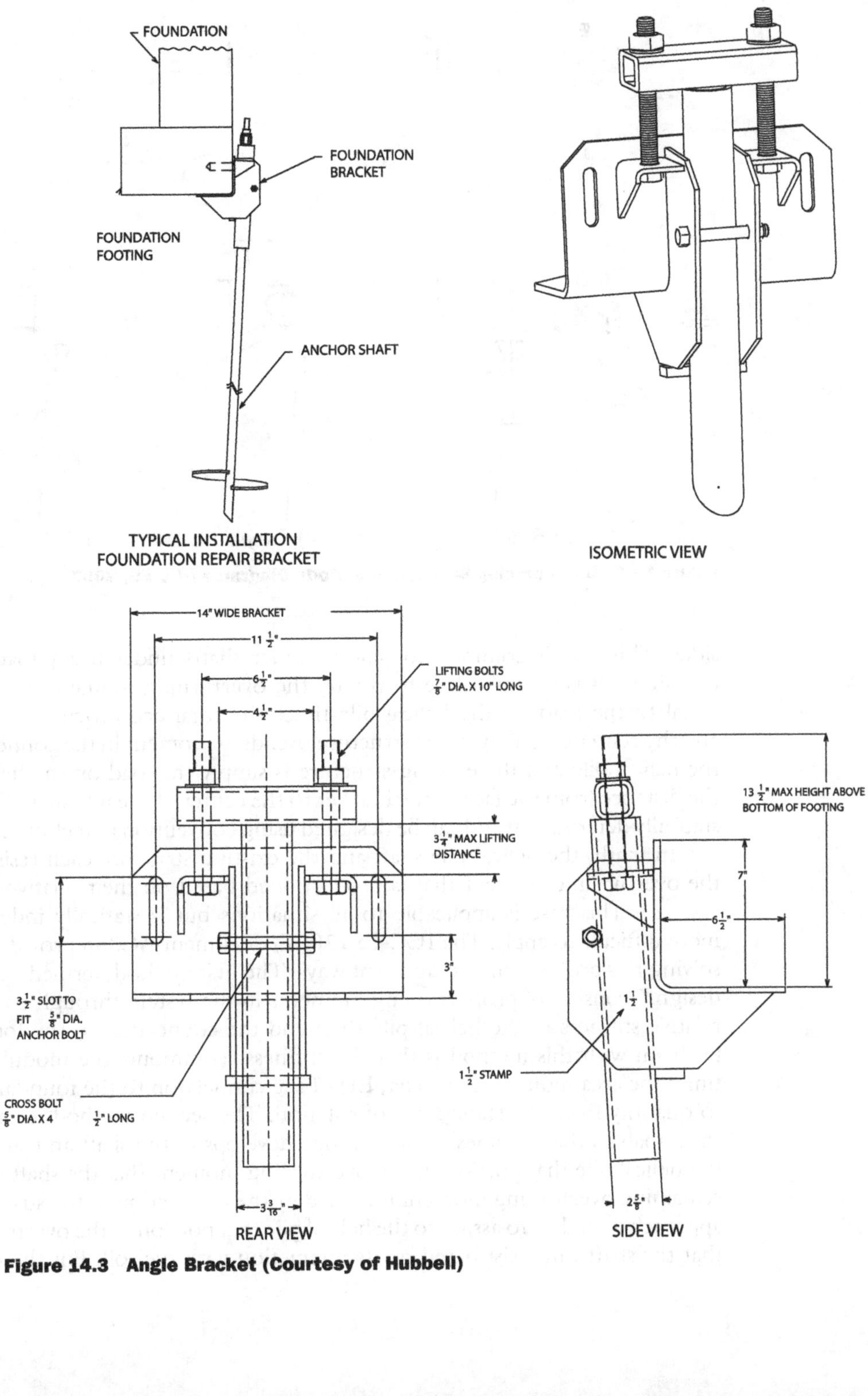

Figure 14.3 Angle Bracket (Courtesy of Hubbell)

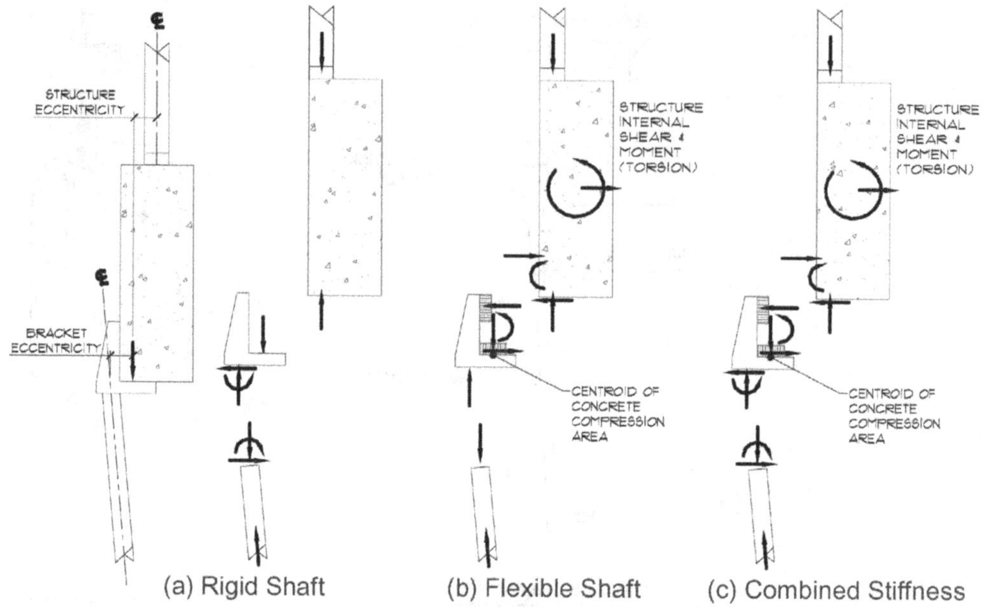

Figure 14.4 Underpinning Bracket Free- Body Diagrams (ICC-ES, 2007)

sides. This case is common for fairly slender shafts under heavy loads. If the helical pile shaft is assumed free to rotate, the overturning moment in the structure is equal to the load on the helical pile times the total eccentricity between the shaft and the center of gravity of the structure. Bending moment in the connection between the helical pile and the existing structure is simply the load on the helical pile times the distance from the face of the bracket to the center of the pile shaft. This case is also statically determinant and can be designed using conventional steel and concrete codes.

In truth, the helical pile shaft and the existing structure each resist a portion of the overturning moment that is proportioned based on their relative stiffness, as in case (c). This case is applicable to all situations but is statically indeterminate and more difficult to apply. The ICC-ES (2007) document AC358 provides guidance for solving this problem in two different ways. The first method, termed "allowable stress design," consists of proportioning moments in the system through considerations of relative stiffness of the helical pile shaft and the connection to the foundation. The problem with this method is that the stiffness (commonly the modulus of elasticity times the area moment of inertia, EI) of the connection to the foundation is difficult to quantify from the standpoint of rotation. The second method is based on a limit state analysis that assumes a plastic hinge develops in the shaft and simply assigns to the helical pile that portion of the overturning moment that the shaft can resist. The remaining overturning moment is carried by the connection to the structure. Another approach would be to assign to the helical pile that portion of the overturning moment that the shaft can resist based on its interaction with the soil. For the latter method, allowable axial force and bending moment on the helical pile shaft can be determined through an iterative approach using L-Pile™ software (discussed in Chapter 10) or other numerical method that considers the shaft-soil interaction.

In any case, the connection between the bracket and the existing foundation has to be evaluated for some lateral load and bending moment. Where angle brackets are not attached to the foundation through any a positive connection such as expansion anchors or epoxy bolts, the lateral resistance in the connection is provided by friction on the horizontal plate. Moment is resisted through the couple formed by friction on the horizontal plate and compression between the upper portion of the bracket and the foundation. Where brackets are attached to the foundation by positive means, the lateral and bending resistance is through combined loads on the connecting anchor bolts and friction, if applicable.

The bottoms of foundations cast on grade are typically very rough. When rough concrete bears on an angle bracket, stress concentrations may develop. The problem in developing a uniform and smooth surface for bearing can be alleviated using a thin layer of quick-set chemical grout on the horizontal bracket plate.

14.3 ROTATIONAL BRACING

In the design of underpinning, it is important to check rotational bracing for that portion of the overturning moment not taken in the helical pile shaft. The combined effect of multiple helical piles located on one side of a foundation needs to be taken into consideration by evaluation of the internal stability of the structure. Figure 14.5 shows an underpinning application wherein multiple helical piles are placed on one side of a grade beam or foundation wall. Unless the helical pile shafts are much more rigid than the structure, each helical pile causes an overturning moment on the structure. The torsional resistance of the foundation element should be checked by comparing the internal strength of the concrete to the total overturning caused by the underpinning pattern.

The design torsional resistance of structural plain concrete can be taken as ACI318 (2005)

$$T_c = \phi 4\sqrt{f_c'}\left(\frac{A_{CP}^2}{P_{CP}}\right) \tag{14.1}$$

Where

ϕ is the resistance factor
f_c' is the compressive strength of the concrete in psi
A_{CP} is the cross-sectional area of the concrete, and
P_{CP} is the perimeter of the concrete cross-section.

Strictly speaking, ACI318 provides this equation for redistribution of factored torsional moment. However, it corresponds well to the ultimate torsional resistance of

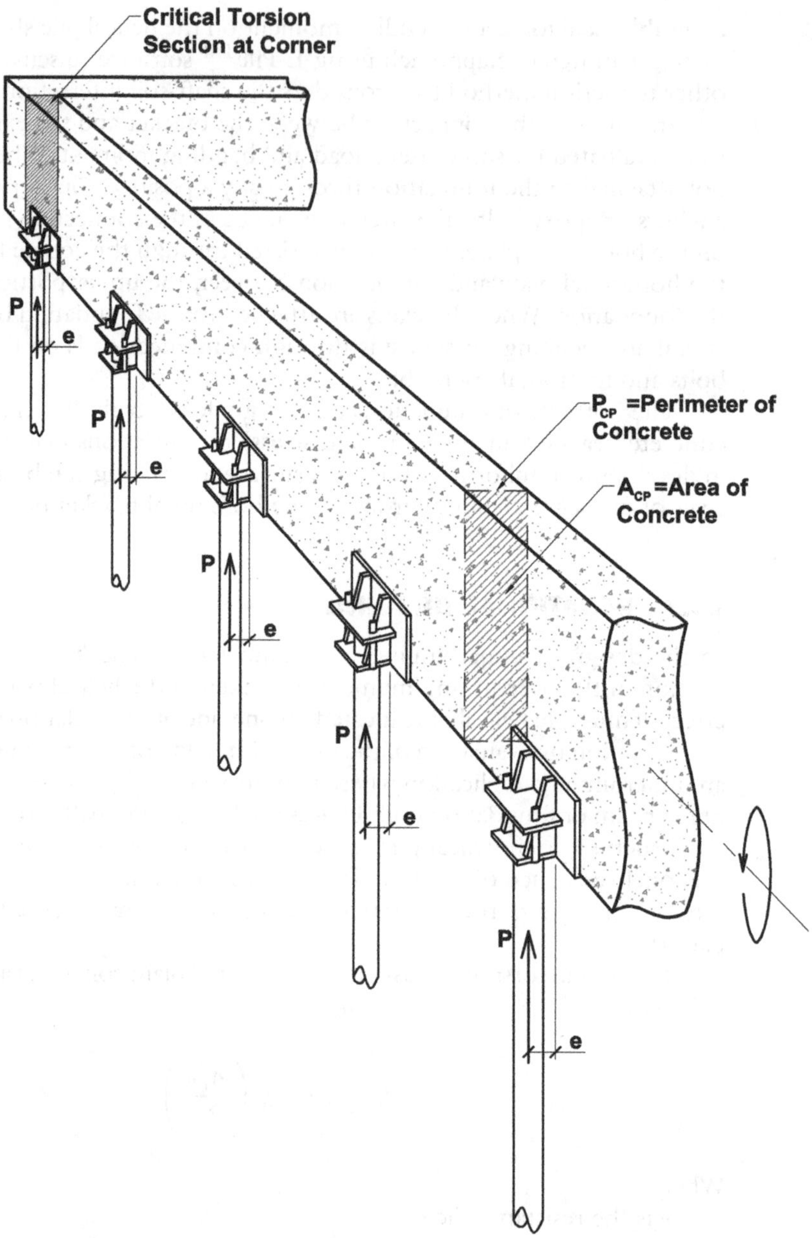

Figure 14.5 Cumulative Grade Beam Torsion

plain concrete rectangular members given by Hsu (1968) and is conservative compared to the cracking resistance given for structural plain concrete by Nilson and Winter (1991). Equation 14.1 is unit sensitive and must be calculated in terms of pounds and inches. According to ACI318 (2005), ϕ shall be 0.75 for torsion.

In the example shown in Figure 14.5, the factored torsion, T_u, is equal to the product of load P on the helical piles, eccentricity e, an appropriate load factor, and the number of piles. If the distribution of live and dead loads is unknown, a load factor of 1.5 is commonly assumed. Factored torsion can be resisted by a corner, as shown in the figure, by an intersecting foundation element, or by a concrete or steel floor or roof system. Anchor bolts in a wood sill plate often provide little resistance to rotation. If underpinning brackets are attached to a footing under a concrete grade beam, their connection should be checked to verify that they can transfer torsion. Otherwise, the area and perimeter used in Equation 14.1 should be based on the footing only.

In preparing an underpinning plan, it is convenient to estimate the torsional capacity of the grade beam or foundation wall to be supported and divide this by the overturning caused by an individual pile to determine the total number of helical piles that can be placed on the same side of the foundation without alternating. An example of this is given next.

Example 14a

Problem: Find the maximum number of underpinning piles that can be placed on the side of a foundation without alternating.

Given: Assume the foundation grade beam pictured in Figure 14.5 has a height of 48 inches [1.2 m] and a width of 8 inches [0.20 m]. It is supported on both ends by corners. The concrete has a compressive strength of 4,000 psi [28 MPa]. Reinforcement in the grade beam is unknown. Each helical pile has a design load of 25 kips [111 kN] and is offset from the resultant foundation force by 3 inches [51 mm]. The helical piles in soil are assumed to be flexible.

Answer: Area of the concrete is $48 \times 8 = 384$ in^2 [0.25 m^2]. The perimeter is $(48 \times 2) + (8 \times 2) = 112$ in [2.8 m]. There are two supports provided by the corners, so the design torsional resistance, Equation 14.1, is given by

$$2(0.75)4\sqrt{4{,}000}\left(\frac{384^2}{112}\right) = 500{,}000\text{ in} - \text{lbs}\,[41{,}600\text{ N-m}] \qquad (14a.1)$$

Each underpinning pile causes an overturning moment of $25{,}000 \times 3 = 75{,}000$inch-pounds [8,500 N-m]. The factored overturning is 1.5 times this, or 112,500 inch-pounds [12,700 N-m]. Hence, 4 ($500/113 = 4.4$) helical piles can be used along the side of the foundation without alternating.

ACI318 states that torsion can be neglected in combined torsion and shear calculations provided that factored torsion is less than 25 percent times the design torsional resistance, Equation 14.1. This amount of torsion is so small that based on the circular interaction between torsion and shear, threshold torsion corresponds to a reduction of only 3 percent in shear. If the applied torsion is above the threshold, combined shear and torsion need to be taken into account using the interaction equation for structural plain concrete shown in Equation 14.3 (Nilson and Winter, 1991):

$$\left(\frac{V_u}{V_c}\right)^2 + \left(\frac{T_u}{T_c}\right)^2 = 1 \tag{14.2}$$

Where

V_u is factored shear

V_C is design shear resistance, and

T_u is factored torsion, The remaining parameter was defined previously.

14.4 FLOOR SLAB SUPPORT

Helical piles are used to underpin existing concrete floor slabs. Several helical pile manufacturers provide preengineered slab support brackets. An example of a common slab support bracket is shown in Figure 14.6. The bracket simply consists of a sleeve that fits over the helical pile shaft and a short section of cannel that supports the underside of the slab. Most brackets also have a threaded adjustment for lifting and transferring the weight of the slab to the helical pile. In order to install the helical pile through the slab, a core hole equal to the size of the helical bearing plates has to be made through the slab. The size of the channel that can be used with this type of slab support bracket depends on the core hole diameter and the amount of excavation under the slab.

A more sophisticated slab support bracket is shown in Figure 14.7. This bracket consists of a collar sleeve that fits over the helical pile shaft and two gusseted support arms that fold out after the unit is inserted below the slab. When folded, the bracket fits through a 10-inch [254-mm] core hole. When expanded, the bracket opens to 19.5 inches [495 mm] wide. The advantage of this bracket is that it provides greater bearing area with respect to the core hole size. Two threaded bars and a fastening strap are provided for expanding the support arms and for transferring loads to the helical pile.

After lifting and releveling a concrete floor slab using a preengineered slab support bracket, many manufacturers require that the void around the slab support bracket be filled with concrete to provide additional strength for long-term support under anticipated live loads. Slab support also can be accomplished using an engineered concrete haunch or pile cap. An example of an engineered concrete haunch system is provided at the end of this section. Before the example is given, it is important to understand the maximum spacing that can be used for support of concrete floor slabs.

The maximum spacing of helical piles for support of concrete floor slabs can be estimated by the direct design method (Section 13.6, ACI318, 2005). According to this method, the total factored static moment, M_o, of a floor slab supported by piles arranged in a square pattern is given by

$$M_o = \frac{wl_n^3}{8} \tag{14.3}$$

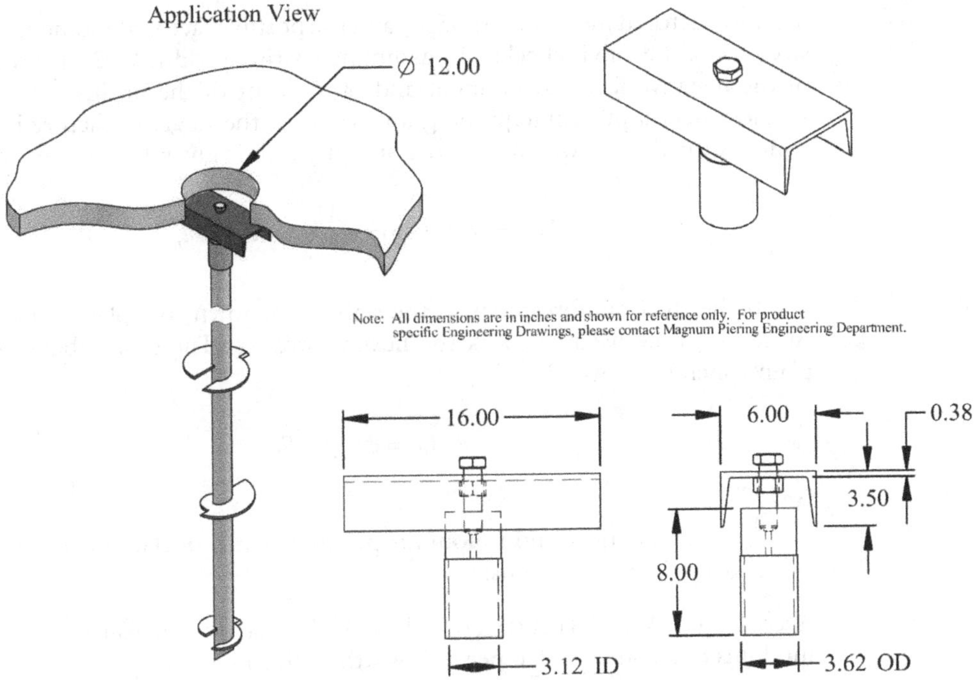

Figure 14.6 Slab Support Bracket (Courtesy of Magnum Piering, Inc.)

Where

w is factored distributed load per unit area, and
l_n is the spacing between piles.

Maximum negative moment will occur directly above each support pile, and maximum positive moment will occur midway between each support pile. In an interior span, the

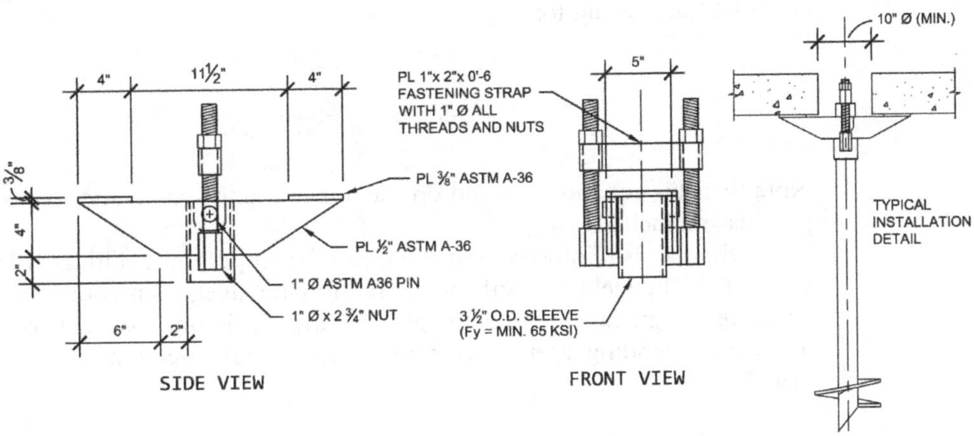

Figure 14.7 Folding Slab Bracket (Courtesy of Ramjack)

negative factored moment is $0.65M_o$ and the positive factored moment is $0.35M_o$. The strength of the slab is checked by assuming a strip of width $l_n/2$ shall resist 75 percent of the negative factored moment and 60 percent of the positive factored moment. In most helical pile underpinning applications, the negative factored moment, M^-, controls the design, which in terms of unit width is given by

$$M^- = 75\% \,(0.65)\,\frac{M_o}{l_n/2} = 0.12 w l_n^2 \tag{14.4}$$

If the steel reinforcement in the slab is unknown, the plain concrete section of ACI318 can be used to check the flexural stresses. The design bending moment of plain concrete is given by

$$\phi M_n = \phi 5\sqrt{f_c'} S_m \tag{14.5}$$

Where

S_m is the elastic section modulus per unit width of slab. Other parameters have been defined previously.

According to ACI318 (2005), ϕ shall be 0.55 for structural plain concrete. The section modulus of a concrete slab per unit width is given by

$$S_m = \frac{t_s^2}{6} \tag{14.6}$$

Where

t_s is the thickness of the slab.

Maximum pile spacing is found by combining Equation 14.4 with Equations 14.5 and 14.6 and solving for l_n.

$$l_n = 2t_s\sqrt{\frac{f_c'^{1/2}}{w}} \tag{14.7}$$

Note that this equation is based on the imperial unit system and has to be worked in pounds and inches.

If the slab is reinforced and the area of steel per unit width is at least 200 t_s/f_y where f_y is the yield strength of the reinforcing steel, then conventional reinforced concrete design can be used to calculate the maximum span between helical piles. The design bending strength of a reinforced concrete section is given by Lindeburg (1997):

$$\phi M_n = \phi f_y A_s \left(d_s - \frac{a}{2}\right) \tag{14.8}$$

Where

A_s is the cross-sectional area of the reinforcing steel per unit width of slab
d_s is the height of the reinforcing steel above the bottom of the slab, and
a is the height of the concrete compression zone.

For simplicity, it is generally conservative to assume $d - a/2 = 0.9d_s$. Per ACI318, ϕ for flexure is 0.9.

Maximum pile spacing is found by combining Equation 14.5 with Equation 14.9 and solving for l_n.

$$l_n = 2.6\sqrt{\frac{f_y A_s d_s}{w}} \tag{14.9}$$

Note that this equation is based on the imperial unit system and has to be worked in terms of pounds and inches. The maximum spacing between helical piles providing continuous support under a floor slab should be limited to about 21 times the slab thickness for deflection and serviceability. Punching shear and bearing at the helical pile supports should be checked following the procedures provided in Chapter 12.

Example 14b

Problem: Find the maximum spacing of helical piles to support a floor slab in a single-family residence.

Given: Floor slab is 4" [102 mm] thick, composed of 2,500 psi [17 MPa] compressive strength concrete, and un-reinforced.

Answer The dead weight of the floor slab is the unit weight of concrete (150 pcf, 2.4 g/cm^3) times the thickness 0.33 ft [10.2 cm] or 50 psf [2.4 kPa]. The live load on a residential floor is commonly taken as 40 psf [1.9 kPa]. The basic ASCE7 load combination is given by

$$w = 1.2(50) + 1.6(40) = 124\ \text{psf} = 0.86\ \text{psi}[5.9\ \text{kPa}] \tag{14b.1}$$

Based on Equation 14.8 for structural plain concrete, the maximum spacing of helical pile supports is given by

$$ln = 2(4)\sqrt{\frac{2{,}500^{1/2}}{0.86}} = 60'' = 5\ \text{ft}[1{:}5\ \text{m}] \tag{14b.2}$$

An example repair plan for a single-family residence is shown in Figure 14.8. This home was constructed on a slab-on-grade foundation with thickened edges over soft fine-grain and organic soils. The soils consolidated over the period of several years, causing considerable damage to the home. The repair plan consists of helical piles spaced along the perimeter of the home and in a square pattern under the floor slab.

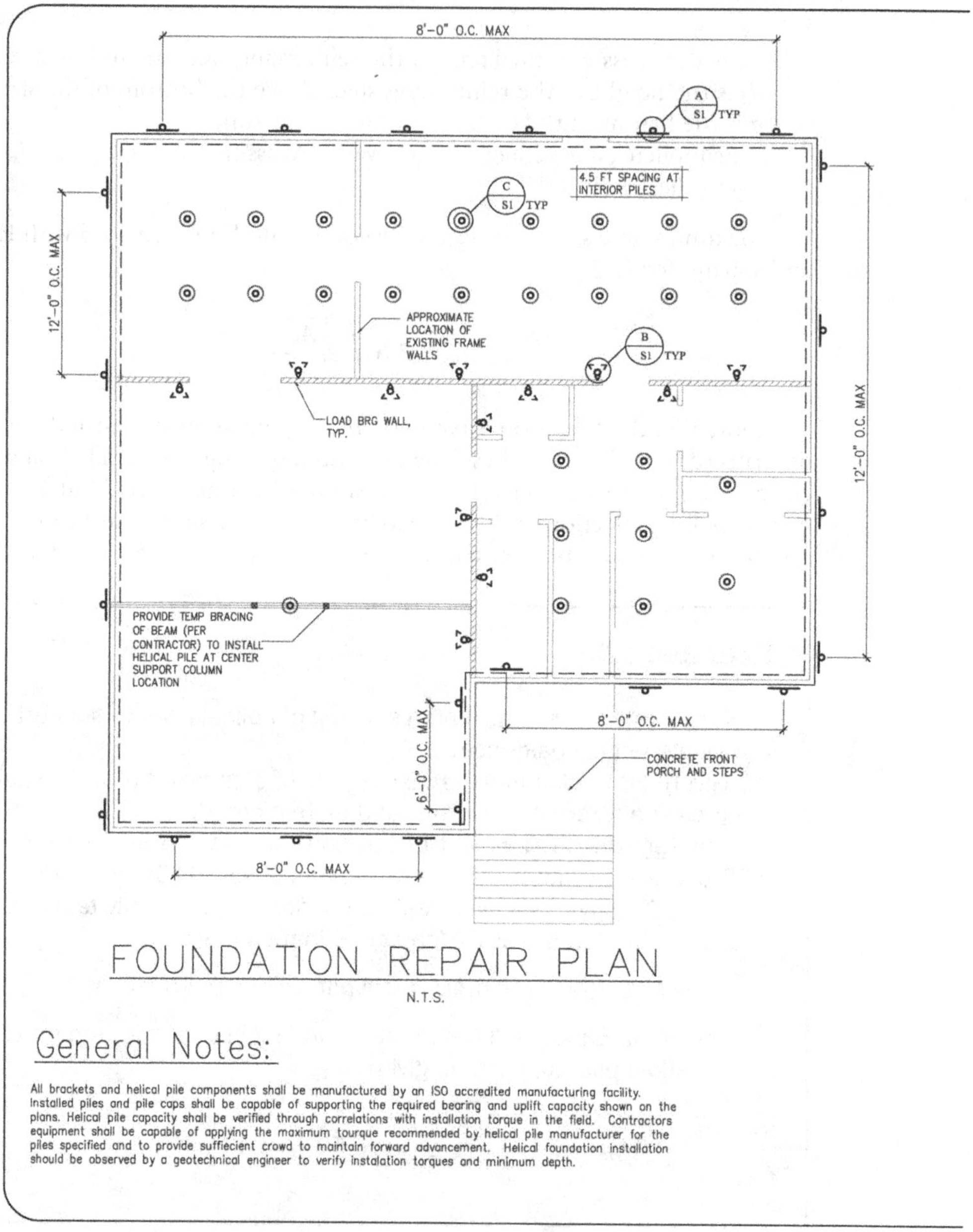

Figure 14.8 Example Repair Plan for Single-Family Residence

HELICAL PILE SCHEDULE					
SYMBOL	REQ'D CAPACITY	REQ'D FINAL TORQUE	BRACKET RYPE	MINIMUM LENGTH	NOTES
◎	5 KIPS	1.25 KIIP–FT	#6 REBAR x 6" LONG	AS REQD [1]	[1] PILE LENGTH SHALL BE IN GENERAL ACCORDANCE WITH THE LOCATION OF RESIDUUM SHOWN IN THE BORINGS. EXTEND ALL PILES TO BEAR ON RESIDUUM
(symbol)	10 KIPS	2.5 KIP–FT	#6 REBAR x 6" LONG	AS REQD [1]	
(symbol)	8 KIPS	2 KIP–FT	PLATE BRACKET	AS REQD [1]	

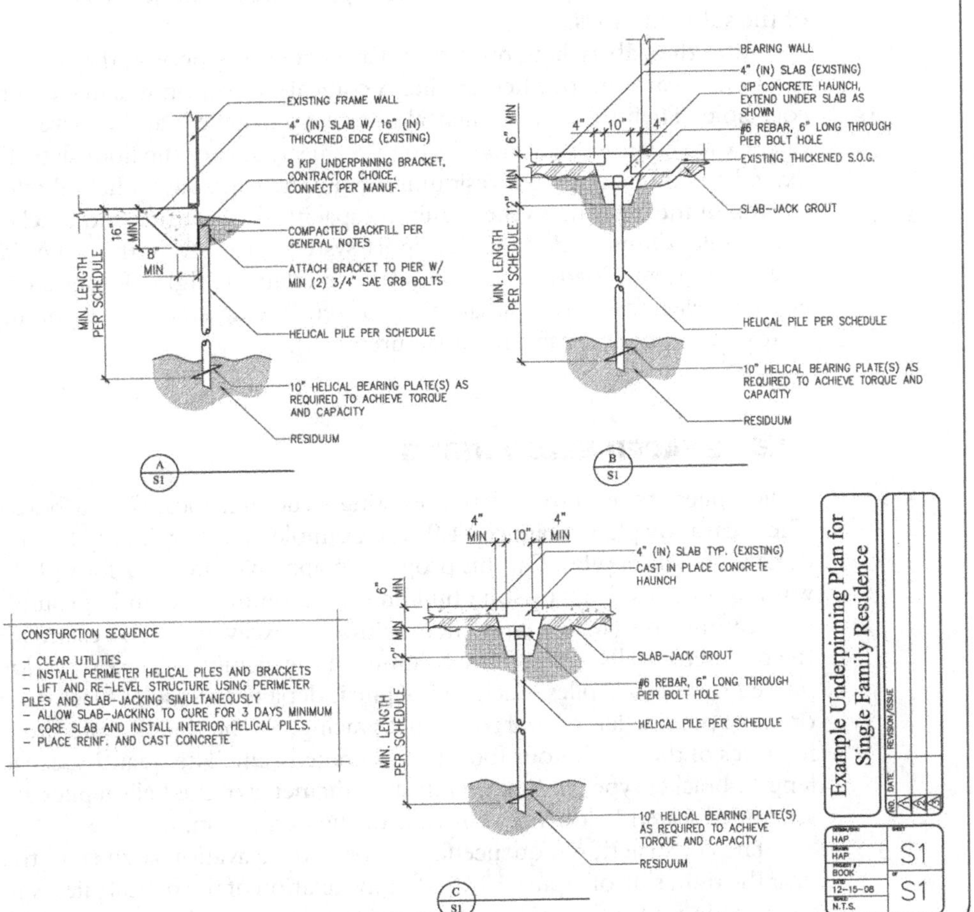

Helical piles also support bearing walls within the interior. The perimeter piles are attached to the foundation using lifting brackets, as shown in section A/S1.

The construction sequence for repairing this home began with installing the perimeter helical piles and attaching them to the foundation with lifting brackets. Then an attempt was made to elevate and relevel the foundation and floor slab using the perimeter piles and slab-jacking on the interior. Slab-jacking is a process whereby small-diameter holes are cored through the floor slab and grout is injected under pressure. Slab-jacking is a quick and effective means to relevel concrete slabs-on-grade, but it may be only a temporary means of support that is subject to additional settlement of the subgrade soils.

After the slab-jacking process and a short curing period, the slab was cored at the location of each interior helical pile. A conical excavation was made in the soil at each core hole. Then the interior helical piles were installed, and concrete haunches were cast in each conical excavation for long-term support of the floor slab. The helical piles extend to more competent residuum below the soft soils. A helical pile schedule near the top of the plan shows the required capacity, final torque, and bracket type for each helical pile. Often in floor slab support systems, the strength of the slab governs the pile spacing, and loads on helical piles are relatively light. Because of the loads, the engineer elected to use a small piece of reinforcing steel bar at the top of each pile instead of more expensive manufactured caps.

14.5 BRACED EXCAVATIONS

Helical piles are used to underpin existing structures located near braced excavations. The excavation plan in Figure 14.9 is an example of underpinning that could have been done with helical piles. On this project, an approximately 14-foot [4.3 m] excavation was required inside an existing building to accommodate underground magnetic resonance imaging (MRI) equipment. Prior to excavation, the example plan calls for the perimeter walls and interior columns in the vicinity of the excavation to be underpinned with helical piles. Each helical pile is depicted as a solid black dot in the plan. In order to provide lateral and rotational bracing, the helical piles are shown on alternating sides of the continuous footing and grade beam. The quantity, capacity, minimum length, bracket type, and minimum shaft diameter of the helical piles are shown in the schedule located below the plan view of the excavation.

The construction sequence for the braced excavation is given in the table located near the right side of Figure 14.9. After installation of the helical piles, small excavations were made under the existing footings in select areas in order to provide additional lateral bracing. Helical anchors were installed at each of the haunch locations. Then reinforcing steel and concrete were placed to connect the footing to the helical anchor. The next stage in construction involved excavating to a depth of approximately 4 feet [1.2 m] below the existing footings, placing a second row of helical anchors and reinforced shotcrete wall facing. The helical anchors were fitted with a manufactured cap with threaded bar that extended through the shotcrete facing. Post-tensioning was performed by tightening a nut and wedge washer against steel plates after the shotcrete was allowed to cure for 48 hours. Plan specifications call for the anchor rod to be lubricated in order to increase the clamping force, as discussed in Chapter 13.

The third and final stage of construction involved excavation to the final depth and completion of the reinforced shotcrete shoring wall facing. Splicing of the shotcrete reinforcing steel was accomplished by driving #5 rebar dowels into the ground between construction Stages 2 and 3 such that the appropriate splice length was left above-ground for Stage 2 and could be uncovered later for incorporation into Stage 3. Details showing the completed braced excavation walls in the area of helical pile underpinning and in areas located away from the underpinning are shown in sections A/S1 and B/S1, respectively. The quantity, design capacity, minimum length, pile cap, and minimum shaft dimensions of the helical anchors are shown in the helical anchor and helical pile schedule. This plan called for 3-inch- [76-mm-] diameter helical piles and anchors because this is the standard for the contractor and manufacturer involved with this project. Other helical pile or anchor sizes could have been specified.

The construction of temporary and permanent excavation shoring requires careful consideration of interim stability. The designer needs to work closely with the builder to minimize the risk of failures during various construction phases. The stability of the braced excavation described in the preceding example was checked at three critical stages. The first critical stage was after removal of the floor slab from inside the building, installation of underpinning piles, and during excavation for the top row of helical anchors. A photograph of this stage of construction is shown in Figure 14.10. The image shows two of the underpinning piles attached to the existing footing. An excavation has been made under the existing footing and a helical anchor is being installed.

The second critical stage was after excavation and installation of the second row of helical anchors. A worker is shown applying the shotcrete facing during Stage 2 in Figure 14.11. Four feet [1.2 m] of the excavation and underpinning pile shafts are standing unsupported during this phase. Helical pile shafts need to be checked for buckling using methods such as those described in Chapter 4. The stability of the soil to stand vertically also needs to be checked. Not all soils have sufficient stand-up time to be conducive to the shotcrete process. One way to check a soil's ability to stand vertically is by using the common geotechnical relationship

$$2c = \gamma H_s \tag{14.10}$$

Where

c is cohesion
γ is the unit weight of soil, and
H_s is the vertical height of soil.

Often c can be estimated from standard penetration tests following techniques given in Chapter 3.

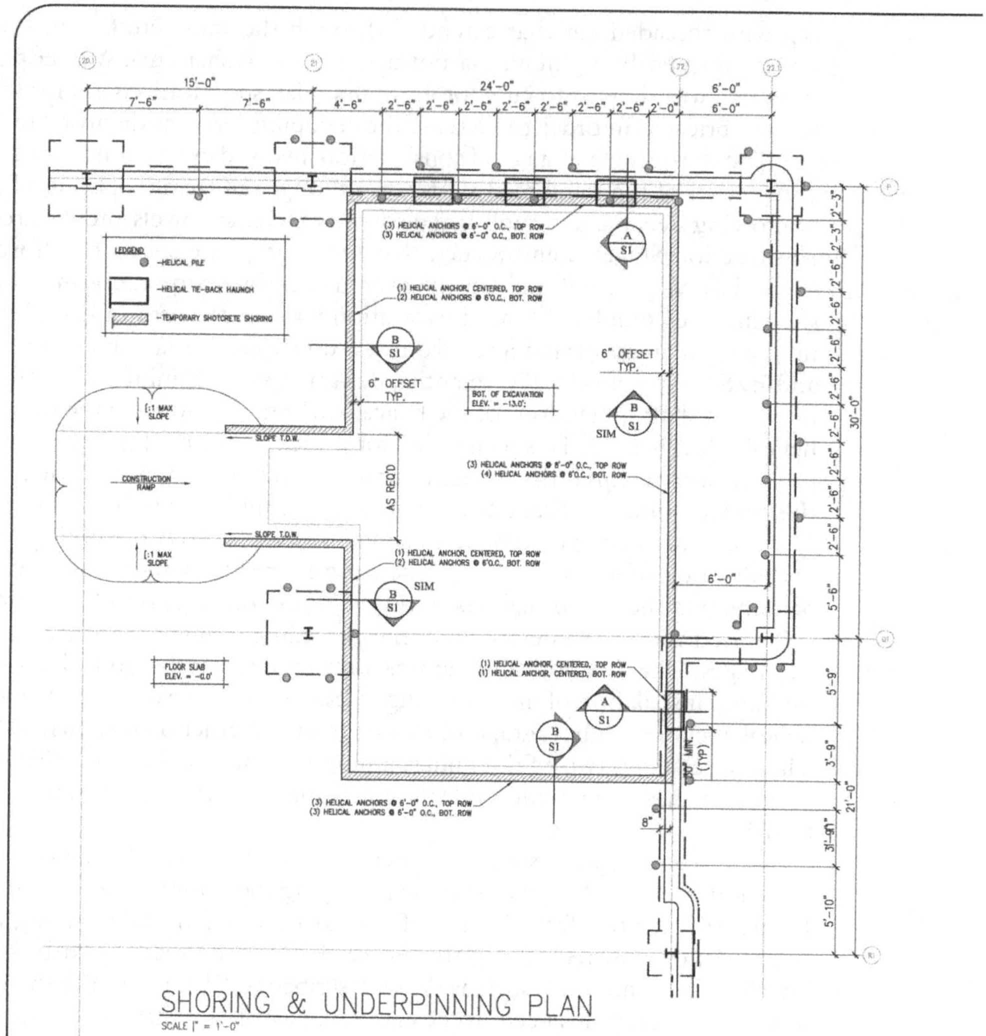

HELICAL ANCHOR AND HELICAL PILE SCHEDULE						
SYMBOL	QUANTITY	DESIGN CAPACITY	MIN. LENGTH *	PIER CAP	SHAFT DIA.	NOTES
⟨1⟩	12	30 KIPS	12' (FT)	TIE-BACK CAP W/ WEDGE WASHERS	MIN. 0.250 WALL 3"~ HSS SHAFT	CLEAN & OIL THREADS & POST TENSION NUT TO 260 lb-ft (MIN.)
⟨2⟩	15	30 KIPS	9' (FT)	TIE-BACK CAP W/ WEDGE WASHERS	MIN. 0.250 WALL 3"~ HSS SHAFT	CLEAN & OIL THREADS & POST TENSION NUT TO 260 lb-ft (MIN.)
●	39	22 KIPS	NA	MANUF. UNDERPINNING BRACKET	MIN. 0.250 WALL 3"~ HSS SHAFT	

* SEE DETAILS FOR LENGTH REQUIREMENTS

Figure 14.9 Example Temporary Excavation Shoring and Underpinning Plan

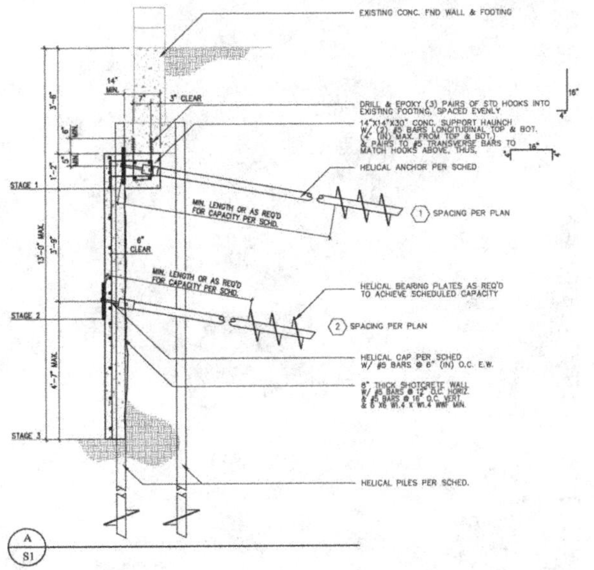

CONSTRUCTION SEQUENCE	
STAGE 1	EXCAVATE TO BOTTOM OF PERIMETER FOOTING, INSTALL ALL HELICAL PILES, EXCAVATE HAUNCH AREAS AMD INSTALL HELICAL TIE-BACKS. CAST CONCRETE HAUNCH PER DETAIL. ALLOW CONCRETE TO CURE 48hrs. INSTALL HELICAL ANCHOR PLATE & SNUG TIGHT HEX NUT.
STAGE 2	EXCAVATE TO BOTTOM OF SECOND HELICAL TIE-BACK ANCHOR PLATE, INSTALL HELICAL TIE-BACKS, PLACE REINFORCING AND SHOTCRETE WALL PER DETAILS. ALLOW CONCRETE TO CURE 48hrs, INSTALL HELICAL ANCHOR PLATE & SNUG TIGHT HEX NUT.
STAGE 3	EXCAVATE TO PLANNED BOTTOM OF EXCAVATION, PLACE REINFORCING & SHOTCRETE WALL PER DETAILS.

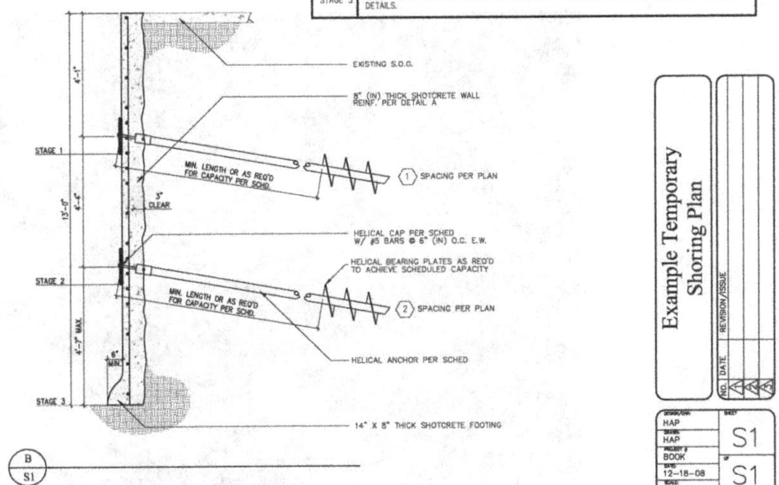

Figure 14.10 Installation of Top Row Helical Anchors (Stage 1)

Figure 14.11 Shotcrete Application (Stage 2)

Figure 14.12 Shotcrete Application (Stage 3)

Figure 14.13 Completed Excavation

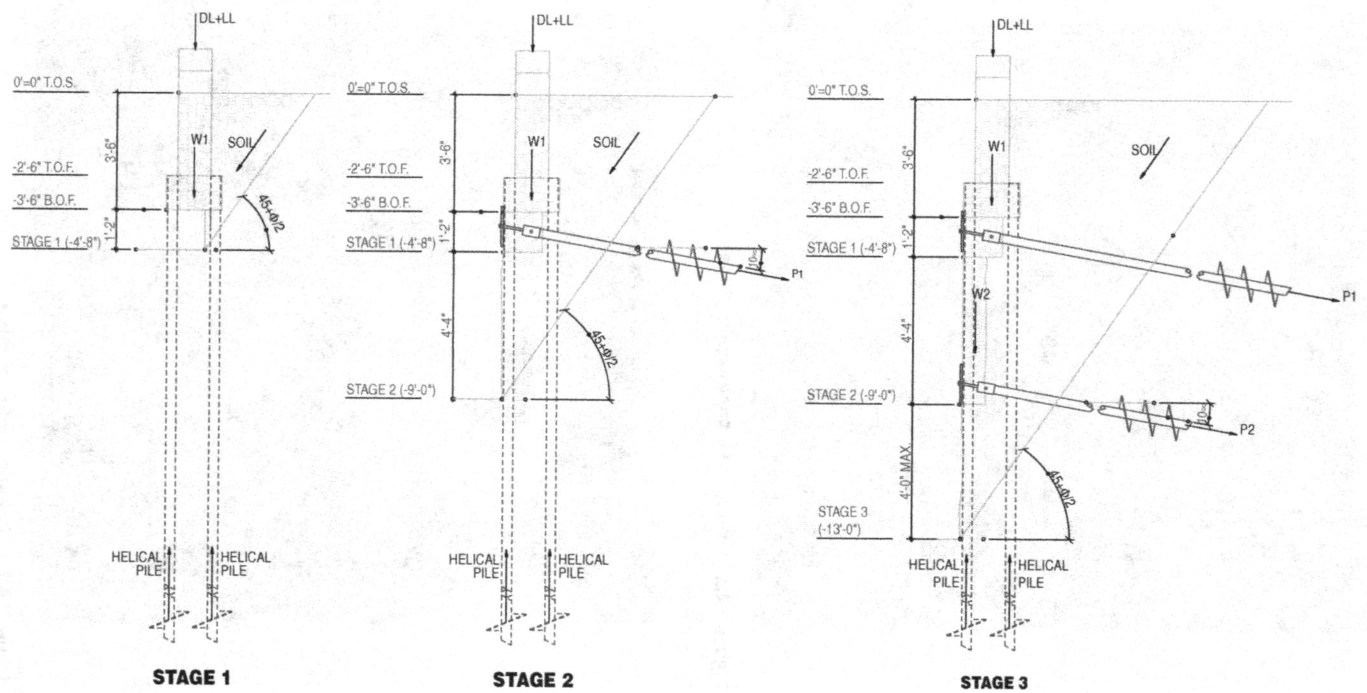

Figure 14.14 Interim-Construction Phase Stability

Upon applying Equation 14.10, one can see that only a small amount of cohesion is required for the soil to stand vertically. The designer is cautioned not to get to aggressive in the design. Occupational Safety and Health Administration (OSHA) standards also should be considered. OSHA allows category A and B soils to stand in vertically in short benches. OSHA category C soils, which include noncohesive sands and gravels, typically will not stand vertically with sufficient stand-up time to allow shotcrete application.

The third critical stage to be considered for interim stability occurred after excavation for the bottom section of reinforced shotcrete. Again, the buckling capacity of the unbraced section of underpinning plies needs to be checked. However, during this stage, the helical piles not only support building foundation loads but also support the weight of the upper portion of shotcrete facing as well as the downward component of force in the helical anchors. The photograph in Figure 14.12 is of workers applying the bottom layer of shotcrete facing. The completed excavation is shown in Figure 14.13.

Interim stability calculations performed at various stages for a similar project are discussed in Perko (2005). Diagrams associated with the different phases of construction are shown in Figure 14.14. A sliding wedge analysis was used by Perko (2005) to determine lateral loads during each of the stages. The angle of the wedge with respect to horizontal was the critical angle determined by Rankine for active earth pressure. The lateral earth pressure could have been reduced significantly by leaving the excavation made along the outside of the grade beam for underpinning pile installation. In the example provided, all excavation on the exterior had to be backfilled immediately for safety reasons. In buckling calculations, Perko (2005) considered the top of underpinning piles to be fixed in translation and rotation. He also considered the piles to be fixed at a depth of 5 feet [1.5 m] below the ground surface per International Building Code (2006) for firm soils.

Chapter 15

Economics

The economics involved in foundation selection is more than just the cost of labor, equipment, and materials. Project timeline is also an important consideration. Site preparation and foundation installation can consume a large fraction of the total time required for construction. The cost of financing, overhead, and lost opportunity must be considered when evaluating factors that affect project timelines.

This chapter discusses the cost and availability of helical piles, economics of different deep foundation types, and typical methods of measurement and payment. Included in this chapter is a brief discussion of basic economic principles including opportunity cost, interest, overhead, and the cost of time. The goal of this chapter is to introduce economic concepts that play a role in foundation selection, especially with respect to helical piles.

15.1 COST AND AVAILABILITY

Helical piles are available through a number of manufacturers in various countries of the world, including the United States, England, Australia, Canada, France, and Japan. Some of the larger manufacturers in the United States include (in alphabetical order) Alpine, Brackett, CantSink, Chance/Atlas, DriveRite, Earth Contact Products, Fasteel, Foundation Supportworks, Griptite, Helical Pier Systems, Ideal, Maclean/Dixie, Magnum Piering, Piertech, and Ramjack. The industry trade Web site, www.helicalpierworld.com, lists more than 50 helical pile manufacturers from around the world.

Several helical pile installation companies can be found in almost every major city of the world. It is typically not difficult to obtain competitive bids for helical pile installation work. Local installers generally can be found on major manufacturer's Web sites and on the World Wide Web.

At the present time, the cost of an installed helical pile foundation varies between geographic regions. In general, a normal-capacity (less than 30 tons [270 kN]) helical pile costs between $600 to $1800 (US$) per pile installed. In some locations, where construction is particularly expensive, such as New York City, a normal-capacity helical pile can cost as much as $3,000 (US$) per pile installed. The unit cost of installed and furnished helical piles can vary with required capacity, depth, and corrosion protection as well as distance to the site, competition, quantity, and schedule. Due to geographic difference in availability of different shallow and deep foundation systems, the question as to the cost of a helical pile foundation cannot be answered quantitatively with any certainty.

An example of the cost for a complete helical pile foundation and other foundation types is shown in Table 15.1. This example is based on a single-family residence located in Denver, Colorado, and a small commercial building in San Francisco, California, United States of America. Foundation costs in Colorado were obtained from verbal quotes from contractors in 2002 and should be considered approximate. Commercial foundation costs in California are based on Means Western Edition (1998) (Perko and Rupiper, 2002).

The costs shown in Table 15.1 are not intended to be absolute. The cost of labor and materials has risen dramatically over the last five years. Rather, the sample costs show how the price of different foundation types relative to each other varies across geographic regions. As can be seen in the table, the cost of a helical pile foundation falls between drilled shafts without casing and those with casing. Costs for a reinforced mat foundation and a driven pile foundation were not available, because these types of foundations were not common for residential construction in Denver, Colorado, at the time of the study. Due to lack of availability, the costs of these foundations for small residential projects would tend to be high. In San Francisco, California, the least expensive deep foundation alternative for the example small commercial building was driven piles. In this geographic area, drilled shafts and cased drilled shafts were infrequently used at the time of this study, and their respective costs were all higher than a helical pile foundation.

Another conclusion that can be drawn from the example cost comparison is that the prices for helical piles are within the general realm of deep foundation costs. Due

Table 15.1 Sample Foundation Costs (Perko and Rupiper, 2002)

Foundation Type	Approximate Cost (U.S. $) in Denver, Colorado[1]	Approximate Cost (U.S. $) in San Francisco, California[2]
Spread footings	2,400	4,952
Reinforced slab-on-grade	N/A	4,200
Post-tensioned slab-on-grade	6,000	18,500
Drilled shafts	3,000	18,800
Helical piles	7,800	17,200
Driven piles	N/A	6,800
Cased drilled shafts	8,000	20,300

to the relative differences between deep foundation costs in various geographic areas, helical pile designers should conduct their own study of their local market to determine how the price of helical piles stacks up against other deep foundation alternatives.

15.2 FOUNDATION ECONOMICS

There is more to the overall cost of a foundation than just the bid cost. The bid cost typically includes all material, labor, and equipment required to furnish and install the foundation. However, the overall cost of a foundation relative to the overall project can include a number of other important factors. General conditions, financing, and even opportunity cost can come into play. Most of these other factors are related to the value of time. The old adage that "time is money" certainly holds true on construction projects, where the cost of an extra day can be on the order of thousands of dollars. This section attempts to address some basic economic concepts that may affect an owner's or engineer's choice in foundation type.

General conditions is the portion of a standard construction contract that deals with project overhead items, such as the project manager, superintendent, job trailer, insurance, permits, inspections, traffic control, utilities, temporary toilets, phones, fencing, and trash removal. In a cost-plus contract, the owner is sometimes required to pay a prorated portion of the general conditions for each day spent in construction. In a lump-sum contract, the general contractor will estimate the cost of general conditions based on a construction schedule. In either case, the cost of general conditions is a function of construction time.

Financing cost generally is thought of as the interest and bank fees paid on a construction loan. Most construction loans are set up where interest is paid only on capital outlays to date, so the cost of financing goes up as the job progresses. However, when the foundation is being installed, there already may be considerable capital outlays such that the daily cost of financing can be appreciable. Some of these early outlays include costs for land acquisition, surveying, engineering, architecture, site clearing, demolition, and deposits to secure labor and materials. In addition to expenses associated with a construction loan, financing costs might also include the opportunity cost of the owner's down payment, which could have been placed in an interest-bearing account.

Opportunity cost is the amount of money lost doing one activity as opposed to another. An example would be an engineer spending time on research instead of working on a billable project. The cost to the company of that time is not the engineer's salary and benefits over that time period. From an economics point of view, the true cost of that activity is the lost billing that could have occurred in lieu of doing the research. In the example of the construction of a commercial building, the opportunity cost of more construction time is the lost rents that could have been received if the building were completed earlier.

An example of general conditions (overhead), financing, and opportunity costs for a \$3 million project consisting of a 15,000-square-foot commercial building is

Table 15.2 Example Daily Cost of Construction

Number of piles	100
Building square footage	15,000
Total cost of construction	$3,000,000
Financing Costs	
Land acquisition cost	$1,125,000
Architecture/engineering cost	$300,000
Demolition/excavation	$75,000
Deposits/materials	$300,000
Total job start cost	$1,800,000
Finance charge (/work day @8% interest)	$552
Opportunity Costs	
Total rents ($21/sf/yr)	$315,000
Expenses (−$15/sf/yr)	($225,000)
Subtotal ($/yr)	$90,000
Opportunity cost (/workday)	$345
Overhead Cost	
Superintendant ($1200/wk)	$171
Employer taxes (17%)	$29
Employee benefits	$39
Construction trailer ($30/day)	$30
Builders' risk insurance ($6,000/yr)	$16
Utilities ($10/day)	$10
Subtotal	$296
Overhead cost (/work day)	$415
Total cost of time (/work day)	$1,313

shown in Table 15.2. In this example, the job start costs prior to installation of the foundation are estimated to be $1.8 million. For a construction loan with an 8 percent annual interest rate, the financing cost on this building with only the job start costs is $144,000 per year, assuming that the interest is paid monthly and not compounded. The financing cost per workday is determined by taking the annual interest divided by 260 workdays per year. For the example, the financing cost at the time of foundation installation is $552 per workday.

The opportunity cost of time can be estimated by taking the total potential annual rents minus the anticipated annual expenses to obtain the pro forma gross profit. For the example in Table 15.2, the gross profit is projected to be $90,000 per year. The opportunity cost for each workday of construction is the gross profit divided by the 260 workdays in a year, or $345 per workday.

The overhead cost or general conditions for the project example includes the salary of the site superintendent and associated payroll taxes and benefits plus expenses for the construction trailer, builders' risk insurance, and utilities. These costs are estimated at $296 per weekday. Multiplying this number by seven weekdays per five workdays (7/5) gives $415 per workday. Hence, the total cost of time for the

Table 15.3 Example Foundation Cost Comparison

Helical Pile Foundation Cost	
Number of Piles	100
Installation Time (days)	4
Labor and Materials ($800/unit)	$80,000
Spoil Removal	$0
Mobilization	$1,200
Subtotal	$81,200
Cost/Pile	$812
Cost of Time	$5,617
Total Cost of Foundation	$86,817
Cost/Pile	$868
Auger-Cast Pile Foundation Cost	
Number of Piles	100
Installation Time (days)	13
Labor and Materials ($700/unit)	$70,000
Spoil Removal	$2,500
Mobilization	$5,000
Subtotal	$77,500
Cost/Pile	$775
Cost of Time	$16,555
Total Cost of Foundation	$96,555
Cost/Pile	$966

example project is the sum of financing, opportunity, and overhead costs, or $1,313 per workday.

In order to determine the true cost of a foundation to a project, the bid cost should be added to the cost of time for the number of workdays required to install the foundation. For example, if a certain type of foundation requires two weeks to install versus another type that requires four weeks to install, the cost of time of each of these alternatives would be $13,130 and $26,260, respectively. One can see that the cost of time for foundation installation can have an appreciable effect on the bottom line of a project.

In the example in Table 15.2, the 15,000-square-foot building requires 100 piles for the foundation. If either helical piles or auger-cast piles could be used on the project, an example of the total cost of these two foundation types is shown in Table 15.3. This table was prepared based on several assumptions. Twenty-five helical piles can be installed per workday. Eight auger-cast piles can be installed per day. The bid price for the helical piles is $812 per unit. The bid price for the auger-cast piles is $775 per unit. Both bid prices include spoil removal, mobilization, and demobilization. At first glance, the cost of the auger-cast piles is approximately 10 percent less than the cost of the helical piles. However, when the cost of time is considered, it can be seen that the helical piles actually cost the owner less than the auger-cast piles in this example.

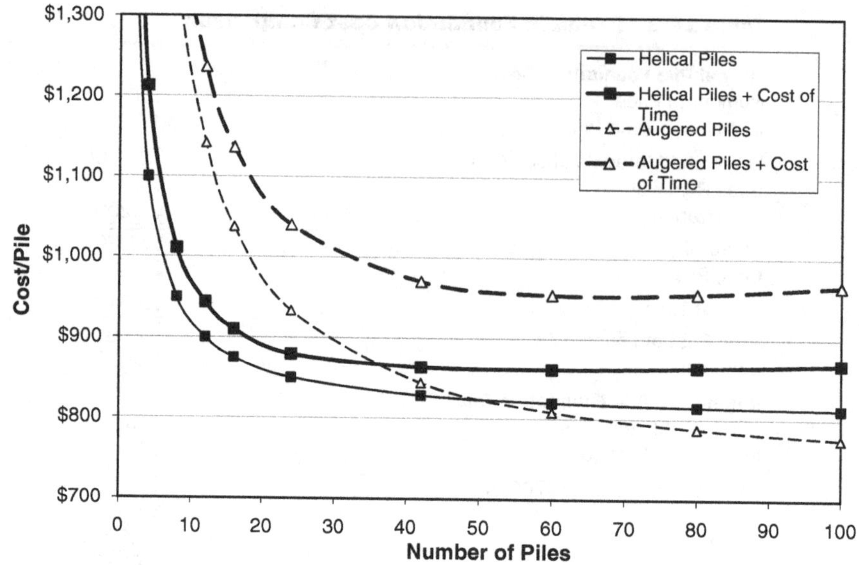

Figure 15.1 Example foundation cost comparison

A similar analysis can be performed for jobs involving a different number of piles following the same assumptions as given in Table 15.3. The results are shown in Figure 15.1 for projects with fewer than 10 piles up to 100 piles. In the figure, the bid price of helical piles is represented by the curve with light line-weight and closed square markers. The bid price of auger-cast piles is shown by the dashed curve with light line-weight and open triangular markers. As can be seen in the figure, these lines intersect at about 40 piles. The bid cost of a helical pile foundation is less than the bid cost of an augered pile foundation when there are fewer than about 40 piles due to the higher mobilization and demobilization costs of augered piles. This is often true for driven piles as well. For this reason, owners and engineers often find helical piles to be more economical than other deep foundation alternatives on smaller projects based on bid price alone.

The total cost of a helical pile foundation, including the cost of time, is represented by the curve with heavy line-weight and closed square markers in Figure 15.1. The total cost of an auger-cast pile foundation is represented by the dashed curve with heavy line-weight and open triangular markers. The results show the total cost of helical piles to a project is less than that of auger-cast piles in all cases. When a helical pile can be used to replace an auger-cast pile on a one-to-one basis, the cost of a helical pile foundation is often low when the cost of time and speed of installation is considered. For reference, the cost of construction time determined in each of these examples is shown in Figure 15.2.

Of course, the foregoing analysis would be drastically different if helical piles are unable to replace augered piles on a one-to-one basis. Normal-capacity helical piles

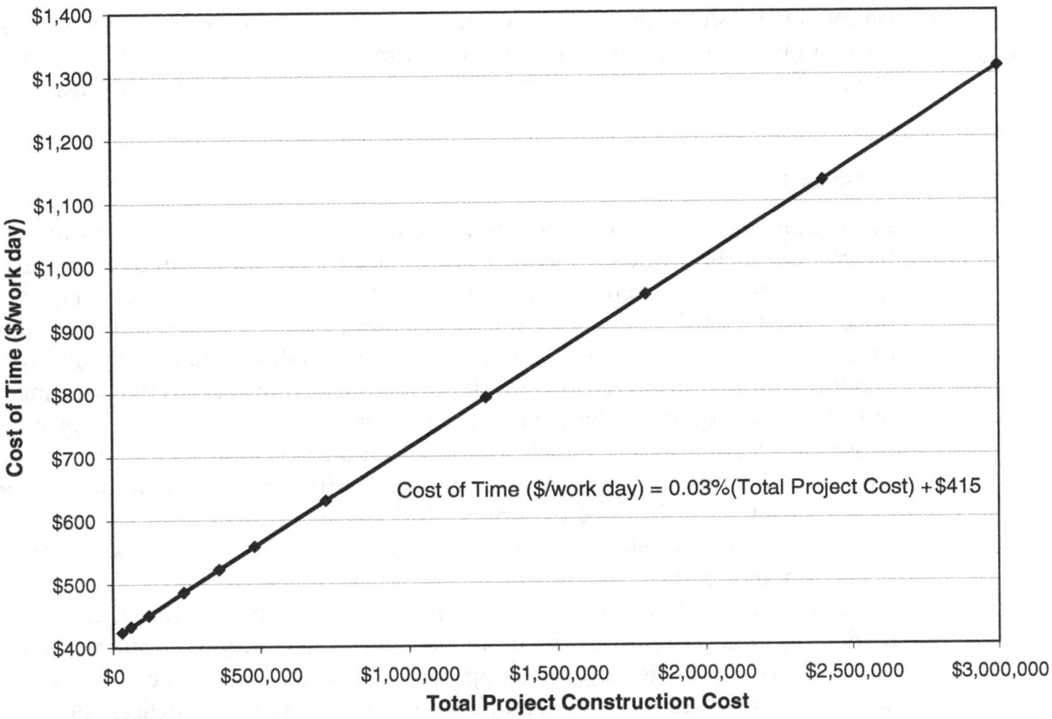

Figure 15.2 Example cost of construction time

have a maximum capacity on the order of 30 tons [270 kN]. Normal-capacity helical piles are well suited to support lighter, low-rise commercial construction. Multiple helical piles or larger-capacity helical piles may be required on projects with heavier column loads. Auger-cast piles can be made larger diameter to support very heavy loads. Auger-cast piles are usually a competitive choice for heavier, high-rise commercial construction.

This section contains several idealized examples intended to demonstrate basic foundation economics concepts. None of the examples should be considered absolute with respect to pricing. There are exceptions to every case. The conclusions presented should be considered at best rules of thumb. Different project requirements, subsurface conditions, and local economic factors can influence the choice of foundation type. The main goal of this section is simply to introduce economic considerations that can influence the selection of foundation type.

15.3 MEASUREMENT AND PAYMENT

Several types of contracts may be employed to secure the furnishing and installation of helical piles. Some of these are described in this section along with potential benefits and drawbacks. The type of contract used for securing helical pile work depends mainly on the relationships among the owner, engineer, and helical pile contractor. There are an unlimited number of methods for contracting helical pile work. This section is intended to introduce the reader to a few of the more common methods.

Lump Sum

Payment includes all labor, materials, and equipment required to furnish and install all helical piles at the locations, inclinations, and elevations shown on the plans and to the specified torque and strength. The benefit of this arrangement to the owner is that the price is fixed regardless of depth, number of helical bearing plates, or time required for installation. This can be a disadvantage to the helical pile installer if difficult installation conditions are encountered or the piles extend to greater depths than assumed in the bid. As a consequence, the helical pile installer may tend to bid the project using conservative assumptions, which may result in an overall greater cost to the owner. Changed condition disputes are more common with lump-sum contracts. In order to reduce problems if this type of contract is desired, the geotechnical engineer, helical pile designer, and installation contractor should agree on a baseline set of design and bid parameters for helical piles.

One of the unique applications of this style of payment is that it can be adapted to design/build contracts, which are becoming increasingly popular in the helical pile industry. In the design/build approach, the structural engineer provides vertical and horizontal column and shear wall load combinations. The helical pile installation contractor is required to provide design and engineering of the helical piles including connection to the concrete pile caps and grade beams. This requires close coordination between the structural engineer and the pile designer but can be beneficial to the owner in that it promotes an efficient use of helical pile spacing, sizes, and capacities. Given the number of different helical pile manufacturers, there is no one-size-fits all for helical piles. The size and strength of helical piles varies widely. A design/build approach allows each installer/manufacturer to best utilize its product selections. It removes some of the liability from the structural engineer but can increase reliability for the owner because the engineers employed by contractors and manufacturers typically specialize in the design of helical piles.

Price per Pile

Payment includes all labor, materials, and equipment required to furnish and install each helical pile at the locations, inclinations, and elevations shown on the plans and to the specified torque and strength. This type of contract is subject to the same benefits and drawbacks as the prior payment option. The only additional benefit of this type of arrangement is that the price for adding or subtracting piles is made by prior agreement. This type of contract can be useful if foundation plans are only partially complete at the bid date.

Price per Foot

Payment includes all labor, materials, and equipment required to furnish and install a unit length of helical pile at the locations, inclinations, and elevations shown on the plans and to the specified torque and strength. This approach is preferred by helical pile installation contractors due to the variability of subsurface conditions. In some cases, the owner may benefit from discounts associated with reduced pile lengths. In other instances, the opposite can be true. In order to keep the field of bids on equal merit, it is important for narrow constraints to be placed on helical pile sizing. A contractor using smaller and fewer number of helical bearing plates often will go to greater depth than another using more larger helical bearing plates. The configuration of the helical piles should be designed based on established baseline geotechnical conditions with some allowance for field modifications should differing conditions be encountered. This can be handled by specifying, in the plans and design documents, the size and number of helical bearing plates, a certain minimum cumulative helix area, or a theoretical bearing capacity.

Time and Materials

Payment shall be made on a unit rate basis at predetermined hourly rates plus the cost of materials and equipment rental plus a fee typically based on a percent of the total cost for overhead and profit. The time and materials method is useful for projects with limited subsurface information and in design/build approaches. The benefit of this arrangement is that the owner, engineer, pile designer, and installation contractor can work together as the project progresses to obtain an optimal use of time and materials. The drawback of this type of contract is that it is more difficult for the owner to estimate total project costs. Sometimes a time and materials contract is written with a guaranteed maximum cost with incentives for completion ahead of schedule and under budget.

Chapter 16

Proprietary Systems

There are more than 160 U.S. patents for different devices and methods regarding helical piles (see Appendix B). Most of these patents have long since become public domain, and anyone can manufacture and install a helical pile. The basic helical pile consisting of a steel shaft with one or more helical bearing plates is nonproprietary and can be used without fear of infringement on older patents. As time has progressed, so too have the technologies used to manufacture, design, and install helical piles. A number of novel approaches have been invented in recent times to improve helical pile penetration and performance. In addition, more and more applications for helical piles are being discovered every day.

This chapter provides a brief description of some of the more recent U.S. patents, many of which are enforced today. These inventions include unique grouting systems, anchoring devices, special helical bearing plate designs, underpinning systems, lateral load enhancement devices, and composite piles. When circumstances merit, the helical pile designer or installer may find the need to call on some of these special designs.

16.1 GROUTING SYSTEMS

As discussed in Chapter 1, T.W.H. Moseley introduced the concept of filling a helical pile shaft with grout around the turn of the 20th century. Franz Dyche showed a helical pile with openings spaced along the shaft through which grout could be pumped to fill the space between each helix. Although it may be the first helical pile with pressure grouting, the Dyche pile is unlike most modern helical piles in that it had a continuous spiral of helical bearing plates. George Ratliff was the first to patent a grouting method using a helical pile that looks more similar to those used today (Figure 16.1). Ratliff placed grout ports at the lead helix and at each coupling. The shaft was tubular and

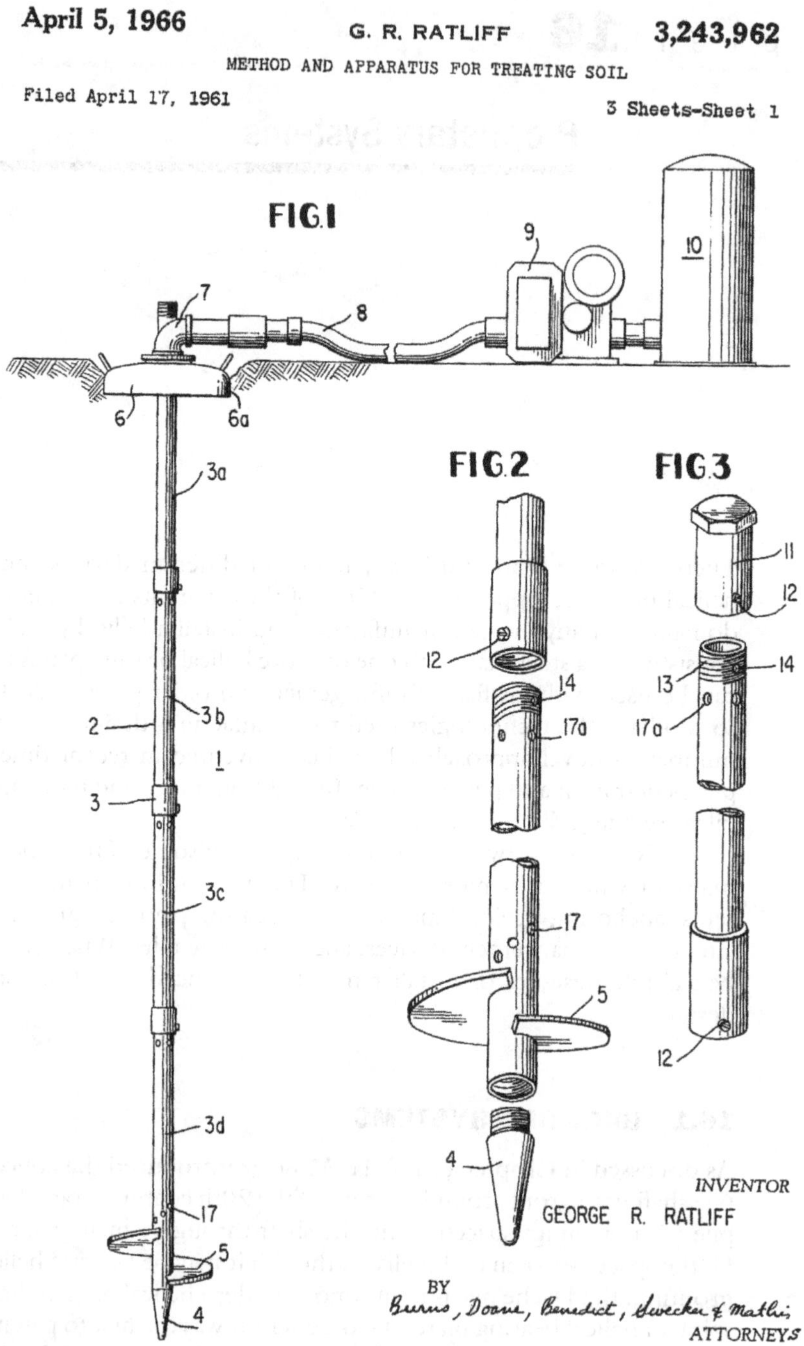

Figure 16.1 Ratliff Soil Treatment Technique

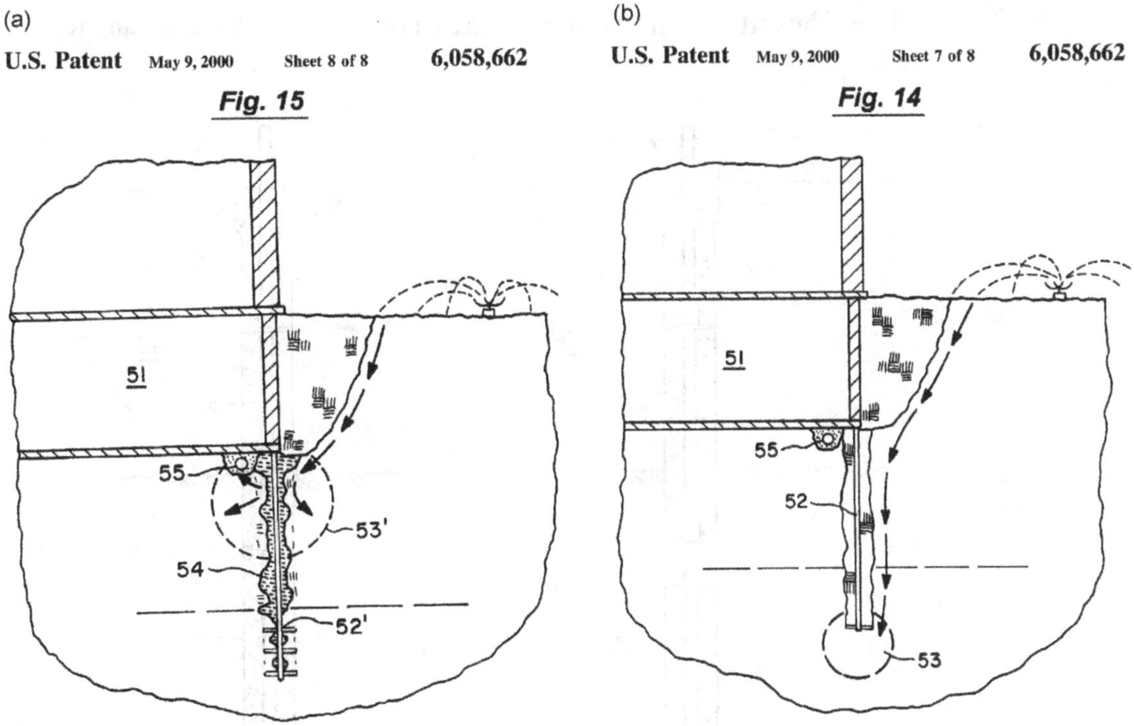

Figure 16.2 Grouting Expansive Soils

the connections were threaded and bolted. Grout is forced through the shaft with an aboveground pump and grout silo.

In 2000, Perko followed the Ratliff technique with a method patent for grouting through ports located just behind the trailing edge of the helices in areas where expansive soils are prevalent. The method is shown in Figure 16.2. The intent of the method is to seal the "tracks" of the helices as they are screwed into the soil and prevent surface water from increasing the zone of wetting and, hence, control soil expansion around a helical pile. As explained in Chapter 9 of this text, increased wetting along a helical pile shaft has not proved to be problematic in expansive soils so the Perko method has experienced limited use.

All of the preceding methods of grouting have involved an open tubular shaft. A modern method for grouting around the exterior of both solid shaft and tubular shaft helical piles is the pull-down method by Vickars, shown in Figure 16.3. In this method, circular disks are placed beneath each coupling. As the couplings are advanced into the ground, the circular disks are pulled down and open an annular space around the helical pile shaft. Grout is gravity fed into the annular space from the ground surface as the helical pile advances. This technique requires fairly soft, low-permeable soils to function properly. The grout enhances the buckling and lateral resistance of a slender shaft helical pile and can increase bearing and pull-out capacity through additional

U.S. Patent Jul. 24, 2001 Sheet 4 of 14 US 6,264,402 B1

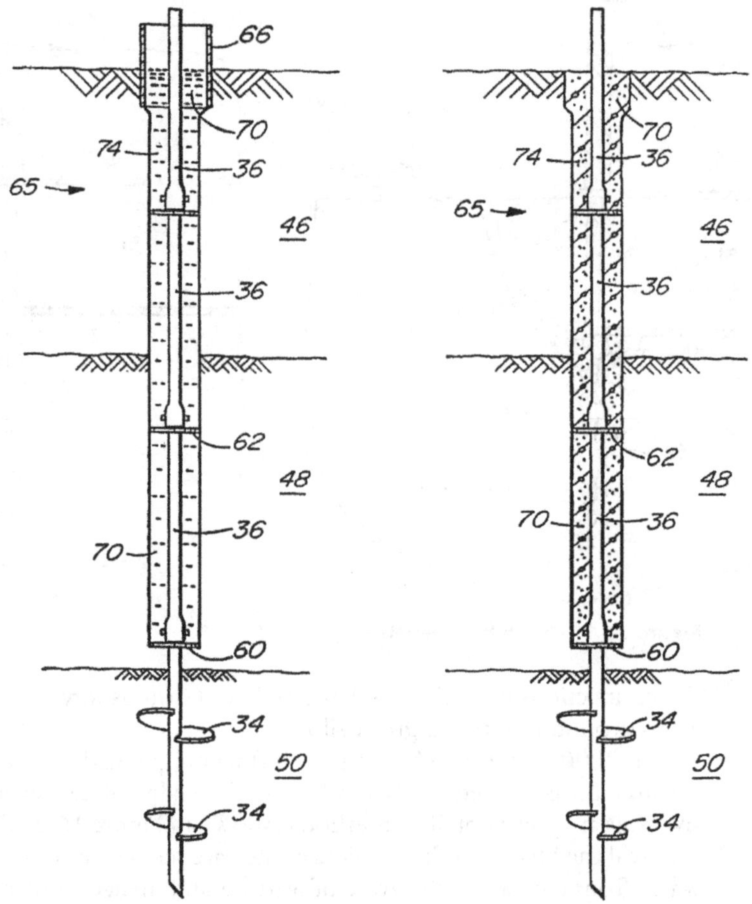

FIG. 4C FIG. 4D

Figure 16.3 Pull-down Pile by Vickars

side shear. Grouting a helical pile through ports along the shaft has been known for a long time and can be used, with few exceptions, by anyone. The pull-down method remains proprietary, and the rights are currently licensed through the A.B. Chance Company.

The capacity of a pull-down pile can be estimated following the methods presented in Chapters 4 and 5. In addition, the adhesion along the grouted shaft is added to the capacity of the pile. This method is described in detail in Clemence and Li (2008).

The capacity of pull-down piles also can be determined using HeliCap^{TM} software described in Chapter 8. Weech (2002) conducted instrumented load tests on pull-down piles and showed that the grouted shaft of a 30-foot-long helical pile in soft marine soils contributed 8 to 21 percent to the ultimate strength of the pile. Wesolek et al. (2005) conducted three tests, two on ungrouted helical piles and one on a grouted helical pile, and showed that the grouted shaft added 10 to 23 percent to the ultimate capacity of the pile. More important, deflection at the allowable load was reduced from 1.1 inches to 0.1 inch, and deflection at ultimate load was reduced from 1.5 inches or more to 0.2 inches.

16.2 GROUND ANCHORS

As discussed in Chapter 1, the first U.S. patent on helical ground anchors was filed back in 1860. Patents have been filed for helical anchors with different shaped installation tools including L-shape, S-shape, square, round, and cruciform shaft sockets. Many patents for helical anchors regard special spade shaped and corkscrew pilot points for penetrating difficult soils. There also are many patents regarding the shape of the helix and its cutting edge. Most of the patents for helical anchors are more than 25 years old and are now public domain.

One of the most recent patents still in effect for helical anchors is the patent filed by Seider and Hamilton for a helical bearing element with thickness decreasing from the shaft outward and from the leading edge backward. Seider's helical anchor is shown in Fig. 16.4. Another aspect of this anchor is that it can be installed using a removable hollow tubular shaft that fits in a socket atop the helix. An anchor rod can be threaded into the helix through the hollow shaft.

16.3 SPECIAL HELIX SHAPES

Another recent innovation in the shape of the helical bearing element is that by Slemons shown in Figure 16.5 and patented in 2006. The Slemons helix is formed by stamping structural ridges radiating outward from the central shaft. These ridges are said to increase the flexural and punching resistance of the helix, hence resulting in increased individual helix bearing capacity and lower deflection. The Slemons helix is currently licensed and sold through Cantsink, Inc., of Atlanta, GA. The helix shown in Figure 16.5 has an octagonal perimeter. The patent covers the ribbed circular helix as well.

Still another recent innovation in helix shape is the dual-cutting edge blade marketed by Magnum Piering, Inc. of Cincinnati, OH shown in Figure 16.6. The dual-cutting edge blade was developed based on claims found in the patent by Perko in 2000. The dual-cutting edge blade has a portion of the helix removed so that the leading edge is approximately two-thirds the radius of a circle inscribing the entire helix. A second cutting edge is located 180 degrees opposed from the leading edge

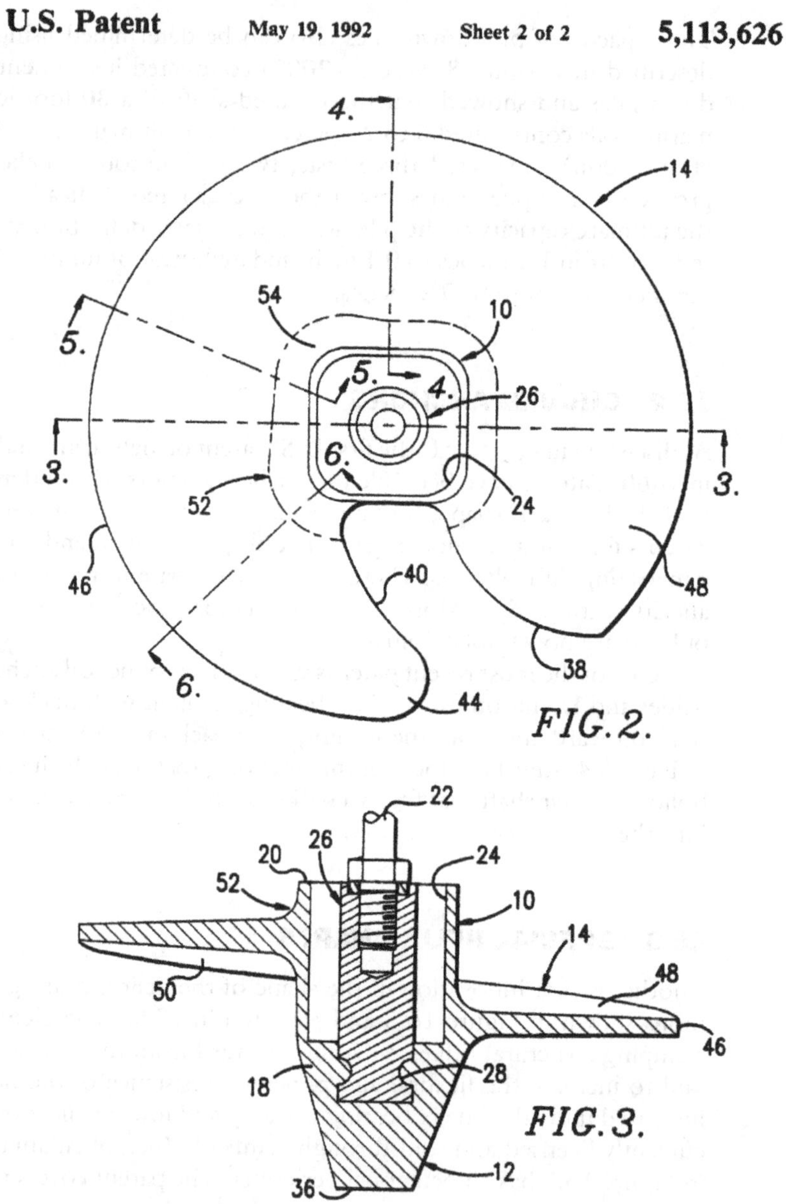

Figure 16.4 Modern Seider anchor

and extends to the full radius of the helix. In this way, cutting forces exerted on the helix during installation are balanced on both sides. The technology is similar to that of a common drill bit. All drill bits have two cutting edges located 180 degrees apart. With only one cutting edge, a drill bit will tend to wobble and walk out of the drill

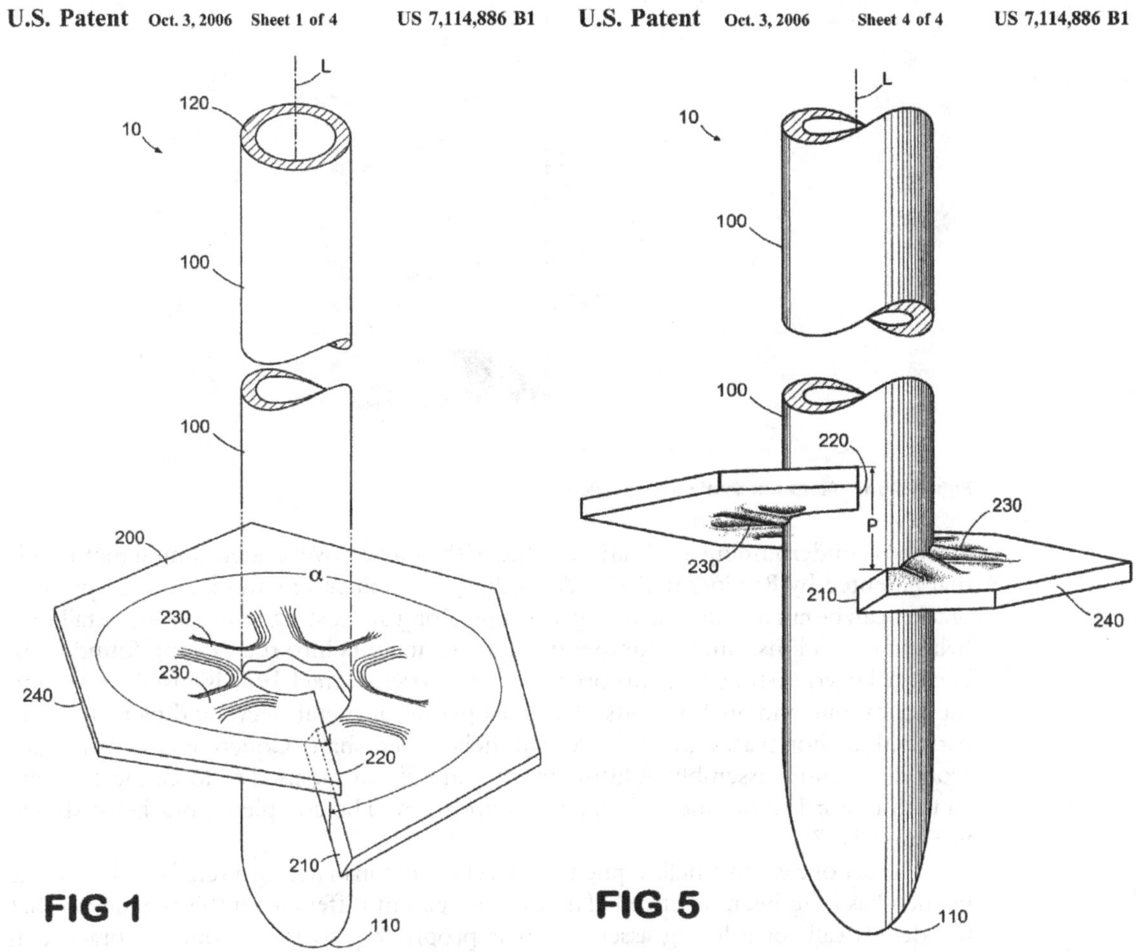

Figure 16.5 Siemons ribbed helix

hole. The dual-cutting edge blade enables the helical pile to install with less wobble and much truer, thereby increasing penetration power and lateral stability of the pile itself. If all things are the same, bearing capacity is reduced slightly by removing a portion of the helix. However, the dual-cutting edge blade will simply penetrate more deeply and bear on more dense material than an equivalent standard circular helix so that capacity of the system is unaffected.

16.4 UNDERPINNING SYSTEMS

As discussed in Chapter 1, the process of underpinning an existing foundation with helical piles is no longer proprietary. As a result, a number of new patents have been filed

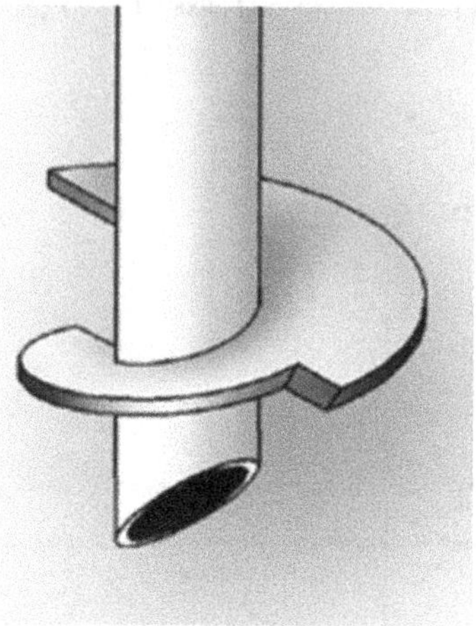

Figure 16.6 Magnum dual-cutting edge helix

for special underpinning applications. One of the more novel underpinning methods is that patented by Rupiper in 2003. According to Rupiper's method, an underpinning bracket can be created by excavating a neat pit along an existing foundation, installing a helical pile, and inserting a number of concrete anchors into the existing foundation. Next, a kit consisting of appropriate-size reinforcing steel bars is installed around the helical pile and anchor bolts. A hollow plastic or metal sleeve and pair of thread bars and anchor plates are slid over the helical pile shaft. Concrete is cast-in-place around the entire assembly. A lifting frame and jack can be used to force the concrete haunch upward to lift and stabilize the foundation. The completed bracket is shown in Figure 16.7.

The art of casting a helical pile in concrete and constructing a reinforced concrete haunch has long been practiced. The Rupiper patent differs from this practice in that the details call for a lifting assembly. It is proprietary to use a concrete bracket in substantial conformance to the one shown in Figure 16.7 to lift and level a structure.

16.5 ENHANCED LATERAL RESISTANCE

As discussed in Chapter 1, there are many methods of enhancing the lateral stability of a slender helical pile shaft in soil. Some of these involve lateral stabilizer plates fit over the top of the helical pile shaft. Of course, another way to enhance the lateral resistance of a helical pile is to make the shaft larger. Many helical pile manufacturers currently produce relatively larger-diameter helical piles for support of lightpoles, sound walls, highway signs, and billboards.

A variation on larger-shaft helical piles is the helical post base patented by Perko et al. in 2004. Shown in Figure 16.8, the post base consists of a hollow steel casing fastened around the upper region of a helical pile. The casing is fitted with a cutting tooth that reduces friction along the length of the casing during installation. A helical

U.S. Patent Jan. 7, 2003 Sheet 1 of 2 US 6,503,024 B2

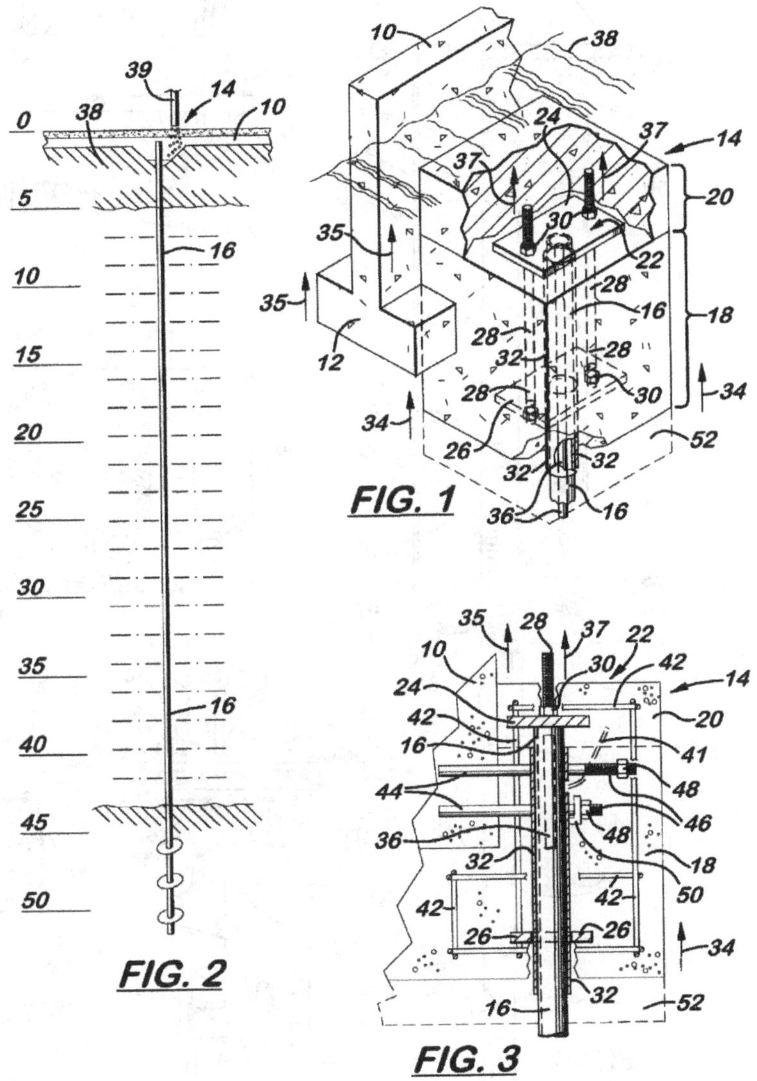

Figure 16.7 Rupiper cast-in-place underpinning bracket

U.S. Patent **Apr. 20, 2004** **US 6,722,821 B1**

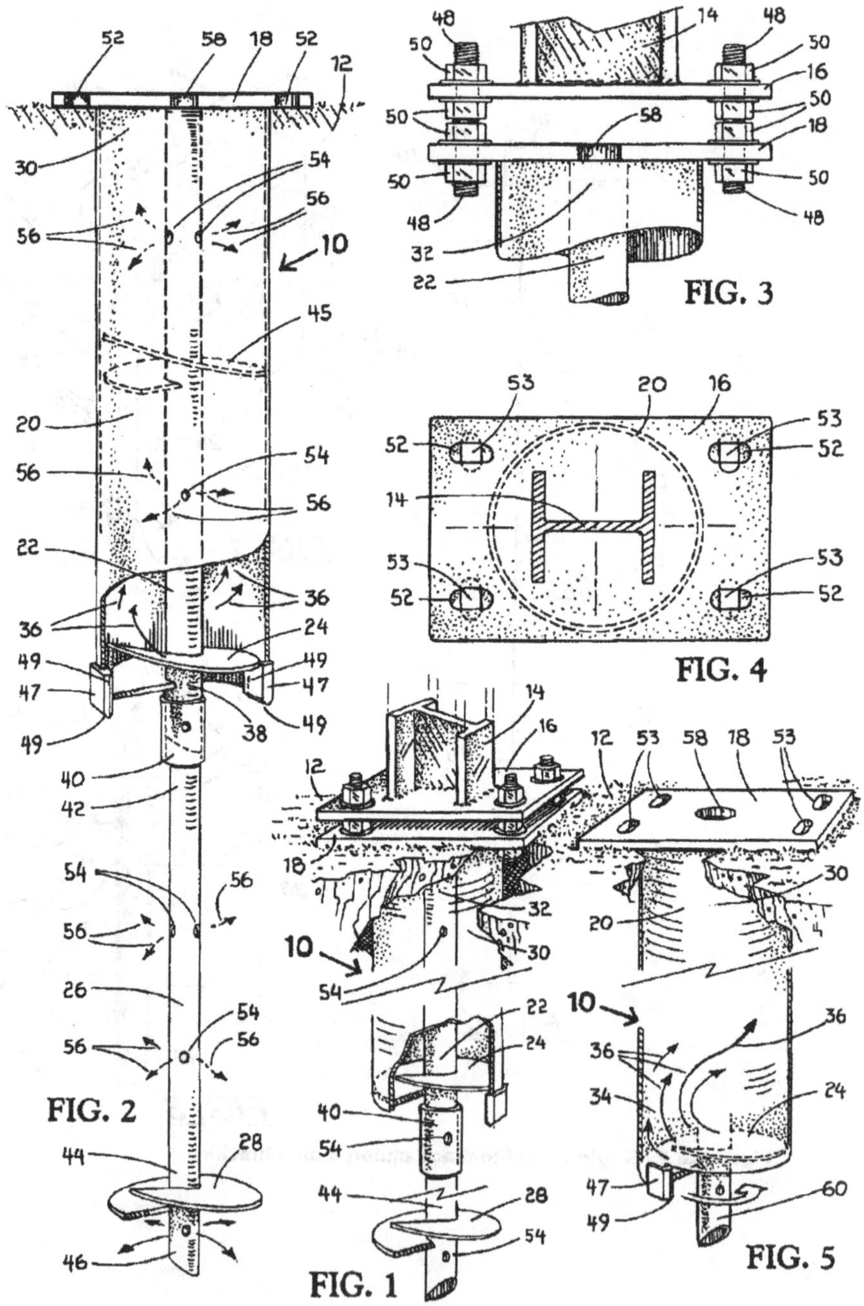

Figure 16.8 Helical post base

bearing plate is fitted to the base of the casing, and soil is forced into the device during rotation. A base plate with slotted holes and adjustable moment connection is rigidly attached to the top of the casing. The helical pile extending from the bottom of the casing can be fitted with one or more additional helical bearing plates for increased load bearing and pull-out resistance. By encouraging soil to enter the device and providing the friction reducing cutting tooth, this post base can be installed in compact soils with considerably less torque. This, in turn, allows much larger diameters. The product is currently available through Secure Piers, LLC in diameters up to 36 inches [914 mm]. It can support structures with much larger lateral loads, such as privacy walls and highway traffic noise barriers.

16.6 COMPOSITE PILES

There have been a number of recent patents regarding composite piles. The latter typically consist of lead sections with square shaft followed by extensions having larger round shafts. Some of these piles can carry larger foundation loads and have improved resistance to buckling and lateral loads. Information about two of these systems can be obtained by contacting the MacLean Dixie organization or the Chance/Hubbell group.

16.7 SPECIAL COUPLINGS

Recently, McLean Dixie has filed for patent on a square engagement coupling system for round shaft helical piles. The coupling is shown in Figure 16.9. The benefit of this

Figure 16.9 Square coupling (Courtesy of MacLean Dixie)

system is that the coupling bolts are not required to resist torque during installation, so it eliminates elongated boltholes and bolt shear during installation. According to company literature, the square engagement coupling system is easy to align which saves time and increases productivity. The two cross bolts increase rigidity. The coupling is also reported to have greater torque capacity compared to other products of the same size. (MacLean Dixie, 2009)

16.8 FUTURE DEVELOPMENTS

According to a survey of professionals in the helical pile industry (Clemence and Li, 2008), current growth areas include foundations for residential and commercial building construction, wind turbines, solar panels, transmission towers, pipelines, and other energy related construction. All respondents agreed that helical pile foundations have become more widely accepted in the past 20 years. Much of the acceptance of helical foundations has stemmed from increased quality and reliability, improved methods of design and capacity prediction, and cost benefits. Advertising, marketing, educational seminars, and technical literature by members of the industry has helped to improve product awareness.

The same survey predicted that in the next 20 years, there will be many new and innovative applications of helical foundations as well as improvements in design. Innovations in helical pile construction may include larger diameter, higher capacity, more rigid couplings, alternative corrosion protection coatings, higher-strength alloy steels, and more use of sacrificial anodes. There may be more use of small helical piles for support of minor structures (i.e., fence posts, highway signs, etc.). If historic trends are an indication of what the future will bring, there will be as many as 140 patents filed for helical piles in the next 20 years. Many of the new inventions may involve combining helical piles with grouting techniques. Improvements in design may include more reliable settlement and capacity prediction methods, widespread use of finite element modeling, adoption of performance design methods, further analysis of lateral resistance, and better understanding of seismic resistance. A flurry of new software packages is likely to be developed for design of helical piles and helical anchors. (Clemence and Li, 2008).

The number of new helical pile patents filed each year is increasing at an exponential rate. Those involved in the industry may find it challenging to keep pace with the changing technologies. By the time this book reaches the consumer, there will be several new patents and novel approaches to the design and construction of helical piles. The application of helical piles to solve problems related to complex ground conditions is limited only by the imagination of the designer.

Chapter **17**

Building Codes

This chapter discusses codes. Building codes are important for the proper design, installation, and application of helical piles in building construction. Building codes help to standardize helical pile and pile cap capacities and thereby give more confidence in capacity determinations. Some building codes also improve manufacturing quality assurance. Building codes are a system for inspection and acceptance of helical pile installations (Perko, 2008b).

17.1 IBC 2006

The *2006 International Building Code* (IBC) Section 1807.2.1 allows for special types of piles. The use of types of piles not specifically mentioned in the code is permitted, subject to approval of the building official. Approval is based on submission of test data, calculations, and other information relating to the structural properties and load capacity of such piles. Allowable stresses shall not exceed the limitations specified in the code. This provision historically formed the basis for allowing the use of helical piles in the construction of buildings under the auspices of the IBC.

Specialty piles are subject to all provisions of Section 1807 for general pile and pier foundations. Some of the criteria in this section include 1807.2.4 Pile or Pier Stability, which says that all piles and piers shall be braced for provide lateral stability. Three or more piles per cap are considered braced, provided the piles are located in radial directions from the centroid of the group at not less than 60 degrees. Piles supporting walls shall be driven at least 1 foot apart and located symmetrically under the center of gravity of the wall (except for single- or two-family dwellings less than 35 feet high).

Section 1807.2.9 Lateral Support states that any soil other fluid soil shall be deemed to afford sufficient lateral support at the top of a pier or pile to prevent buckling. Piles standing unbraced in air, water, or in fluid shall be designed as columns. Such piles supported in firm ground can be considered fixed and laterally supported at a depth of 5 feet below ground surface. In soft material, the piles may be considered

Figure 17.1 Building codes

fixed and laterally supported at a depth of 10 feet below ground surface. The 2006 IBC does not define the terms "firm" or "soft soils."

17.2 IBC 2009

In September 2008, the International Code Council (ICC) formally adopted a proposal submitted by the Deep Foundation Institute (DFI) Helical Foundations and Tie-Backs Committee to include specific provisions for helical piles in the building code. A copy of the language to be included is shown in the excerpt. Due to its historic significance to the helical pile industry, a photograph of the code hearing at which the provisions for helical piles were adopted is shown in Figure 17.2.

Figure 17.2 Presentation of helical pile codes at ICC hearing

Excerpts from IBC2009—Helical Pile Codes

1802.1 Definitions. The following words and terms shall, for the purposes of this chapter, have the meanings shown herein.

HELICAL PILE. Manufactured steel deep foundation element consisting of a central shaft and one or more helical bearing plates. A helical pile is installed by rotating it into the ground. Each helical bearing plate is formed into a screw thread with a uniform defined pitch.

Table 1810.3.2.6 Allowable Stresses for Materials Used in Deep Foundation Elements

Material Type and Condition	Maximum Allowable Stress[a]
3 Structural steel in compression	
Cores within concrete-filled pipes or tubes	$0.5\ F_y \leq 32{,}000$ psi
Pipes, tubes, or H-piles, where justified in accordance with Section 1810.3.2.8	$0.5\ F_y \leq 32{,}000$ psi
Pipes or tubes for micropiles	$0.4\ F_y \leq 32{,}000$ psi
Other pipes, tubes, or H-piles	$0.35\ F_y \leq 16{,}000$ psi
Helical piles	$0.6\ F_y \leq 0.5\ F_u$
5 Structural steel in tension	
Pipes, tubes, or H-piles, where justified in accordance with Section 1810.3.2.8	$0.5\ F_y \leq 32{,}000$ psi
Other pipes, tubes, or H-piles	$0.35\ F_y \leq 16{,}000$ psi
Helical piles	$0.6\ F_y \leq 0.5\ F_u$

1810.3.3.1.9 Helical Piles. The allowable axial design load, Pa, of helical piles shall be determined as follows:

$$Pa = 0.5\,Pu \qquad \text{(Eq: 18-4)}$$

where Pu is the least value of:

Pu = sum of the areas of the helical bearing plates times the ultimate bearing capacity of the soil or rock comprising the bearing stratum

Pu = ultimate capacity determined from well documented correlations with installation torque

Pu = ultimate capacity determined from load tests

Pu = ultimate axial capacity of pile shaft

Pu = ultimate axial capacity of pile shaft couplings

Pu = sum of the ultimate axial capacity of helical bearing plates affixed to pile

(Continued)

1810.3.1.5 Helical Piles. Helical piles shall be designed and manufactured in accordance with accepted engineering practice to resist all stresses induced by installation into the ground and service loads.

1810.3.5.3.3 Helical Piles. Dimensions of the central shaft and the number, size and thicknesses of helical bearing plates shall be sufficient to support the design loads.

1810.4.10 Helical piles. Helical piles shall be installed to specified embedment depth and torsional resistance criteria as determined by a registered design professional. The torque applied during installation shall not exceed the maximum allowable installation torque of the helical pile.

1810.4.1011 Special inspection. Special inspections in accordance with Sections 1704.8 and 1704.9 shall be provided for driven and cast-in-place deep foundation elements, respectively. Special inspections in accordance with Section 1704.10 shall be provided for helical piles.

1704.10 Helical Pile Foundations. Special inspections shall be performed continuously during installation of helical pile foundations. The information recorded shall include installation equipment used, pile dimensions, tip elevations, final depth, final installation torque, and other pertinent installation data as required by the registered design professional in responsible charge. The approved geotechnical report and the documents prepared by the registered design professional shall be used to determine compliance.

17.3 PRODUCT EVALUATION REPORTS

The International Building Code requires that all manufactured products used in construction, including helical foundations, have a current product evaluation report. ICC-Evaluation Services, Inc. (ICC-ES), the product evaluation arm of the ICC, publishes these reports. Evaluation reports contain product capacity information as well as design and installation guidelines.

There are currently over 50 helical foundation manufacturers in the world. Manufacturing quality varies considerably. Most do not have ICC-ES product evaluation reports. A small number of manufacturers have a “legacy” evaluation report. Legacy reports are those published prior to the formation of ICC-ES by old code agencies, such as the International Conference of Building Officials (ICBO) or Building Officials and Code Administrators International (BOCA). However, a standard guideline for evaluation of helical foundations did not exist when these reports were written. As a result, they contain limited information. Manufacturing quality is not calibrated to one standard, and they do not address connections to structures, soil interaction, buckling, or corrosion. Instructions for installation vary considerably.

Without a detailed product evaluation report, engineers and building officials need to review an excessive amount of information on every job to assess product acceptance on a case-by-case basis. This information may include weld inspections, mill run certificates, structural designs, material testing reports, field test reports, and manufacturing processes.

Evaluation reports also establish a quality assurance plan for the manufacture of helical piles, which promotes improved manufacturing quality assurance. An example evaluation report is contained in Figure 17.3. Evaluation reports contain a general description of the product, material specifications, and information on design and installation. The basis for evaluation reports is derived from Section 104.11 of the IBC, which reads:

> The provisions of this code are not intended to prevent the installation of any materials or to prohibit any design or method of construction not specifically prescribed by this code, provided that any such alternative has been approved. An alternative material, design or method of construction shall be approved where the building official finds that the proposed design is satisfactory and complies with the intent of the provisions of this code, and that the material, method or work offered is, for the purpose intended, at least the equivalent of that prescribed in this code in quality, strength, effectiveness, fire resistance, durability and safety.

Similar provisions are contained in the Uniform Codes, the National Codes, and the Standard Codes.

17.4 AC358 CRITERIA DEVELOPMENT

The ICC voted unanimously in June 2007 to adopt AC358 Acceptance Criteria for Helical Foundations and Devices. These new criteria establish a much-needed standard for evaluation of helical piers and foundation brackets. AC358 takes into account not only pier material strengths but also connections to structures, buckling, corrosion, and soil interaction. Ultimately, the improved evaluation reports resulting from AC358 will standardize capacities, give more confidence to engineers and building officials, and improve manufacturing quality assurance.

In order to address the lack of clear standards, the Ad Hoc Committee of Helical Foundation Manufacturers (CHFM) formed in early 2005. The committee organized under the rules of parliamentary procedure and consisted of nine independent manufacturing companies: Chance Civil Construction, Magnum Piering, Techno Metal Post, RamJack, Grip-Tite, Earth Contract Products, Cantsink, MacLean Dixie, and Fasteel. The goal of the CHFM was to propose and present universal guidelines for helical foundation product evaluation.

The CHFM met about every two months over the course of two years to put together the criteria that the ICC-ES eventually adopted. Two outside engineering

LEGACY REPORT **PFC-5551***

Reissued April 1, 2001

ICC Evaluation Service, Inc.
www.icc-es.org

Business/Regional Office ▪ 5360 Workman Mill Road, Whittier, California 90601 ▪ (562) 699-0543
Regional Office ▪ 900 Montclair Road, Suite A, Birmingham, Alabama 35213 ▪ (205) 599-9800
Regional Office ▪ 4051 West Flossmoor Road, Country Club Hills, Illinois 60478 ▪ (708) 799-2305

Legacy report on the 1997 *Uniform Building Code*™

DIVISION: 02—SITE CONSTRUCTION
Section: 02350—Piles and Caissons

DIXIE ANCHORING HELICAL FOUNDATION SYSTEM

MACLEAN DIXIE, LLC
3098 PELHAM PARKWAY
PELHAM, ALABAMA 35214

1.0 SUBJECT

Dixie Anchoring Helical Foundation System.

2.0 DESCRIPTION

2.1 General:

The Dixie Anchoring helical foundation system is used to underpin foundations of existing structures, to form deep foundations for new structures, to retrofit or remediate deficient foundations of existing structures, and to provide bearing for new foundations. The foundation anchors are used with foundation attachment brackets, designed and manufactured by others, to connect the foundation of the structure to the installed helical foundation anchor.

2.2 System Components:

2.2.1 Helical Steel Piles (Foundation Anchors): The foundation anchor lead sections consist of one or more helical-shaped circular steel plates welded to a central steel hub. The depth of the foundation anchors in the soil is typically extended by adding one or more steel hub extension shafts mechanically coupled together to form one long, continuous steel pile. Extensions can be with or without attached helical-shaped steel plates. After fabrication, the helical steel piles have a Class B1, hot-dipped, galvanized coating, complying with ASTM A 153, as applied.

Each helical steel plate is $^3/_8$ inch (10 mm) or $^1/_2$ inch (13 mm) thick and has an outer diameter ranging from 6 to 15 inches (152 to 381 mm), and an inner annulus either $1^1/_4$, $1^1/_2$ or $1^3/_4$ inches (31.7, 38 or 44.5 mm) square. Each plate is formed with all radial sections normal to the central longitudinal axis, ± 3 degrees. The helix pitch is 3 inches (76 mm).

The central steel hub of lead sections and extension sections is round cornered square (RCS), solid steel bars. RCS bars are either $1^1/_4$, $1^1/_2$ or $1^3/_4$ inches (31.7, 38 or 44.5 mm) square.

Each lead section of a foundation anchor has provisions at the top for a connection to an extension, and has an earth-penetrating pilot at the bottom. Each extension has provisions for a coupler at one end and a connection at the other. The coupler is an integrally forged socket that slips over an RCS hub of the same size. Each socket has a transverse hole so that lead sections and extensions can be connected with a bolt and nut.

For all foundation anchor lead sections and some extension shafts, helical plates are welded to their respective hubs. Nominal spacing between helical plates is not less than 2.36 times the diameter of the lower helix. Figure 1 illustrates the foundation anchor lead sections and extension shafts.

2.2.2 Foundation Attachments (Brackets): The foundation attachments (brackets) are used to address foundation settlement. The brackets are supplied by others and their design is not considered in this report. Figure 2 provides a typical illustration of the foundation attachments.

2.3 Material Specifications:

2.3.1 Helical Plates: Hot-wrought carbon steel for the helical plates conforms to ASTM A 36, with a minimum yield strength of 36,000 psi (248 MPa) and a minimum tensile strength of 58,000 psi (400 MPa).

2.3.2 Pile Hubs (Lead Sections and Extensions): The $1^1/_4$-, $1^1/_2$- and $1^3/_4$-inch (31.7, 38, and 44.5 mm) square RCS hubs are formed from hot-wrought carbon steel conforming to ASTM A 576, Grade C-1045, having minimum yield and tensile strengths of 45 and 82 ksi (310 and 565 MPa), respectively. The hubs are also available in steel conforming to ASTM A 576, Grade 1530M, having minimum yield and tensile strengths of 60 ksi and 110 ksi (414 MPa and 758 MPa), respectively.

2.3.3 Bolts: Coupling bolts are $^5/_8$ inch (15.9 mm), $^3/_4$ inch (19.1 mm) or $^7/_8$ inch (22 mm) in diameter and shall comply with ASTM A 325 Type 1, and shall have a Class C hot-dipped zinc coating that complies with ASTM A 153. The corresponding nut shall conform to either ASTM A 194 Grade 2H or ASTM A 563 Grade DH. Nuts shall have a Class C hot-dipped zinc coating that complies with ASTM A 153.

2.4 Design:

2.4.1 General: Structural calculations must be submitted to the building official for each project, and must be based on accepted engineering principles. The design of the steel components must be in accordance with the Allowable Stress

***Revised January 2006**

Page 1 of 5

Figure 17.3 Example evaluation report

Figure 17.4 Meeting of ad hoc committee of helical foundation manufacturers

consultants assisted with writing the guideline. In addition, each manufacturer involved its in-house engineers. After considerable effort, the CHFM presented a working draft of the helical foundation acceptance criteria to the ICC-ES in Whittier, CA. From there, engineers with the ICC-ES took over development. The criteria underwent much scrutiny, and eventually the ICC reviewed and adopted the document. The adopted criteria are described in the next section.

From this point forward, ICC-ES will use AC358 to prepare all new evaluation reports and for reevaluation of legacy reports. AC358 is perhaps the most comprehensive criteria ever established for a manufactured foundation system. The higher quality of helical foundation evaluation reports and the resulting increased reliability in capacity should serve as a catalyst for increasingly broad acceptance of helical foundations.

The adoption of AC358 is a significant milestone for the helical foundation industry. Having thorough, standardized evaluation reports will make it easier to specify helical foundations and may even increase their use. Engineers and building officials can expect higher quality and reliability of helical foundations by insisting on ICC-ES evaluation reports based on the new acceptance criteria.

17.5 NEW EVALUATION CRITERIA

The resulting AC358 includes new construction, foundation augmentation, slab support, and tension anchor product applications. Four primary strength components are considered, as shown in Figure 17.5. These include brackets (P1), shafts (P2), helical

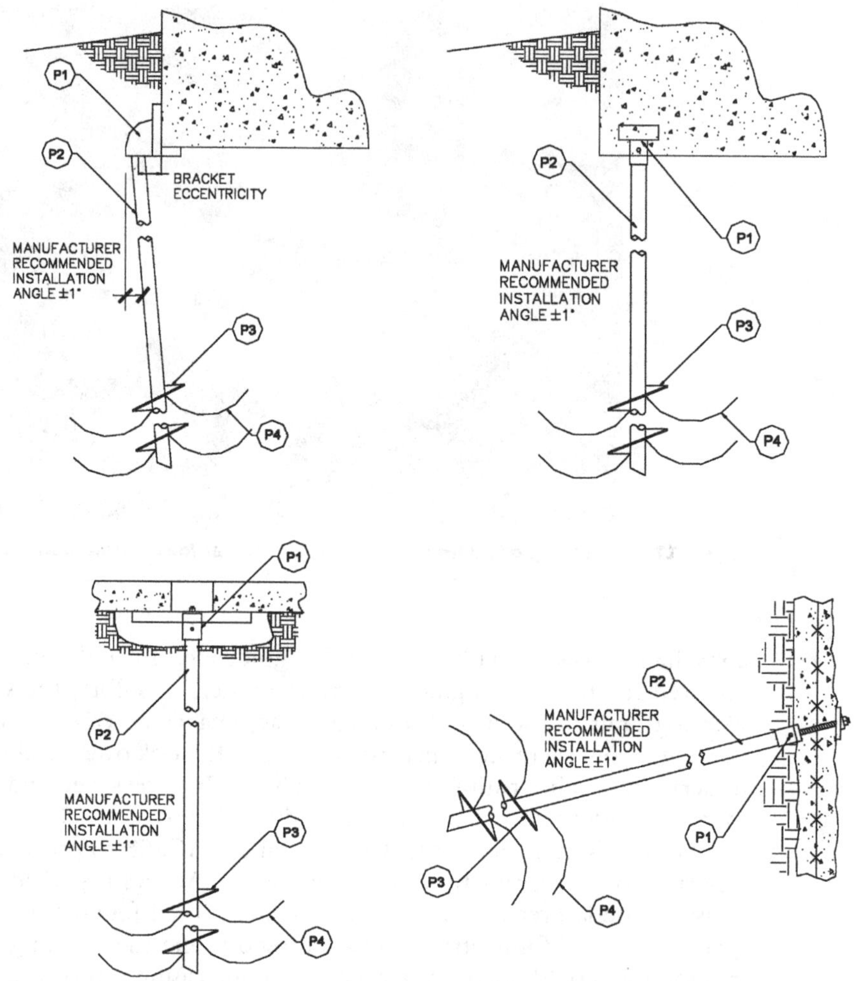

Figure 17.5 Strength components of helical piles (ICC-ES, 2007)

bearing plates (P3), and soil interaction (P4). Brackets are evaluated for their connection to structures, internal strength, and connection to pier shafts. Shaft evaluation includes tension, compression, and flexure of couplings, as well as shear and torsion. Helical bearing plates are evaluated for punching, weld shear, and torsion. Soil capacity evaluation considers compression, tension, and lateral load resistance.

Engineers preparing an evaluation report shall examine each of the four primary strength components through a combination of design calculations and product testing. All design calculations need to be traceable to the International Building Code and referenced standards. All product tests need to be conducted by an International

Accreditation Service (IAS)–accredited laboratory following a published standard. In addition, AC358 introduces a number of new tests, such as coupling rigidity, shaft and helix torsion, and helix punching shear.

One of the features of helical foundations that contributes greatly to their popularity and quality is that field capacity can be determined from final installation torque. AC358 goes beyond any previous building codes and considers the relationship between torque and capacity. The criteria delineate between conforming and nonconforming helical foundations. "Conforming" means the product generally matches the design and configuration of previously manufactured helical foundations, which were used in past research to establish the well-documented capacity-to-torque relationship. In order to be conforming, products must follow a number of rules regarding helix thickness, pitch, diameter, shape, and geometry. A list of the criteria for conforming helical piles from AC358 is given in Chapter 6. Nonconforming products need to undergo considerably more testing to establish capacity-to-torque ratios. The important point is that future evaluation reports should contain a capacity-to-torque ratio that can be used during installation to verify capacity for specific helical foundation products.

Most product evaluation criteria published by the ICC-ES are generated with the assistance of one or possibly two manufacturers. Consequently, many are heavily biased toward a particular product and not necessarily considered an industry standard. The president of the ICC-ES commented after the adoption of AC358 that due to the extensive involvement of numerous manufacturers and engineers, the criteria for helical foundations is a unique example of how the product evaluation portion of the building codes is supposed to work. AC358 is truly the state of the art with respect to helical foundation design, installation, and manufacturing.

Acceptance criteria AC358 and product evaluation reports are readily available to engineers, architects, contractors, and the general public and can be downloaded free of charge from the Web site www.icc-es.org. Acceptance criteria are issued to provide all interested parties with guidelines for demonstrating compliance with performance features of the applicable codes(s) referenced in the acceptance criteria. ICC-ES may consider alternate criteria, provided an evaluation report applicant submits valid data demonstrating that the alternate criteria are at least equivalent to the criteria set forth in AC358.

17.6 FORTHCOMING CODES

At the time of this writing, the New York City Department of Buildings (DOB) was working on a section of code dealing with helical piles. A request for proposal was sent out in March 2008. The consulting firm of Mueser Rutledge was awarded the contract. A report with sample calculations was completed in late summer. The DOB is in the process of reviewing the report and writing the code language. By the time this text is published, provisions for helical piles should be incorporated into the New York City Building Code.

The North Carolina Department of Transportation (DOT) is working on a specification for anchored temporary shoring. The specification recognizes and allows for the use of helical anchors. It requires ICC-Evaluation Report in conformance with AC358 or calculations and test information has to be submitted to DOT for approval. Design loads are kept below 60 percent of ultimate strength. Anchors have to be installed within 3 degrees of planned inclination and 12 inches of planned position. The document references AASHTO (2004) for Performance Tests.

Appendix **A**

Common Symbols and Abbreviations

Ø	Drafting symbol for diameter
#4	No 4 reinforcing steel bar (≈ 4/8 Ø)
#7	No 7 reinforcing steel bar (≈ 7/8 Ø)
9d	9 penny nails (d = penny, unit of nail size)
A36	Common ASTM steel grade (Min $f_y = 35$ ksi)
AASHTO	American Association of State Highway and Transportation Officials
AB	After boring completion
AC	Acceptance criteria
ACI	American Concrete Institute Code
ACIP	Auger cast in-place pile
ADSC	Association of Drilled Shaft Contractors
AIR	Percent air entrainment in concrete (6% Typ)
AISC	American Institute of Steel Construction Code
APPROX	Approximately
ASCE7	Code for determining loads on structures
ASD	Allowable stress design
ASTM	American Society for Testing and Material
BGS	Below ground surface
BM	Survey elevation reference (Benchmark)
BOCA	Building Officials and Code Administrators International
BOE	Bottom of excavation elevation
BOGB	Bottom of grade beam elevation
BOF	Bottom of footing/foundation elevation
BOS	Bottom of slab elevation
BRCS	Claystone bedrock

BRSS	Sandstone bedrock
BRIB	Interbedded sandstone and claystone bedrock
CADD	Computer aided design and drafting
CEC	Certified consulting engineer
CF	Cubic feet
CF16	Typically 16" wide continuous spread footing
CF24	Typically 24" wide continuous spread footing
CH	High plasticity clay
CIP	Cast-in-place
CL	Low plasticity clay
CONC	Portland cement concrete
CPT	Cone penetration test
CRS	Cold rolled steel
CY	Cubic yard (27 CF)
CT	Count
D	Diameter
DCP	Dynamic cone penetrometer
DFI	Deep Foundation Institute
DIAM	Diameter
DF#1	Douglas fir, grade 1 (select structural)
EB	Expansion bolt
ELEV	Elevation
EG	Existing grade
EST	Estimate
EW	Each way
F3	Typically 3′ × 3′ footing pad
F5.5	Typically 5.5′ × 5.5′ footing pad
FHWA	Federal Highway Administration
FF	Finish floor elevation
FG	Finish grade elevation
FT	Feet
F'c	Minimum 28-day compressive strength
Fy	Minimum yield strength
GB	Grade beam
GC	General contractor (also clayey gravel)
GM	Silty gravel
GP	Poorly graded gravel
GR	Material grade
GW	Well graded gravel
HD	Heavy duty Magnum helical pile
HF#2	Hemlock fir, grade 2 (common)
HORZ	Horizontal
HP	H-pile
HSS	High strength steel tube

ICBO	International Conference of Building Officials
ICC	International Code Council
ICF	Insulated concrete form
ID	Inside diameter
ISO9001	Standard of quality for manufacturing
IN	Inch
KIP	Kilo-pound (1,000 lb)
KN	Kilo-newton (225 lb)
KSF	Kilo-pound per square foot (1,000 psf)
KSI	Kilo-pound per square inch (1,000 psi)
LB	Pound
LF	Lineal feet
LL	Liquid limit from Atterberg test
LONG	Longitudinal
LVL	Laminated veneer lumber
MAX	Maximum
M	Meter (3.2808 ft)
MF	Matt foundation
MH	High plasticity silt
MIL	Thousandth of an inch (0.001 in)
MIN	Minimum
ML	Low plasticity silt
MM	Millimeter (0.0394 in.)
N	SPT blow count (blows/ft.)
NEMA	National Electrical Manufacturers Association
NIL	Negligible
OC	On-center spacing
OD	Outside diameter
OSHA	Occupational Safety and Health Administration
PCC	Portland cement concrete
PE	Professional engineer
PI	Plasticity index (LL-PL)
PLF	Quantity per lineal foot
PL	Plastic limit from Atterberg test
PLS	Professional land surveyor
PSF	Pounds per square foot
PSI	Pounds per square inch
QTY	Quantity
R	Radius
RCP	Reinforced concrete pipe
RCS	Round-corner square solid bar
SBCCI	Southern Building Code Congress International
SC	Clayey sand

SCH	Schedule
SD	Standard duty Magnum helical pile
SF	Square feet
SM	Silty sand
SP	Poorly graded sand
SPT	Standard penetration test
SS150	1.5" × 1.5" solid square shaft Chance helical pile
SS250	2.5" × 2.5" solid square shaft Chance helical pile
STA	Position station (example STA 3 + 50 = 350 ft.)
SW	Percent soil swell when wet (also southwest and well graded sand)
T	English ton (2,000 lb)
T/B	Top and bottom
TJI	I-joist (by Truss Joist Inc.)
TOF	Top of foundation elevation
TOGB	Top of grade beam elevation
TOS	Top of slab elevation
TRANS	Transverse
TSF	Tons per square foot
TYP	Typical
UC	Unconfined compressive strength
UNO	Unless noted otherwise
VAR	Variable
VERIFY	Check in field
VERT	Vertical
WD	While drilling
WOR	Weight of rod in SPT test
WOH	Weight of hammer in SPT test
W10 × 50	Wide flange steel beam (10" depth × 50 plf)

Appendix **B**

Summary of Prior Art

Patent No.	Date	Inventor	Assignee	Place	Title	Claims Unique
30,175	1860	Ballard		Iowa	Earth-Borer for Post-Holes	Helical set of plates mounted to solid round shaft
101,379	1870	Parker		Massachusetts	Improvement in Screw-Piling	Helical pipe pile with inner and split outer helix with different pitch
108,814	1870	Moseley		Massachusetts	Metal Screw Piles	Hollow metal helical pile, can have grout, sub-pile and/or pilot bit
160,648	1875	Clarke		Iowa	Imporvement in Earth-Augers	Earth anchor with detacheable pipe shaft
172,917	1876	Durrin		Wisconsin	Improvement in Picket-Stakes	Earth anchor with rod, screw flange, and guy ring
200,217	1878	Mudgett		Iowa	Improvement in Fence Posts	Angle iron fence post with screw-type pilot point
269,548	1882	Stephenson		Ontario	Post-Hole Auger	Earth anchor with pipe shaft and unique shaped helix
303,263	1884	Burton	Screw Post Co.	New York	Fence-Post	T-Shape fence post with screw-type pilot point
314,870	1885	Nendel & Knowlton		Iowa	Metalic Fence Post	Round fence post with conical screw-type pilot point

(*Continued*)

Patent No.	Date	Inventor	Assignee	Place	Title	Claims Unique
315,593	1885	Boehmke & Bohlken		Ohio	Ground Auger Anchor	Gimlet-pointed screw with spiral blade and swivel guy loop
414,700	1889	Gray		Illinois	Pile	Threaded adjustable helix and riveted angle iron shaft
495,471	1893	Brown	WK Clyne	Ohio	Fence-Post	Angle iron fence post with screw-type pilot point
505,811	1893	Brown		Ohio	Fence Post	Metal angle iron fence post with lower helical portion
575,374	1897	Reck		Ohio	Fence Post	Formed sheet steel fence post with twisted helical portion
598,003	1898	Oliver		Illinois	Fence Post	Pipe shape fence post with screw pilot and X-shape lateral stabilizer
624,724	1899	Alter		Indiana	Fence Post	X-shape fence post with screw pilot and cylindrical lateral stabilizer
795,405	1905	Moore		California	Fence Post	L-shape fence post with screw-type pilot point and wire support tabs
816,631	1906	Widmer		Ohio	Guy-Anchor	Guy anchor installed with removable wrench
818,061	1906	Toy, et al.		Ohio	Land-Anchor	Guy anchor with S-shaped socket
832,565	1906	Wilson		Nebraska	Portable Fence Post	Pipe shape fence post with wire screw pilot point
839,822	1907	Dunnington		Kansas	Tent-Stake	Helical tent stake with spring shaft
888,917	1908	Lucas		Ohio	Ground-Anchor	Guy anchor with T-shaped wrench
1,030,411	1912	Larsen		Wisconsin	Metal Fence Post	Multi-piece rivited fence post with twisted helical portion
1,049,549	1913	Tesreau & Sundman		Missouri	Metal Fence Post	Cast cylindrical fence post with cast screw-type pilot point
1,109,020	1914	Skiff & Westphall		Iowa	Fence Post	Improved metal fence post with half-moon helical plates
1,193,725	1916	Smith		Missouri	Guy Anchor	Guy anchor with tubular wrench
1,455,163	1922	Blackburn		Missouri	Expanding Screw Anchor	Three part helix that expands about square hub
1,791,185	1931	Birkenmaier	WN Matthews	Missouri	Anchor	Hand installed guy anchor
1,791,368	1931	Mullet		Ohio	Anchor Post	Pipe post with bow-tie helical plates
1,797,637	1931	Chance		Missouri	Screw Anchor	Guy anchor with cruciform socket

1,800,504	1929	Chance		Missouri	Earth-Anchor Tool	Anchor installation wrench with cammed engagement
1,849,268	1932	Birkenmaier	WN Matthews	Missouri	Anchor	Guy anchor with square wrench
1,883,477	1932	Bash		Missouri	Guy Anchor	Guy anchor with pyramidal pilot point
1,894,401	1933	Hollos		Ohio	Anchor Post	Post with spiral auger and extendable scissor-like anchor plates
1,940,938	1933	Chance		Missouri	Earth Anchor	Guy anchor where guy rod is used as wrench
2,063,052	1936	Robins		Pennsylvania	Self-Securing and Propelling Anchor	Double threaded screw anchor with spiral pilot point
2,234,907	1941	Williams	Malleable Fittings	Florida	Earth Anchor	Guy anchor with tapered helix
2,569,528	1951	Kandle		Chicago	Screw Anchor Adapter	Guy anchor with laterally confined eyelet
2,603,319	1949	Dyche		Texas	Ground Anchor	Anchor with tapered spiral auger with lubrication ports
2,643,843	1953	Brown		New York	Sand Anchoring Device	Helical pile beach umbrella support
2,864,633	1958	Mackie		Missouri	Methods and Apparatus for Anchoring Pipe Lines. . .	Method of using helical anchors for pipe lines
2,999,572	1961	Hinckley		Louisiana	Earth Anchor	Earth anchor with hinged helical flanges, conical tip, and driven by fluid
3,011,597	1961	Galloway & Galloway		Indiana	Supporting Post	Helical pile with three triangular wings for lateral loads
3,016,117	1962	Peterson		California	Earth Anchor	Helical pile with curved leading edge, dual spade pilot point
3,148,510	1964	Sullivan	Chance	South Carolina	Method of Installing Earth Anchors	Method of installing helical anchors using reciprocable dogs
3,148,739	1964	Mattingly, et al.	Walter Happe	Missouri	Boring Apparatus with Screw Anchor	Screw anchor installation adaptor for boring machine
3,243,962	1966	Ratliff		Mississippi	Method and Apparatus for Treating Soil	Slurry injected thru helical shaft conduit with threaded couplings
3,293,809	1966	Daline		Minnesota	Easily Installable Post for Fences, Docks and the Like	Hollow tubular fence post with tapered screw-type lead
3,295,274	1967	Fulton	De Kalb Toys	Illinois	Combination of Pole Anchor with a Gym Set. . .	Pole with helix and link chain
3,318,058	1967	Sullivan	Chance	South Carolina	Apparatus for installing Earth Anchors	Drive tool with outer housing, guide, and reciprocable dogs

(*Continued*)

Patent No.	Date	Inventor	Assignee	Place	Title	Claims Unique
3,377,077	1968	Hollander & Lewis	Chance	Missouri	Power Installed Screw Anchor Wrench	Drive tool with positionable dogs
3,427,812	1969	Hollander	Chance	Missouri	Method and Apparatus for Anchoring Offshore Pipelines	Underwater pipe anchoring device with oppositely rotating anchors
3,525,225	1968	Yager, et al.	Chance	Missouri	Method of Installing Earth Anchors	Method of helical anchor installation using a shear pin device
3,645,055	1972	Roza	Joslyn	Illinois	Screw Anchor	Guy anchor with spiral cutting edge
3,662,436	1972	Roza	Joslyn	Illinois	Screw Anchor	Guy anchor with multiple helix and v-notch neck in shaft
3,688,454	1972	Wofcarius		Belgium	Landmark	Helical survey marker
3,710,523	1973	Taylor		Texas	Earth Anchor	Guy anchor with convoluted helix and auger screw lead
3,793,786	1974	Jahnke	Joslyn	Illinois	Screw Anchor	Guy anchor with crowned helix and pointed spiral lower end
3,810,364	1974	Johnson		British Columbia	Ground Anchor	Pipeline support coupling, jagged helix, predrill with freezing fluid
3,828,562	1974	Petres	Joslyn	Illinois	Method and Apparatus for Installing Anchors	Method of installing anchor with mechanical crowd device
3,830,315	1974	Love	Wiley	Oklahoma	Apparatus for Implacement of Subterranean Screw Anchors	Drive tool with rotary kelly drive element
3,832,860	1974	Jahnke	Joslyn	Illinois	Method and Apparatus for Installing Anchors	Drive tool with spring actuated crowd system
3,841,032	1974	Grannis	Chance	Missouri	Article and Screw Anchor-Supported, Load-Bearing Pad...	Helical pile supported fiberglass enclosure for utilities
3,871,142	1975	Abbott	Windtie	Indiana	Device for Holding Down Mobile Homes	Steel strap attached to small helical anchor
3,903,626	1975	Ford		Ohio	Earth Anchors	Garden plant marking device
3,952,523	1976	Gale		Texas	Method of Installing a Screw-Type Anchor	Method of installing helical anchors using removable outer mandrel
4,068,445	1978	Bobbitt	Chance	Missouri	Lightweight, screw anchor supported foundation...	Foam filled conical housing attached to multiple helical piles
4,280,768	1981	Pardue, et al.	Dixie	Tennessee	Anchor Drive Coupler	Torque limiting friction plate drive coupler
4,290,245	1981	Pardue, et al.	Dixie	Tennessee	Earth Anchor	Helix with lagging leading edge, end cap with re-bar, dowel coupling

4,316,350	1982	Watson		Texas	Wing Screw Earth Anchor	Cast wing shaped helical blade, 4" pitch
4,334,392	1982	Dziedzic	Chance	Missouri	Modular Screw Anchor Having Lead Point	Increased hub diameter at blade for longer welds
4,389,034	1983	Suttles	Anchoring Intnl.	Texas	Underwater Pipe Anchoring Device	Helical anchor pipeline support that prevents metal to metal contact
4,405,262	1983	Nagashima		Japan	Method for Erection of a Temporary Bridge, and a Pile Means...	Continuous flight metal screw pile
4,467,575	1984	Dziedzic	Chance	Missouri	Internally Driven Earth Anchor Having Small Diameter...	Modular anchor with removable square driving tube
4,471,588	1984	Schirm		Deleware	Tie Rod	Helical support for pallets
4,492,493	1985	Webb		Oklahoma	Pipeline Anchor Hook	Helical anchor with hook-like cap for restraining pipelines
4,499,698	1985	Hoyt, Bobbitt, et al.	Chance	Missouri	Method and Apparatus for Anchoring Retaining Walls...	Threaded helical anchor cap, dog-type coupling system
4,533,279	1985	van den Elzen, et al.	Fundeman-tum	Netherlands	Method for Making a Foundation Pile	Grout behind helical lead section with smaller diameter shaft
4,561,231	1985	Hoyt, et al.	Chance	Missouri	Tower Foundation Anchor Interconnection	Tower leg is inserted into hollow tubular shaft of helical pile
4,580,795	1986	Burtelson & Riley	Joslyn	Illinois	Apparatus for Installing Anchors	Socket drive with sliding lock pins
4,598,511	1986	Severs & Leonard	Cooper Industries	Pennsylvania	Earth Anchor	Ovular helix that is offset from hub axis
RE32076	1986	Dziedzic	Chance	Missouri	Modular Screw Anchor Having Lead Point Non-Integral with Helix...	Earth anchor with seperable pilot point
4,650,372	1987	Gorrell	Dow Chemical	Texas	Drive Screw Pile	Driven screw pile with steeply pitched helices
4,659,259	1987	Reed & Gibson	Chevron	California	Method and Device for Mixing Stabilizing Chemicals...	Helical lead with chemical ports through top and sides of cast helix
4,707,964	1987	Hoyt, et al.	Chance	Missouri	Method of Providing Support for an Elongated Tower Leg	Helical pile with solid lead and groutable hollow extension
4,742,656	1988	Farmer	Dixie	Tennessee	Earth Anchor with Multi-Sided Blade	Lineal and arcuate cutting edges, generally octagonal shaped helix
4,756,129	1988	Webb		Oklahoma	Ground Anchor and Apparatus to Set and Remove Same	Helix with bent lead edge, external drive tool pushes on back of helix

(*Continued*)

Patent No.	Date	Inventor	Assignee	Place	Title	Claims Unique
4,833,846	1989	McFeetors & Wilson		Manitoba	Ground Anchor System for Supporting an Above Ground Structure	Support anchor with three wings hammered down for lateral loads
4,923,165	1990	Cockman		South Carolina	Stabilized Post Anchor	Helical pile with bracket for wood post
4,979,341	1990	Norman, et al.	Dixie	Tennessee	Integral Earth Anchor	Rectangular hub, spade point, square blade with rounded corners
4,981,000	1991	Hamilton & Odom	Chance	Missouri	Penetration of Power Installed Anchor	Earth anchor with flat pilot point that directs soil over helix
4,996,806	1991	Platz		Canada	Lead Point for Helical Earth Anchor	Removable hollow drive tool extends over anchor rod
5,011,336	1991	Hamilton, et al.	Chance	Missouri	Underpinning Anchor System	Helical pile with underpinning bracket
5,066,168	1991	Holdeman	Chance	Missouri	Cylindrical Foundation Support. . .with Removable Helix	Large diameter casing inserted with removable helical pile
5,113,626	1992	Seider and Hamilton	Chance	Missouri	Earth Anchor Apparatus Having Improved Load Bearing Element	Thickness of helix decreases out from hub and from lead edge back
5,120,163	1992	Holdeman, et al.	Chance	Missouri	Foundation Underpinning Bracket and Jacking Tool Assembly	Helical pile underpinning bracket with T-pipe
5,139,235	1992	Kilmer		Kansas	Corner Fence Post System	Helical pile fence post base will two battered helical pile braces
5,139,368	1992	Hamilton, Hoyt, et al.	Chance	Missouri	Method of Underpinning Existing Structures	Method of underpinning with helical piles
5,171,107	1992	Hamilton, Hoyt, et al.	Chance	Missouri	Method of Underpinning Existing Structures	Method of underpinning and lifting with helical piles
5,213,448	1993	Seider, et al.	Chance	Missouri	Underpinning Bracket for Uplift and Settlement Loading	Helical pile underpinning bracket with uplift resistance
5,224,310	1993	Edwards & Holdeman	Chance	Missouri	Hand-Installed Landscape Foundation	Helical pile support for yard light with hole for utility line
5,286,142	1994	Hoyt & Hamilton	Chance	Missouri	Reduced Moment Anchor Hub	Earth anchor with pivoting hub
5,295,766	1994	Tikkainen		Finland	Apparatus and Method for Building a Foundation for Uprights. . .	Helical pile with solid leand and tapered tubular extension

5,408,788	1995	Hamilton & Seider	Hubbell	Missouri	Hollow Hub Helical Earth Anchor with. . .Spade/Pilot Point	Earth anchor with pilot tip having diametrically opposed cutting teeth
5,482,407	1996	Raaf	Atlas	Missouri	Helical Outrigger Assembly. . . for an Underpinning Drive. . .	Pair of helical piles used as a reaction for hydraulically driven pier
5,575,122	1996	Hamilton, Hoyt, et al.		Missouri	Earth Screw Anchor Assembly Having. . .Penetrating Capability	Singe blade with spade shaped pointed end
5,575,593	1996	Raaf	Atlas	Missouri	Method. . .for Installing a Helical Pier with Pressurized Grouting	Tubular helical pile with grout holes arranged along shaft
5,607,261	1997	Odum, Hoyt, et al.	Hubbell	Missouri	Clamshell Power Installed Anchor Screw	Earth anchor with C-shaped hub socket
5,683,207	1997	Mauer		Maryland	Pier Assembly and Method for Installing Same	Helical anchor with conical end cap that is drawn into ground
5,707,180	1998	Vickars, et al.	Vickars Develop.	California	Method and Apparatus for Forming Piles In-Situ	Method of forming cast concrete pile simultaneously with helical pile
5,791,820	1998	Rempel		Alberta	Method and Apparatus for Implanting Screw-In Pilings. . .	Method for determining crowd and capacity of lightly loaded helical pile
5,800,094	1998	Jones		Colorado	Apparatus for Lifting and Supporting Structures	Helical pile underpinning bracket with two hydraulic jacks
5,833,399	1998	Bullivant	Global Innova-tions	Great Britain	Apparatus for Use in Forming Piles	Install lead helical pile, extend and grout with removeable larger shaft
5,904,447	1999	Sutton, Rupiper, et al.	IST	Clorado	Drive Device Used for Soil Stabilization	Helical pile with grout ports and helix at each coupling
5,919,005	1999	Rupiper	IST	California	Ground Anchor Device for Penetrating. . . Rock Formation	Hollow helical pile with down hole rock hammer for drilling penetration
5,934,836	1999	Rupiper & Ludwig	IST	California	Ground Anchor Device	Thin wall helical pile driven with removeable internal kelly bar
5,980,162	1999	McCrown		Kansas	Seismic Shock Absorbing Pier	Helical pile underpinning bracket with coil spring and I-beam haunch
6,032,880	2000	Verrills & Verrills		Great Britain	Ground Spike for a Sun Umbrella	Hollow post with spiral flange
6,050,740	2000	Dziedzic	Dixie	Tennessee	Combined Lockdog and Kelly Bar Adapter	Torque sensitive spring actuated lock for driving helical piles

(*Continued*)

Patent No.	Date	Inventor	Assignee	Place	Title	Claims Unique
6,058,662	2000	Perko	Magnum Piering	Colorado	Earth Anchors and Methods for Their Use	Moment balanced helix, other unique helix geometries
6,066,015	2000	Brown		Alabama	Method and System for Anchoring a Buoy. . .	Earth anchor attached to buoy and installed with quick disconnect drill
6,079,905	2000	Ruiz, et al.	Ruiz, LLC	Missouri	Bracket Assembly for Lifting and Supporting a Foundation	Helical pile underpinning bracket with adjustable lifting cross beam
6,094,873	2000	Hoffman, et al.		Indiana	Foundation for Manufactured Homes	Support beams on helical piles with concrete pad arround upper shaft
6,128,867	2000	MacKarvich		Georgia	Ground Anchor with Stabilizer Cap	Helical earth anchor with short cylindrical cap forced into soil
6,142,710	2000	Holland & Holland		North Carolina	Apparatus and Method for Raising a Foundation	Underpinning bracket with side-to-side and front-to-back pivots
6,142,712	2000	White, et al.		Alberta	Hollow Screw-In Pile	Hollow pipe pile installed with removeable auger
6,193,443	2001	Trudeau, et al.		California	Anode Installation Apparatus and Method	Small helical pile used as an anode driven with larger helical casing
6,202,368	2001	Wallace		North Carolina	Earth Anchoring System	Helical pile with tubular post slid over shaft extending from ground
6,264,402	2001	Vickars, et al.	Vickars Develop.	California	Method and Apparatus for Forming Piles in Place	Method of forming telescoping concrete pile with helical pile
6,272,798	2001	Cockman		South Carolina	Anchor with Pivotal Attachment	Helical earth anchor with cylindrical cap and pivot plate for guy
6,352,390	2002	Jones		Colorado	Apparatus for Lifting and Supporting a Foundation under Tension. . .	Underpinning bracket with diagonal straps extending up foundation
6,352,391	2002	Jones		Colorado	Piering Device Having a Threaded Shaft and Helical Plate	Helical pile with all-thread shaft
6,394,704	2002	Sacki, et al.	Nippon Steel	Japan	Screwed Steel Pile and Method of Construction Management. . .	Helical pile with split helix at leading end of pipe shaft
6,412,235	2002	Pylant		Texas	Removable Screw-Type, In-Ground Anchor Device	Conical lead with helical flights and conical tip

6,412,236	2002	Johnson		Texas	Easily Installable Fence Post	T-Shape fence post with screw-type pilot point
6,435,776	2002	Vickars, et al.	Vickars Develop.	California	Method and Apparatus for Forming Piles in Place	Method of forming concrete pile with helical pile and grout thru shaft
6,503,024	2003	Rupiper		California	Concrete Foundation Pierhead and Method of Lifting a Foundation. . .	Reinforced, cast-in-place concrete underpinning bracket
6,539,685	2003	Bell & Rietveld		Illinois	Apparatus and Method for Lifting Sunken Foundations	Gusseted plate on concentric tubular sleeve with clamp to foundation
6,615,554	2003	Rupiper		California	Helice Pier Coupling System Used for Soil Stabilization	Helical pile coupling system with diagonal pipe joint
6,641,332	2003	Alvarado	Appalachian Systems	North Carolina	Foundation Support and Process for Structures	Helical pile with elongated pilot forced into ground as a guide for pile
6,652,195	2003	Vickars, et al.	Vickars Develop.	California	Method and Apparatus for Forming Piles in Place	Method of forming concrete pile with helical pile and remove shaft
6,659,692	2003	May		Arizona	Apparatus and Method for Supporting a Structure. . .	Underpinning bracket with long sleeve forced over helical pile
6,682,267	2004	Jones		Colorado	Piering Device with Adjustable Helical Plate	Helical pile with helix adjustably attached to shaft with key and notch
6,702,239	2004	Boucher		California	Apparatus and Method for Supporting the Trunk of a Tree	Vertical helical anchor tree suppport
6,722,821	2004	Perko & Rupiper	Secure Piers	Colorado	Helice Pier Post and Method of Installation	Large diameter casing with helix and cutting tooth at leading end
6,814,524	2004	Peterson		Florida	Method and Apparatus for Lifting and Stabilizing Subsided. . .	Helical pile bracket that assembles under slabs
6,814,525	2004	Whitsett		Louisiana	Piling Apparatus and Method of Installation	Helical lead with grouted hollow cylindrical upper sections
6,817,810	2004	Jones		Colorado	Piering Device with Adjustable Helical Plate	Helical pile with threaded square shaft and adjustable helix
6,840,714	2004	Vache		Texas	Foundation Repair Bracket	Three-piece bracket with tubular sleeve and I-beam haunch
6,872,031	2005	May		Arizona	Apparatus and Method of Supporting a Structure with a Pier	Underpinning bracket with rotatable shelf

(Continued)

Patent No.	Date	Inventor	Assignee	Place	Title	Claims Unique
6,986,495	2006	Pinkleton & Showler		Michigan	Walkway Bracket for use with Helical Anchor	Vertical support bracket with clip for tension anchor
7,004,683	2006	Rupiper		California	Helice Pierhead Mounting Plate and Bolt Assembly	Pair of circular plates separated by thread bars for moment connection
7,018,139	2006	Slemons	Cantsink	Georgia	Structural Helical Plate	Helix with structural ribs
7,037,045	2006	Jones		Colorado	Modular Tubular Helical Piering System	Rectangular tubular hub with helix attached to square sleeve
7,040,842	2006	Stotzer	Schroben-hausen	Denmark	Method and Device for Making a Foundation Member	Uses helical pile to mix soil and grout, grouts through central shaft
7,044,686	2006	May		Arizona	Apparatus and Method for Supporting a Structure with a Pier	Underpinning bracket with vertically adjustable rotatable shelf
7,090,437	2006	Pinkleton		Michigan	Modular Helical Anchor	Square shaft with helix attached to square tubular sleeves
7,094,003	2006	Faires, et al.	Dixie	Tennessee	Bracket Assembly for Lifting and Supporting a Foundation	Underpinning bracket with bolt that adjusts for pile angle
7,112,012	2006	Whitsett		Louisiana	Piling Apparatus and Method of Installation	Helical lead with grouted swedged cylindrical upper sections
7,114,886	2006	May	Cantsink	Georgia	Structural Helical Pile	Helix with structural ribs that do not extend to edge
7,163,357	2007	Peterson		Florida	Method and Apparatus for Lifting and Stabilizing Subsiding…	Helical pile bracket that assembles under slabs
7,195,426	2007	May		Arizona	Structural Pier and Method for Installing Same	Underpinning bracket with adjustable threaded diagonal braces
7,220,081	2007	Gantt		South Carolina	Concentric Load Bearing Piping with Liner for Foundation Anchor	Helical pile with internal and external overlapping sleeves
7,338,232	2008	Nasr		Alberta	Method for Installing a Screw Pile	Helical pile with internal gorut pipes for high pressure grouting

Appendix C

Load Test Results

List of Site References

a. Norman, OK, Victor and Cerato (2008)
b. Norman, OK, Cerato (Pending)
c. Adams and Klym (1972)
d. Mooney et al. (1985)
e. lab compacted soils, Narasimha Rao et al. (1991)
f. Windsor, CO, CTL|Thompson
g. West Chester, OH, Magnum Piering, Inc.
h. Colorado State University, CTL|Thompson
i. lab compacted soils, Ghaly, et al. (1991)
j. University of Cincinnati, Magnum (2001)
k. Syracuse, NY, Mitsch (1985a)
l. lab compacted soils, Mitsch (1985a)
m. Chicago, IL, Deardorff (2007)
n. Edmonton, AB, Tappenden (2007), Zhang (1999)
o. Bruderheim, AB, Tappenden (2007), Zhang (1999)
p. Ft. McMurray, AB, Tappenden (2007), Schmidt (2004)
q. Slemons (2008)
r. AB Chance (1994)
s. Wesolek (2005)
t. Huang et al. (1995)
u. Atlas (1995)
v. Fort Collins, CO, CTL|Thompson
w. Cincinnati, OH, Magnum Piering, Inc.
x. Dallas, TX, Hargrave (1992)
y. Denver, CO, Park Range Construction
z. San Jose, CA, Rupiper (1989)
a1. Columbia, SC, Terratech
b1. Irving, TX, Witherspoon (2006)
c1. Colebrook, BC, Weech (1996)
d1. Denver, CO, Terracon
e1. Long Island City, NY, Superior
f1. Brooklyn, NY, Tauberer
g1. Arverne, NY, Soil Mechanics
h1. New York, NY, Loftus
i1. Ft. Saskatchewan, AB, Tappenden (2007)
j1. Lamont, AB, Tappenden (2007)
k1. Hythe, AB, Tappenden (2007)
l1. Ft. St. John, BC, Tappenden (2007)
m1. Saskatoon, SK, Tappenden (2007)
n1. Bronx, NY, Soil Mechanics
o1. New York, NY, Tauberer
p1. Fort Collins, CO, Dixie Electrical Manufacturing
q1. Windsor, CO, Dixie Electrical Manufacturing
r1. Denver, CO, CTL|Thompson

Axial Load Tests

Pile No.	Shaft Shape	Shaft Dim in.	Pitch in.	Helices in.							Site Ref.	Depth ft.	Avg Spt-N bpf	Friction Angle deg	Soil Type	Final Torque ft-lb	Load Test Deflctn in.	Load Test Load kips	Notes
TENSION																			
V1*	SQR	1.75	3.0	8	10						a	10			SM	3,900	−1.30	12	Shallow ground water
V2*	SQR	1.75	3.0	8	10	12	14				a	10			SM	3,400	−2.10	36	Long term test
V3*	SQR	1.75	3.0	8	10						a	25			Shale	2,500	−1.69	39	
V4*	SQR	1.75	3.0	8	10						a	14			SM	3,700	−1.02	44	Dynamically tested
V1	SQR	1.75	3.0	8	10	12	14				b	10		35	SC/SW	3,197	−2.00	36	
V2	SQR	1.75	3.0	8	10						b	10		35	SW	3,530	−1.30	12	Shallow ground water
V3	SQR	1.75	3.0	8	10	12					b	14		35	SW	4,231	−1.80	48	Dynamically tested
V4	SQR	1.75	3.0	8	10	12					b	21		35	SM	5,295	−1.60	34	Dynamically tested
V5	SQR	1.75	3.0	8	10	12	14				b	16		35	SW	4,593	−1.40	49	Dynamically tested
V6	SQR	1.75	3.0	8	10						b	22			Shale	3,660	−1.10	44	Dynamically tested
V7	SQR	1.75	3.0	8	10						b	25			Shale	2,532	−1.30	39	
V8	SQR	1.75	3.0	8	10	12	14				b	23			Shale	2,994	−2.00	61	Dynamically tested
SR1				10	10	11	11				c	134						49	
SR2				10	11	14	15				c	141						66	
SR3				2.0	2.5	2.8					d				CL/ML			0.13	
SR4				2.0	2.5	2.8					d				CL/ML			0.19	
SR5				2.0	2.5	2.8					d				CL/ML			0.19	
SR6	SQR	1.5	3.0	8	10	12					d				CL/ML			11.5	
SR7	SQR	1.5	3.0	8	10	12					d				CL/ML			11.0	
SR8	SQR	1.5	3.0	8	10	12					d				CL/ML			10.9	
1-P1	RND	1.73	2.3	4	4						e	25			CL		−1.26	1.18	PL = 25%, WC = 40%
2-P2	RND	1.73	2.3	4	4	4					e	25			CL		−1.18	1.97	PL = 25%, WC = 40%
3-P3	RND	1.73	2.3	4	4	4	4				e	25			CL		−2.17	1.97	PL = 25%, WC = 40%
4-P1	RND	1.73	2.3	4	4						e	25			CL		−1.57	1.57	PL = 25%, WC = 45%
5-P2	RND	1.73	2.3	4	4	4					e	25			CL		−1.97	1.77	PL = 25%, WC = 45%
6-P3	RND	1.73	2.3	4	4	4	4				e	25			CL		−2.05	1.97	PL = 25%, WC = 45%
7-P1	RND	1.73	2.3	4	4						e	25			CL		−1.97	1.18	PL = 25%, WC = 50%

8-P2	RND	1.73	2.3	4	4	4					e	25			CL		−2.36	1.38	PL = 25%, WC = 50%
9-P3	RND	1.73	2.3	4	4	4	4				e	25			CL		−1.97	1.97	PL = 25%, WC = 50%
1-P4	RND	2.36	2.3	6	6						e	25			CL		−2.56	0.33	PL = 25%, WC = 45%
2-P5	RND	2.36	2.3	6	6	6					e	25			CL		−2.76	0.38	PL = 25%, WC = 45%
3-P6	RND	2.36	2.3	6	6	6	6				e	25			CL		−1.97	0.39	PL = 25%, WC = 45%
1-P7	RND	0.98	1.0	3	3						e	39			CL		−1.18	0.16	PL = 16%, WC = 26%
2-P8	RND	0.98	1.0	3	3	3					e	39			CL		−0.98	0.19	PL = 16%, WC = 26%
3-P9	RND	0.98	1.0	3	3	3	3				e	39			CL		−1.38	0.20	PL = 16%, WC = 26%
1-P10	RND	0.98	1.0	3	3						e	39			CL			0.65	PL = 16%, WC = 26%
2-P11	RND	0.98	1.0	3	3	3					e	39			CL			0.71	PL = 16%, WC = 26%
1-P7	RND	0.98	1.0	3	3						e	39			CL			1.52	PL = 23%, WC = 35%
2-P8	RND	0.98	1.0	3	3	3					e	39			CL			1.86	PL = 23%, WC = 35%
3-P9	RND	0.98	1.0	3	3	3	3				e	39			CL			2.13	PL = 23%, WC = 35%
4-P10	RND	0.98	1.0	3	3						e	39			CL			1.19	PL = 23%, WC = 35%
5-P11	RND	0.98	1.0	3	3	3					e	39			CL			1.48	PL = 23%, WC = 35%
	RND	3	3.0	8	10	12					f	17	45		SP/GP	9,860	−1.61	76	
	RND	3	3.0	8	10	12					g	18			CL	8,330	−1.56	56	
	RND	3	3.0	8	10	12					h	16	27		CL	5,355	−1.50	52	
	RND	3	3.0	8							h	22	40		Claystone	11,560	−1.25	48	
	RND	3	3.0	8							h	22	40		Claystone	11,560	−1.20	50	
	RND	3	3.0	8							f	27	150		Sandstone	13,000	−0.80	120	
	RND	3	3.0	10	12	14					g	17			CL	12,325	−1.83	72	
	RND	3	3.0	10	12						f	17	45		SP/GP	6,205	−1.73	69	
	RND	3	3.0	10	12						g	18			CL	6,545	−1.60	50	
	RND	3	3.0	10	12						h	16	27		CL	4,505	−1.60	39	
	RND	3	3.0	12							f	16	42		SP/GP	8,245	−1.80	66	
	RND	3	3.0	12							f	18	48		SP/GP	10,285	−1.80	65.2	
	RND	3	3.0	12							f	17	45		SP/GP	10,285	−1.80	63	
	RND	3	3.0	12							h	22	40		Claystone	9,435	−1.75	48	
	RND	3	3.0	12							h	22	40		Claystone	8,670	−1.65	41	
	RND	3	3.0	14							h	16	27		CL	4,505	−1.95	30	
	RND	3	3.0	14							g	18			CL	7,395	−1.90	38	
	RND	3	3.0	14							f	27	150		Sandstone	13,000	−1.40	120	

(Continued)

Pile No.	Shaft Shape	Shaft Dim in.	Pitch in.	Helices in.							Site Ref.	Depth ft.	Avg Spt-N bpf	Friction Angle deg	Soil Type	Final Torque ft-lb	Load Test Deflctn in.	Load Test Load kips	Notes
	RND	2.875	3.2	8	10	12					f	17	45		SP/GP	5,610	−1.52	30	
	RND	2.875	3.2	8							f	18	48		SP/GP	5,610	−1.30	29.5	
	RND	2.875	3.2	8							f	26	75		SP/GP	6,600	−1.21	84	
	RND	2.875	3.2	8	10	12					f	13	34		SP/GP	4,900	−1.20	84	
	RND	2.875	3.2	8	10	12					h	7	26		CL	5,400	−1.07	60	
	RND	2.875	3.2	8	10	12					f	7	21		SP/GP	7,700	−1.06	43	
	RND	2.875	3.2	8	10	12					h	7	26		CL	5,610	−1.05	39	
	RND	2.875	3.2	8							f	13	34		SP/GP	7,000	−0.89	39	
	RND	2.875	3.2	10	12						f	17	45		SP/GP	5,610	−1.40	26.5	
	RND	2.875	3.2	10	12						h	15	27		CL	5,610	−1.21	42	
	RND	2.875	3.2	10	12						f	9	21		SP/GP	3,800	−1.18	43	
	RND	2.875	3.2	10	12						h	6	21		CL	5,200	−1.16	52.5	
	RND	2.875	3.2	12							h	18	40		Claystone	5,610	−1.33	40	
	RND	2.875	3.2	12							h	17	40		Claystone	5,610	−1.31	34	
	RND	2.875	3.2	14							f	11	27		SP/GP	7,100	−1.55	72.5	
	RND	2.875	3.2	14							h	14	27		CL	5,610	−1.51	45	
	RND	2.875	3.2	14							h	7	26		CL	3,800	−1.44	31.5	
	RND	0.70	0.4	2							i				SW	25	−0.59	0.94	Research models
	RND	0.70	0.6	2							i				SW	29	−0.59	0.93	Research models
	RND	0.70	0.8	2							i				SW	30	−0.59	0.92	Research models
	RND	0.70	0.6	2							i				SW	22	−0.59	0.73	Research models
	RND	0.70	0.6	2							i				SW	17	−0.59	0.47	Research models
FL	RND	3	3.0	8	12						j	15	41		SC	3,700	−3.34	30	Dual edge helix
	SQR	1.5	2.6	8	10	12					k	10		37	SP	1,500		31	
	SQR	1.5	2.6	8	10	12					k	10		37	SP	2,000		36	
	SQR	1.5	2.6	8	10	12					k	10		37	SP	1,750		38	
	SQR	1.5	2.6	8	10	12					k	10		37	SP	2,500		42	
	SQR	1.5	2.6	8	10	12					k	14		37	SP	3,000	−2.50	52	
	SQR	1.5	2.6	8	10	12					k	14		37	SP	3,500	−2.50	52	
	SQR	1.5	2.6	8	10	12					k	14		37	SP	1,750	−2.50	43	

	SQR	1.5	2.6	8	10	12					k	14		37	SP	2,500	−2.50	50	
	SQR	1.5	2.6	8	10	12					k	10		37	SP	1,500		31	
	SQR	1.5	2.6	8	10	12					k	10		37	SP	2,700		33	
	SQR	1.5	2.6	8	10	12					k	10		37	SP	2,000		32	
	SQR	1.5	2.6	8	10	12					k	14		37	SP	1,900		45	
	SQR	1.5	2.6	8	10	12					k	14		37	SP	2,000		45	
	SQR	0.5	0.9	3	3	4					l	1		46	SP	115	−0.20	1.53	Shallow failure
	SQR	0.5	0.9	3	3	4					l	1		46	SP	132	−0.20	1.58	Shallow failure
	SQR	0.5	0.9	3	3	4					l	1		46	SP	112	−0.20	1.54	Shallow failure
	SQR	0.5	0.9	3							l	1		46	SP	25	−0.20	0.24	Shallow failure
	SQR	0.5	0.9	3	3	4					l	2		46	SP	165	−0.20	2.74	
	SQR	0.5	0.9	3	3	4					l	2		46	SP	175	−0.20	2.95	
	SQR	0.5	0.9	3	3	4					l	2		46	SP	160	−0.20	2.94	
	SQR	0.5	0.9	3							l	2		46	SP	75	−0.20	1.08	
	SQR	0.5	0.9	3	3	4					l	1		35	SP	20	−0.20	0.53	Shallow failure
	SQR	0.5	0.9	3	3	4					l	1		35	SP	21	−0.20	0.52	Shallow failure
	SQR	0.5	0.9	3	3	4					l	1		35	SP	19	−0.20	0.52	Shallow failure
	SQR	0.5	0.9	3							l	1		35	SP	9	−0.20	0.13	Shallow failure
	SQR	0.5	0.9	3	3	4					l	2		35	SP	24	−0.20	0.86	
	SQR	0.5	0.9	3	3	4					l	2		35	SP	22	−0.20	0.86	
	SQR	0.5	0.9	3	3	4					l	2		35	SP	25	−0.20	0.94	
	SQR	0.5	0.9	3							l	2		35	SP	15	−0.20	0.38	
3	RND	3.5	2.7	8	10	12					pl	21	40		Claystone	9,200	−1.41	57.0	
2	RND	3.5	3.0	10	12						pl	21	40		Claystone	8,000	−1.35	42.5	
1	RND	3.5	2.6	14							pl	14	26		CL	6,150	−1.80	56.6	
10	RND	3.5	2.4	8	10	12					ql	10	4		SP	4,400	−1.21	22.0	
8	RND	3.5	2.7	8							ql	10	4		SP	2,270	−1.13	11.0	
6	RND	3.5	2.7	8							ql	22	28		GP	9,800	−1.29	61.0	
9	RND	3.5	2.8	10	12						ql	11	9		SP	4,100	−1.26	20.6	
7	RND	3.5	2.7	14							ql	20	34		GP	7,400	−2.00	61.0	
3	RND	2.88	2.5	10	12						ql	13	9		SP	2,400	−1.00	16.0	
4	RND	2.88	2.6	8	10	12					ql	13	9		SP	3,400	−1.25	39.0	
5	RND	2.88	2.6	8							ql	9	4		SP	740	−0.25	4.3	

(*Continued*)

Pile No.	Shaft Shape	Shaft Dim in.	Pitch in.	Helices in.							Site Ref.	Depth ft.	Avg Spt-N bpf	Friction Angle deg	Soil Type	Final Torque ft-lb	Load Test Deflctn in.	Load Test Load kips	Notes
1	RND	2.88	2.6	8							ql	10	4		SP	1,000	−0.25	9.0	
2	RND	2.88	2.4	14							ql	15	9		SP	4,400	−1.30	24.0	
1	RND	2.88	2.6	14							pl	20	40		Claystone	8,500	−1.50	30.0	
2	RND	2.88	2.6	10	12						pl	12	26		CL	5,500	−1.20	56.0	
L1	RND	4.5		12	14	16					m	57	26.5		CL	22,300	−2.40	70.00	
L2	RND	4.5		10	12	14	14				m	62	26.5		CL	24,200	−1.00	82.00	2.25"stinger lead
B1	RND	4.5		10	12	14	14				m	67	28		CL	24,200	−1.00	70.00	2.25"stinger lead
4	RND	5.5		19	19	19	19				rl	43			Bedrock	46,000	−2.00	340	
T1	RND	8.62	3.0	14	14	14					n	16	17		CL	16,300	−0.39	47.2	
T2	RND	8.62	3.0	14	14	14					n	10	14		CL	15,000	−0.39	31.5	
T3	RND	8.62	3.0	14	14						n	16	17		CL	16,900	−0.47	47.2	
T4	RND	8.62	3.0	14	14	14					o	16	6		SM/SP	37,500	−1.57	81	
T5	RND	8.62	3.0	14	14	14					o	10	11		SM/SP	31,500	−0.55	43	End bearing on OC crust
T6	RND	8.62	3.0	14	14						o	16	6		SM/SP	35,300	−1.97	81	
T7	RND	7	3.0	30							p	19	32		CL	60,000	−2.40	180.0	
T8	RND	8.62	3.0	30	30						p	20	32		CL	90,000	−2.60	298.0	
T9	RND	7	3.0	30							p	16	209		SP	190,000	−1.97	455.0	Jack capacity exceeded
COMPRESSION																			
1	RND	2.875		8							q	44	120		ML	4,000	0.80	25	Hexagonal helix
2	RND	2.875		8	8	8					q	24	13		ML	1,450	0.80	12	Hexagonal helix
3	RND	2.875		14							q	10	21		ML	3,000	1.40	25	Hexagonal helix
4	RND	2.875		14							q	38	51		ML	4,000	1.40	40	Hexagonal helix
5	RND	2.875		8	14						q		85		ML	6,785	1.10	70	Hexagonal helix
6	RND	2.875		14	14						q	10	21		ML	4,000	1.40	43	Hexagonal helix
7	RND	2.875		14	14						q	34	43		ML	6,195	1.40	70	Hexagonal helix
	SQR	1.5	3.0	8							r					2,000	0.31	45	
	SQR	1.75	3.0	8	10	12					r					6,000	0.34	52	

	RND	3.5	3.0	8	10	12					r						10,000	0.34	52	
	SQR	1.5	3.0	8	10	12					r						1,600	0.48	16	
	SQR	1.5	3.0	8	10						r						3,000	0.49	44	
	SQR	1.5	3.0	8	10	12					r						3,000	0.90	32	
	SQR	1.5	3.0	8	10	12					r						4,000	1.03	28	
	SQR	1.5	3.0	8							r						4,000	1.09	60	
	SQR	1.5	3.0	8	10	12					r						3,500	1.13	24	
	SQR	1.5	3.0	8	10	12					r						4,500	1.30	30	
	SQR	1.75	3.0	8	10	12	14	14	14	14	r						10,000	1.35	71	
	SQR	1.5	3.0	10							r						1,000	0.22	34	
	SQR	1.5	3.0	10							r						1,500	0.22	34	
	SQR	1.5	3.0	10							r						1,000	0.28	35	
	RND	3.5	3.0	10	12						r						5,000	0.34	52	
	RND	3.5	3.0	10	12						r						5,000	0.34	52	
	SQR	1.5	3.0	10							r						1,250	0.43	33	
	SQR	1.5	3.0	10							r						2,400	0.72	35	
	SQR	1.5	3.0	10	12						r						2,000	0.97	33	
	SQR	1.5	3.0	14							r						2,600	0.15	20	
1	RND	1.73	2.3	4	4						e	25				CL		0.08	0.32	PL = 25%, WC = 40%
2	RND	1.73	2.3	4	4	4					e	25				CL		0.22	0.39	PL = 25%, WC = 40%
3	RND	1.73	2.3	4	4	4	4				e	25				CL		0.31	0.44	PL = 25%, WC = 40%
4	RND	1.73	2.3	4	4						e	25				CL		0.16	0.23	PL = 25%, WC = 45%
5	RND	1.73	2.3	4	4	4					e	25				CL		0.31	0.31	PL = 25%, WC = 45%
6	RND	1.73	2.3	4	4	4	4				e	25				CL		0.31	0.33	PL = 25%, WC = 45%
7	RND	1.73	2.3	4	4						e	25				CL		0.12	0.19	PL = 25%, WC = 50%
8	RND	1.73	2.3	4	4	4					e	25				CL		0.12	0.21	PL = 25%, WC = 50%
9	RND	1.73	2.3	4	4	4	4				e	25				CL		0.35	0.24	PL = 25%, WC = 50%
1-P4	RND	2.36	2.3	6	6						e	25				CL		0.59	2.38	PL = 25%, WC = 45%
2-P5	RND	2.36	2.3	6	6	6					e	25				CL		0.47	2.60	PL = 25%, WC = 45%
3-P6	RND	2.36	2.3	6	6	6	6				e	25				CL		0.59	2.71	PL = 25%, WC = 45%
1-P7	RND	0.98	1.0	3	3						e	39				CL		0.39	0.18	PL = 16%, WC = 26%
2-P8	RND	0.98	1.0	3	3	3					e	39				CL		0.28	0.21	PL = 16%, WC = 26%
3-P9	RND	0.98	1.0	3	3	3	3				e	39				CL		0.28	0.22	PL = 16%, WC = 26%

(*Continued*)

Pile No.	Shaft Shape	Shaft Dim in.	Pitch in.	Helices in.							Site Ref.	Depth ft.	Avg Spt-N bpf	Friction Angle deg	Soil Type	Final Torque ft-lb	Load Test Deflctn in.	Load Test Load kips	Notes
1-P10	RND	0.98	1.0	3	3						e	39			CL			0.15	PL = 16%, WC = 26%
2-P11	RND	0.98	1.0	3	3	3					e	39			CL			0.16	PL = 16%, WC = 26%
	SQR	1.75	3.0	8	10						s	55			CL	9,690	1.45	73	
	SQR	1.75	3.0	8	10						s	60			CL	9,085	1.75	65	
	SQR	1.75	3.0	8	10						s	62			CL	12,114	2.00	40	
	SQR	1.5	3.0	8	10	12					t	14				5,500	0.50	47	
	SQR	1.5	3.0	10							t	24				5,500	0.75	75	
	SQR	1.75	3.0	8							u					10,000	0.30	35	
	RND	2.875	3.0	12	14						u					8,000	1.00	42	
	RND	2.875	3.0	12	14						u					3,000	1.00	31	
	RND	3	3.0	8							f	27	75		Sandstone	4,000	0.80	65	
	RND	3	3.0	8							f	24	75		Sandstone	8,700	0.80	38	
	RND	3	3.0	8	10	12					h	10	26		CL	6,171	1.12	51	
	RND	3	3.0	8	10	12					g	17			CL	5,355	1.13	27	
	RND	3	3.0	8	10	12					v	8	20		SP	5,355	1.35	49	
	RND	3	3.0	8							v	9	33		SP	4,505	1.65	45	
	RND	3	3.0	8	10	12					g	16			CL	6,205	1.78	35	
	RND	3	3.0	10	12						v	8	20		SP	5,780	1.40	55	
	RND	3	3.0	10	12						g	16			CL	5,780	1.53	36	
	RND	3	3.0	10	12						h	10	26		CL	5,348	1.57	51	
	RND	3	3.0	10	12						g	15			CL	5,780	1.60	40	
	RND	3	3.0	12							v	8	33		Sandstone	6,582	1.60	64	
	RND	3	3.0	12							v	10	67		Sandstone	5,760	1.60	52.5	
	RND	3	3.0	12							v	10	67		Sandstone	6,994	1.65	65	
	RND	3	3.0	12							v	10	67		Sandstone	3,291	1.90	101	
	RND	3	3.0	14							g	16			CL	5,780	1.19	42	
	RND	3	3.0	14							f	27	75		Sandstone	5,500	1.40	65	
	RND	3	3.0	14							f	23	75		Sandstone	8,000	1.40	35	
	RND	3	3.0	14							g	15			CL	5,780	1.53	42	
	RND	3	3.0	14							h	10	26		CL	4,937	1.75	42	

	RND	2.875	3.2	8							f	10	24		SP/GP	1,700	0.84	23	
	RND	2.875	3.2	8							f	14	37		SP/GP	6,000	0.90	38	
	RND	2.875	3.2	8							h	18	40		Claystone	5,610	0.95	34	Eccentrically loaded
	RND	2.875	3.2	8							h	23	40		Claystone	6,600	0.98	42	
	RND	2.875	3.2	8							f	17	45		SP/GP	5,610	0.98	69	
	RND	2.875	3.2	8							f	26	75		Sandstone	6,600	0.99	38	
	RND	2.875	3.2	8	10	12					f	12	31		SP/GP	7,300	1.12	53	
	RND	2.875	3.2	8	10	12					h	20	40		Claystone	3,600	1.14	38	
	RND	2.875	3.2	8	10	12					f	6	21		SP/GP	5,865	1.20	57	
	RND	2.875	3.2	8	10	12					h	8	26		CL	5,610	1.38	38	
	RND	2.875	3.2	10	12						h	15	27		CL	3,300	1.23	47	
	RND	2.875	3.2	10	12						f	12	31		SP/GP	7,100	1.23	57	
	RND	2.875	3.2	10	12						f	5	21		SP/GP	5,865	1.25	60	
	RND	2.875	3.2	10	12						h	6	26		CL	5,610	1.42	45	
	RND	2.875	3.2	14							f	10	24		SP/GP	2,700	1.43	15	
	RND	2.875	3.2	14							f	12	31		SP/GP	6,600	1.51	52	
	RND	2.875	3.2	14							h	22			Claystone	6,600	1.52	29	
	RND	2.875	3.2	14							h	3	50		CL	5,270	1.72	57	
	RND	2.875	3.2	14							h	3	50		CL	5,610	1.82	54	
FL2	RND	3	3.0	12							w	11	41		CL	7,300	2.25	74	Dual edge helix
	SQR	1.5	3.0	10							x	19				1,450	0.80	18	
	SQR	1.5	3.0	10							x	19				875	0.80	9	
	SQR	1.5	3.0	10							x	19				1,800	0.80	20	
	SQR	1.5	3.0	10							x	19				1,800	0.80	22	
	SQR	1.5	3.0	10							x	19				2,050	0.80	23	
	SQR	1.5	3.0	10							x	19				2,100	0.80	22	
	SQR	1.5	3.0	10	12						x	13				2,100	0.80	24	
	SQR	1.5	3.0	10	12						x	13				1,650	0.80	18	
	SQR	1.5	3.0	10	12						x	13				2,000	0.80	24	
	SQR	1.5	3.0	10	12						x	13				1,700	0.80	22	
	SQR	1.5	3.0	10	12						x	13				1,800	0.80	26	
FL3	RND	3	3.0	8	12						j	15			SC	3,700	2.30	36	Dual edge helix (DE)
	SQR	1.75		8	10						y	56				9,459	0.69	87	

(*Continued*)

Pile No.	Shaft Shape	Shaft Dim in.	Pitch in.	Helices in.							Site Ref.	Depth ft.	Avg Spt-N bpf	Friction Angle deg	Soil Type	Final Torque ft-lb	Load Test Deflctn in.	Load Test Load kips	Notes
	SQR	1.5	3.0	14							z					2,160	0.15	20	
SC2	RND	3	3.0	10	12						al	52	45		Bedrock	12,000	1.45	60	DE, 15 deg out-of-plumb
SC1	RND	3	3.0	10	12						al	52	45		Bedrock	12,000	1.81	103	DE
31	RND	3.5	3.2	10	12						bl	30	12		CL/CH	5,097	0.23	28	Possible eccentric load
30	RND	3.5	3.2	10	12						bl	32	13		CL/CH	5,097	0.30	28	Possible eccentric load
29	RND	3.5	3.2	10	12						bl	27	13		CL/CH	5,097	0.36	27	Possible eccentric load
32	RND	3.5	3.2	10	12						bl	29	12		CL/CH	5,097	0.43	28	Possible eccentric load
28	RND	3.5	3.2	12							bl	45	13		CL/CH	5,097	0.31	24	Possible eccentric load
26	RND	3.5	3.2	12							bl	26	13		CL/CH	5,097	0.38	26	Possible eccentric load
25	RND	3.5	3.2	12							bl	24	13		CL/CH	5,097	0.49	25	Possible eccentric load
27	RND	3.5	3.2	12							bl	34	14		CL/CH	5,097	0.68	30	Possible eccentric load
PP1	RND	3.5	3.0	14	14	14	14	14			cl	30	soft		CL/ML		3.00	17	6"diam pull down piles
TP1	RND	3.5	3.0	14	14	14	14	14			cl	30	soft		CL/ML		3.00	18	6"diam pull down piles
TP2	RND	3.5	3.0	14	14	14	14	14			cl	30	soft		CL/ML		2.00	20	6"diam pull down piles
TP3	RND	3.5	3.0	14	14	14	14	14			cl	30	soft		CL/ML		3.00	17	6"diam pull down piles
TP4	RND	3.5	3.0	14	14	14	14	14			cl	30	soft		CL/ML		3.00	17	6"diam pull down piles
TP5	RND	3.5	3.0	14	14	14	14	14			cl	30	soft		CL/ML		3.20	15	6"diam pull down piles
TP6	RND	3.5	3.0	14	14	14	14	14			cl	30	soft		CL/ML		3.50	14	6"diam pull down piles
1/11	RND	5.5		16							dl	26			Claystone	51,000	0.51	560	
#1	RND	10.75		24							el	50	RQD = 95		Schist	68,000	0.52	480	Verification, not ultimate
#2	RND	10.75		24							el	50	RQD = 95		Schist	68,000	0.54	480	Verification, not ultimate
TP1	RND	5		18							fl						0.81	220	Verification, not ultimate
#8	RND	5									gl		20		SP		0.91	120	
#9	RND	5									gl		20		SP		0.90	120	
TP2	RND	7		14	14						hl	36	35		SM		1.70	221	
C1	RND	8.62	3.0	14	14	14					n	16	17		CL	15,000	0.83	40.5	
C2	RND	8.62	3.0	14	14	14					n	10	14		CL	11,500	0.75	36.0	
C3	RND	8.62	3.0	14	14						n	16	17		CL	14,400	0.59	47.2	
CP2	RND	8.62	3.0	14	14						n	16	17		CL	15,000		47.2	
C4	RND	8.62	3.0	14	14	14					o	16	6		SM/SP	33,000	1.57	106	
C5	RND	8.62	3.0	14	14	14					o	10	11		SM/SP	30,000	1.97	94	
C6	RND	8.62	3.0	14	14						o	16	6		SM/SP	33,000	1.57	85	
C7	RND	7	3.0	18							il	15			CL	18,900	0.71	48	
C8	RND	8.62	3.0	18							il	15			CL	25,700	0.67	60	

C9	RND	7	3.0	20	18						i1	18			CL	23,200	0.79	84	
C10	RND	9.61	3.0	20	18						j1	31			Shale	87,500	0.91	265	
C11	RND	10.75	3.0	30							p	19	32		CL	63,000	3.54	246	
C12	RND	10.75	3.0	30	30						p	20	32		CL	72,000	2.36	309	
C13	RND	8.6	3.0	16							k1	31	164		Sandstone	87,500	0.71	265	
C14	RND	12.8	3.0	36	36						l1	34			CL	58,300	1.20	143	
C15	RND	5.5	3.0	20	20	20					l1	20			CL	14,500	1.20	61	
C16	RND	4.5	3.0	18	18						l1	16			CL	9,990	0.75	55	
C17	RND	4.5	3.0	18							l1	13			CL	5,920	0.87	38	
C18	RND	4.5	3.0	18							m1	16	7		ML	8,140	1.10	46	
C19	RND	4.5	3.0	16							m1	16	7		ML	7,030	1.20	33	
C20	RND	4.5	3.0	18	16						m1	19	7		ML	7,030	0.91	45	
5	RND	3.5	2.6	10	12						p2	9	4		SP	2,270	1.10	36.0	
4	RND	3.5	2.7	8							p2	12	7		SP	2,570	0.80	14.0	
6	RND	3.5	2.7	8	10	12					p2	7	5		SP	3,490	1.00	34.5	
3	RND	3.5	2.7	14							p1	14	27		CL	4,900	1.40	48.0	
1	RND	3.5	2.7	14							p1	21	40		Claystone	12,000	1.40	60.0	
5	RND	3.5	2.7	10	12						p1	14	27		CL	4,900	1.10	55.5	
2	RND	3.5	2.7	8							p1	25	40		Claystone	10,400	0.80	60.5	
4	RND	3.5	2.7	8	10	12					p1	18	27		CL	7,700	1.00	56.6	
1	RND	2.88	2.6	8							q1	18	34		GP	6,450	0.80	83.0	
2	RND	2.88	2.4	10	12						q1	10	4		SP	2,100	0.25	23.4	
3	RND	2.88	2.4	8	10	12					q1	10	4		SP	1,350	0.25	14.0	
1	RND	2.88	2.6	8							p1	18	40		Claystone	8,000	1.00	56	
2	RND	2.88	2.6	8							p1	16	27		CL	3,200	1.00	37	
3	RND	2.88	2.6	10	12						p1	18	27		CL	6,700	1.20	57	
4	RND	2.88	2.6	8	10	12					p1	14	27		CL	3,800	1.10	50	
5	RND	2.88	2.6	14							p1	17	40		Claystone	8,500	1.50	60	
1	RND	4.5		16							r1	40			Bedrock	26,000	1.60	192	
2	RND	4.5		16							r1	41			Bedrock	30,600	2.00	140	
3	RND	4.5		16							r1	41			Bedrock	31,000	1.60	160	
1	RND	5.5		19							r1				Bedrock		3.25	240	
3	RND	5.5		19							r1	44			Bedrock	46,000	3.50	310	
2	RND	5.5		19							r1	44			Bedrock	45,000	2.50	320	
L2	RND	4.5		10	12	14	14				m	62	26.5		CL	26,900	1.30	136	2.25" stinger lead
B1	RND	4.5		10	12	14	14				m	70	33		CL	26,900	1.38	136	2.25" stinger lead
#44	RND	8.625		14							n1	44	RQD = 53-100		Gneiss	69,000	0.35	160	Verification, not ultimate
#44A	RND	8.625		14							n1	28	RQD = 53-100		Gneiss	69,000	0.18	160	Verification, not ultimate
1P	RND	8.625		10	12	14	14				o1		18		SM		0.29	160	

Lateral Load Tests

Site	Soil Condition	Average N-Value	Design Capacity Theoretical (kip)	Design Capacity Measured (kip)	Pile Depth (ft)	Shaft Diameter (in)	Shaft Thickness (in)	Lead Configuration	Ratio
University of Cincinnati, Ohio	Silty clay fill	7	590	1,250	16	3	0.25	12"DCE	2.12
University of Cincinnati, Ohio	Silty clay fill	7	583	850	10	3	0.25	10"& 12" circular	1.46
Dwyer Yard	Sandy clay	16	1,488	1,800	na	3	0.25	unk	1.21
Dwyer Yard	Sandy clay	16	1,478	1,500	18	3	0.25	8", 10", 12"DCE	1.01
Dwyer Yard	Sandy clay	16	1,520	1,050	12	3	0.25	8", 10", 12" circular	0.69
Colorado State Univ	Clay	34	2,200	2,000	10	2.875	0.203	8" circular	0.91
Colorado State Univ	Clay	34	2,200	2,500	10	2.875	0.203	8" circular	1.14
Colorado State Univ	Clay	34	2,200	2,000	10	2.875	0.203	8" circular	0.91
Colorado State Univ	Clay	34	2,200	3,000	10	2.875	0.203	8" circular	1.36
Colorado State Univ	Clay	34	2,200	2,700	10	2.875	0.203	8" circular	1.23
Colorado State Univ	Clay	34	2,200	3,000	10	2.875	0.203	8" circular	1.36
Colorado State Univ	Clay	34	2,200	1,300	10	2.875	0.203	8" circular	0.59
Colorado State Univ	Clay	34	2,200	2,000	10	2.875	0.203	8" circular	0.91
Ontario Hydro (Puri et al. 1984)	Stiff clay	stiff	1,878	2,400	14.5	3.5	*0.3*	10", 11", 13.5"	1.28
Ontario Hydro (Puri et al. 1984)	Stiff clay	stiff	7,050	7,000	14.5	8.625	*0.322*	unk	0.99
West Chester Lot	Clay	21	2,213	2,955	12	3	0.25	8" circular	1.34
West Chester Lot	Clay	21	2,213	3,485	12	3	0.25	8" circular	1.57
West Chester Lot	Clay	21	2,213	5,000	12	3	0.25	8" circular	2.26
West Chester Lot	Clay	21	2,213	3,430	12	3	0.25	8" circular	1.55
West Chester Lot	Clay	21	2,213	2,200	12	3	0.25	8" circular	0.99
West Chester Lot	Clay	21	2,213	4,400	12	3	0.25	8" circular	1.99
West Chester Lot	Clay	21	2,213	2,265	12	3	0.25	8" circular	1.02
West Chester Lot	Clay	21	2,213	3,185	12	3	0.25	8" circular	1.44
ESSA (Puri et al. 1984)	Sand	*loose*	2,785	2,750	9.5	8.625	*0.322*	unk	0.99
Georgia Power (Puri et al. 1984)	Sand	*dense*	6,100	6,250	15	8.625	*0.322*	unk	1.02
Houston, TX (Puri et al. 1984)	Clay	*soft*	2,450	2,500	31.5	8.625	*0.322*	unk	1.02
Alberta (Schmidt, 2004)	Clay	*stiff*	10,320	8,800	20	10.75	*0.365*	30"	0.85
Alberta (Schmidt, 2004)	Clay	*stiff*	10,320	8,300	20	10.75	*0.365*	30", 30"	0.80
Edmonton, AB, Zhang (1999)	Clay	14	7,050	4,495	16	8.625	0.26	14", 14", 14"	0.64
Edmonton, AB, Zhang (1999)	Clay	14	7,163	4,945	16	8.625	0.32	14", 14", 14"	0.69
Bruderheim, AB, Zhang (1999)	Silty sand	17	4,750	6,970	16	8.625	0.26	14", 14", 14"	1.47
Bruderheim, AB, Zhang (1999)	Silty sand	17	4,850	7,305	16	8.625	0.32	14", 14", 14"	1.51
Dwyer Yard	Sandy clay	16	11,000	10,000	18	24*	24*	8", 10", 12"DCE	0.91
Dwyer Yard	Sandy clay	16	11,000	8,800	18	24*	24*	8", 10", 12"DCE	0.80
Dwyer Yard	Sandy clay	16	11,000	14,000	18	24*	24*	8", 10", 12"DCE	1.27
Dwyer Yard	Sandy clay	16	11,000	12,000	18	24*	24*	8", 10", 12"DCE	1.09

Appendix **D**

Nomenclature

α	=	adhesion between soil and a helical pile shaft
α_o	=	shaft adhesion coefficient in expansive soils
α_s	=	ratio between side shear and penetration stress
δ	=	pile head deflection
H	=	helical pile–soil interaction coefficient
θ	=	ratio of helix radius to shaft radius
Θ	=	angle of soil influence cone from vertical
γ	=	unit weight of soil
γ'	=	effective unit weight of soil
γ_w	=	unit weight of water (62.4 pcf) [1 g/cm^3]
λ_c	=	column slenderness parameter
λ_{CPT}	=	ratio between SPT blow count and CPT tip resistance
λ_δ	=	deflection fitting constant
λ_k	=	capacity to torque ratio fitting factor (22 in$^{0.92}$/ft, 1433 mm$^{0.92}$/m)
λ_s	=	effective helical pile shaft length
λ_{SPT}	=	SPT correlation factor (0.065 tsf/blow, 6.2 kPa/blow)
λ_t	=	ground disturbance factor (0.87)
λ_ϕ	=	correction factor for overburden (0.048 blow/psf, 1 blow/kPa)
μ	=	coefficient of friction
ν	=	Poisson ratio
η	=	group efficiency
π	=	Pi (3.141593 rad)
ρ	=	redundancy factor
σ'_n	=	effective confining stress
σ_M	=	standard deviation
T	=	shear strength
Φ	=	angle of internal friction
ϕ	=	resistance factor

A = surface area of a helical pile
A_{CP} = area enclosed by outside perimeter of concrete section
A_n = area of a helical bearing plate n
A_g = gross cross-sectional area of helical pile shaft
A_1 = area of the lowest helical bearing plate; also, bearing area of pile base plate
A_s = cross-sectional area of reinforcing steel per unit width of slab
A_T = area of the uppermost helical bearing plate
a = batter angle; also, height of concrete compression zone

B = width of spread footing or pile cap
B_n = bearing strength
b_o = shear perimeter

COV = coefficient of variation
C_S = seismic response coefficient
C_1 = slope of linear regression line in load test interpretation
C_2 = y-intercept of linear regression line in load test interpretation
C_u = disturbance factor for lateral load determination
c = cohesion

d = side dimension of a square shaft or outside diameter of a round shaft
d_1 = inside diameter of a round shaft
d_b = anchor rod or bolt diameter
d_c = cohesion depth factor
d_q = overburden depth factor
d_γ = friction depth factor
d_{eff} = effective shaft diameter
d_s = depth to reinforcing steel
d_w = depth of active zone below the top of helical piles in expansive soils
D = diameter of helical bearing plate
D_T = diameter of uppermost helical bearing plate
D_{AVG} = average diameter of helical bearing plates on a given pile

e = eccentricity of an applied axial or lateral load
E = modulus of elasticity
E_f = current efficiency of a sacrificial anode
E_H = horizontal component of earthquake loads
E_{peak} = modulus of elasticity of competent, intact rock
E_s = stress-strain modulus for soil
E_V = vertical component of earthquake loads
E_r = Soil Penetration Test (SPT) hammer energy ratio

F = crowd force
F_a = seismic site coefficient
F'_b = bending stress in wood design
F_{cr} = critical buckling stress

F_s = factor of safety; also, tension or compression in seismic ties
F_y = yield strength
f = Broms' shaft length resisting lateral shear
f'_c = compressive strength of concrete
f_y = yield strength of reinforcing steel

G_M = arithmetic mean
g = Broms' shaft length resisting overturning moment

H = total length of shaft above top helix; also, height of retained earth
H_{eff} = effective shaft length for adhesion calculations
H_s = vertical height of soil
h = thickness of concrete above or below a helical cap plate
h_w = height of water above the tip of a helical pile
h_y = number of hours in a year (8766)

i = number of helical piles in a group
i_o = electrical current density
I = area moment of inertia
I_{req} = required electrical current in sacrificial anode system

K = scaling parameter used to compute shape and depth factors for bearing
K_a = active earth pressure coefficient
k = effective length factor
k_c = Cone Penetration Test (CPT) bearing capacity factor
K_o = at-rest lateral earth pressure coefficient
K_p = passive lateral earth pressure coefficient
K_T = ultimate capacity to installation torque ratio for helical piles

L_s = length of spread footing
L = helical pile length
L_t = life span of a sacrificial anode
l = unsupported column length in buckling, also spacing between H-pile flanges for lagging design
l_n = maximum spacing between helical piles under a floor slab
m = integer place holder; number of helical bearing plates cutting separate path through soil
m_1 = width of a group of piles in plan view
m_2 = breadth of a group of piles in plan view
M = overturning moment
M_n = nominal flexural strength
M^- = negative factored moment
M_o = total factored static moment
M_{MAX} = maximum bending or overturning moment

N = Standard Penetration Test (SPT) blow count, N-value
N_T = relative embedment ratio (H/D_1)

N_c = cohesion bearing capacity factor
N_q = overburden bearing capacity factor
N_γ = friction bearing capacity factor
N'_c = combined cohesion bearing, shape, and depth factor
N'_γ = combined overburden bearing, shape, and depth factor
N'_γ = combined friction bearing, shape, and depth factor
N'_u = empirical uplift capacity factor
N_k = empirical cone factor
N_{55} = Standard Penetration Test (SPT) blow count (55% energy ratio)
N_{60} = Standard Penetration Test (SPT) blow count (60% energy ratio)
N_{70} = Standard Penetration Test (SPT) blow count (70% energy ratio)
n = integer place holder; number of helical bearing plates; sample population; number of years to zinc depletion

P = axial load applied to a helical pile or pile group
P_a = allowable capacity of a helical pile (a.k.a. working capacity)
P_c = clamping force; also, higher column load for seismic ties
P_{CP} = outside perimeter of concrete cross section
P_{dead} = dead load applied to a helical pile
P_L = lateral load on a helical pile
P_{max} = maximum uniform pressure on timber lagging
P_n = nominal capacity of pile shaft
P'_o = effective overburden stress at depth of sampler or pile tip
P_u = ultimate axial capacity of a helical pile
P_s = swell pressure
P_{ue} = axial capacity of eccentrically loaded helical pile
P_{ug} = ultimate axial capacity of a helical pile group
P_{ut} = corrected ultimate pullout capacity of a helical anchor
p = helix pitch
p_f = combined probability of poor performance
p_1 = probability of poor performance based on limit state methods
p_2 = probability of poor performance based on torque correlations

q' = effective overburden stress at bearing depth
q_u = unconfined compressive strength of soil or rock
q_{ult} = ultimate bearing pressure of soil or rock
q_c = Cone Penetration Test (CPT) cone tip resistance
q_{ca} = equivalent Cone Penetration Test (CPT) cone tip resistance
q_s = Cone Penetration Test (CPT) unit skin friction
q_{peak} = unconfined compressive strength of competent, intact rock

RQD = rock quality designation
r = radius of gyration of helical pile shaft
r_s = effective radius of helical pile shaft
R = radius of helical bearing plate; reaction force

s = spacing along the shaft between successive helical bearing plates
s_u = undrained strength of fine-grain soil
s_c = cohesion shape factor
s_q = overburden shape factor
s_γ = friction shape factor
S = on-center spacing between helical pile shafts
S_m = elastic section modulus per unit width of slab
S_S = maximum considered short period earthquake acceleration
S_{MS} = adjusted maximum earthquake acceleration
S_{DS} = design earthquake acceleration

T = installation torque
T_c = design torsional resistance of concrete
T_d = design thickness
T_h = theoretical current output from a sacrificial anode
T_n = nominal thickness
T_s = sacrificial thickness
T_u = factored torsional load
t = helix thickness
t_d = design life span (years)
t_o = thickness of zinc coating
t_s = slab thickness

U_f = utilization rate of sacrificial anode (85%)

V = shear
V_c = design shear resistance of concrete
V_{HC} = highest conceivable value
V_{LC} = lowest conceivable value
V_n = nominal shear strength
V_u = factored shear load

W = weight of soil influence cone; weight of the structure; also weight of sacrificial anode
w = distributed load; also, uniform surcharge pressures

y_n = value for a select test in a population

z = depth
z_{avg} = average depth of helical bearing plates
z_n = depth of *n*th helical bearing plate

Glossary of Terms

Aeolian—Windblown deposit (a.k.a. eolian).

Adhesion—Similar to cohesion except adhesion is the shear strength between soil and another material, such as a pile shaft or concrete foundation.

Angle bracket—Side load bracket having a horizontal bearing plate that extends below and supports an existing foundation or other structural element.

Angle of internal friction—Soil shear strength parameter that governs strength due to confining stress.

Batter angle—Inclination angle from vertical at which helical piles are installed; can be given as percentage, ratio of vertical depth to horizontal run, or angle in degrees.

Battered piles—Helical piles installed at an angle other than vertical; typically used for lateral stability or in areas where the pile butts are constrained together and the pile tips splay apart to avoid group effects.

Bearing plate—See *helix*.

Bedrock—Continuous rock layer that forms the Earth's crust; may outcrop at the ground surface or be buried under a thick layer of soil, cobble, and boulders.

Bentonite— High-swelling clay soil derived from chemical alteration of volcanic ash.

Bracket—Manufactured steel cap or assembly that attaches to the helical pile butt and is used to transfer loads to new or existing foundation elements; examples include angle bracket, plate bracket, new construction bracket.

Buttress—Section of reinforced concrete or masonry that is constructed inside of a structure and is oriented perpendicular to the primary wall system in order to resist lateral movement.

Caliche—(pronounced "cal-lee-chee") Soil with grains cemented by carbonates below-ground in semiarid conditions.

Collar—Larger-diameter outer sleeve of a helical pile coupling.

Conforming helix—Term that describes helical bearing plates with true helix shape that are normal to the shaft such that leading and trailing edges are parallel (see *AC358*).

Consolidation—Change in volume of soil involving the drainage of pore water over time under applied pressures; typically associated with fine-grain soils having low permeability.

Cohesion—Soil shear strength parameter that is constant with respect to confining pressure; basically, the forces within soil that hold it together.

Counterfort—Section of reinforced concrete or masonry that is constructed outside of a structure and is oriented perpendicular to the primary wall system in order to resist lateral movement.

Coupling—Pinned, bolted, or welded connection of two helical pile shafts.

Daylights—Verb indicating that the end of a helical anchor, foundation drain, or other underground structure extends out of the ground surface.

Deadman—Type of anchor wherein a steel plate, concrete block, wood beam, pile, or other object is buried in the earth behind a retaining system and connected via a threadbar, rod, strap, or strand to the wall facing.

Deviations—Angle, percent slope, or linear distance differences between final and planned helical pile location, butt elevation, inclination, orientation, batter angle, or plumbness.

Disk—Circular plate with square hole that fits over a solid square shaft under the knuckle in the proprietary helical pile grouting process known as the pull-down pile used by A. B. Chance

Direct load bracket—Manufactured bracket for attaching a helical pile to a structure such that applied compressive or tensile loads are concentric with the primary axis of the helical pile shaft.

Down drag—The phenomenon where soft soils surrounding a pile consolidate and produce a downward force on the pile tending to cause additional settlement. Consolidation can be triggered by soil self-weight, periodic changes in groundwater, placement of fill, or other surcharge loads on the ground surface.

Drained—Refers to the draining of pore water pressures within a soil during slow loading and is often associated with coarse-grain soils because of high permeability.

Drift pin—Tapered cylindrical tool used to align bolt holes when connecting helical pile couplings during installation.

Drive pin—Pin used to connect a helical pile butt to a drive tool on a torque motor; typically a high-strength smooth round pin, although any dowel of appropriate strength can be used.

Drive tool—Adapter for transferring torque between the torque motor and the helical pile consisting of a hex or other shape Kelly bar socket and a round or square collar sleeve.

Duckbill—Type of earth anchor consisting of a plate attached by a hinge to an anchor rod; the plate is driven into the earth to a desired depth, and the anchor rod is post-tensioned, which causes the plate to turn broadside into the soil (commercial names include Manta-Ray and Sting-Ray).

Dunnage—Spare pieces of scrap timber, steel, or other material used to build up an area and form a temporary structure, such as a footing pad or construction platform.

End bearing—Term used to describe a pile that generates a majority of its capacity from the pile tip, such as piles that bear directly on bedrock; essentially all helical piles with single helix are end bearing.

Extension—Sections of central shaft with or without helical bearing plates used to extend the lead section of a helical pile to greater depth and form a continuous column of steel in the ground. Extensions with helical bearing plates often are required in very soft soils or in soil nail applications.

Flange—Pair of often-identical plates that form the main structural components of a wide flange beam, H-pile, or I-joist; flanges are separated and held apart by the web.

Friction—The shear force caused by normal stress of concrete on an angle bracket; see also *Angle of Internal Friction.*

Glacial Outwash—Material deposited by streams that drained from the front of a melting glacier typically composed of gravel, sand, and silt.

Glacial till—Material deposited by a glacier without subsequent transport by water; typically a heterogeneous mixture of soil and rock varying in size from clay to boulders.

Gumbo—Very plastic, typically very soft to medium stiff, usually wet, dark- colored clays found in Louisiana and surrounding area.

Groundwater—1. (noun) Water within soil or bedrock (e.g., Helical piles penetrate groundwater without difficulty.). 2. (adjective) Pertaining to water within soil or bedrock (e.g., Groundwater levels do not preclude helical pile installation.).

Hardpan—Stratum of hard soil that is difficult to excavate, drill, or install a helical pile through.

Helical pile—Manufactured steel foundation consisting of one or more helix-shaped bearing plates affixed to a central shaft that is rotated into the ground to support structures.

Helix—Generally circular steel plate pressed in a spiral shape with uniform pitch. The helix is welded to the shaft and used to install the helical pile in a screwing action into the ground and to transfer the load from the shaft into the surrounding soil or rock material.

Homogeneous—Term indicating that a particular physical property of soil or rock, such as color, density, moisture content, or consistency, is the same in all directions.

In situ—(pronounced "in sit-you") State of being undisturbed and in place.

Isotropic—Term indicating that a particular mechanical property of soil or rock, such as strength, bearing capacity, or permeability, is the same in all directions.

Kelly bar—Short, typically round or hexagonal drive shaft extending from a torque motor, auger drive, or soil/rock drill.

Knuckle—Slang term for the forged upset couplings on a square-shaft helical pile.

Lateral load—Force acting on a helical pier in a direction that is transverse (perpendicular) to the central shaft.

Lateral resistance—Capacity of a helical foundation system or device to resist lateral loads.

Lead—First section of a helical pile to enter the ground; consists of a pilot point, central shaft, and one or more helical bearing plates.

Lifting bracket—Also referred to as foundation bracket or underpinning bracket, a manufactured steel bracket used by foundation repair contractors to connect a helical pier to an existing foundation for lifting or stabilizing.

Loess—(pronounced "lus") Very porous aeolian- (wind-) deposited silt soil cemented by calcium carbonate or clay; can stand on vertical slopes, and often collapses when wet.

Marl—Calcareous silt or clay of marine origin; ranges from very soft to stiff, typically very moist to wet.

Micropile—Small-diameter (typically 4-inch to 8-inch [101 mm to 203 mm] drilled pile with central reinforcing steel bar surrounded by cement grout; One of the main distinctions of micropiles compared to other types of drilled piles is the use of smaller drilling equipment, with short mast and segmental drill stem.

Mooring—Underwater anchoring system for securing a boat or ship. Typically consists of a helical anchor, cable or chain, and buoy.

Net deflection—Total axial deflection of the pile head under static load test minus elastic shortening/elongation of the helical pile shaft.

New construction bracket—Helical pile bracket typically consisting of a collar sleeve that fits over the pile butt and a square plate that is significantly larger in area than the pile shaft; the entire bracket is cast in the concrete pile cap, grade beam, or wall and increases punching shear resistance in order to transfer higher loads.

Oedometer—Machine used to conduct swell/consolidation tests.

Outcrop—Bedrock exposed at the ground surface.

Overconsolidated—Refers to the stress history of soil; indicates that the soil has previously experienced significantly higher confining pressures than its current state.

Pile butt—Trailing end of a helical pile that is attached to the torque motor drive tool; name is derived from driven pile industry wherein the pile butt is the end of the pile that is impacted by the hammer.

Pile cCap—Reinforced concrete structure of variable thickness and geometry placed over one or a group of helical piles and used to transfer loads to a column, grade beam, wall, or other structure.

Pile freeze—Generally defined as the gradual increase in pile capacity with time and pore water pressure dissipation.

Pile head—Aboveground end of a helical pile that is cast in a pile cap or affixed to a bracket; synonymous with pile butt.

Pile tip—Leading end or belowground end of a helical pile that is often the sharp end of the pilot point on a lead section; term is most often used when referring to the elevation of the deepest end of the pile.

Pile toe—Synonym for pile tip.

Pilot hole—Predrilled hole made by a conventional drill rig or down-hole hammer and used in difficult ground conditions to break through obstructions, guide the helical pile shaft, and improve penetration; the pilot hole is typically slightly larger than or the same size as the helical pile shaft.

Pilot point—Short section of shaft with tapered tip extending below the lead helix; facilitates positioning and proper installation.

Pitch—Distance between the leading and trailing edges of a helix; also, the approximate distance a helical pile should advance with proper installation.

Plate bracket—Side load bracket having a vertical plate that is mounted against the side of an existing foundation or other structural element and used to transfer loads to the pile.

Plumbness—Deviation of a helical pile shaft from vertical.

Plunging—Continuous deflection of helical pile under constant load.

Proof test—Abbreviated load test on a helical anchor, which typically consists of applying 100% of the allowable load in one increment and waiting until movement stops.

Performance test—Full-scale load test on a helical anchor, which typically consists of applying 150% to 200% of the allowable load in increments while monitoring displacement.

Raker—Diagonal brace extending from the base of an excavation to a waler in excavation shoring design.

Reaction bar—Also referred to as a torsion bar, long tube or I-beam used as a counter resistance to the torsional force of a portable, hand-operated, free-standing torque motor during helical pier installation; the reaction bar is attached to a torque motor at one end, and the other end is braced against an immovable structure.

Settlement—Relatively immediate change in volume of soil as a result of mechanical restructuring under applied loads; normally associated with free-draining coarse-grain soils.

Shaft—Central tubular or solid steel column that makes up the core of a helical pile and transfers loads from the helical bearing plate(s) to the pile butt; typically square or round cross-section, although other shapes may be used.

Shear strength—(a) For soil or rock, the resistance to movement along a plane and is generally the sum of cohesion and normal force times the tangent of the angle of internal friction; (b) for components of a helical pile, the maximum stress that can be exerted on the cross-section of a shaft or bracket; (c) for concrete, the resistance to movement along a plane typically given as some fraction of the compressive strength.

Side load bracket—Manufactured bracket for connecting a helical pile to a structure such that applied compressive or tensile loads are eccentric with respect to the primary axis of the pile shaft.

Sidesway braced—Indicates that a connection to a helical pile prevents lateral translation but not necessarily rotation of the pile butt or point along the shaft.

Slab support bracket—Direct load bracket designed specifically for underpinning and support of existing floor slabs.

Soil nail—Helical anchor used in earth retention with helical bearing plates spaced along the entire length of the shaft so as to reinforce an entire zone of retained earth; also a term used with grouted anchors and other systems with continuous bond length.

Soldier—Vertical pile used in earth retention.

Spoil—Excess soil or rock produced during excavation or drilling activities; helical piles do not produce drill spoil.

Spotter—Member of the helical pile installation crew responsible for observing from a vantage point perpendicular to that of the operator; the spotter helps direct installation by checking plumbness, positioning, and alignment of the torque motor.

Static load test—Pile load test wherein load increments are applied to a helical pile at a slow rate to simulate static conditions. The quick test method is considered a type of static load test; loads may be applied with deadweight blocks or by hydraulic ram; the word "static" is used to differentiate this test from cyclic, dynamic, impulse, or statinamic load testing.

Swaged—Adjective describing helical pile couplings formed by a manufacturing process whereby a tubular shaft is enlarged to form a socket, thus eliminating the need for a collar.

Tension anchor racket—Direct load bracket designed specifically for tension applications; typically consists of a collar sleeve for connection to the helical pile and a continuously threaded bar, washer plate, and hex nut.

Tie-back—Helical anchor used in earth retention with helical bearing plates located a significant distance past the active zone of retained earth and a central shaft that extends through the active zone to a rigid wall facing; also a term used for grouted anchors, duck-bill anchors, and deadman anchors where the bond zone is outside of the active zone of retained earth.

Tuff—Compact fine-grained slightly cemented volcanic soil and ash; common in the Pacific Northwest of the United States.

Undrained— Refers to the generation of pore water pressures within a soil during rapid loading and is often associated with fine-grain soils because of low permeability.

Underpinning—Method of installing helical piles alongside or underneath an existing structure, connecting the piles to the structure, and transferring support from the existing foundation elements to the helical piles; may be associated with lifting and releveling of structures.

Underconsolidated—Refers to a soil that is not in equilibrium with its self-weight and is currently undergoing consolidation over time.

Unsupported length—Length of helical pile shaft standing unbraced in air, water, or fluid soils.

Waler—Horizontal beam spanning between tie-backs, rakers, or braces in earth retention systems.

Web—Central plate component oriented perpendicular to and separating the flanges of a wide flange beam, H-pile, or I-joist; the depth of a beam is the distance from the top of one flange to the bottom of the other measured along the length of the web

Varved clay—Lacustrine (lake) deposit with alternating clay and sand/silt strata;, extremely nonisotropic and can be very weak if pore pressure builds in permeable zones.

Bibliography

AASHTO 2004. *LRFD Bridge Design Specifications.*, Washington, DC: American Association of State Highway Transportation Officials.

A.B. Chance Co. 1992. *Anchor Corrosion Reference and Examples*, Bulletin 01-9204. Centralia, MO: A. B. Chance Co.

———. 1993a. *Helical-Pier Foundation System.* Manufacturer technical support document. Centralia, MO: A.B. Chance.

———. 1993b. *Tension Anchor System for Tieback Applications.* Manufacturer technical support document. Centralia, MO: A.B. Chance.

———. 1995. *Sample Calculations for Helical Pier Application.* Manufacturer technical support document. Centralia, MO: A.B. Chance.

———. 1996. *Buckling of Helical Anchors in Underpinning Applications.* Bulletin 01-9602, manufacturer technical support document. Centralia, MO: A.B. Chance.

———. 2003. *Helical Screw Foundation Design Manual for New Construction.* Manufacturer technical support document. A.B. Chance: Centralia, MO

———. 2006. *Corrosion—An Overview.* Version 1. 0, manufacturer technical support document. Centralia, MO: A. B. Chance.

ACI318. 2005. *Building Code Requirements for Structural Concrete and Commentary.* ACI Standard, ACI Committee 318. Farmington Hills, MI: American Concrete Institute.

Adams, J.I. and D.C. Hayes. 1967. "The Uplift Capacity of Shallow Foundations." *Ontario Hydro Research Quarterly*, Vol. 19, No. 1, pp. 1–13.

Adams, J.I. and T. W. Klym. 1972. "A Study of Anchors for Transmission Tower Foundations." *Canadian Geotechnical Journal*, Vol. 9, No. 1, pp. 89–104.

American Galvanizers Association (AGA). 2000a. *Hot-Dip Galvanizing for Corrosion Protection of Steel Products.* Englewood, CO: American Galvanizers Association.

———. 2000b. *Zinc Coatings.* Englewood, CO: American Galvanizers Association.

AISC. 2001. *Manual of Steel Construction* 3rd ed. Chicago, IL: American Institute of Steel Construction.

American Society of Civil Engineers (ASCE). 2006. *Minimum Design Loads for Buildings and Other Structures (ASCE7).* Reston, VA: American Society of Civil Engineers.

ASTM International. 2008. "ASTM A123 Standard Specifications for Zinc (Hot-Dip Galvanized) Coatings on Iron and Steel Products." *Annual Book of Standards.* West Conshohocken, PA: ASTM International.

———. "ASTM A153 Standard Specifications for Zinc (Hot-Dip Galvanized) Coatings on Iron and Steel Hardware." *Annual Book of Standards.* West Conshohocken, PA: ASTM International.

———. "ASTM B633 Standard Specifications for Electrodeposited Coatings of Zinc on Iron and Steel." *Annual Book of Standards.* West Conshohocken, PA: ASTM International.

———. "ASTM A653 Standard Specifications for Continuous Sheet Galvanizing." *Annual Book of Standards.* West Conshohocken, PA: ASTM International.

———. "ASTM B695 Standard Specifications for Coatings of Zinc Mechanically Deposited on Iron and Steel." *Annual Book of Standards.* West Conshohocken, PA: ASTM International.

———. "ASTM D1143 Standard Test Method for Piles under Static Axial Compressive Load." *Annual Book of Standards.* West Conshohocken, PA: ASTM International.

———. "ASTM D1586 Standard Test Method for Standard Penetration Test (SPT) and Split-Barrel Sampling of Soils." *Annual Book of Standards.* West Conshohocken, PA: ASTM International.

———. "ASTM D2166 Standard Test Method for Unconfined Compressive Strength of Cohesive Soil." *Annual Book of Standards.* West Conshohocken, PA: ASTM International.

———. "ASTM D2487 Standard Classification of Soils for Engineering Purposes (Unified Soil Classification System)." *Annual Book of Standards.* West Conshohocken, PA: ASTM International.

———. "ASTM D2488 Standard Practice for Description and Identification of Soils (Visual-manual Procedure)." *Annual Book of Standards.* West Conshohocken, PA: ASTM International.

———. "ASTM D3441 Standard Test Method for Mechanical Cone Penetration Tests of Soil." *Annual Book of Standards.* West Conshohocken, PA: ASTM International.

———. "ASTM D3689 Standard Test Method for Individual Piles under Static Axial Tensile Load." *Annual Book of Standards.* West Conshohocken, PA: ASTM International.

———. "ASTM D3966 Standard Test Method for Individual Piles under Lateral Loads." *Annual Book of Standards.* West Conshohocken, PA: ASTM International.

———. "ASTM D4318 Standard Test Methods for Liquid Limit, Plastic Limit, and Plasticity Index of Soils." *Annual Book of Standards.* West Conshohocken, PA: ASTM International.

———. "ASTM D6913 Standard Test Method for Particle-Size Distribution (Gradation) of Soils Using Sieve Analysis." *Annual Book of Standards.* West Conshohocken, PA: ASTM International.

Atlas Systems, Inc. 2000. *Technical Guide*, ed. 1.4. Centralia, MO.

Bassett, R.H. 1978. "Underreamed Ground Anchors." *Bulletin of Engineering Geology and the Environment*, Vol. 18, No. 1, December Springer, Berlin/Heidelberg, pp. 11–17.

Beer, F.P. and E.R. Johnston, Jr. 1981. *Mechanics of Materials.* New York: McGraw-Hill.

Black, D.R. and J. S. Pack. 2001. "Design and Performance of Helical Screw Piles in Collapsible and Expansive Soils in Arid Regions." In *Proceedings of the 36th Symposium, Engineering Geology and Geotechnical Engineering*, University of Nevada, Las Vegas, pp. 567–576.

———. 2002. "Design and Performance of Helical Screw Piles in Collapsible and Expansive Soils in Arid Regions of the United States." *Proceedings of the 9th International Conference on Piling and Deep Foundations*, Nice, France. Presses Ponts et Chaussees, Paris, pp. 469–476.

Bobbitt, B.E. and S.P. Clemence. 1987. "Helical Anchors: Applications and Design Criteria." *Proceedings of the 9th Southwest Asian Geotechnical Conference*, Bangkok, Thailand, Vol. 2, pp. 105–120.

Bowen, G. 2009. Personal Communication.

———. (In press). "A Static Based Theory for the Capacity to Torque Factor for Helical Piers in Compression", *Proceedings of Helical Foundations and Tie-Backs Seminar*, Deep Foundation Institute, University of Alberta, Canada, June 2009.

Bowles, J.E. 1988. *Foundation Analysis and Design* 4th ed. New York: McGraw-Hill.

Bradka, T.D. 1997. *Vertical Capacity of Helical Screw Anchor Piles*, M.S. report, Department of Civil Engineering, University of Alberta, Edmonton, Alberta.

Brinch-Hansen, J. 1961. "The Ultimate Resistance of Rigid Piles against Transversal Forces." *Danish Geotechnical Institute Bulletin No. 12*, pp. 5–9.

———. 1963. "Discussion on Hyperbolic Stress-Strain Response: Cohesive Soils." *Journal of the Soil Mechanics and Foundation Division*, Vol. 89, pp. 241–242.

———. 1970. "A Revised and Extended Formula for Bearing Capacity." *Danish Geotechnical Institute Bulletin No. 28*, pp. 5–11.

Broms, B.B. 1964a. "Lateral Resistance of Piles in Cohesive Soils." *Journal of the Soil Mechanics and Foundation Division*, Vol. 90, pp. 123–156.

———. 1964b. "Lateral Resistance of Piles in Cohesionless Soils." *Journal of the Soil Mechanics and Foundation Division*, Vol. 90, pp. 27–63.

Bustamante, M. and L. Gianeselli. 1982. "Pile Bearing Capacity Prediction by Means of Static Penetrometer CPT." *Proceedings of the 2nd European Symposium on Penetration Testing*, EWOPT-II, Vol. 2, pp. 493–500.

Carville, C.A. and R.W. Walton. 1994. "Design Guidelines for Screw Anchors." In *Proceedings of the International Conference on Design and Construction of Deep Foundations*, Orlando, FL, Vol. 2, pp. 646–655.

———. 1995. "Foundation Repair Using Helical Screw Anchors." In *Foundation Upgrading and Repair for Infrastructure Improvement* pp. 56–75. Reston, VA: American Society of Civil Engineers.

Cerato, A.B. 2007. "Dynamically Loaded Helical Anchors for Small Wind Tower Guyed Cable Foundations." Presented at the Proceedings of the Helical Foundations and Tie-Backs Seminar, Deep Foundation Institute, New Orleans.

Cerato, A.B. and R. Victor. 2008. "Effects of Helical Anchor Geometry on Long-Term Performance for Small Wind Tower Foundations Subject to Dynamic Loads." *Journal of Deep Foundations Institute*, Vol 2, pp 30–41.

———. (in press). "Effects of Long-Term Dynamic Loading and Fluctuating Water Table on Helical Anchor Performance for Small Wind Tower Foundations." *Journal of Performance of Constructed Facilities.*

Chapel, T.A. 1998. "Field Investigation of Helical and Concrete Piers in Expansive Soils." Masters thesis, Colorado State University, Fort Collins, CO.

Chapel, T.A. and J.D. Nelson. 2002. "Field Investigation of Helical and Concrete Piers in Expansive Soils." In *Proceedings of Geotechnical. Engineering Conference*, Rio de Janeiro, Brazil.

Chen, F.H. 1988. *Foundations on Expansive Soils (Developments in Geotechnical Engineering)*. Amsterdam: Elsevier Science Publishers.

Chuan, H.S. 2006. "Uplift Capacity of Helical Anchor in Sand." undergraduate thesis, University of Technology, Malaysia.

Clemence, S.P. 1985. "Uplift Behavior of Anchor Foundations in Soil." In *Proceedings of a Session Sponsored by the Geotechnical Engineering Division of ASCE*, Detroit, MI.

Clemence, S.P., L.K. Crouch, and R.W. Stephenson. 1994. "Uplift Capacity of Helical Anchors in Soils." *Proceedings of the 2nd Geotechnical Engineering Conference*, Cairo, Egypt, Vol. 1, pp. 332–343.

Clemence, S.P. and Y. Li. 2008. "Review of Helical Foundations over Twenty Years." Presented at the Proceedings of Helical Foundations and Tie-Backs Specialty Seminar, Deep Foundation Institute, Los Angeles, CA.

Clemence, S.P., and F.D. Pepe Jr. 1984. "Measurement of Lateral Stress around Multihelix Anchors in Sand." *Geotechnical Testing Journal*, Vol. 7, No. 3, pp. 145–152.

Clemence, S.P., and A. P. Smithling. 1984. "Dynamic Uplift Capacity of Helical Anchors in Sand." In *Proceedings of the 4th Australia-New Zealand Conference, Geomechanics*, No. 1, pp. 88–93.

Clemence, S.P. and C.J. Veesaert. 1977. "Dynamic Pullout Resistance of Anchors in Sand." *Proceedings of the International Symposium on Soil-Structure Interaction*, Roorkee, India, pp. 389–397.

Cole, W.H. 1978. "An Innovative Use for Multi-Helix Anchors." Presented at the EEI T&D Subcommittee, Key Biscayne, FL (unpublished).

Craig, B.D. 1995. *Handbook of Corrosion Data*. Materials Park, OH: ASM International.

Curle, R. 1995. "Screw Anchors Economically Control Pipeline Buoyancy in Muskeg." *Oil and Gas Journal*, Vol. 93, No. 17, pp. 49–54.

Dai, S.H. and M.O. Wang. 1992. *Reliability Analysis in Engineering Applications*. New York: Van Nostrand, Reinhold.

Das, B.M. 1990. *Principles of Geotechnical Engineering* 2nd ed. Boston: PWS-Kent Publishing Company.

Davisson, M.T. 1972. "High Capacity Piles." In *Proceedings of Lecture Series on Innovations in Foundation Construction*, ASCE, Illinois Section, Chicago (March), pp. 81–112.

Davisson, M.T. and H.L. Gill. 1963. "Laterally Loaded Piles in Layered Soil System." *Journal of Soil Mechanics and Foundation Division*, ASCE, Vol. 89, No. SM3, pp. 63–94.

Deardorff, D. 2007. "Torque Correlation Factors for Round Shaft Helical Piles." In *Proceedings of the 32nd Annual Conference of the Deep Foundation Institute*, Colorado Springs, CO, pp. 439–450.

DeBeer, E.E.1967/1968. "Proefondervindlijke Bijdrage Tot de Studie van Het Grensdraag Vermogen van Zand onder Funderingen op Staal." *Tijdshift der Openbar Verken van Belgie*, No. 6 (1967), and Nos. 4, 5, and 6 (1968).

———. 1970. "Experimental Determination of Shape Factor and Bearing Capacity Factors of Sand." *Geotechnique*, Vol. 20, No. 4, pp. 387–411.

Decourt, L. 1999. "Behavior of Foundations under Working Load Conditions." In *Proceedings of the 11th Pan-American Conference on Soil Mechanics and Geotechnical Engineering.*, Foz DoIguassu, Brazil, August, Vol. 4, pp. 453–488.

Dejong, J.T., M.F. Randolph, and D.J. White. 2003. "Interface Load Transfer Degradation during Cyclic Loading a Microscale Investigation." *Soils and Foundations*, Vol. 43, No. 4, pp. 81–93.

Dejong, J.T., D.J. White, and M.F. Randolph. 2006. "Microscale Observation and Modeling of Soil-Structure Interface Behavior Using Particle Image Velocimetry." *Soils and Foundations*, Vol. 46, No. 1, pp. 15–28.

Diewald, G.A. 2003. *A Modified Soil Suction Heave Prediction Protocol: With New Data from Denver Area Expansive Soil Sites*, MS Thesis, University of Colorado at Denver, Denver, CO.

Dobry, R., E. Vicente, M.J. O'Rourke, and J.M. Roesset. 1982. "Horizontal Stiffness and Damping of Single Piles." *Journal of the Geotechnical Engineering Division*, Vol. 108, No. GT3, pp. 439–459.

Downey, S. 2008. "A Quality Foundation." Presented at the Proceedings of Helical Foundations and Tie-Backs Specialty Seminar (November) Deep Foundation Institute, Los Angeles, CA.

Duncan, J.M. 2000. "Factors of Safety and Reliability in Geotechnical Engineering." *Journal of Geotechnical and Geoenvironmental Engineering*, Vol. 126, No. 4, pp. 307–316.

Duncan, J.M. and S.G. Wright. 2005. *Soil Strength and Slope Stability*. New York: John Wiley and Sons.

Duzceer, R. and A. Saglamer. 2002. "Evaluation of Pile Load Test Results." *Proceedings of the 9th International Conference on Piling and Deep Foundations* Nice, France. Paris: Presses Ponts et Chaussees, pp. 469–476.

El Marsafawi, H., Y.C. Han, and M. Novak. 1992. "Dynamic Experiments on Two Pile Groups." *Journal of Geotechnical Engineering*, Vol. 118, No. 4, pp. 576–592.

Elias, V. 2000. "Corrosion/Degradation of Soil Reinforcement for Mechanically Stabilized Earth Walls and Reinforced Soil Slopes." Report No.FHWA-NHI-00-044, Federal Highway Administration.

Fellenius, B.H. 2001a. "We Have Determined the Capacity, then What?" *Fulcrum* (Deep Foundation Institute) (Fall), pp. 23–26.

———. 2001b. "What Capacity Value to Choose from the Results of a Static Load Test." *Fulcrum* (Deep Foundation Institute) (Winter), pp. 19–22.

Fleming, W.G.K., A.J. Weltman, M.F. Randolph, and W.K. Elson. 1985. *Piling Engineering*. Glasgow: Surrey University Press.

Gazetas, G. and R. Dobry. 1984a. "Horizontal Response of Piles in Layered Soils." *Journal of Geotechnical Engineering*, Vol. 110, No. 1, pp. 20–40.

———. 1984b. "Simple Radiation Damping Model for Piles and Footings." *Journal of Engineering Mechanics*, Vol. 110, No. 6, pp. 937–956.

Ghaly, A.M. and S.P. Clemence. 1998. "Pullout Performance of Inclined Helical Screw Anchors in Sand." *Journal of Geotechnical and Geoenvironmental Engineering*, Vol. 124, No. 7, pp. 617–627.

———. 1999. "Closure to Pullout Performance of Inclined Helical Screw Anchors in Sand." *Journal of Geotechnical and Geoenvironmental Engineering*, Vol. 125, No. 12, pp. 1102–1104.

Ghaly, A.M. and A.M. Hanna. 1991a. "Experimental and Theoretical Studies on Installation Torque of Screw Anchors." *Canadian Geotechnical Journal*, 28, No. 3, pp. 353–364.

———. 1991b. "Stress Development in Sand Due to Installation and Uplifting of Screw Anchors." *Proceedings of the 4th International Conference on Piling and Deep Foundations*, Vol. 1, pp. 565–570.

Ghaly, A. and A. Hanna. 1992. "Stresses and Strains around Helical Screw Anchors in Sand." *Soils and Foundations*, Vol. 32, No. 4, pp. 27–42.

———. 1993. "Model Investigation of the Performance of Single Anchors and Groups of Anchors." *Canadian Geotechnical Journal*, Vol. 31, pp. 273–284.

Ghaly, A.M. and A.M. Hanna. 1994a. "Model Investigation of the Performance of Single Anchors and Groups of Anchors." *Canadian Geotechnical Journal*, Vol. 31, pp. 273–284.

———. 1994b. "Ultimate Pullout Resistance of Single Vertical Anchors." *Canadian Geotechnical Journal*, Vol. 31, No. 5, pp. 661–672.

Ghaly, A., A. Hanna, and M. Hanna. 1991a. "Uplift Behavior of Screw Anchors in Sand. I: Dry Sand." *Journal of Geotechnical Engineering*, Vol. 117, No. 5, pp. 773–793.

———. 1991b. "Uplift Behavior of Screw Anchors in Sand. II: Hydrostatic and Flow Conditions." *Journal of Geotechnical Engineering*, Vol. 117, No. 5, pp. 794–808.

———. 1991c. "Installation Torque of Screw Anchors in Dry Sand." *Soils and Foundations, Japanese Society of Soil Mechanics and Foundation Engineering*, Vol. 31, No. 2, pp. 77–92.

Ghaly, A.M., Hanna, A., Ranjan, G. and Hanna, M. 1991. "Helical Anchors in Dry and Submerged Sand Subjected to Surcharge." *Journal of Geotechnical Engineering*, Vol. 117, No. 10, pp. 1463–1470.

Gill, H.S. and J.J. Udwari. 1980. "Pullout Tests: Multi-Helix Screw Anchors." report prepared for the Virginia Electric and Power Company, Richmond, VA.

Gregory, S. and R.M. Hoyt. 2005. "Strength Assessment of Helix Plates for Helical Piles." *Proceedings of the Helical Foundations and Tiebacks Seminar*. Torrence, CA: Deep Foundation Institute.

Hanna, A. and A. Ghaly. 1992. "Effects of K_0 and Overconsolidation on Uplift Capacity." *Journal of Geotechnical Engineering*, Vol. 118, No. 9, pp. 1449–1469.

———. 1994. "Ultimate Pullout Resistance of Groups of Vertical Anchors." *Canadian Geotechnical Journal*, Vol. 31, No. 5, pp. 673–682.

Hanna, T.H., E. Sivapalan, and A. Senturk. 1978. "The Behavior of Dead Anchors Subjected to Repeated and Alternating Loads," *Ground Engineering*, Vol. 2, No. 4, pp. 28–34.

Hardesty, R. 2007. "Helical Piers vs. Drilled Concrete Piers in Highly Expansive Soil Areas." Hardesty Consulting, LLC, Denver, CO, www.helicalpierworld.com.

Hargrave, R.L. and R.E. Thorsten. 1992. "Helical Piers in Expansive Soils of Dallas, Texas." *7th International Conference on Expansive Soils*, Vol. 1, pp. 125–130. Lubbock, TX: Texas Tech University Press.

Helmers, M.J., J.M. Duncan, and G.M. Filz. 1997. "Use of Ultimate Load Theories for Design of Drilled Shaft Sound Wall Foundations." Report of the Virginia Transportation Research Council.

Hendrickson, R. 1984. *The Ocean Almanac*. New York: Doubleday.

Housner, G.W. 1959. "Behavior of Structures during Earthquakes." *Journal of the Engineering Mechanics Division*, Vol. 85, No. EM4 pp. 109–129.

Hovland H.J. 1993. "Discussion of Helical Anchors in Dry and Submerged Sand Subjected to Surcharge." *Journal of Geotechnical Engineering*, Vol. 119, No. 2, pp. 391–392.

Hoyt, R. 2007. *Engineering Manual: Ram Jack Helix Screw Anchors, Ram Jack Steel Piers*. Ada, OK: Ram Jack Systems Distribution, LLC.

Hoyt, R.M. and S.P. Clemence. 1989. "Uplift Capacity of Helical Anchors in Soil." *Proceedings of the 12th International Conference on Soil Mechanics and Foundation Engineering*, Rio de Janeiro, Brazil Vol. 2, pp. 1019–1022.

Hoyt, R.M., G. Seider, L.C. Reese, M. Hon, and S. Wang. 1995. "Buckling of Helical Anchors Used for Underpinning." In *Foundation Upgrading and Repair for Infrastructure Improvement*, pp. 89–108. Reston, VA: American Society of Civil Engineers.

Hsu, T.T.C. 1968. "Torsion of Structural Concrete-Plain Concrete Rectangular Sections." *Torsion of Structural Concrete*, Portland Cement Association, Bulletin D134, pp. 203–238.

Huang, F.C., I. Mohmood, M. Joolazadeh, and G.W. Axten. 1995. "Design Considerations and Field Load Tests of a Helical Anchoring System for Foundation Renovation." *Foundation Upgrading and Repair for Infrastructure Improvement*, pp. 76–88. Reston, VA: American Society of Civil Engineers.

ICC-Evaluation Services. 2003. "AC228 Acceptance Criteria for Corrosion Protection of Steel Foundation Systems Using Polymer (EAA) Coatings." www.icc-es.org.

———. 2004. "Dixie Anchoring Helical Foundation System.", ICC-ES Legacy Report 21-47, www.icc-es.org.

———. 2007. "AC358 Acceptance Criteria for Helical Pile Foundations and Devices." www.icc-es.org.

Industrial Galvanizers America. 1999. "Product Galvanizing Brochure. International Business Publishers, Atlanta, GA.

International Code Council. 2006. *International Building Code (IBC)*. Washington, DC: International Code Council.

Johnston, G.H. and B. Ladanyi. 1974. "Field Tests of Deep Power-Installed Screw Anchors in Permafrost." *Canadian Geotechnical Journal*, Vol. 11, No. 3, pp. 348–359.

Jones, D.A. 1986. *Principles and Prevention of Corrosion* 2nd ed. London: Prentice-Hall.

Jones, F., H. H. Ryffel, E. Oberg, and C. J. McCauley. 2004. *Machinery's Handbook* (27th ed.). New York: Industrial Press.

King, R.A. 1977. Corrosion Nomograph, TRRC Supplementary Report, *British Corrosion Journal*

Kishida, H. 1996. "Damage to Reinforced Concrete Buildings in Niigata City with Special Reference to Foundation Engineering." *Soils and Foundations* (Tokyo), Vol. 6, No. 1, pp. 71–88.

Klym, T.W., H.S. Radhakrishna, et al. 1986. "Helical Plate Anchors for Tower Foundations." In *Proceedings of Symposium on Anchor Systems in Geotechnical Engineering*. Toronto: Canadian Geotechnical Society.

Levadoux, J.N. and M.M. Baligh. 1980. "Pore Pressures During Cone Penetration in Clays." Research Report R80-15, Department of Chemical Engineering, MIT, Cambridge, MA.

Lindeburg, M.R. 1997. *Civil Engineering Reference Manual*, 6th ed. Belmont, CA: Professional Publications.

Lunne, T. and A. Kleven. 1981. "Role of CPT in North Sea Foundation Engineering." In *Cone Penetration Testing and Experience*, Geotechnical Engineering Division, ASCE National Conference, St. Louis, MO, pp. 76–107.

Lunne, T., P.K. Robertson, and J.J.M. Powell. 1997. *Cone Penetration Testing in Geotechnical Engineering Practice*. London: Blackie Academic & Professional.

Lutenegger, A.J. 2003. "Helical Screw Piles and Screw Anchors—An Historical Prospective and Introduction." In *Proceedings of the Helical Foundations and Tie-Backs Seminar*, Deep Foundation Institute, Cincinnati, OH.

Lutenegger, A.J., B.L. Smith, and M.G. Kabir. 1988. "Use of In Situ Tests to Predict Uplift Performance of Multi-Helix Anchors." *Special Topics in Foundations*, ASCE, pp. 93–110.

MacLean Dixie. 2009. *Product Information Bulletin No. 008-0*, Manufacturer Product Literature, MacLean Dixie, Franklin Park, IL.

MacNab, A. 2002. *Earth Retention Systems Handbook*. New York: McGraw-Hill.

McCann, M. 2006. "Heavy Equipment and Truck-Related Deaths on Excavation Work Sites." *Journal of Safety Research*, Vol. 37, No. 5, pp. 511–517.

McKeen, R.G. and L.D. Johnson. 1990. "Climate-Controlled Soil Design Parameters for Mat Foundations." *Journal of Geotechnical Engineering*, 116(7), pp. 1073–1094.

McOmber, R.M. and R.W. Thompson. 2000. "Verification of Depth of Wetting for Potential Heave Calculations." In *Advances in Unsaturated Geotechnics: Proc. Sessions of Geo-Denver*, C. Shackelford, S. Houston, and N.Y. Chang, eds., pp. 409–422,

Means 1998. *Building Construction Cost Data*, Western ed., Kingston, MA: R.S. Means.

Meyerhof, G.G. 1951. "The Ultimate Bearing Capacity of Foundations." *Geotechnique*, Vol. 2, No. 4, pp. 301–331.

———. 1976. "Bearing Capacity and Settlement of Pile Foundations." *Journal of Geotechnical and Geoenvironmental Engineering*, Vol. 102, No. 3, pp. 195–228.

Meyerhof, G.G. and J.I. Adams. 1968. "The Ultimate Uplift Capacity of Foundations." *Canadian Geotechnical Journal*, Vol. 5, No. 4, pp. 224–244.

Michaelides, O., G. Gazetas, G. Bouckovalas, and E. Chrysikou. 1997. "Approximate Non-linear Dynamic Axial Response of Piles." *Geotechnique*, Vol. 48, No. 1, pp. 33–53.

Miller, F.E., J.E. Foss, and D.C. Wolf. 1981. *ASTM STP* 741, American Society for Testing and Materials, p. 19.

Mitsch, M.P. and S.P. Clemence. 1985a. "The Uplift Capacity of Helix Anchors and Sand." *Uplift Behavior of Anchor Foundations in Soil, ASCE*, pp. 26–47.

———. 1985b. "Uplift Behavior of Anchor Foundations in Soil." *Journal of Geotechnical Engineering*, pp. 26–47.

Mizuno, H. 1987. "Pile Damage during Earthquake in Japan (1923–1983)." *Dynamic Response of Pile Foundations—Experiment, Analysis and Observation*, ASCE Special Geotechnical Publication No. 11, T. Nogami, Ed., pp. 53–78.

Mooney, J.S., S. Adamczak Jr., and S.P. Clemence. 1985. "Uplift Capacity of Helix Anchors in Clay and Silt." *Uplift Behavior of Anchor Foundations in Soil*, ASCE pp. 48–72.

Mudali, U.K., H.S. Khatak, and B. Raj. 2007. "Anodic and Cathodic Protection." *Encyclopedia of Electrochemistry*. Hoboken, NJ: Wiley Interscience, pp. 401–442.

Mylonakis, G., and Gazetas, G. 1999. "Lateral Vibration and Internal Forces of Grouped Piles in Layered Soil." *Journal of Geotechnical and Geoenvironmental Engineering*, Vol. 125, No. 1, pp. 16–25.

Narasimha Rao, S. and Y.V.S.N. Prasad. 1993a. "Behavior of Model Screw Anchors in Soft Clays." *Geotechnique*, Vol. 43, No. 44, pp. 605–614.

———.1993b. "Estimation of Uplift Capacity of Helical Anchors in Clays." *Journal of Geotechnical Engineering*, Vol. 119, No. 2, pp. 352–357.

Narasimha Rao, S., Y.V.S.N. Prasad, and C.V. Prasad. 1990. "Experimental Studies on Model Screw Pile Anchors." *Proceedings of the Indian Geotechnical Conference, Bombay*, pp. 465–468.

Narasimha Rao, S., Y.V.S.N. Prasad, and M.D. Shetty. 1991. "The Behavior of Model Screw Piles in Cohesive Soils." *Soil and Foundations*, Vol. 31, No. 2, pp. 35–50.

Narasimha Rao, S., Y.V.S.N. Prasad, M.D. Shetty, and V.V. Joshi. 1989. "Uplift Capacity of Screw Pile Anchors." *Geotechnical Engineering*, Vol. 20, No. 2, pp. 139–159.

Narasimha Rao, S., Y.V.S.N. Prasad, and C. Veeresh. 1993. "Behavior of Embedded Model Screw Anchors in Soft Clays." *Geotechnique*, Vol. 43, pp. 605–614.

Nasr, M.H. 2004. "Large Capacity Screw Piles." In *Proceedings of International Conference: Future Vision and Challenges for Urban Development*, Cairo, Egypt.

Nasr, M.H. 2007. "Use of Large Capacity Screw Piles—High Pressure Grouted Piles—Barbados." *Proceedings of the Helical Foundations and Tie-Backs Seminar*, Deep Foundation Institute, New Orleans.

Nasr, M.H. 2008. "Behavior of High Pressure Grouted Screw Piles in Santa Clara, CA." www.helicalpiersystems.com/pdf/HPS-Paper%20Santa%Clara.pdf.

NCDOL. 2008. *A Guide to OSHA Excavations Standard.* B.R. Davis, ed. Division of Occupational Safety and Health, North Carolina Department of Labor, Raleigh, NC.

Nelson, J.D., and D.J. Miller. 1992. *Expansive Soils: Problems in Practice in Foundation and Pavement Engineering.* John Wiley and Sons, New York.

Nelson, J.D., D.D. Overton, and D.B. Durkee. 2001. "Depth of Wetting and the Active Zone." *Proc. of Expansive Clay Soils and Vegetative Influence on Shallow Foundations*, C. Vipulanandan, Marshall Addison, and Michael Hansen, Eds., ASCE, pp. 95–109.

Nelson, J.D., K.C. Chao, and D.D. Overton. 2007. "Definition of Expansion Potential for Expansive Soil." *Proc. 3rd Asian Conf. on Unsaturated Soils*, UNSAT-ASIA, Nanjing China, Z.Z. Yin, Y.P. Yuan, and A.C.F. Chin, Eds., Science Press.

Nilson, A.H. and G. Winter. 1991. *Design of Concrete Structures*, 11th ed. New York: McGraw-Hill.

Noe, D.C. 2007. *A Guide to Swelling Soil for Colorado Homebuyers and Homeowners*, 2nd ed. Denver, CO: Colorado Geological Survey Special Publication 43, DNR.

Nogami, T., J. Otani, K. Konagai, and H.L. Chen. 1992. "Nonlinear Soil-Pile Interaction Model for Dynamic Lateral Motion." *Journal of Geotechnical Engineering*, Vol. 118, No. 1, pp. 89–106.

Novak, M. 1991. "Piles under Dynamic Loads." *Proceedings of the 2nd International Conference on Recent Advances in Geotechnical Earthquake Engineering and Soil Dynamics*, March 11-15, St. Louis, MO, Paper No. SOA14, pp. 2433–2456.

Novak, M., and F. Aboul-Ella. 1978. "Stiffness and Damping of Piles in Layered Media." *Proceedings of the ASCE Geotechnical Engineering Division Specialty Conference on Earthquake Engineering and Soil Dynamics*, June 19-21 1978, Pasadena, CA, Vol. 2, pp. 704–719. Reston, VA: ASCE Press.

Novak, M., and B. El Sharnouby. 1983. "Stiffness Constants for Single Piles." *Journal of Geotechnical Engineering*, ASCE, Vol. 109, No. 7, pp. 961–974.

National Park Service (NPS). 2007. *List of National Historic Landmarks*. www.nps.gov. U.S. Department of the Interior.

Occupational Health and Safety Administration (OSHA). 2008. *Excavations* (Document 2226). Washington, DC: Occupational Safety and Health Administration.

Pack, J.S. 2000. "Design of Helical Piles for Heavily Loaded Structures." In *New Technological and Design Developments in Deep Foundations*, Reston, VA: ASCE Press, pp. 353–367.

———. 2004. *Practical Design and Inspection Guide for Helical Screw Piles and Helical Tension Anchors*, 3rd ed. Aurora, CO: IMR, Inc.

———. 2006. "Performance of Square Shaft Helical Pier Foundations in Swelling Soil." *Proceedings of Geo-volution*, ASCE and AGU Joint Conference Denver, CO ASCE Geotechnical Practice Publication No. 4, pp. 76–85.

———. 2007. "Design, Specification and Installation of Square Shaft Helical Piers in Expansive Soils." *Proceedings of the 32nd Annual Conference of the Deep Foundation Institute*, Colorado Springs, CO, pp. 321–330.

Pack, J.S. and K.M. McNeill. 2003. "Square Shaft Helical Screw Piles in Expansive Clay Areas." *Proceedings of the 12th Pan American Conference on Soil Mechanics and Geotechnical Engineering*, pp. 1825–1832.

Parry, R.H.G. 1977. "Estimating Bearing Capacity of Sand from SPT Values." *Journal of Geotechnical Engineering Division*, Vol. 103, No. GT 9, pp. 1014–1019.

Peck, R.B. 1969. "Deep Excavations and Tenneling in Soft Ground," *Proceedings of 7th International Congress of Soil Mechanics and Foundation Engineering, State-of-Art Volume*, pp. 225–290.

Peck, R.B., W.E. Hanson, and T.H. Thornburn. 1965. *Foundation Engineering*, New York: John Wiley and Sons.

Perko, H.A. 1999. "Summary of Earth Retaining Methods Utilizing Helical Anchors." *Magnum Piering Technical Reference Guide, Engineering Analysis*, Section 3, Cincinnati, OH: Magnum Piering, Inc.

———. 2001. "Energy Method for Predicting the Installation Torque of Helical Foundations and Anchors." In *New Technological and Design Developments in Deep Foundations*, pp. 342–352. Reston, VA: American Society of Civil Engineers.

———. 2003. "Lateral Capacity and Buckling of Helix Pier Foundations." In *Proceedings of the Helical Foundations and Tie-Backs Specialty Seminar*, Deep Foundation Institute, Cincinnati, OH.

———. 2004a. "Introduction to Corrosion and Galvanizing of Helix Foundations." In *Proceedings of the Helical Foundations and Tie-Backs Specialty Seminar*, Deep Foundation Institute, Tampa, FL.

———. 2004b *Helipost*™ *Engineering Handbook*, Rev. 2. Fort Collins, CO: Secure Piers, LLC.

———. 2005. "Underpinning and Shoring for Underground MRI Research Facility at Ohio State University." In *Proceedings of Underground Construction in Urban Environments*, Specialty Seminar by ASCE Metropolitan Section Geotechnical Group at the Geo-Institute of ASCE, New York, NY.

———. 2006a "Geotechnical Techniques Used in Planetary Exploration." Keynote address, *Proceedings of Geo-volution*, ASCE and AGU Joint Conference, Denver, CO, pp. 109–119.

———. 2006b "ICC-ES Acceptance Criteria for Helical Foundations and Devices." In *Proceedings of the Helical Foundations and Tie-Backs Specialty Seminar*, Deep Foundation Institute, Newark, NJ.

———. 2006c. "Installation Torque as a Predictor of Helical Pier Axial Capacity." Electronic publication, www.helicalpierworld.com.

———. 2007a. "Lateral Resistance of Helical Foundations for Hurricane-Prone Coastal Areas." In *Proceedings of the Helical Foundations and Tie-Backs Specialty Seminar*, Deep Foundation Institute, New Orleans, LA.

———. 2007b. "Evidence of Seismic Resistance of Helical Foundations." In *Proceedings of the Helical Foundations and Tie-Backs Specialty Seminar*, Deep Foundation Institute, New Orleans, LA.

———. 2007c. "Creating Acceptance for Helical Foundations." Code Updates, *Structure Magazine*. December, pp. 49–50.

———. 2008a "Helical Pile Foundation for Alexan Broadway Parking Structure." *Proceedings of Case Histories in Deep Foundations*, Deep Foundation Institute, Cincinnati, OH.

———. 2008b "Helical Piles in the Building Codes." *Proceedings of the Helical Foundations and Tie-Backs Specialty Seminar*, Deep Foundation Institute, Los Angeles, CA.

Perko, H.A. and J.B. Boulden. 2008. "Lateral Earth Pressure on Lagging in Soldier Pile Wall Systems." *DFI Journal*, Vol. 2 (November), pp. 52–60.

Perko, H.A., and R.A. Doner.(in press). "Full-Displacement, Augered Friction Piles and a Method for Estimating Capacity." *Proceedings of the 34th Annual Conference of the Deep Foundation Institute*, Kansas City, MO.

Perko, H.A. and S. Rupiper. 2002. Helical Pile Engineering Handbook. Manufacturer technical Literature. Larkspur, CO: Precision Pier USA.

Plog, B.A.,et al. 2006. *Strategies to Prevent Trenching-Related Injuries and Deaths.* Silver Spring, MD: Center to Protect Workers' Rights.

Prasad, Y.V.S.N. and S. Rao. 1994. "Pullout Behavior of Model Piles and Helical Pile Anchors Subjected to Lateral Cyclic Loading." *Canadian Geotechnical Journal*, Vol. 31, No. 1, pp. 110–119.

Prasad, Y.V.S.N. and S.N. Rao. 1996. "Lateral Capacity of Helical Piles in Clays." *Journal of Geotechnical Engineering*, Vol. 122, No. 11, pp. 938–941.

Prayer, J.H.,et al. 1980. Material Performance (as referenced by Jones, 1986).

Puri, V.K., R.W. Stephenson, E. Dziedzic, and L. Goen. 1984. "Helical Anchor Piles under Lateral Loading." *Laterally Loaded Deep Foundations: Analysis and Performance*, ASTM STP 835, Langer, Mosley, and Thompson, Eds. West Conshohocken, PA: American Society for Testing and Materials, pp. 194–213.

Radhakrishna, H.S. 1972. "Helix Anchor Tests in Stiff Fissured Clay." Ontario Hydro Research Division Report No. 72-12-H. Nanticoke, Ontario, Ontario Hydro.

Radhakrishna, H.S. 1976. "Helix Anchor Tests in Sand." Ontario Hydro Research Division Research Report No. 76-130-K, pp. 1–33.

Rao, S.N. and Y.V.S.N. Prasad. 1993. "Estimation of Uplift Capacity of Helical Anchors in Clays." *Journal of Geotechnical Engineering*, Vol. 119, No. 2, pp. 352–357.

Rao, S.N., Y.V.S.N. Prasad, and C.V. Prasad. 1990. "Experimental Studies on Model Screw Pile Anchors." *Proceedings of the Indian Geotechnical Conference*, Vol. 1, Bombay, pp. 465–468.

Rao, S.N., Y.V.S.N. Prasad, and M.D. Shetty. 1991. "Behavior of Model Screw Piles in Cohesive Soils." *Soils and Foundations*, Vol. 31, No. 2, pp. 35–50.

Rao, S.N., Y.V.S.N. Prasad, and C. Veeresh. 1993. "Behavior of Embedded Model Screw Anchors in Soft Clays." *Geotechnique*, Vol. 43, No. 4, pp. 605–614.

Read, A.A.L. and S. Sritharan. 1993. "Reconnaissance Report on the Ormond Earthquake—August 10, 1993." *Bulletin of the New Zealand National Society for Earthquake Engineering*, Vol. 26, No. 3, pp. 292–308.

Revie, R.W. (ed.). 2000. *Uhilig's Corrosion Handbook*, 2nd ed. Electrochemical Society Series. New York: Wiley Interscience.

Robinson, K.E. and H. Taylor. (1969). "Selection and Performance of Anchors for Guyed Transmission Towers." *Canadian Geotechnical Journal*, Vol. 6, pp. 119–135.

Rodgers, T.E. Jr. 1987. "High Capacity Multi-Helix Screw Anchors for Transmission Line Foundations." *Foundation for Transmission Line Towers*, ASCE Reston, VA, ASCE Press, pp. 81–95.

Romanoff, M.[1957] 1989. *Underground Corrosion*. National Bureau of Standards No. 579. Reprinted by NACE, Houston, TX.

———. 1972. "NBS Papers on Underground Corrosion of Steel Piling." National Bureau of Standards Monograph 127. Gaithersburg, MD: NIST.

Rupiper, S. 1990. "Helix Piers Are Solutions for Column Reactions." In *Abstracts of the Proceedings of the 8th Structures Congress.* Baltimore, MD: ASCE.

———. 1994. "Helical Plate Bearing Members, A. Practical Solution to Deep Foundations." In *Proceedings of the International Conference on the Design and Construction of Deep Foundations*, Vol. 2, Hawthorne, NJ: Deep Foundation Institute, pp. 980–991.

———. 2000. Personal communication, San Jose, CA.

Rupiper, S. and W.G. Edwards. 1989. "Helical Bearing Plate Foundations for Underpinning." In *Proceedings of Foundation Engineering Congress/SCE/CO Division*, Evanston, IL, June 25–29.

Sailer, D. and B. Soth. 2004. "Helical Pier Foundations for Problem Sites." *Journal of Light Construction* (May).

Schmidt, R. 2004. "Screw Piles: Uses and Considerations." *Structure Magazine* (June), pp. 25–31.

Seider, G. 1993a. "Eccentrically Loaded Helical Pier Systems," Bulletin 01-9303. Centralia, MO: A.B. Chance Company.

———. 1993b. "Eccentric Loading of Helical Piers for Underpinning." In *Proceedings of the 3rd International Conference on Case Histories in Geotechnical Engineering*, St. Louis, MO, Vol. 1, pp. 139–145.

———. 2004. "Helical Foundations: What an Engineer Needs to Know." *Structure Magazine*, Vol. 11, No. 6, pp. 27–28.

Skempton, A.W. 1951. "The Bearing Capacity of Clays." In *Proceedings of the Building Research Congress*, Vol. 1, pp. 180–189.

Slemons, P.E. 2008. "A Rational Approach to Calculating Torque to Capacity in Cohesive Soils." *Proceedings of the Helical Foundations and Tie-Backs Specialty Seminar.* Deep Foundation Institute, Los Angeles, CA, Hawthorne, NJ: Deep Foundation Institute.

Spencer, White, and Prentis, Inc. 1986. "Lagging Design." Sample Calculations provided by Tom Tuozzolo, Moretrench Geotec, Rockaway, NJ.

Stagg, K.G. and O.C. Zienkiewicz. 1968. *Rock Mechanics in Engineering Practice*, New York: John Wiley and Sons.

Sun, K. and J.A. Pires. 1993. "Simplified Approach for Pile and Foundation Interaction Analysis." *Journal of Geotechnical Engineering*, Vol. 119 No. 9, pp. 1462–1479.

Tabesh, A. and H.G. Poulos. 1999. "The Effect of Soil Yielding on Internal Pile Response." In *Proceedings of the 2nd International Conference on Earthquake Geotechnical Engineering*, ed. Seco e Pinto vol. 1, 327–333. Rotterdam: A.A. Balkema.

Tappenden, K.M. 2004. "Predicting the Axial Capacity of Screw Piles Installed in Western Canadian Soils." Master's thesis, University of Alberta, Edmonton, Alberta.

———. 2006. "Screw Piles: Use and Design." www.almita.com/_html/technical/html.

Terzaghi, K. 1943. *Theoretical Soil Mechanics.* New York: John Wiley and Sons.

Terzaghi, K. and R.B. Peck. 1967. *Soil Mechanics in Engineering Practice.* New York: John Wiley and Sons.

Thompson, R.W., W. Rethamel, and H.A. Perko. 2006. "Comparison of Constant Volume and Oedometer Swell Pressures." In *Proceedings of Unsat 2006*, American Society of Civil Engineers, Phoenix, AZ.

Tomlinson, M.J. 1986. *Foundation Design and Construction*, 5th ed. New York: John Wiley and Sons.

Trofimenkov, J.G., and L.G. Maruipolshii. 1965. "Screw Piles Used for Mast and Tower Foundations." In *Proceedings of the 6th International Conference on Soil Mechanics and Foundation Engineering*, Vol. 2, pp. 328–332.

Udwari, J.J., T.D. Rodgers, and H. Singh. 1979. "A Rational Approach to the Design of High Capacity Multi-Helix Screw Anchors." *Proceedings of the 7th Annual IEEE/PES, Transmission and Distribution Exposition*, Piscataway, NJ: IEEE Publishing, pp. 606–610.

Uhlig, H.H. and R.W. Revie. 1985. *Corrosion and Corrosion Control*, 3rd ed. New York: John Wiley and Sons.

U.S. Navy. 1988. "Military Handbook: Seawalls, Bulkheads, and Quaywalls." MIL-HDBK-1025/4.

Velez, A., G. Gazetas, and R. Krishnan. 1983. "Lateral Dynamic Response of Constrained Head Piles." *Journal of Geotechnical Engineering*, Vol. 109, No. 8, pp. 1063–1081.

Vesic, A.S. 1973. "Analysis of Ultimate Loads of Shallow Foundations." *Journal of Soil Mechanics and Foundation Design*, Vol. 99, No. SM 1, pp. 45–73.

Vickars, R.A. and S.P. Clemence. 2000. "Performance of Helical Piles with Grouted Shafts." In *New Technological and Design Developments in Deep Foundations*, pp. 327–341. Reston, VA: American Society of Civil Engineers.

Victor, R. and A. Cerato. 2008. "Helical Anchors as Wind Tower Guyed Cable Foundations." In *Proceedings of the BGA International Conference on Foundations.* Dundee, Scotland: HIS BRE Press.

Vyazmensky, A.M. 2005. Numerical Modeling of Time Dependent Pore Pressure Response Induced by Helical Pile Installation." Masters. thesis, University of British Columbia, Vancouver.

Walsh, K.D. et al. 2009. "Method for Evaluation of Depth of Wetting in Residential Areas", *Journal of Geotechnical and Geoenvironmental Engineering*, February Edition, ASCE, pp. 169–176.

Weech, C.N. 2002. "Installation and Load Testing of Helical Piles in Sensitive Fine-Grained Soil." Master's thesis, University of British Columbia, Vancouver.

Weikart, A.M., and S.P. Clemence. 1987. "Helix Anchor Foundations—Two Case Histories." *Foundations for Transmission Line Towers*, pp. 72–80. Reston, VA: American Society for Civil Engineers.

Wesolek, D.A., F.C. Schmednecht, and G.L. Seider. 2005. "Helical Piers/Anchors in the Chicago Building Code." In *Proceedings of the 30th Annual Conference on Deep Foundations.* Chicago, IL

Witherspoon, T.W. 2006. *Underpinning Systems in Expansive Clay Environment*, Ph.D. diss., University of Texas at Arlington.

Yokel, F.Y., R.M. Chung, and C.W.C. Yancey. 1981. "NBS Studies of Mobile Home Foundations." U.S. National Bureau of Standards Report NBSIR 81-2238. Gaithersburg, MD: NIST.

Zaki, A. 2006. *Principles of Corrosion Engineering and Corrosion Control.* Oxford, U.K: Butterworth-Heinemann.

Zhang, D.J.Y. 1999. "Predicting Capacity of Helical Screw Piles in Alberta Soils." Master's thesis, University of Alberta, Edmonton.

Index